THE MODERN PLASTERER

W. Verrall

With an introduction by Tim Ratcliffe and Jeff Orton

Volumes I and II

Routledge
Taylor & Francis Group
LONDON AND NEW YORK

First published by Caxton Publishing Company Ltd
Published by Donhead Publishing Ltd 2000

Published 2015 by Routledge
2 Park Square, Milton Park, Abingdon, Oxon OX14 4RN
711 Third Avenue, New York, NY 10017, USA

Routledge is an imprint of the Taylor & Francis Group, an informa business

A CIP catalogue for this book is available from the British Library

ISBN 13: 978-1-873394-45-8 (hbk)

Introduction to 2000 edition

In the late 1920s and early 1930s Caxton Publishing brought out a series of books on the building trades, including *The Modern Carpenter and Joiner, The Modern Painter and Decorator, The Modern Bricklayer* and, of course, *The Modern Plasterer.* This was a significant period in the development of the building trade. There had been a shortage of labour since the end of the Great War, and there was a general view that it was no longer possible to rely just on site based learning to ensure tradesman were properly trained. Emphasis was being put on the need for some college teaching, with students being encouraged to take the City and Guilds examinations.

Caxton's decision to publish a co-ordinated series of books was in response to the consequent need for reference material on the different trades. Along with these books they issued booklets of coupons which entitled the purchaser to write to an Information Bureau with enquiries about their particular trade. We don't know to what extent Caxton's Information Bureau was made use of, but its existence demonstrates a clear commitment to the furtherance of knowledge and skills in the building trades.

In recent years there has been a distinct lack of literature on plastering, generally, and on traditional materials and methods in particular. In 1998, Donhead, therefore, decided to republish William Millar's 1897 book *Plastering - Plain and Decorative,* which records the skills and knowledge that existed at the end of the nineteenth century. In 1999, The Plasterers Craft Guild persuaded another publisher to update and re-issue the 1970s; *Plastering - An Encyclopaedia* by Stagg and Pegg. Verrall's book. *The Modern Plasterer* stands as a very important bridge between these two definitive works. Whereas Mr Millar's book assumes a certain amount of existing experience and understanding, Verrall starts at a level which is more akin to the understanding of the modern reader. This really is, therefore, an excellent reference book for anyone wanting to rediscover and learn about traditional plastering.

The period when Caxton published their series was significant because, even though new methods and materials were being introduced, traditional materials and skills were still in common usage. *The Modern Plasterer,* for example, explains how to use cement in renders and mouldings, but gives just as much, if not more, emphasis to traditional materials such as lime plasters. The book also makes intriguing references to compositions like magnesium oxychloride floors, which appear to have been a relatively new invention, but have since disappeared.

The Modern Plasterer is a book that is always useful when trying to either understand or describe a traditional detail, or when needing to overcome particular technical problems. Whilst it may have been written with apprentice and trainee plasterers in mind it is clear, both from the level of information included in the book, and from publicity produced by Caxton's, that the book was aimed at a wider audience of plasterers. It gives basic descriptions of the principles of plastering, but also goes into considerable detail about the intricacies of how to achieve particular finishes. We believe that it will prove to be useful to everybody from relative beginners through to experienced traditional plasterers. It should, also, we believe, provide a lot of invaluable information to designers and specifiers who want to understand plastering methods and materials better.

Mr Verrall had a great gift for explaining technical issues, in a way that is clear, without ever being simplistic. One of the most useful things is that, as far as possible, he illustrates everything he describes with a drawing or diagram. There are drawings of basic plasterers tools, through to illustrations of how to use a tramel to set out an ellipse.

The Modern Plasterer describes the characteristics of different types of plaster and their uses, and the varieties of lathing available, before describing how to apply and finish flat lime plaster. This includes a description of how to achieve a flat and level surface using dots and screeds and the best way to finish angles. Although the emphasis of the book is on quality, Verrall is always keen to discuss how best to make economic use of plant and the easiest ways to execute work.

The book goes into great detail explaining how to set up and run mouldings of different types, such as ellipses and segment arches, using different materials. It then explores the casting of mouldings and the uses of fibrous plaster. There are a lot of other materials and methods described, including exterior rendering and cement floors, and traditional materials like gesso and artificial marbles. The range of topics should mean that this book will be useful to a wide range of plasterers, not just those involved in conservation and repair work.

Verrall taught plastering at the North London Polytechnic and was committed, throughout his life, to the continuing education of plasterers. He wrote a number of smaller books on the subject, but *The Modern Plasterer* stands as his seminal work. It was a standard reference for the City and

Guilds exams in plastering for decades, and is still held in high regard by those who teach on modern plastering courses. Its strength and appeal to those of us involved in repair and conservation work is that it is written by someone for whom traditional skills and materials were still common place. Verrall explains that the techniques and skills that he knew so well in a way that is accessible to modern readers. We hope that it will prove as useful to others as it has been to us.

Tim Ratcliffe and Jeff Orton
November 1999

Tim Ratcliffe (Architect) BA DipArch RIBA
Tim Ratcliffe is an architect working on historic buildings in and around the Midlands and North Wales. He joined Donald Insall Associates in 1997, having previously worked for Rodney Melville and Partners. In 1987 he was awarded a travelling scholarship by the Society for the Protection of Ancient Buildings and has also worked as a labourer for a number of specialist plastering and conservation contractors. He sits on the Technical Panel of the SPAB.

Jeff Orton (Plasterer) CRP MPCG
Jeff Orton served a full plastering apprenticeship from 1963-1968 with a traditional family firm, and gained a distinction in the Advanced Craft City and Guilds Examination. He is registered with the Worshipful Company of Plasterers (CRP) and in 1981 he attended the San Servolo Architectural Conservation Course for Craftsmen in Venice. Since 1987 he has worked for Trumpers Ltd, repairing and reinstating plasterwork in historic buildings for private clients as well as The National Trust and English Heritage.

Tim Ratcliffe and Jeff Orton are both members of the Plasterers' Craft Guild and the Building Limes Forum. They give lectures and demonstrations together on traditional plastering techniques. The Plasterers' Craft Guild was established, almost 50 years ago, to pursue the ideal of craft excellence within the plastering trade; through education and the ongoing development of skills in the industry. The Building Limes Forum was formed in 1992 to encourage the development of expertise and understanding in the use of lime in building.

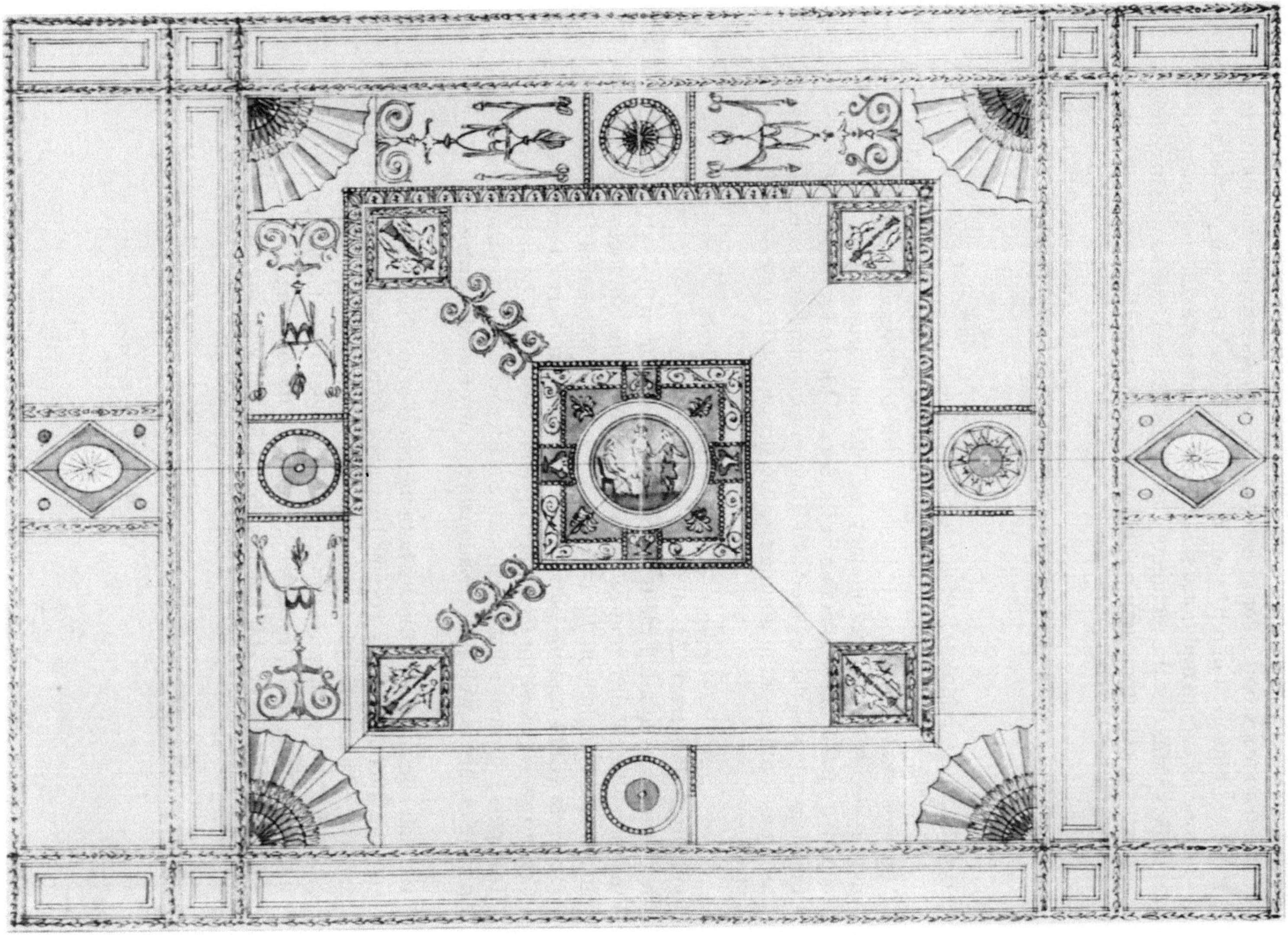

ROBERT ADAM CEILING.

The general scheme carried out in low tones, painted on the flat plaster or with ornament in slight relief in composition, the panels in strong colours.

[*Frontispiece*

THE MODERN PLASTERER

A PRACTICAL WORK ON PLASTERING, DECORATIVE PLASTERWORK, ROUGH-CASTING, CEMENT-WASHES, PLASTIC FLOORS, ARTIFICIAL MARBLES, ETC.

BY

W. VERRALL

TEACHER OF PLASTERING, NORTHERN POLYTECHNIC, LONDON; FIRST-CLASS HONOURS AND MEDALLIST IN PLASTERING, CITY AND GUILDS OF LONDON INSTITUTE, ETC.

WITH SPECIAL CONTRIBUTIONS BY

HERBERT JEANS

AUTHOR OF "THE PERIODS IN DECORATION"

AND

J. TURNER

MEMBER OF THE NATIONAL ASSOCIATION OF PLASTERERS, GRANOLITHIC AND CEMENT WORKERS

EIGHT COLOURED PLATES BY

H. GUY HOLT, F.R.I.B.A.

WITH EIGHTEEN HALF-TONE PLATES AND 343 DIAGRAMS IN THE TEXT

VOLUME I

CAXTON PUBLISHING COMPANY, LIMITED
CLUN HOUSE, SURREY STREET, LONDON, W.C.2

PREFACE

Much has been written on decorative plaster work, mainly from the historical and art points of view, very little on the strictly technical and practical sides of the craft.

This book has been planned and written to satisfy the need for the modern treatment of the subject, widely felt alike among architects and members of every branch of the building trade, particularly by plasterers themselves.

It is a comprehensive work, covering the whole field in a way to assist craftsmen and interest everyone concerned with the use of plaster and cement under present-day conditions. It will be seen that great care has been taken with the very large number of drawings and photographs which elucidate and adorn the text. A number of the full-page plates, illustrating fibrous plaster ceilings and mural ornament and fibrous plaster work, represent work carried out by The Associated Crafts Limited, Cricklewood, and by Messrs. Humphreys Limited, Knightsbridge, London.

The General Editor.

CONTENTS

VOLUME I

CHAPTER I

INTRODUCTORY

CHAPTER II

ANCIENT AND ASIATIC PLASTER

CHAPTER III

MODERN DECORATIVE PLASTER

CHAPTER IV

CEILINGS: DECORATIVE AND OTHERS

CHAPTER V

PLASTERING MATERIALS

CONTENTS

CHAPTER VI

PORTLAND CEMENT

CHAPTER VII

HOW TO MAKE CONCRETE

CHAPTER VIII

SELECTION OF MATERIALS

CHAPTER IX

PREPARING THE BASE FOR PLASTERING

CHAPTER X

LATHING

CHAPTER XI

PLASTERERS' TOOLS

CONTENTS

CHAPTER XII

SCAFFOLDING AND PLANT

CHAPTER XIII

INTERIOR PLASTERING

CHAPTER XIV

RUNNING CORNICES AND MOULDINGS

LIST OF PLATES

VOL. I

THE MODERN PLASTERER

CHAPTER I

INTRODUCTORY

What is Plaster ?—Why we Plaster—Sequence of Plastering Operations : Rendering ; Floating ; Setting ; Special Finish—Precautions : Well-slaked Lime ; Clean Sand ; Impure Water—Defects : Crazing and Cracking—The Revival of Stucco—Cost of Upkeep—Stucco in Dirty Atmospheres—Variety of Treatment.

PLASTERING is the process of covering the surfaces of a building, or parts of a building, with pure lime or mortar. Usually this mortar consists of mixtures of lime or gypsum and sand, gauged with pure water or milk of lime, with, for certain purposes, the addition of sugar, water-proofing agents, cow's hair or other fibrous material. But in modern practice cement mortars are also used, while the various processes of rough-casting form an important branch of plastering greatly in demand in the present day.

Another development of plastering is in the use of cast plaster, generally in the form of slabs (particularly for curved surfaces) and ornament, in which lime and fine sand, pure plaster of Paris (gypsum) or cement is cast in prepared moulds. A further development of this is seen in fibrous plaster castings, in which the plaster slabs or castings have a backing or a core of strong canvas, or have fibres incorporated in the plaster mixture.

WHY WE PLASTER

It is desirable that the reasons for adopting the expensive processes of plastering should be understood. There are several reasons, but they all tend both to the improvement and the preservation of structures. The conservative services of plastering, though often relegated to a secondary place, are really the most important.

If a wooden or brick structure is left uncovered, exposed to the weather, it suffers from the effects of sunrays during the summer months, of rain and frost in autumn and winter, and in certain conditions is subject to abrasion, as when exposed to very high winds, especially if dust or sand laden, or, and this applies especially to interiors, the rubbing and knocks by persons or objects carried by persons. In a plastered or stuccoed surface (the term "stuccoed surfaces" applies mainly to the smooth, polished plastering on exterior work), the wood or brick is preserved, and the plaster itself, if well laid with good materials, is very durable, and, moreover, can be protected by means of two or more coats of paint. The repairing of stucco is usually cheaper than the raking out of jointing, washing down, repointing and possibly cutting out and replacing of decayed, soft bricks. Plastering, and in this we include rough-casting, has an economic protective function even on concrete slab, block and mass constructions.

Secondly, plaster, when properly prepared and applied, being compact, of close texture and without joints, makes a building weather-tight. It offers a considerable opposition to the infiltration of moisture, whether in the form of rain or damp vapour. In slight structures, frame or of other types, it is a draught excluder. These properties make plaster an effective insulating medium, so that a plastered house is likely to be cooler in the summer and warmer in the winter, as well as drier, as the result of such protection. In a series of experiments carried out by the Building Research Board on heat transmission through walls, concrete and plasters, the following results were obtained. Conductivity in gram-calories per square centimetre per second for 1 centimetre thickness and 1° C. difference of temperature:

Roman mortar (ground bricks and lime)	·00100
Sand 2, lime 1	·00113
Sand 4, cement 1	·00128
Coke breeze 4, cement 1	·00140
Concrete granite 4, cement 1	·00176
Concrete Portland stone 4, sand 2, cement 1	·00186
Concrete York stone 4, sand 2, cement 1	·00242

Thus it will be seen that while a brick-dust lime mortar gave the best insulating results, both lime and cement mortars are much superior insulators to the best concretes. This was confirmed by experiments carried on in Switzerland and America, as shown below:

In a series of experiments connected with a lengthy investigation into the phenomena of transmission of heat through walls, it was found that the loss due to the degree Centrigade between inside and outside temperature (taking stability as 0) was for bricks set in cement (the wall about 10″ thick), 2·4; for red bricks in mortar (about 9″ thick), 2·1; for hollow brick wall, from 1·2 to 1·5, according to the system of construction, the greater the amount of air space, but the more it is split up into cells, the less the loss; and for masonry, 2·1 (Prof. A. Dumas, Chief Building Materials Experimental Laboratory, Lausanne).

In a set of American experiments it was found that a 13″ brick wall, plastered inside with cement, exposed to a wind blowing 15 miles per hour, allowed a passage of 0·111 cubic feet of air per minute. The leakage through an ordinary glazed window was at the rate of 47·5 cubic feet per minute; but if the window was provided with draught excluder, the passage of air was reduced to 7·8 cubic feet per minute.

Thirdly, we apply plaster in order to obtain a surface which is smooth, easily kept clean, and suitable to be white-washed, painted or papered. Where the surface is to be left white (usually smoothed and polished with the trowel) or painted with oil colours, a hard-drying cement, such as Keene's or Sirapite, is advisable ; but if it is to be distempered, treated with washable water-colours or papered, the softer lime plasters are preferable.

Fourthly, we apply plaster to walls and ceilings as a medium for decoration, as when we cover it with decorative painting, impress it with ornament (as in pargeting), or produce relief designs with it by moulding, modelling or carving. In this connection of decoration, it should be pointed out that the setting or finishing coat in white or coloured lime or cement plasters can be finished smooth and given a high degree of polish, as in best trowelled work on Keene's, or given a distinctive texture, ranging from the sand-floated finish to the rough, foam-sprinkled effects of sponge finish, all suitable for interior as well as exterior work, besides which there are the various forms of dash for exterior treatment.

So much for the why and wherefore of plastering. Let us now pass on to a consideration of the first principles of the craft.

Sequence of Plastering Operations

Plastering may be applied to wood in various forms (beams, planks, laths), to metal laths, brickwork, stone-masonry or concrete. It is only in the cheapest form of work that plaster is applied in one coat. As a rule, two or three coats, and in the best work four coats, are applied, each of moderate thickness, and one applied upon the other before the bottom one is hardened.

Rendering.—We commonly say we plaster laths and render brickwork or masonry ; but the first coat is rendering, and it is carried out with coarse stuff, a mortar of lime, sand and hair. It is usually applied ½″ thick, on a well-brushed and wetted surface. This is left rough from the trowel and then scored over with a bunch of pointed laths to provide a key.

Floating.—The floating coat is usually fine stuff, composed of pure slaked lime and fine sand, but no hair, applied to the damp scored rendering, and in its turn left fairly rough and scored, often with circular, not too deep lines. In two-coat work, this is the finishing coat, and then the plaster commonly consists of gauged stuff, a mixture of coarse stuff with setting stuff, or with plaster of Paris.

Setting.—The setting coat requires special care, because, as a rule, it is the finishing coat. For this, setting stuff, also known as plasterer's putty, is employed. Like fine stuff, it consists of pure slaked lime, often with a little fine sand, both passed through fine hair sieves. It is important that the lime should have no lumps. It is finished by polishing with the trowel and wooden floats.

Special Finish.—Ceilings, cornices and sometimes walls, when left with a plaster finish, receive a fourth coat. This is usually very thin and consists of a gypsum plaster (plaster of Paris), or one of the close-tectured, hard-drying compositions, such as Keene's, Parian, Sirapite, etc. This treatment is frequently applied to cornices, friezes, etc., where very fine work and high finish are demanded.

There are, of course, variations on the above, and we have not taken into consideration the fixing of moulded and fibrous work, running mouldings, modelling, etc., but these elaborations can be left for later description. We may, however, touch upon a few points of importance in obtaining satisfactory results.

PRECAUTIONS

Plastering demands care from start to finish, in the selection and preparation of materials as well as in carrying out the work. We do not at the present stage propose to deal with the nature of materials or the composition of plasters, but to give a few general hints and suggest how certain defects occur and how they may be avoided or remedied.

Well-slaked Lime.—We shall see presently that the old master-plasterers took extraordinary precautions to use only thoroughly well-matured lime. They were quite justified in doing so, for good work can only be achieved if the lime is pure and well-slaked. As long as the lime is in the quick, unquenched state, is hot, it will absorb moisture, and in so doing expand and produce vapour. The consequence is that if quicklime is present in the mortar, instead of setting normally and hardening gradually into a compact, cohesive mass, the mortar, owing to the quicklime absorbing moisture and thereby expanding, and by emitting heat and vapour, is disrupted, opens out into bubbles, cracks, or tiny hair-like fissures appear, and it loses cohesion. The effect of all this in plastering is that the plaster cracks, blisters and crumbles away. So the lime must be well slaked. It is well to remember that this cannot be ensured if the lime is lumpy, for when this is the case, the chances are that water has not got to every particle, so that the interior of the lumps is likely to contain quick, unquenched lime. So lime for plaster work must be sieved and all lumps rejected ; put aside for crushing and more or less regular slaking to be fitted for ordinary bricklayers' coarse stuff. Fine sieving is particularly necessary for floating and setting stuff. The poorer the slaking the more is the lime likely to effloresce, show forth a white crumbly powder, which will work its way even through two or three coats of oil paint.

Clean Sand.—The sand used should be clean—that is to say, free from earth, dust or common salt. If the sand is clean, a chemical action will set up in the mortar and closer adhesion result. But if it contains earth, dust or salts (other than calciums), it simply adulterates the lime, there follows reduction in adhesion and, in bad cases, crumbling. If you demolish old Tudor plaster work it offers considerable resistance and comes away in big flakes; but if you attack inferior rendering on brickwork or plaster on laths, it crumbles amidst clouds of fine dust.

Impure Water.—If impure water, containing organic matter or chloride of sodium (ordinary salt), is used for gauging, the lime is adulterated, the sand fouled and again poor adhesion results, whilst the chloride will assuredly effloresce and spoil the surface.

Imperfections

Some of the imperfections to which plaster is liable, and the ways to avoid them, have been discussed above. But there are two other defects, cracking and crazing, which demand more detailed discussion.

Cracks.—By cracking we mean the disfigurement of a plastered surface with long, deep cracks, on occasion developing to $\frac{1}{16}''$ in width. These are most liable to occur on partitions, but are also seen on ceilings. This phenomenon may be due to defective plaster; either plaster containing quicklime or an insufficient allowance of lime. The remedy, of course, is to see that good materials are used, properly apportioned and well mixed. Cracks of this kind are also caused by structural defects. For instance, too-light wood laths, warped laths, also laths insufficiently nailed and so too springy, will cause cracks. So they will if too dry, by robbing the plaster of its moisture by absorption. The cause may be more deep rooted, in poor studding and framing of partitions, too light or unstrutted ceiling joists. These deeper causes are beyond the domain of the plasterer. But the choice of sound laths and their secure fastening are matters that rest with him.

Where diagonal cracks, intersected by others, horizontal or vertical, are numerous, deep and wide, stripping and replastering, after inspection and repair of foundation, will be called for. In ordinary cases it is only necessary to cut back the plaster all along both sides of the crack and after damping refilling with a rich mixture. On moulded work Keene's or Parian is the best material.

Crazing.—Crazing may vary from tiny fissures to the more unsightly and serious shrinkage cracks. Hair-like fissures are generally the result of the plaster having been too rich or too wet. They are also occasionally due to too thin coatings on surfaces subject to vibration; for instance, two-coat work on ceiling laths nailed direct to the underside of light flooring joists. These fissures can usually be filled up with a rich lime wash in which a little sugar or size has been melted. Shrinkage cracks are the result of too rapid setting, possibly due to dry or absorbent laths, more likely to insufficient water in gauging or too dry an atmosphere. If the plaster shows signs of rapid setting, sprinkling will prevent damage. This particularly applies when cement is used in place of lime.

Crazing may also be set up by undue disturbance of the setting plaster. Again this applies with special force to cements. Plaster may stand in need of consolidating and polishing with the trowel or float, but there is a limit to this. If too great and prolonged pressure is applied, the setting may be disturbed and crazing will appear. This is a serious defect, and can only be hidden, not remedied, by lime washing.

Revival of Stucco

Recently Mr. W. D. Caroë, F.S.A., architect to the Ecclesiastical Commission, made what may be called a laudatory reference to the stucco work of Palladio and of our own Georgian period. These remarks of so learned a practitioner deserve more than passing note, for they coincide with a trend in some quarters to reintroduce stucco on an extensive scale. This reversion from plain to covered brickwork, it is interesting to record, has already made considerable progress in the United States of America, where it began with the opening years of the century.

As has been shown in the above discussion, such a revival can be fully justified from the practical as well as from the art point of view.

In America, where stucco is used as a surface dressing on brick, concrete, hollow slab and metal lath and plaster structures, the Bureau of Standards found much inferior material being used. In a series of experiments carried out on fifty-six stucco panels, approximately 15′ × 10′, exposed to the weather for six months, only 40 per cent. proved satisfactory, and only two out of the whole lot were rated as perfect—that is to say, entirely free from cracks, blotches or unevenness. All the panels had been given a smooth sand-float finish. With rougher finish there is less tendency to fissure and blotch. It is quite clear that stucco should not be applied until the brick walls, or concrete, are quite dry. Then, if the work is carefully done, a smooth, strongly adhering surface coat can be obtained, which will prove to provide weather-tight walls. This is important in these days when it is difficult to ensure good interior plastering to cope with the demand. Yet brick walls do stand in need of efficient protection inside and outside.

When stucco is to be applied, surface finishing of the brick walls is not necessary, so that the smooth surfacing is not all added cost. Although time must be allowed for drying, the increase of labour involved by this method of finishing is not much greater than in good painting, the actual time of work involved being very nearly the same.

Cost of Upkeep

Once done, the surface only requires to be kept in fair repair to ensure weather-tight walls, which look well. The periodical painting (or cement washing in the case of rough-dressed stucco) takes little time, and, in the long run, will prove less costly than repointing the brick, entailing the preliminary process of raking out joints and cutting out soft crumbling bricks and painting the stone dressings. The painted stucco can be washed, when occasion necessitates, much more quickly, cheaply and with less damage than unprotected brick can be cleaned; which is a point in its favour, especially for town buildings, where the use of stucco seems to be specially indicated.

We are promised purer atmosphere for our cities in the future, but we cannot very well get rid of dampness which injuriously affects even polished granite. Granite is not appropriate for every building, and experience has shown us that even granite cannot give us walls so easily kept presentable as good stone.

Stucco in Dirty Atmospheres

"Astragal," in *The Architects' Journal* of August 18, 1926, wrote:

"Manchester is a rich city, and very large sums are expended upon the new buildings which align the streets. Expensive materials—the best stone, the best wood, and costly metal-work—are employed in their construction, and yet the complaint is made that in a few years the façades will be covered with grime. Obviously the remedy is paint, and there is one material suitable for facing buildings which also is a perfect background for paint. That material is stucco. The initial cost of a stucco-fronted building is so much less than one of Portland stone that the interest on the difference between the capital expenditure involved in these two methods of wall construction will more than cover the occasional expense of painting the walls. In advocating the liberal use of stucco in Manchester, I am not guilty of suggesting an architectural innovation foreign to the traditions of the place, for in walking through the streets of that city one may still come upon examples of modest Regency façades which, although over a hundred years old, still look pleasant and fresh, and are a most welcome relief from their sooty neighbours. The owners of the buildings just referred to are not princes of industry and commerce, but for the most part have quite small establishments, and if they can afford to enliven with a coat of paint that part of the street frontage for which they are responsible, so much more ought it to be possible for their more prosperous and important commercial confrères to adorn their façades in that manner. Perhaps when the present craze for Portland stone shall have run its course, the glories of stucco will be revived. No section of the community will stand to gain so much from such a development than would the inhabitants of our great, but grimy, industrial cities."

Variety of Treatment

If it is hardly to be expected that stucco will be selected for the outside dressing of the more pretentious business premises, it possesses merits that should make it attractive to those erecting smaller business houses and private residences in towns, or in the vicinity of manufacturing centres. It should be noted that the treatment of stucco surfaces can be sufficiently varied to allow for the expression of individuality of an architect and to give character to different buildings. It is at least as " plastic " in the artistic sense as brick, and perhaps more so than granite. So there should be no objection to it on the score of " sameness," which is one of the criticisms so often brought against stucco in its days of degeneracy, the early Victorian period.

CHAPTER II

ANCIENT AND ASIATIC PLASTER

Egyptian Plaster—Grecian Plaster and Stucco—Roman Plaster Work : Slaking Lime for Stucco—Foundations for Plaster ; Plastering ; Contrast with Modern Methods—Preservation from Damp of Mural Paintings and Decoration : Greek Method ; Vitruvius' Treatment of Damp Walls ; Examples of Work—Byzantine Plaster—Saracenic Plaster—Other Asiatic Plaster : Chunam.

THE art of plastering is very ancient, so ancient that it is lost in the mist of time. But there is evidence that it took its origin from the wattle-and-daub method of construction, a method common to many primitive people and persisting from prehistoric times to our own days. It was a natural transition to replace the mud or puddled clay daubed on interlaced wattle or small timber walls with a more permanent material and excluder of the elements compounded of burnt and slaked lime mixed with sand. What happened in Arabia affords an illustration of the slow growth. We learn that the mosque at Medina in A.D. 622 had partitions of plastered wattles, and about 250 years later we are told that it was surrounded by arcades, the arches resting on plastered pillars.

EGYPTIAN PLASTER

The property of burnt and slaked lime, or gypsum, to harden and, either alone or mixed with sand and pebbles, to form a tough, adherent body capable of receiving a high polish, was early discovered. Gypsum, and what could be done with it, was known in Egypt at least 3,000 years before our era, and though we have no extensive examples of its use in construction, we do possess plenty where a gypsum stucco was employed as a coating (true plastering) on stone and wood. Many of the ancient sarcophagi (or coffins) are plastered both inside and out. We have plastered walls in the pyramids and many of the temples. The very tools the plasterers used in those far-off days have come down to us and show that in essentials little difference has been made. The gypsum was burnt, and from what we see of results, it must have been very pure ; probably left long to mature before use. It was applied as three-coat work, and some process must have been adopted to harden it. Much of the plaster is very thin, but extremely hard, with highly polished surface. Slabs were cast, apparently for ceilings as well as for pavements, such as the beautiful example at Tel-el-Amarna. This plaster was elaborately painted with pictures, hieroglyphics, name and title, cartouches and sacred symbols. So well has the plaster and its decoration lasted, that it has given us a large measure of our knowledge of the history and manners of the people. But we cannot be quite sure how they produced that hard, close-textured, brilliantly polished thin skin of gypsum plaster which we find on the sarcophagi and certain other articles, and it remains a marvel and puzzle to us, as does the secret of the splendid colouring. All we do know is that for fine work on wood they first glued a groundwork of linen, on which they laid a coat of gypsum ground in size; then they applied size in which they laid threads of parchment; over this a second coat of gypsum gauged with size. Finally they mixed their pigments with size.

Partly as the result of weathering and partly through damage during excavations, etc., the method of decorative work has been revealed. It appears that a plastered surface to be decorated was covered by vertical and horizontal lines drawn in red chalk at regular intervals. By means of these squares the design to be copied was easily transferred by the first artist, who drew in outline. Such drawings appear to have been supervised, for they are often altered and corrected. These outlines were filled in by painting or modelling, and again corrections were carried out, for in some cases we see where limbs or ornament have been wiped out, and new painting or modelling introduced. No doubt this accuracy was needed where decoration was mostly ritualistic, or in nature of hieroglyphics, and so not suitable for free-hand rendering.

That some traditions of Egyptian plasterers must have lingered in Northern Africa we shall see when considering the fine Saracenic work, which had so widespread an influence in mediæval times.

Grecian Plaster

Grecian stucco was a lime plaster, which had been brought to a high state of perfection at a very early date, as samples of the Mycenean age is very close textured, white and applied in a quite thin sheathing. It was used extensively internally and externally on temples and public buildings 500 years before the Christian era. It was stained and decorated. Much of the lime, no doubt, was obtained from marble. The plaster, often little thicker than a sheet of paper such as used in this book, spread for artists to paint on, Pliny tells us consisted of "stucco (refined lime) mixed with milk and saffron, and polished with spittle rubbed with the ball of the thumb." A good deal of saffron must have been used, because the plaster was strongly scented with it. Saffron was probably introduced to produce a golden stain, which would act as a foil for the painting, showing through the washes of colour and giving them brilliance. Later, the Byzantine artists gilded their plaster to secure a luminous foil. This stucco was so hard that it was used for pavements in houses and courtyards, and was beautifully decorated in colour. There is no evidence that it was modelled, but we do know that casts were made of it, for sculptors even took casts of their sitters.

The Greeks carried their art of fine plastering to their colonies in Grecia Magna, where it spread to the Etruscans, who were mighty plasterers, and probably influenced the Romans.

Roman Plaster Work

Our knowledge of Roman plaster work is very copious and definite, both from the abundance of examples and written records, for Vitruvius, Julius Cæsar's military engineer, has left us a very valuable and detailed work on architecture, composed about the year 16 B.C. Certain sections of his book are devoted to the subject of plaster, and from these we can gather a fair idea of the materials and methods employed in ancient Rome to produce the marvellously fine work which has endured to our own days, giving evidence of its close texture, hardness, polished surface and often superb modelling in low and medium relief. It must be remembered that much of this work has endured for upwards of 2,000 years, been subject to the ravages of the elements and the still more destructive activities of man, yet the fragments are still to be seen in perfect condition.

Unlike the Egyptians, the Romans used lime, not gypsum, and it underwent most elaborate preparation for the production of plaster, as the following directions will show.

Slaking Lime for Stucco.—Vitruvius tells us that for stucco "the lime should be of the best quality, and tempered a long time before it is required for use, so that if any of it be not burnt enough, the length of time taken in slaking it may bring the whole mass to the same consistency." To secure this uniformity in the material, the burnt and water-slaked lime was chopped with iron hatchets and turned over. This was to be repeated at intervals if necessary until the choppers became glutinous with the unctuous lime, which indicated that it was rich through slaking.

Foundations for Plasters

For lathing he recommends "Greek reeds bruised and tied with Spanish broom," or in default "marsh reeds tied with cords." This is most interesting, as it connects up with the primitive wattle and daub, and shows us a method which persisted even in England to within 300 years ago. Great stress is laid on the necessity of selecting woods which will not rot or shrink for ceiling joists and cradling purposes, so cypress, heart of oak, olive, box and juniper are recommended.

Plastering.—The plaster was a mixture of the unctuous lime and sand. It was laid on in two coats of lime and a little sand, and then a little lime and more sand, and when dry polished with chalk or marble. "The first coat on walls," says Vitruvius, "is to be laid on as roughly as possible, and while drying the sand coat spread thereon. When the work has dried, a second and a third coat are laid on. The sounder the coat is, the more durable will the work be. The coat of marble dust follows, and this is to be so prepared that when used it does not stick to the trowel. While the stucco is drying, another thin coat is to be laid on; this is to be well worked and rubbed, and then still another thin coat is laid on, finer than the last. So with three sand coats and the same number of marble dust coats, the walls will be solid and not liable to crack." Experience has, indeed, proved him to be quite correct. It is interesting to note that the fact that the richer, fatter, the lime, the more plastic and capable of being worked it became. This was, and is, a matter of immense importance when it comes to a case of modelling. The importance attached to thoroughly working and rubbing is seen from the following passage: "The wall that is well covered with plaster and stucco, when well polished, not only shines, but reflects to the spectator the images falling on it. The plasterers of the Greeks not only make their stucco work hard by adhering to these directions, but when the plaster is mixed cause it to be beaten with wooden staves by a great number of men, and use it only after this preparation. Thus, some people cutting slabs from ancient walls use them for tables and mirrors." It will be acknowledged that the test was a severe one, both as to stability and high degree of polish. From these sentences it will be seen that a distinction is made between plaster, or rough stuff for rendering, and the stucco for the setting or finishing coat. But it is obvious that for it to be possible to cut slabs capable of being used as tables, every coat must have been perfect of its kind, and worked into a homogeneous, or at least closely adhering, mass, which was only possible with good material and skilful workmanship.

The variations of plasters for special purposes, and the directions as to treatment for ornamental purposes, etc., show evidence of sound knowledge and practice. For such reasons the passages relating to these subjects are worth careful study, for they have lessons useful to the present-day plasterer.

Contrast with Modern Methods.—The methods adopted are not very different from that of our own day, though we stir, but do not beat or hammer, the thickening plaster. The writer's own experience in a sculptor's studio is that you can "perish" the plaster by over-stirring and beating. The sculptor of to-day finds that a very liberal use of water produces the result obtained by "beating" in a more certain and satisfactory manner. Plaster and cement are like the dog and the walnut tree, only, instead of beating, they are made better by ample liquid refreshment.

Preservation from Damp of Mural Paintings and Decoration

But of course the water having developed all the adhesive properties of the cement and plaster, which is only another form of cement, if the walls are to be beautified with paintings, like the splendid work done in

Greece and Italy, and later in our own country, then it is essential that the water must be thoroughly dried out.

The Greek Method.—Here Vitruvius is again interesting. In the first process "trullisatio," the wall was coated with trowelled plaster made up of coarse sand, brick and tile well powdered, and lime—1 part lime to 2 parts sand, and 1 part brick and tile. This left a rough surface, so that the following materials would stick fast to it.

Then followed what they called "arenatum" which we call sand mortar (1 part lime, 2 parts wet sand). Of this three coats were applied while the previous coatings were still damp. This was very solidly done, for everything depended upon it.

The final coating was of mortar made up with marble, and was known as "marmoratum." This lime and marble dust (both fine and coarse) was laid on, the coarse being used first, the two later coats being fine. The last coat was applied after the two coats had dried. It was composed of 2 parts marble dust to 1 part of lime.

Experience showed that if this elaborate process was curtailed the finished plaster-work surface of the wall or bath was apt to crack, and the colours of wall decoration were not so brilliant, nor was the whole of so lasting a character.

The thickness of the best work in the baths of Titus is given as 1¼", the surface being as hard as marble.

Vitruvius' Treatment of Damp Walls.—In our recent national effort to supply houses for the people after the building stoppage during the War, a wall with an inner space enclosed in it was much talked of. Vitruvius was long before us. Rightly or wrongly—for our own experiments have not been very conclusive—he advises an inner wall for dampness, that is, an air space or channel.

Examples of Work.—As we have stated, numerous examples of Roman plaster work of different periods still exist in Italy and elsewhere. Roman mortar was an extraordinarily fine material, and the Romans carried their secret of making it wherever they worked, for their lime mortar masonry is as good in Britain or in Africa as it was in Rome itself. The same may be said of the plaster. But it is the stucco which shows to greatest advantage. As we have seen, it was finished smooth and highly polished, to be left white or painted, but it was also elaborately modelled. There is a specimen of low-relief plaster work of Vitruvius' time in the Capitol Museum of Rome. Another example of first-century work on a vaulted ceiling shows the space divided up into rectangular panels by beaded mouldings, containing beautiful figure work modelled in low relief. This shows Greek tasteful restraint. We have other examples of the more exuberant Roman style, in which we find rich mouldings for rectangular panels enclosing oval or circular panels, filled in with human figures, birds and animals, the spaces between filled up with portrait medallions, candelabra, scrolls and the fantastic "arabesques" of leaves, curved and merging into heads, torsos and weird creatures. These are of peculiar interest, because they gave rise to the decorative splendour and extravagances of the Renaissance.

As in Greece, the Romans employed hardened plaster for pavements, although for this purpose their incomparable mosaics predominated.

Modelled plaster ornament was applied to walls as well as to ceilings. It was sometimes left white, usually vividly coloured and brilliantly polished.

Byzantine Plaster

When Constantine the Great transferred the capital of the Roman Empire, A.D. 330, to Byzantium (Constantinople), he took with him Roman and Græco-Roman artists and artisans, among them plasterers and stucco workers. In that great city on the verge of Europe, the influence of the glowing East was felt, particularly in art. To the already rather extravagant form of decoration prevalent in Rome was superadded the luxuriant fancy and love of rich colouring innate in Asiatic art, and this was seen reflected in the plaster work of the Byzantine craftsman. The ornamentation was exuberant, rather coarsely modelled, easily slipping into the grotesque and richly coloured. The " arabesque " designs of metamorphosed vegetation, with scrolls blossoming forth into animal forms, took on an Oriental character in which evidence of paganism is not far to seek. But the Latin Court at Byzantium was Christian, and this soon had a sobering effect on art.

Religious subjects and symbolism were introduced, and were inspired by a mystical feeling, which was translated in art by means of a primitive style of design, in which the flat, almost shadowless and simplified form of perspective was adopted. This brought about the abandonment of modelling in favour of flat painting and mosaic work, which lent themselves admirably to flat treatment and gorgeous colouring with gold and paint or rare marbles. Therefore plaster speedily fell into abeyance, so far as decorative stucco work is concerned. But the plastering remained exquisite, providing splendid grounds for painting and mosaic inlays. For both the common plaster and the fine stucco, lime was used, and apparently the same elaboration for maturing after slaking was observed.

Saracenic Plaster

That the Arabs were great plasterers from an early age there can be no doubt, but what we know of their work reflects the influence of Byzantine art, which spread far East, and through Arabia to Northern Africa, and thence to Spain.

Saracenic plaster work is fine, hard and durable. It was mostly coloured, except when used in coarse form, as a rendering on brick, mud or masonry walls. It was highly worked, modelled, but not in the sense that we in Europe understand the word modelling. The plastered surfaces of walls, arches and ceilings were often covered with foliage, flowers, scrolls, geometric patterns and Arabic script in medium relief, the designs being very regular and giving the impression of being stamped on the wet plaster, which was generally the case.

Side by side with this surface embossed decoration was evolved a peculiar form of ornamentation, known as the stalactite. Longish prisms of plaster were formed, and given different shapes, usually having the lower half scooped out as diminutive alcoves. These prisms were set in the plaster covering walls, arches and vaults, so as to form a mass of regularly disposed steps, easily adapted to the architectural forms prevalent. It is in the nature of planted ornament, the damp plastered wall or ceiling being cut into to receive the prisms, which are then cemented with semi-liquid plaster. Pendants and hanging cornices are supported by reinforcements generally in the form of iron chains, fixed to the wall and ornaments and heavily plastered. Cradling is also used for cornices.

Another interesting use of plaster is in the forming of panels, sometimes on wood frames, which are pierced while wet, the holes being glazed with large or small pieces of coloured glass set in the plaster. Such panels are employed as screens or windows, and may be flat or curved for dome work. The system of building up this kind of plaster ornament is well shown by Owen Jones, who says that these tiny blocks permitted combinations to be made " as various as the melodies which may be produced from the seven notes of the musical scale. With these stalactite plaster bricks they formed cornices, arches, domes and pendants with the greatest facility." The surfaces of the larger blocks were often decorated with fine lace-like geometric patterns and scrollings. All was painted in strong colours with gilding.

While they recognised and thus made use of the plastic qualities of good plaster, they were masters in producing smooth, flat-plastered surfaces for interior walls, domed ceilings, etc.

Other Asiatic Plaster

In Asia we see the influence both of the Byzantine and Saracenic plasterers. Much of the plaster work throughout Asia, from Persia to India and China, is over-worked, a mass of modelled decorated mouldings, beadings, cornices and friezes loaded with sculptured subjects, now purely floral and geometric, anon, as in polytheistic India, a medley of grotesque forms, half human, half bestial. Much of the floral work is very beautiful, and so is the geometric patterning ; but it is overdone.

In Persia a good deal of plastering was in the nature of perishable *kahgill*, a compound of mud and chopped straw. But for better-class work they employ *gatch*, a gypsum plaster, which hardens well and takes on a high polish. There is little casting, the modelling being done by hand in the mass. However, the stucco is applied in a peculiar manner. It is made as plastic as possible by constant kneading. Largish balls are formed, and after working up are forcibly pressed into the wet plaster rendering, and then flattened by hand to the desired shape. Modelling follows rapidly, and another ball is stuck in, flattened and modelled, and so on until the whole surface is covered. This method is applied to walls, ceilings, arches, pillars and all other architectural features. It is thus seen to be closely allied with the stalactite style of ornamentation. It will, however, be understood that preliminary floating is necessary, and the fact is that the flat, smooth-plastered surfaces in Persia are as fine as they are in India.

In India the plaster used is made of lime and known as *chunam*. In the best work the lime is very pure and is mixed with fine sand. It produces a good stucco, lending itself readily to the elaborate modelling work and also to the polishing process much in favour. For plain wall floating and pavement work, a mortar of quicklime and sand is used, but the working up is very prolonged, so that slaking takes place before actual plastering. When fine sand is used, this form of plaster takes a high polish and proves durable.

TUDOR STYLE OF COVED CEILING, WITH PENDANTS; FIBROUS PLASTER.

PANELLED AND RIBBED CEILING IN FIBROUS PLASTER.

CHAPTER III

MODERN DECORATIVE PLASTER

Italian Developments : Gesso Grosso ; Gesso Sottile ; Gesso Bolognese ; Stucco—French Work : Papier Mâché ; Carton Pierre ; Pâte Coulante ; Paste Composition—English Plaster Work : Pargeting ; Early Plaster Work—Exterior Work.

TENDENCIES in the Roman Empire after the second century of our era, both as seen in the West at Rome, Herculaneum and Pompeii, and in the East at Byzantium and other large cities, reduced plastering to a subordinate place. The mission of the plasterer was chiefly to prepare good surfaces for painting on, or producing smooth, slow-setting matrices for mosaic work. But the craftsmanship shown in preparing the plaster and in applying it was of a very high order, as existing examples clearly testify. There must also have been a great demand for plain plaster, both in the West and East, for interior and exterior rendering, as smooth surfaces over rough masonry and slighter forms of construction were needed.

In Europe on the break-up of the Empire much tended to keep the plasterer to plain straightforward craftsmanship, for important buildings during the Romanesque or Norman period of architecture (which in England lasted from the eleventh to the twelfth century) were mostly of stone or brick, both of which were decorated with carvings. Apart from the ecclesiastical edifices and the castles, however, there were the humbler domestic buildings, and on these plastering was applied with technical skill and artistic appreciation, in England certainly to a remarkable extent.

ITALIAN DEVELOPMENTS

It is generally assumed that the revival of the art of the plasterer dates from the Renaissance, which began about 1493, though in the applied arts not very evident until 1500 or a little later. That lime plaster, or *stucco*, was extensively used before that there can be no reasonable doubt, though we have no direct mention of it. But Cennino Cennini, a painter, born about 1360, wrote a book in 1437 on art as practised in his days and by his predecessors. In this he has much to say about *gesso*, a gypsum plaster, used, among other purposes, for decoration in relief and taking moulds. He mentions three kinds, *gesso grosso*, (or coarse), *gesso sottile* (or fine) and *gesso bolognese*, or *volterrano*.

Gesso Grosso.—Of gesso grosso he says it is prepared by mixing ground alabaster (gypsum) with glue. It is to be kept warm and applied with a spatula to panels or slabs to be painted on. "Then," he continues, "take some of this ground plaster, warm it, take a soft hog's hair brush and give a coat on the cornices and foliage (the raised ornament), and on the even surfaces with the spatula. Give three or four coats on the other parts of the cornices." This, it will be seen, was for ground work.

Gesso Sottile.—For all fine work gesso sottile was to be used. He tells us: "This is made from the same plaster as the last, but it must be well washed and kept moist in a large tub for at least a month; stir it up well every day until it almost rots and is completely slaked, and it will become as soft as silk. Throw away the water, make it (the plaster) into cakes and let it dry. . . . It is used for gilding, for working in relief and other fine work." Thus he provides for both water and air slaking.

As to tempering, Cennini directs that "having laid on the gesso grosso, rubbed down the surface, and polished it well and delicately, put some cakes of the gesso sottile into a pipkin of water, and let them absorb as much as they will. Put a small portion of it at a time on the porphyry slab, and, without adding any water to it, grind it to an impalpable powder. Put it then on a piece of linen-cloth, strong and white. When you have ground as much of it as you want (for you must consider what quantity you will want, that you may neither have to make two portions of tempered plaster nor to throw away any good plaster), take some of the same glue with which you tempered the gesso grosso. You must make sufficient at one time to temper both kinds of gesso. The gesso sottile requires less tempering than the gesso grosso; the reason for this is, that the gesso grosso is the foundation of all your work, and that how much soever you press the gesso grosso, a little water will still remain in it. For this reason make the same kind of glue for both. Take a new pipkin which is free from grease, and if it be glazed so much the better. Take a cake of this gesso sottile, and scrape it fine with a knife, as you would cheese, and put it into the pipkin. Put some of the glue on it, and stir the gesso as you would a paste for making fritters, smoothly and evenly, until there are no longer any lumps. Procure a cauldron of water, and make it very hot, and put into it the pipkin containing the tempered gesso. Thus the gesso will become warm, but will not boil; for if it should boil, it would be spoiled. When it is warm, take your picture, and a large and very soft pencil of hog's bristles dipped in the pipkin, and taking up a proper quantity at a time, neither too much nor too little, spread it evenly over the level surfaces, the cornices, and the foliage. It is true that in doing this the first time you should spread and rub the gesso with your fingers and hand wherever you can, and this will incorporate the gesso grosso with the gesso sottile. When you have done this, begin again, and spread it with the brush, without touching it with the hand. Let it rest a little, but not so long as to dry thoroughly; then pass over it a third time with the brush, and let it dry as usual. Then give it a coat on the other side; and in this manner, always keeping your gesso warm, give the panels eight coats. Foliage and relievos require less, but you cannot put too much on cloths. This is on account of the rasping or rubbing down, which is done afterwards."

He adds: "There are, nevertheless, some persons who grind gesso sottile with water and glue. This is proper for grounds where no gesso grosso is used, which require to be more diluted. This same kind of gesso is good for raising foliage and other works in relief, which are frequently necessary to be done. But when you are going to execute works in relief with this gesso, add to it enough Armenian bole to give it a little colour."

Gesso Bolognese.—Gesso bolognese or volterrano was prepared from white alabaster found in quarries near Bologna and Volterra. This Cennini recommends for taking mouldings of the human face. It was to be fresh burnt, sifted and tempered with water.

Further interesting information about gesso volterrano is given by Vasari. In his Life of Andrea Verrocchio (1435–1488) he says: "Andrea (Verrocchio) took much pleasure in making models of that kind of plaster (gesso) which is made of soft stone found in Volterra and Siena, and in many other parts of Italy; which stone, burnt in the fire, and then pounded and made into a paste with cold water, becomes so supple that you may make what you please of it; and afterwards it becomes so hard and firm that it may be used for making casts of whole figures. Andrea then used to form models of natural objects, namely, hands, feet, knees, legs, arms and bodies, from such casts, for the convenience of having them always before him, so that he might imitate them. In his time began the practice of taking casts of dead persons, at a small expense; so that there might be seen in the passages, doors, windows and cornices of every house in Florence an infinite number of these casts so well and naturally done that they appeared alive. And from that time forward the practice was, and is, still followed; and very useful has it been to us in procuring many of the portraits which are placed in the palace of the Duke Cosmo; and for this we are greatly indebted to the skill of Andrea, who was the first who made use of it."

Instructions are given by Cennini for producing raised ornament, such as foliage, etc., in pictures, by applying gypsum tempered with glue, with a paint brush.

He then goes on: "There are many different ways of working in relief, therefore I will teach you some of them. With the same kind of gesso or with a portion of stronger glue, you may cast heads of lions or of anything, modelled in earth or in chalk. Oil the mould with lamp-oil (*olio da brucciare*), fill it with the gesso well diluted, and let it cool; then remove the gesso with the point of a knife, and blow (*souffare*) it strongly. It will come out quite clean; let it dry. Afterwards, when ornamenting anything with the same gesso, you must proceed in this manner, with the same gesso which you used for the ground, and with the same casts, first oiling the part with the pencil where the heads are to be fixed, then press them with the finger, and fix them in the usual manner. Afterwards, with a pencil of minever lay a coat or two of the same kind of gesso on the parts you mean to appear in relief, and which you have previously marked out. Afterwards remove with the knife any irregularities."

We shall have more to say about gesso in the course of our remarks.

Stucco.—Cennini when giving directions for taking gesso casts for gesso moulds states that if a little brickdust is added to the gesso for casting it will be improved. This shows that he was aware that a pure gypsum plaster would not answer for all purposes. It becomes still more evident when he gives directions for working mortar in relief on walls. "I shall," he says, "teach you how to raise designs in relievo on walls. In the first place, there are certain parts of the wall that are either circular or enriched with foliage, on which the mortar cannot be spread with the trowel. Take some lime and sand, both well sifted. Put them into a basin, and, with a large hog's-hair pencil, make them into a paste with water, and apply several coats of this mortar with the same pencil on these places. Then polish the parts with the trowel, and the work will be done. You may paint on it in fresco or in secco, as I directed you when speaking of fresco painting. Grind a little of the before-mentioned lime on the stone, then make what parts you please in relief on the walls, as I have told you to do with regard to pictures, especially when the lime is rather fresh."

Further direct reference to lime plaster or stucco is found in a Marciano Manuscript in the Library of St. Mark, Venice, in which is given a recipe "as tried by Messere Jacopo de Monte San Savino, the Sculptor." It is described as "Admirable Stucco for making and modelling figures and for colouring and resisting water." "Take of finely powdered travertine 5 lb., and if you would have it fine and more delicate, take fine marble instead of travertine, and 2 lb. of slaked lime, and stir and beat them well together like a fine paste and execute what work you will with it, either by forming it with your hands, or in moulds, and dry it in the shade. And if you wish to colour it white (when the work is dry enough to be nearly firm, but not quite dry, grind white lead with water in same way as colours are ground, and the flour of sifted lime, and apply it with a brush, and it will be very white and resist water. And if you wish to colour it with other colours, let the work dry perfectly and then colour it; but these colours will not resist water like the white. If then you wish the colours to resist water, apply on the work the above-mentioned composition and paint with oil colours." This manuscript was written about 1503. Travertine is a calcium carbonate rock formed by deposits from springs and rivers, principally in volcanic regions. An interesting detail is the manipulation of the stuccos apparently much in the same way as prevails among Persian plasterers.

It is with stucco of this kind that the great masters of the Renaissance did most of their splendid work. But there were experimenters. We have seen that Cennini recommended the addition of brickdust to lime stucco. Donatello, who died in 1466, has left us some remarkable high-relief modelling in a mixture of lime plaster and brickdust, one of these being in the Victoria and Albert Museum. This mixture dried so hard that an example at Padua was long considered to be of terracotta. Another important advance is recorded by Vasari, who states that Bramante, the celebrated architect, used a mixture of lime to decorate the exterior of houses with festoons and friezes of foliage. One of the houses so decorated was that of his nephew, Raffaello. Now, Raffaello and Giovani da Udine were commissioned by the Pope Leo X to open the Golden House of Nero at Rome. While Raffaello was enraptured with the decorative painting, da Udine paid special attention to the perfectly preserved ancient plaster. It was to da Udine and Jacapo Sansavino, working under the direction of Raffaello, Giulio Romano and other artists, that we owe the extraordinarily lavish decoration of public buildings in Rome, Florence, Venice and elsewhere with stucco modelled in high relief, magnificently painted and gilded. The chief thing to note is that da Udine professed to be guided by a study of the ancient Roman stucco and the directions of Vitruvius. So he, and his followers, used well-slaked lime from selected stone or marble, mixed with fine white sand.

P. Ligorio, who was joint architect with Michelangelo at the Vatican, and had employed da Udine, has left a record, preserved in the Bodleian Library, Oxford, of the stucco then used. He directs: "Take three parts of pounded Parian marble, easily got from among the ruins of Rome and from broken statues; add one part of lime which is to be perfectly slaked by letting it lie in a heap covered with pozzalana and exposed to the sun and rain for at least a year. The lime is to be made from pure white marble, not from travertine, or any other stone which is full of holes and yellowish in tint; mix a day before with sufficient water on a tile floor. The first coat to be mixed with coarsely pounded, the finishing coat with finely pounded white marble." Pozzalana is a volcanic rock. Some craftsmen added a small quantity of burnt gypsum to the lime putty.

These artist plasterers produced a very plastic material, with which they plastered walls and ceilings, and blocking out in rough projecting masses, proceeded to model and carve with the freedom and ambitions of true sculptors. Not content with modelling flowers, foliage and arabesques, they proceeded to figure work, now in miniature, often enough life size, and all this in conjunction with the most elaborate architectural details, mouldings, pediments, complete entablatures, covings, covering flat walls, pillars with their capitals, vaults, soffits of arches and staircases, the framings of doors and windows with magnificent modelled plaster work.

While the general practice was to render the walls and then add plaster to be modelled by hand, wooden stamps were also used to impress in the plaster *in situ,* the pressed-out ornament then being finished by hand.

In the seventeenth and eighteenth centuries, the use of ornament became excessive, and as the whole space, except the panels left for painting, was covered, the stucco ornament became larger and even more projecting; and being given a very highly polished smooth finishing coat, it aped the tortured marble sculptures of the baroque period. But however much one may criticise the art of these centuries, the craftsmanship continued to be of the best, and from the middle of the eighteenth century to the first decades of the nineteenth, many of the craftsmen, finding no scope for their efforts in the politically disturbed peninsula, migrated and carried their knowledge to France, Germany, Spain and England. Wherever these Italians went, however, they experienced difficulties as regards materials, missing their favourite marble dust and the finest gypsum. At home, well into the nineteenth century, the habit of long maturing the water and air slaked lime was maintained. In many works, three or more large tanks were kept, each in turn coming into use, but not before they had had many years to mature.

French Work

We have no detailed records of native plaster work in France prior to the sixteenth century, though we know that the gypsum quarries round Paris and lime deposits elsewhere were actively exploited. Much decorative work before the sixteenth century was carried out in *papier mâché* and *carton pierre,* and these materials have persisted, among the French (and to some extent in the seventeenth and eighteenth centuries among the Germans, to a much less extent with us) ever since. Fine examples of papier mâché work, including huge trophies of arms and heraldic devices, are to be seen in the Louvre at Paris, notably in the Council Chamber of Henri II and also at St. Germain.

Papier Mâché.—The methods for the preparation of papier mâché varied considerably from time to time. The most primitive way was to soak paper in glue and press successive layers in oiled moulds until the desired thickness was obtained. These casts were taken out, planted in position and painted or gilded. Another way is to form a paste by tearing up soft paper and steeping in thin glue water, stirring frequently and when soft squeezing out the water and mixing the paste with glue. This paste is pressed into moulds, and then backed by paper soaked in glue. Large casts are usually backed by wood frames or cradles. Long lengths of friezes, and also large panels, are cast in this material. It possesses the great advantages of being very light and tough; it can be nailed in position; but unless varnished, it is subject to the influence of moisture and heat. With proper precautions, however, it is very enduring and will receive any kind of decoration.

Carton Pierre.—Carton pierre is an improvement on the latter kind of papier mâché. It is a preparation of paper, glue, flour and a little whiting or gypsum. The best paper is that made from rags, but any tissue paper, even such as that used for wrapping oranges, does capitally. It is weighed dry, torn up and soaked in water. Good fish glue, 2 lb., is dissolved in two quarts of water; to this ½ lb. of flour is stirred in carefully so that there shall be no lumps. The paper pulp is squeezed dry and added. The whole is boiled for an hour, and then simmered for another hour or more, when it is spread out on a marble slab and dusted with finely powdered whiting, the whole being kneaded until sufficiently stiff for pressing into moulds. The French as a rule brought the paper pulp to the boil, then adding the dissolved glue with rye flour and whiting. Plaster of Paris was used as a stiffening. Like papier mâché, it is very light and tough, can be moulded into quite thin casts, and is easily fixed into position by nailing or otherwise. It can be treated with distemper colours, oil paints, gilded and varnished. It does not require elaborate cradling.

Pâte Coulante.—Pâte coulante, or liquid paste, was another substance largely used in France, being employed chiefly for elaborate moulded work, the substance being poured into the oiled moulds. This material was prepared by mixing dry equal proportions of plaster of Paris and whiting, and then gauging with dissolved glue to produce a freely running paste, a little alum being added to harden the casts. In some cases tissue-paper pulp was added, in about the proportion of one-half the quantity of plaster.

Paste Composition.—Paste composition was also much in favour in France, being introduced in the reign of Louis XIV. It is of particular interest to us because it was adopted with slight modifications by Robert Adam, and, as his much-talked-about "compo" or "composition," was much used by him for the decoration of ceilings, walls, chimney-pieces, woodwork, etc. There are various ways of preparing the material, which, when fresh, has the appearance and feel of a stiffish dough. The following may be taken as a standard composition. Take 2½ lb. of refined glue, 2 lb. of resin and 1 pint of linseed oil. Mix the whole in a warm water bath, and when thoroughly compounded pour out on a marble or hard wood slab. Knead, and roll out into a firm but pliable dough. It is warmed before use. It has been found that by adding about ¼ lb. of Burgundy pitch, shrinkage of the paste during drying is avoided. An equal quantity of Venice turpentine will also prevent shrinkage, and have the additional good effect of increasing toughness. This "compo" can be cast in sulphur, metal or wood moulds. The florid work of the Louis XV period will show how plastic such material can be and how fine casts can be made from it, whether in the form of mouldings, scrollings or other ornament in high relief.

Stucco.—That the use of gypsum plaster was well understood in France early in the eleventh century, we have plenty of evidence. But the use of stucco as a decorative material was introduced into France by Italian workers not much earlier than the opening years of the sixteenth century. But it was not until 1536, when Francis I imported Italian workers under Romano and Primaticcio for work at Fontainebleau, that the Italian style made headway. After that, however, the old flat incised and low-relief decoration of the native workers was abandoned for the more massive work. Under the new conditions the stucco was used very boldly in two ways, structurally, as in the forming of coving and coffering on ceilings, the casing of piers and modelling of capitals, and in the decorative modelling. In the better kind of work, high-relief, large-scale modelling was the rule, and much figure work was introduced. Classical details, as to mouldings and friezes, were brought into fashion, and during the Regency and Louis XV periods blended with the extravagant scrolling, rock and shell ornaments. A tall, slim, elegant type of the human form, especially the female form, was made fashionable by Primaticcio early in the sixteenth century, and had a lasting effect on French decorative art. They were very beautiful, but carried out on somewhat too large a scale for general harmony, and led directly to the exaggeration of the heavy trophies and other ornaments introduced towards the end of the reign of Louis XIV, which had a very bad effect on art in Germany and was not without its influence in England in the early part of the eighteenth century.

Later French work has been classical in style, very severe under the First Empire, then with an admission of the realistic, especially in floral work.

English Plaster Work

In England the preparation of lime plaster was well understood at an early period, possibly as a tradition from the old Roman artificers. At all events we have examples of Norman and early Gothic work of the very best description, practically all plain plastering for painting, as carved work was carried out in stone and brickwork. Not content with lime plaster, however, about 1160 plaster of Paris, or gypsum, was introduced over here, and thenceforward was in constant use in connection with lime plaster, wherever great smoothness and whiteness was demanded. In the archives of the Guildhall, London, there is an agreement dated 1317 whereby Adam le Plasterer bound himself to find plaster of Paris and therewith " plaster and complete the hall " of John de Bretagne, Earl of Richmond. In the same archives for 1350 we find the City Regulations as to the wages to be paid to different classes of craftsmen, and here both " plasterers " and " daubers " are mentioned. This is important as showing that the craft was well developed, the several descriptions of work being provided for. Walls were rough-cast, plastered (mostly for painting), wrought with plaster of Paris and pargeted or white limed.

The rough-casting was carried out by the daubers and was much in the nature of the kind of work known as dinging. Plastering was three-coat work, the finishing coat being brought to a beautiful close-texture and smoothness, so that painting on it has lasted 800 years and more. Plastering and white lime were largely used in domestic architecture of the framed building type, our charming half-timbered work so largely used in town and country, which lasted well into the Jacobean period. The spaces between the timbers were filled with bricknogging or stone rubble, or merely with lath-and-plaster, the surface smoothed and limewashed at intervals.

The wrought work in plaster of Paris was smooth plastering for interiors, to be left smooth, or run as mouldings and perhaps modelled with low-relief ornament.

But it was in the pargeting and white liming that the native genius showed itself for at least three centuries, down to Jacobean times.

Pargeting.—Pargeting in its best form is essentially an English art craft. In its old form it constituted a simple but very effective way of decorating plaster. A smooth finish was obtained, and while this was still soft it was impressed with designs, usually of a repeat order. Two methods were used. Designs were cut out on wood blocks and these were used as stamps. For small designs, however, suitable for borders and banding for example, the patterns were cut out on cylinders, mounted loosely on handles to form rollers, which could be rolled up, down or along the damp surface. Variations were introduced in carving the designs, by scooping out (which would produce embossed ornament), or in relief (which would produce depressed or sunken ornament). The craftsman had to use his judgment in selecting, applying and combining his ornaments. In many cases we find an "all over" repeat pattern. In others there are bandings, borders, corner pieces, and well-filled panels. A certain amount of freehand work is also traceable, apparently carried out with the edge of the trowel or a small raking implement, to produce curved lines, often supplemented by dots. A clever development of this freehand style in modern work is to be seen at Becontree, where some of the cement-plastered houses are ornamented by means of the trowel, pointed laths and wire brushes with combinations of curves, which look very well.

Reverting to old pargeting, in a series of articles on local variations in construction, Mr. Edwin Gunn, A.R.I.B.A., writes: "The distinctive scratched or impressed plaster patterns which enliven the protective coating in East Anglia . . . are deserving of careful study. This was evidently a local art, like thatching, where each plasterer contrived his own tools—simple combs, rakes and the like—and displayed his own preference in their use. There is a certain similarity in the framework surrounding the patterns which appear to have little reference to any structural timbers behind, but aim at dividing the surface between door and window openings into manageable panels. It may be surmised that these margins, which are usually defined by shallow flush mouldings, were applied as screeds, and the intervening spaces plastered and patterned singly after the framework had been laid on." The designs are all-over chevrons, wave patterns, basket weave, etc.

In other parts of the country the work was often more formal. Small flowers, geometric patterns, leaf designs and a kind of strapwork. The patterns were often painted, usually a shade of brown or yellow, but were distinctive enough without that.

Early Plaster Work.—Mention has already been made of the early use of plaster in England as well as the introduction of gypsum plaster in the middle of the eleventh century. By the beginning of the twelfth we have evidence that the fire-resisting and sanitary qualities of lime plaster were thoroughly appreciated, for King John ordered that all houses on London Bridge and by the river-side were to be plastered inside and out, as well as all houses where bread was baked and beer was brewed. Plaster was also used in ecclesiastical buildings and palaces, for traces of it are still extant, and we know from records that painting on ceilings and walls was carried out extensively and often at great cost, which, of course, implies good plaster grounds, for it is obvious from the accounts that much of the painting was not on wood.

We have seen that the pargeters were at work in quite early days. Their guild received a charter of incorporation from Henry VII. It seems unquestionable that these men developed their craft on art lines, and from executing repeat patterns by stamping and freehand scratch-work they came to more studied work of impressing and modelling, such as we see in the very fine strapwork plaster designs of the Tudor period. This strapwork is, of course, an evolution from the spiral and ribbon designs which we associate with Celtic art, only in the Tudor designs the ribbons are wider, flatter and drawn in more formal patterns, suggesting broad leather thongs rather than the flimsier style of the Celtic spirals. This suggestion of leather work is also to be found in the split and upturned scrollings constantly associated with the plain straps. With bands of this kind very fanciful and handsome designs can be produced, the flat bands forming, with occasional interlacing, frames, curves, cutting up ceilings in square, oblong, oval, circular and diamond-shaped panels, which could be filled with stamped or raised ornament, while the flat bands themselves could be embellished with running patterns, embossed or incised as well as painted. At the crossings of the straps medallions were placed, usually containing an armorial shield, a crest, badge or some symbol. They were largely floral or geometrical in shape, direct copies from Gothic ornament, and this Gothic element was also seen in the large pendants which frequently replaced the medallions at the intersections. Such pendants were often in the shape of half-open buds or flowers, or purely architectural in form, as with the stone pendants in Gothic edifices. Ceilings of this description, with a considerable admission of the heraldic in the decoration, became very elaborate during the reigns of Elizabeth and James I, and, although usually carried out in stucco rather than plain lime plaster, were clearly a native art, developed on distinctive lines, though often in association with Italian workmen, and with Italian-compounded materials. It is well, however, to insist on the strength of the position of the native plasterer, and his inclination to develop along national lines, for this accounts for a good deal in the later progress of decorative plaster work in England.

Stucco.—Henry VII, who knew something about the revolution in arts and crafts brought about by the Renaissance on the Continent, provided in his Charter to the " Plasterers and Pargeters " for the study of and research as to plaster and plastering, which shows that the spirit of inquiry was astir within the corporation. It was left to his successor to carry the matter further. Henry VIII was a lavish patron of the arts, very curious as to what the French and Italians were doing, nevertheless fundamentally English in his outlook. Thus, while he built a splendid palace at Cheam and was anxious to embellish it with the finest Italian work, Nonesuch was English in planning and construction. It was a half-timbered building, with panels filled with bricks and plaster, and had a main body and two wings. For the wrought plaster work he wanted Italians, so he brought over the brothers Luca and Bartolommeo Penni and others, who, contemporary foreign travellers tell us, adorned the King's side with stucco figures in medium relief, but larger than life, representing the Labours of Hercules, while the Queen's side was adorned with nude female figures of mythological origin, also larger than life.

It is significant that the modelling of the figures was kept low, possibly in accordance with the English tradition and taste, but also, no doubt, owing to difficulties in the compounding of the stucco. The Italians then and at later times had much difficulty in this, finding that the marble dust was not abundant, and that the imported gypsum did not answer all their desires. For much of the work carried out at Hampton Court Palace, composition was used instead of plaster for decorative purposes. A compound of gypsum, rye flour and glue was used in some cases; evidently a suggestion from French practice.

Although the Penni brothers modelled huge figures in stucco, figures did not play a large part in Tudor plaster work, in spite of a constant flow of Italian craftsmen into England. We hear of master men of that nation over here under Edward, Mary and Elizabeth. We see small figures, and occasionally portraits, introduced in a type of Renaissance arabesque and Gothic tracery friezes, but rarely otherwise. While a good deal of experimenting was going on, as foreshadowed in Henry VII's Charter, and classical elements crept into plaster decorations, it is true to say that, in the main, decorative plaster work in England under the

Tudors was of the strapwork and Gothic character. This persisted well into the Jacobean period. Though there has been a lamentable amount of destruction, happily we possess a great number of exceptionally interesting and beautiful specimens of plaster work of these periods, mostly of ceilings, more rarely mural. The Eastern Counties and the Midlands are especially rich in these treasures. In London very few remain, but in the neighbourhood of London (as at Hatfield) there are some splendid specimens of Jacobean work. The subject can also be studied at the Victoria and Albert Museum.

In Jacobean plaster the human figure, boldly modelled, appears much more frequently, in alliance with classical arabesques and architectural features.

As regards this Tudor work, it should be said that much of it was painted and gilded, the heraldic motifs lending themselves well to this treatment. Often the plain surfaces, even the flat spaces on the strapwork, were left white, only the ornament being decorated with colour and gilding. The plaster work was usually carried down to form a deep, heavily wrought cornice, but rarely farther, for wainscoting and the hanging of walls with tapestry or other needlework was the fashion, while door and window frames were adorned with carved wood.

With the rise of Inigo Jones, the Gothic and Renaissance elements fell into disuse, and we had a rather florid classic style instead. As regards ceilings, this meant large, flat, painted panels with heavily wrought plaster frames, the cornices being broad and projecting, the friezes deep. Most of the enriched plaster work was of the repeat order of decoration. It was nearly always cast and then fixed in position, not modelled *in situ*. Jones planned on a grandiose scale, establishing the fashion for stateliness and formality in building and decoration, and we find his favourite ornaments for plaster work to be unadorned classic mouldings, and foliage or flowers treated in a formal manner. He brought the plaster down on the walls, simply as mouldings on cornices, and over doors or framings to doors and windows and sometimes for overmantels.

His successors were less restrained. From smooth sweeps and rounded edges the acanthus became rugged, the flowers and leaves more varied and prominent, military trophies and symbolic attributes of the arts quite common. So the plaster work grew in volume, and became heavy and clumsy.

Better taste, with a gayer note, is seen in the work of Sir Christopher Wren, who with all his love for the classic style had an eye for striking ornament drawn with much freedom though carefully grouped and rigidly balanced. Wren greatly admired the French plaster work of the period, which was a florid style, with great bunches and garlands of flowers and fruit, serrated acanthus leaves and figure work, carried out in a rich stucco, given a whiteness and hardness by the addition of " marble meal." While his plaster work was chiefly confined to the ceilings, he did introduce plaster ornament over doors, windows, mantelpieces, and on the soffits of staircases and the walls of staircase wells. Pillars and pilasters, with handsome capitals, were also carried out in the brilliant hard stucco. The style is heavier than that of Inigo Jones, the details being modelled in high relief, almost undercut, and suggest the work of the woodcarver rather than that of the plasterer. But it harmonised with the formal lines of his structures and their lavish surface decoration. The beautiful example of the wreath of flowers and fruit on the ceiling of St. Clement's Church, Eastcheap, illustrated on Plate VII, Vol. II, is a good example of Wren's careful plaster work. Not only are the details carried out in high relief suitable for their position, but the design is well balanced in every way. Although from the west door and the altar the broad wreath appears truly circular, it is an ellipse, so planned in order to correct optical illusion, and it does so quite effectually.

The tendency for plasterers under Wren to become carvers, and to treat flowers, leaves, fruit, and, to a limited extent, the human figure realistically, was much exaggerated under his successors, notably William Kent the architect, decorator and designer of furniture. He was building mansions in London, its neighbourhood and all over the country in the grand classical style, and his plaster work was very heavy. Ceilings were occasionally divided into panels, more often left flat with a broad framing or deeply modelled plaster, but not brought down on the walls. The cornices were greatly reduced, thin mouldings being placed at top and bottom of the coved surroundings. On the other hand, he used massive stucco ornament over doors, windows and chimneypieces. His floral treatment was naturalistic, and this was combined sometimes with the human form, terminal figures, sirens and sphinxes, and on other occasions with heraldic devices or trophies. He was fond of the pediment, both closed and open, which usually had plain outer mouldings, with a number of receding decorated mouldings, including the egg and tongue and the dentil. Such pediments were placed over doors, windows and fireplaces. Below them was often placed a dentil moulding, and below this a heavy wreath of oak or bay leaves (frequently measuring one-fourth of the height of the pediment), tied with ribbons. Such entablature was supported by a deep frame with heavy mouldings, classical fluted pillars with decorated capitals, and occasionally by terminal figures, while the sloping sides of the pediments supported the form of nude boys, in high relief or carved in the round. While he employed the flowing lines of scrolled acanthus leaves, wave forms and vine tendrils, of the Rocaille style, he adopted little more from that school than the large escallop shell, chiefly introduced in place of a keystone.

Like Wren, Kent added marble dust to his stucco; he also used eggs as a binding in large quantities. From quite an early period both beer and eggs had been used by English plasterers to secure a more plastic material, which would adhere strongly, dry hard and glossy. We hear of beer, milk, eggs, gluten, sugar, pitch and wax being added to the mixtures, even before the time of Wren, but in later times eggs and milk, or skim milk, were the usual plasticisers used in lime plasters. The work produced was durable, notwithstanding its great bulk and elaborate modelling, which frequently approached working in the round.

But Kent's plaster, at all events on ceilings, was mainly in the nature of an ornamental framing for painting, either direct on the plaster or on canvas stretched thereon. This had a powerful influence on his successors, for they abandoned the heroic in plaster work and produced designs which they described as suitable for execution in stucco, carving or painting. It was the commencement of the degradation of decorative plaster, which was accentuated under Robert Adam, and increased rapidly under Sir William Chambers and in the hands of Collins and Rose, the master plasterers of the eighteenth century who organised big and busy workshops.

In justice it must be said that Robert Adam had two distinct manners. When building or remodelling great mansions he could use stucco massively, for heavy modelled borders round ceilings, broad cornices and deep friezes, accompanied by broad bands or plaster modelled in high relief over doors, windows and chimneypieces, forming panels on walls, especially in halls and on staircases. There is much very grand work of his in this style in different parts of the country. But his second manner was more used and perhaps more characteristic, and this favoured considerable flat, smooth spaces, with bands and medallions delicately modelled in low relief and usually on a small scale. For his grand manner he favoured deep, plain mouldings, boldly massed acanthus leaves, oak and bay wreaths and swags, and certain classic patterns, such as the curled wave, key and honeysuckle. This bold work shows both the influence of the Palladian school and his own "Etruscan" studies. At Harewood House, Yorkshire, for instance, there is much very strong decorative plaster which Rose carried out for Adam. On the ceiling of the state bedroom are large figures of Bacchus, Ariadne, Diana, Venus, Endymion, Adonis, Caphalus and Pocris. In the great gallery the stucco work is in the architect's favourite "Etruscan" style, with griffins, urns, foliated candelabra, fuchsia cup festoons, shield-shaped pateræ and mouldings forming rectangular panels exquisitely painted.

Before dwelling on Adam's second manner it is necessary to point out that some of his precursors and contemporaries, architects, decorators and plasterers, were disciples of the rococo style in its various phases. We see both on ceilings and walls (chiefly over doors and chimneypieces) fairly heavy plaster modelled into the tortured scrolls, reeds, shells, massive mouldings, with small nude cupids, doves, lovers' knots and other details of Louis XV decoration, gradually growing more eccentric until it developed into what is generally known as Chinese rococo or

Chinese Chippendale This was characterised by deeply serrated and contorted scrolls, tall reeds, waterfalls, pagados and weird beasts and birds. This was often richly gilded, or partly gilded and partly painted, and constantly formed framings for paintings of mythological or Arcadian subjects.

Reverting to Robert Adam and his second manner, we find that he used narrow mouldings in low relief to make small rectangular, odd-shaped, but principally circular, oval, and semicircular or semioval panels, which he filled in either with painting or with modelled ornament, frequently figure work. He was fond of the fan palm (or traceried fan), the Greek honeysuckle ornament, oak leaves and acorns, fuchsia drop, classical urns, the key and wave patterns. His figures were of the idealised Greek type familiarised to us by John Flaxman's modelling, so widely diffused by his designs in bas-relief on Wedgwood jasper ware. Unlike the old Greeks as seen in their work on the painted vases and architectural sculpture, Flaxman avoided the nude, covering his figures more or less in flowing draperies. He set a lasting fashion in this type. The ornamental ceilings were harmonised with the room as a whole by the cornices and deep friezes, supplemented by more modelled work, not only over doors, windows and chimneypieces, but as panels on the walls.

Most of this work was coloured, the tints being pale and low in tones, with the exception of browns; the yellows verging on the lemon rather than the primrose or buttercup; the blues being milky; the pinks also soft. In many cases the figures moulded in low relief were left white, as in the jasper pottery just referred to; but the painted panels (sometimes by foremost painters of the day) were in the nature of miniatures usually in strong colours. Joseph Rose, who had an extensive business and employed a large number of hands, did much of the plaster and "compo" work in the second manner, while Jackson, the founder of the firm of George Jackson, was another.

George Richardson followed as a designer of plaster work, principally ceilings, which were mostly carried out by A. Collins and his staff. Here, again, the relief is very low and the mouldings narrow, the ornament and figure work rather niggling. Their work lacked the fine balance and dignity of the Adam designs. Their work, too, was coloured in low tones. Much of the ornament was painted.

Sir William Chambers designed with rather more strength, and some of the plaster work carried out for him by W. Collins (for instance, at Somerset House, in the Strand) is worthy of note. Gibbs, another fashionable architect of the late eighteenth century, also paid attention to plaster, but his work is poor compared with the sixteenth- and seventeenth-century masterpieces.

Many of the English plasterers of this period showed decided individuality, though complying with the general fashion. Thus, William Rhodes, of whose work we have a fair amount at Drayton, Northamptonshire, while using very slender decoration in low relief, has a dainty touch, a true sense of beauty and balance. His work is mostly of the nature of laurelling, slim festoons of wreaths, formed with a series of consecutive loops about pine-capped rods, with dainty figures holding flowing scarves. This kind of ornament he applied to ceilings, narrow friezes and framed panels on the walls. Then, again, the small portrait medallions on the ceilings of Sir Robert Taylor, in the Bank of England, are most delicately modelled and full of character.

During the early years of the nineteenth century decorative plaster fell still lower, and there set in the demand for the plain ceilings, with central cast roses (often of cast metal or composition) and insignificant cornices. After the first half of the century fibrous plaster was introduced, and although this seemed to presage a further fall in craftsmanship, it really acted as a stimulus. Fine old examples were reproduced and good designs specially prepared. Side by side with this there was a revival of craftsmanship, and towards the end of the century much excellent work was done, though it was not widespread.

In this rapid review of plaster work in England the general trend has been kept in view, rather than the exceptional examples.

Composition.—Mention has been made of the use of composition both in ceilings and walls. Workers in the reign of Henry VIII were using a paste of rye meal with gypsum. A century or two later fat lime was gauged with milk and eggs for fine work, and gesso was occasionally introduced. Robert Adam relied extensively on his "compo," prepared as a secret by Rose and by Jackson. Its use persisted well into the nineteenth century, especially for decoration on wood.

Gesso.—We have mentioned that gesso was introduced here in the eighteenth century. As shown elsewhere in this volume, gesso is something much more than a name for plaster of Paris, though the Italians used it in this narrow sense. It has long been known as a description for plaster plus glue, size or other medium which delays the hardening or "setting" of plain plaster, and so gives the craftsman an excellent medium for modelling decoration in low relief upon furniture, wainscot panels, ceilings, etc., with a brush or with wood or steel tools. Whiting is to-day often used instead of plaster or clay.

Much attention was given to gesso by the British pre-Raphaelite School—Burne-Jones, Walter Crane and others. The last-named gives as his recipe for "gesso"—one part of powdered resin in four parts of linseed oil, and six parts of melted glue boiled and well mixed.

Gesso can be worked upon almost any surface either by modelling tools or with a brush. It can also be impressed with a stamp.

Common flour mixed with plaster of Paris is also a well-known material for gesso ornamentation or "painting in relief" as it has also been called.

The temptation is to *overdo* decoration by these fascinating processes. As an example, in Walter Crane's famous ceiling at Coombe Park, Sevenoaks, the outer panels are very beautiful and restrained in the very important matter of the quantity of the decoration. This is mostly composed of a series of figures.

In the great central panel, in the writer's opinion, far too many elements or *objects* are given in figure and in floral ornaments. The two fan pattern ornaments at each end of the oblong, and the floral designs as a whole, are, in fact, very " shoddy."

Saracenic decoration in early times was very intricate and profuse, but they seldom failed to leave plain places of an ample kind to throw up the marvellous beauty of their mosques. They also used leading lines and forms like the horseshoe arch, which dominated the whole and held the mind of the onlooker, as it were, in a leash.

Exterior Work

As we have already seen, lime plaster was extensively used for external work, even in London, in the reign of King John, and we have shown that it was an important element in half-timbered construction, which was the main type for domestic buildings in England well into the seventeenth century. All the work that it has been possible to examine has proved to be three-coat lime plaster work, with chopped straw in cheap, and ox or horse hair in better-class work for the rendering and sometimes the middle coats. It would appear that in the better-class work beer was often used, probably with the idea of retarding setting and increasing binding effects. The plaster was lasting even when applied to bruised reeds or to riven laths.

Such work was generally whitewashed, occasionally painted. But the usual methods of decorating was by means of incised lines and dots —scratched ornament—or by impressing stamps, which could give either sunken or embossed ornament, according to the cutting of the dies. These were the methods adopted by the pargeters. Their work was often quite elaborate, the ornament sometimes being coloured. There was little actual modelling in this style, beyond the running of flat bands to form panels, these being run in the fashion of screeds.

In the Perpendicular Gothic of the Tudor period, however, outside decorative plastering became more self-assertive. Nonesuch Palace built for Henry VIII had the exterior embellished with modelled and carved plaster, including large-sized human figures, by Italian and English craftsmen. This decorative plaster was of the nature of stucco, containing gypsum or marble dust. The same kind of ornament was adopted by the Elizabethan builders and was elaborated by the Jacobeans, who were very fond of figure work in low and medium relief, either single figures or groups, sometimes processional in arrangement, so having a truly decorative value. Work of this period, though scarce, may still be seen *in situ*, while at the Victoria and Albert Museum there is the charming façade of Sir Paul Pindar's house, once a landmark, but not an exceptional one, of London's streets.

As the classical taste of Inigo Jones asserted itself, this kind of thing was abandoned for better-class houses, and the art gradually died out. The influence of Sir Christopher Wren, William Kent, Gibbs, Sir William Chambers and others was against, at all events, the decorative use of plaster on exteriors, though plain plastering was never quite abandoned and probably was encouraged by Wren's experiments to improve plaster. Even Kent, however, employed stucco on exterior walls. For instance, Lord Burlington's splendid Palladian " villa " at Chiswick was plastered with stucco externally, and we learn that during the carrying out of this work the neighbourhood was denuded of eggs and buttermilk used in tempering the stucco. Other examples of this could be cited, but this practice was not generally adopted, though Palladio's practice in Italy justified such treatment.

It was not, however, until about the middle of the eighteenth century that stucco began to be extensively used for the outside protection of walls, especially in towns. Robert Adam is generally credited with this revival of an old practice. It lent itself with special force to his treatment of terrace-houses, as we see, for instance, in the Adelphi Terrace, London, overlooking the Thames, which presents the appearance of a great building, though including many houses. As he used exterior stucco, there was panelling by broad bands, occasionally vertical, but more commonly in the form of string courses, decorative panels, well-defined cornices, pediments over doorways and windows, often heavy architraves, with capped pilasters and other architectural refinements. In this way not only was he able to produce a uniformity in the design of his elevations, but he was able to bring this into harmony with the treatment of the interior, even when the inside ornament was in " compo " of low projection. Adam adopted this method in Edinburgh as well as in London.

So successful was this exterior stucco work that it was long continued. It was claimed that on brick or rubble it provided weather-tight, warm houses, easy to decorate and keep in order. Stucco was generally oil-painted, and so lasted well, only requiring periodical washing and occasional repainting. The routine repainting was probably inimical to the system in the end, because only too often one coat of paint was added upon another, which, of course, tended to blur, where it did not entirely obliterate, any raised plaster ornament, and so gave a cheap appearance to the finish. But during the late Georgian and early Victorian era stucco was very generally used for exterior finish. Nash, among others, made good use of it, as could be seen until lately in the old harmonious Regent Street, and now still to a large extent in Portland Place and its neighbourhood, but particularly in the stately terraces surrounding Regent's Park. There it will be seen that stucco is not used merely as a smooth surface, but modelled to give decorative pilasters, cornices, etc. Later Victorian and neo-Georgian architects preferred exposed materials of construction, even when these were a mere shell, or veneer, for the cheaper material behind. Concrete construction has again suggested the rendering of exterior walls with cement plaster, rough-cast and so on, and this is once more bringing stucco into some prominence. The use of cement and refined lime with gypsum favours this revival, which seems to have a promising future, both in town and country, though particularly in the former, where it presents many advantages.

In many regions of France, Italy and Spain, and many other parts of the Continent, the exterior plastering of better-class houses, when not built of dressed stone, is the rule. It is often treated as stone, rules being run so as to form joints in the wet plaster, the basement or pedestal "blocks," quoins and other "dressings" in many cases being pressed with engraved wood or metal plates to imprint vermiculations or other types of face-dressed stones. Cornices, string courses and framings for all openings are usually carried out in moulded or modelled stucco. Painting in oil-colour is uncommon on smoothly finished stucco, but in the country districts and smaller towns a moderately coarse finish is much in vogue and this is generally painted, often very fancifully, including pictorial panels. It is the very best kind of finish for inferior brick and rubble walls, whether treated in severe plainness or ornamentally.

In America during the "Colonial" style of architecture in the eighteenth century, lime and gypsum stuccoing was common, while in California and the other Spanish regions lime plastering of the superior type of adobe buildings was the rule. Of late years there has been a pronounced revival in the use of pure white lime stucco on outside walls, including decorative modelling and the running of ornamental mouldings, but, as mentioned in another chapter, smooth finish is by no means the only rule, various types of rough texture being much in favour.

CHAPTER IV

CEILINGS: DECORATIVE AND OTHERS

Designing a Decorative Ceiling—Selection of Ornament—Some Authorities—Styles of Decoration: Ancient Work—Foreign Developments: Saracenic Plaster, Persian Plaster, Italian Work, French Plaster Work—Ceiling Work in England—Types of Ceilings—Flat Ceilings —Brandered and Strapped Ceilings—Provision for Lighting—Cornices—Panelled Ceilings—Coffered Ceilings—Coved Ceilings—Vaulted Ceilings—Barrel Vaulted Ceilings—Curved Surfaces —Groined Ceilings—Colouring—Pugging.

A cursory study of the architecture of the day goes to show that the mid-nineteenth century idea of "plain walls and ceilings," of "ceilings as smooth as mirrors," with their hard, commonplace central rosettes, has been outgrown, and that we are returning to the older, more sane plan of decorating our ceilings both with modelled ornament and colour. This is a distinct gain to architecture, whether concerned with public buildings or domestic interiors, and again opens out great opportunities for the skilled plasterer.

Designing a Decorative Ceiling

Much of the plaster work on modern ceilings is simple enough, plain panelling, linked up with the upper cornice of the architrave; coffering between beams; perhaps coving; but there is also a fair amount of more decided decorative work. Often the more elaborate decorative work, particularly when period styles are in demand, is carried out by means of pre-cast fibrous plaster, of which we shall have more to say in a special chapter. Even, however, when fibrous plaster is used, the plasterer's skill is called for in fixing; in many cases we find handwork, modelling and the planting of cast ornament associated with fibrous plaster.

When discussing decorative ceilings, we must never lose sight of the fact that the treatment of the ceiling must be considered in conjunction with the treatment of the walls. This holds good with plastering as with any other type of decoration. And when we come to study old rooms, we find that the elaborate plaster moulding on the ceiling corresponds with plaster work or wood carving on the walls, or at least over door and window openings and the chimneypiece. As a rule, too, the heavier the work on the ceiling itself, the more distinct is the architrave crowning the walls, with its lower and upper cornices and deep frieze, with their mouldings and enrichments. This is perfectly right, because the ceiling should not be uninspiringly flat and colourless, nor should it become the focal point. If there is to be a focal point, that should be the chimneypiece, the main doorway or some other feature on the vertical plane. Therefore, ornament on a ceiling should never be overdone; it should harmonise with the treatment of the walls.

It is also to be observed that square panels, or oblong panels carried across a long room, tend to restrict apparent area, while oblong panels carried lengthwise help to increase the appearance of length. A number of small coffers also reduce apparent area. Heavy projecting ornament, especially drops, tend to dwarf the apparent height; so, too, will heavy architraves. Having due regard to wall treatment, ornament on a ceiling should be distributed fairly evenly, to avoid a strong focal point being produced overhead. That is to say, if we have an elaborate central ornament, the upper cornice should advance well on to the ceiling, and small panels, mouldings or scattered sprays, etc., should link up the central ornament with the corners of the rooms. In this way a good percentage of plain surface can be retained without producing the patchiness that the central ornament isolated from the cornice would naturally suggest.

Lighting has also to be taken into account. A ceiling with heavy projected ornament, or with elaborately delicate decorations, demands as nearly equal distribution of natural and artificial light as possible. If the lighting is strong in one quarter and poor elsewhere, ornament should be restricted. A coved ceiling is easier to light than one with deep coffers, and a square or circular one than one of great length. Exceptional height in relation to floor area also presents a difficult problem in lighting, though heavy ceiling ornament, again having regard to the walls, is rather indicated in order to give it the necessary emphasis, and at the same time assist in restoring the sense of proportion.

Selection of Ornament

Size of ornament, as well as projection, must have reference to size of room, character of room, or doors, architraves, chimneypiece, etc.

Choice of ornament must, of course, depend upon the character of the architecture, or at least the style of decoration adopted for the room. This leads to the need for the study of styles as exemplified by old examples. A plasterer will often be called upon to produce an exact replica of a given ceiling, or more frequently of a special period style. As a rule, he is supplied with detailed drawings by the architect or decorator, but frequently he is given a simple sketch and general instructions, the carrying out of details being left to him, and this means that if he is to do himself justice, he must possess a general knowledge of the history of plastering as an art, and also know where to go to find details of each period. In the following pages it is only possible to give a wide survey of the subject. For more detailed study of the various periods readers may refer to such books as:

Some Authorities

Sebastiano Serlio, *Il Terzio Libro di S.S.*, 1544; Andrea Palladio, *Architettura*, 1642; Colin Campbell, *Victruvius Britannicus*, 1717–25; William Kent, *The Designs of Inigo Jones*, 1727; Robert and James Adam, *The Decorative Work of Robert and James Adam, being a reproduction of the plates illustrating decoration, etc., from their Works in Architecture*, 1901; George Richardson, *A Book of Ceilings*, 1776; Sir William Chambers, *A Treatise on the Decorative Part of Civil Architecture*, 1826; Owen James, *Grammar of Ornament*, 1865. All these books are valuable, some of them rare, but can be found in good libraries.

Visits to the Victoria and Albert Museum, South Kensington, and to the Architectural Courts at the Crystal Palace, Sydenham, will also be found extremely useful.

There are several excellent coloured plates of ceilings in *The Modern Painter and Decorator*.

Styles of Decoration

Ancient Work.—It will not be necessary to dwell at any great length on the decoration of ancient ceilings, though some reference to the subject is advisable, as the early steps explain developments and often underlie present-day practice. It is certainly interesting to know that the early Egyptians employed a peculiarly hard and durable stucco to cover their ceilings, mostly flat and formed of stone slabs. This stucco was either painted on the flat, or modelled in low relief and then painted. It is to them that we owe the conception of blue ceilings, either sprinkled with yellow or gilded stars, or painted with the symbols of the Zodiac. They were also fond of depicting on their ceilings protecting gods and goddesses in the form of birds, or their symbolic feather sceptres and fans. These symbols were repeated in stucco over doorways. Divinities in human form with animal heads, or animals with human heads, also appear, singly, in processional array or in groups, alone or in conjunction with conventional ornament, some of which are still with us.

Other decorative schemes included conventionalised representations of reed matting, plant forms, such as the lotus bud, the papyrus bloom, the palm leaf, and intricate patterns of curved and crossed lines. The colouring was bold and vivid, bright blues, reds and greens, with black or yellow used as a harmonising medium.

The Greeks and Romans employed timber, boards, large thin slabs of stone and tiles for ceiling purposes. The primitive type of timbers with boarding was represented by projecting cross members with square or oblong depressions between, called *lacunaria*, our coffering. These lakelets or coffers were filled with painted stone slabs, encaustic tiles, painted boards, but quite as often with stone or wood covered with stucco painted and often sculptured in relief with rosettes. In Greece the flat ceiling generally prevailed in public buildings, but the barrel vault was common in dwellings. Red and blue and gold were the prevailing colours.

In Rome, especially under the Empire, ceilings were very elaborate. They were flat, with or without *lacunaria*, barrel vaulted and coved, while the octagon or circular cupola was much used. Some of these ceilings were of rare woods, such as ebony, encrusted with ivory and precious metals. Encaustic tiles were also employed, but the prevailing material was a fine variety of stucco, richly moulded with heavy cornices, panels and ornaments in low or high relief. These ornaments consisted of geometrical patterns, conventionalised foliage and flowers, and grotesques, which latter embraced intersecting foliage scrolls combined with or merging into human and bestial forms, shells, war trophies and an endless variety of objects. The colouring was lavish and vivid. The star-spangled sky was also represented. This coloured stucco work is of peculiar importance, as it was the direct inspiration for the revival of plaster under the Italian Renaissance in the sixteenth century. At a late period mosaic was introduced, the patterns generally being geometric or floral, though figures were also introduced, the tesseræ being black and white, or of coloured marbles, or with cubes of coloured clay.

In Byzantine architecture the vaulted ceiling largely prevailed. The surface was covered with a rich, smooth lime plaster, and either painted on the flat or encrusted with mosaic. The colouring was lavish, the drawing conventional, with little or no attempt at perspective; the pictorial panels, surmounted by elaborate ornamentation of geometric and foliated patterns, often framing small miniatures. The mosaic was very rich, including coloured marbles and rare stones and vitrified cubes. Gold was a favourite background for the figure paintings.

Foreign Developments

Before dealing with the decorative development of the ceiling in England, it will be well to pass in rapid review the progress of the art abroad, for this largely influenced our own practice.

Traditions set by the Egyptians and Romans, though suffering an eclipse during the Middle Ages, did not quite die out. Fine schools of plastering were evolved by the Saracens and the Persians, while a much impoverished version of Roman and Byzantine practice lingered in various parts of Italy. For the most part, however, mediæval architecture was chiefly a matter of stone and brickwork, plastering being relegated to a very secondary part.

Saracenic Plaster.—In the early part of the seventh century the mosque at Medina was built of wattled walls plastered. Two hundred years later a beautiful mosque in Cairo had the walls covered with plaster decorated with hand-modelled rosettes and flower patterns, and from this a well-defined scheme of decorative design was evolved. Most Saracenic plaster work consists of involved lines and highly conventionalised flowers and leaves, carried out in low relief, forming crowded patterns having a lace-like appearance.

The involved line ornament is best seen as flat ribbons, interlaced like our Tudor strapwork, but more angular in contour, the patterns often designed to form star shapes and horse-shoe arched ends; the flat surface delicately worked with lace and scrolled figures. Such work covered whole panels of ceilings. Patterns in ceilings, especially of the domed variety, are often pierced and sometimes filled with coloured glass. But they also formed pierced screens of plaster, the patterns being geometrical and conventionalised floral. Particularly characteristic is the prism motif, sometimes called the stalactite; a long cube, hexagon or octagon in form, and frequently with the lower part of one facet shaped out with a coved head. These prisms were built up honeycomb fashion to form coved cornices to cover the intradoses of arches, or the whole of a vaulted ceiling. Arabic script is also much used. All this elaborate plaster work is richly coloured in green, red, blue, yellow, with occasional browns and a little white. In the Turkish form white and pale yellow is more common, and the ornament is on a larger scale, more open, sparsely distributed so that extensive plain spaces are seen.

Persian Plaster.—Persian decorative plaster early reached a high pitch of excellence. It is mostly gypsum plaster, and is profusely worked on ceilings. The ornamental work was, and is, carried out in a peculiar way. The plaster is gauged in quite small quantities, and rather stiff; it is then freely manipulated, being pressed and kneaded with the hands until the mass attains a certain plasticity. In this state it is daubed on the wall or ceiling, more and more being added on a narrow space in such a manner as to rough out the design, the ornamentation being finished by modelling and impressing or stamping out. While the designs are Saracenic in feeling, the flowers are less conventionalised, the patterns more open than the Moorish, but more regular than the Turkish. The ribbon pattern is absent, and in early work animals, and even the human form, are introduced, though sparingly. Blues and greens of the turquoise tints are the predominating colours, with a brownish red and a little pale yellow.

Italian Work.—As already stated, the art of plastering, so finely employed by the Romans, lingered on in out-of-the-way corners of Italy during the Middle Ages, to be revived about the middle of the thirteenth century, though we do not discover the sumptuous ceiling work until the early part of the sixteenth century. As we have seen in an earlier chapter, the Italians worked with a plaster formed of old lime putty enriched with pure white marble dust, which gave it a sparkle, a brilliancy unseen in ordinary lime plaster, and a wonderful hardness. This *stucco duro* was used on ceilings of all shapes to produce those quaint arabesques which Raeffel and others evolved from a study of decadent periods of Roman art as revealed in many of the ancient buildings unearthed in Rome and other cities. At first the ceiling ornamentation was confined largely to arabesques, foliated forms merging into animals, human torsos or heads, candelabras, cartouches, enriched mouldings and the like.

Plaster work on ceilings at this period formed a gorgeous surround for carefully finished pictures, large and small, very splendid as regards composition, execution, and full of colouring, but conceived more as easel paintings than decoratively, so that unless shown on the friezes of an architrave or pendentives of a vaulted ceiling, they are often difficult to look at and provide focal points which prove rather distracting. The craftsmanship of the plaster work is superb, with its boldness of modelling, its careful finish and extraordinary variety of ornament and treatment. Ceilings of this period are splendid, and, apart from the pictures, truly decorative and in harmony with, not dominating, the general architectural design.

Fresh discoveries of Roman art were being made, mostly of the ostentatious period of decadence under the later emperors, and as these discoveries of statuary and bas-reliefs were made, gradually a kind of exaltation supervened, and forgetting the proper function of decorators, to embellish an architectural conception, not to overshadow it, the plasterers grew bolder and introduced on a large scale cherubs, cupids and adult human forms, heads and masks, together with the luxuriant acanthus leaf, armorial shields and insignia, together with much else. Often the modelling was very strong, almost approaching carving in the round, the projection being considerable.

Raeffel Sanzio, with his pupils and compeers, in the decoration of the Vatican and other buildings used stucco lavishly, but kept to sound proportions, as their work in the loggias at the Papal Palace, the Villa Madama and elsewhere testifies. Serlio, writing in 1566, says that ceilings in small rooms should be ornamented with small designs, not overloaded with detail, and in low relief. For large apartments and high ceilings, more elaborate designs in bold relief are prescribed, but in both cases he advocates richly ornamental cornices. Unlike the masters of his day, he liked monochrome for small designs, and for others not more than two colours, unless blue could be used as a background to act as a foil and harmonising agent. Raeffel and his fellows, however, painted and gilded their stucco ceilings with lavish hands, using bright reds, blues, greens and yellows, often in thin washes, which permitted the brilliance of the pure white stucco beneath to pierce through. This love of the sensuous line and colouring increased with the passing of time. The grandeur merging into extravagance of decorative plaster ceilings is seen in work of Alexandro Vittoria, acting under Sansovino in Venice. Flat and vaulted ceilings are a mass of projecting plaster ornament, carved on a large scale, surrounding rectangular, circular and oval panels most beautifully painted in brilliant colours by the foremost artists of the day. The plaster work framing and surrounds include floral displays, figures, armorial devices, with some geometrical designs, but little of the older arabesques. It was highly coloured and heavily gilded.

Curiously enough, the revolt against the extravagance of Renaissance architecture and decoration, which was crystallised in the work of Andrea Palladio at the opening of the seventeenth century, showing a rather cold and formal rendering of the earlier classic style, was hardly reflected in decorative plaster. The designs were less extravagant as regards admixture of subjects, but ceilings and cornices were loaded with classic or mythological figures, sculpturesque in character, large in scale and too heavy to be restful or altogether pleasing. We find war trophies being introduced. The bold, self-assertive plaster work frequently formed a framing for large painted panels, mostly of an allegorical type, though sometimes pastoral in character.

The Italian plasterers were masters of their work, recognised all over Europe, and were received everywhere, introducing their style and methods. But it must be said that they worked with a heavy hand, and certainly overloaded the ceilings, though usually supported by bold architraves, which continued the work well down the walls, thus in a measure preserving balance. Such ceilings, however, rather suggested parade and grandeur than homeliness, peace and comfort.

In the eighteenth century the Neo-classic generally prevailed, represented by classic mouldings, scrollings, festoons and medallions in low relief and painted in low tones. In the nineteenth century decorative plaster work was mainly confined to rich cornices and good mouldings round the ceilings, while the broad, flat plaster surfaces were painted with clouds, scenery and figure subjects, or rather tame imitations of lace curtains. Plain white plastered ceilings were the rare exceptions.

French Plaster Work.—Both lime and gypsum plaster were used for ceilings in France from early times, but were mostly applied so as to obtain smooth surfaces for painting. Ornamentation to the end of the fifteenth century was mostly carried out in *papier mâché* and later in *carton pierre*, both of which materials have already been described. Modelled stucco was scarcely known until the opening years of the sixteenth century, when it was introduced by Italian plasterers, whose efforts, however, were modified by French taste. The Renaissance work in France was quite as ponderous as in Italy, but less extravagant. Arabesques soon gave way to the acanthus leaves, classical mouldings, nude and partially draped human figures, beautifully modelled armorial insignia and floral garlands. Military trophies, helmets and breastplates were much in favour. The colouring was vivid and gilding heavy. Under the Regency and Louis-Quinze the Rococo style was introduced, in which we see the ceilings covered with foliated and twisted scrolls, associated with the escallop shell, slight garlands, cupids and female figures. The whole work was much less heavy, the modelling being on a smaller scale and usually in medium relief. Panels and plain spaces were often filled in with delicate and fanciful trellis work, and ribbons were used to tie up garlands, musical trophies, shepherd's crooks and so on. The colours are clear but subdued; pinks, pale blues, lemon yellows, with tender greens largely prevailing. While the ceilings are not so overloaded, the designs are out of balance and somewhat niggling. With the next reign there was a greater soberness, a return to classic garlands, mouldings and figures on a rather larger scale, but the colouring is still lower in tone, and the human figure often left white. The Empire saw a return to flatter projection, the introduction of such symbols as lyres, eagles, sphinxes, Roman helmets and breast-plates, and other ornament from Rome and Egypt. The colouring was brilliant, a rich red and a deep blue prevailing.

Ceiling Work in England

Although in England, as elsewhere, the Romanesque architecture and early Gothic tended to minimise the importance of plastering, there is evidence that plastered ceilings were common in domestic architecture, usually being smooth and painted. We have little evidence of how it was used in this connection during the Early English (1200–1300), Decorated (1300–1400), and Third Pointed Period (1400–1460) of Gothic work, it is only with the Perpendicular style that it comes into prominence, especially with the Tudor period.

As we have seen, under Henry VII the pargeters formed a powerful guild. They covered ceilings with strapwork, geometrical patterns and fanciful figures, mostly stamped, and in the best class of work largely coloured. Henry VIII introduced Italian plasterers who introduced classic decoration on the purely English type of florid Gothic. They attempted modelling and carving on the native plaster, but did not find it sufficiently plastic, so used for their finer work a paste of rye meal, glue and fat lime. This was used to obtain casts from moulds for repeat ornaments, or bossed out on the ceilings and then carved. All this was painted.

Figure work on ceilings during the Tudor period was the exception, and when used was mostly in medium relief. Indeed, medium relief was the rule, except as regards the knots and pendants, which we shall come to presently. A characteristic ornament of this period is the strapwork which is seen in extraordinary variety on ceilings of Elizabethan mansions and manor houses. This strapwork consists of flat, broad ribbons, twisted in many geometric patterns, including rectangles and curves, often edged, the surface decorated with filigree work, with bosses or rosettes at the intersections. Such patterns often covered the whole ceiling, with mouldings at the cornices; but are also seen in conjunction with geometrical mouldings, ribbing and great pendants. This style lasted well into the Jacobean period. While most of the ornament was in medium relief, the ribbing, in imitation of Gothic mouldings on groining and fan spandrels, was usually raised, rebated and enriched with tracery. Pendants, at the intersections of interlacings and ribbings, sometimes took the form of floral bosses, which, like those in Gothic detail, might enclose figures, armorial shields, crests or badges; at other times they were like the great leaf buds of finials, or were architectural in character and might project 6″ or more. Much of this work was left white, with a little gilding, though the heraldic shields and other symbols were painted in their proper pure strong colours, forming a good contrast on its own plane and with the wainscoted or tapestry-covered walls.

Geometrical patterns persisted in the Jacobean period, with the addition of scroll-work, an attenuated form of acanthus leaf, mouldings, heraldic motifs, and in the cornices or friezes small figures of animals or the human form. More distinct floral designs were introduced, but highly conventionalised, and we find "jewelling" among the strapwork (bossed diamonds, carbuncles or oval cartouches, circles, etc.).

With Inigo Jones, the Palladian style of architecture came in, and also the classical sculptured plaster work for ceilings. There were regular entablatures, with lower and upper cornices, the latter infringing on the ceiling and deep friezes. Such entablatures were enriched with plaster sculpture, including human forms, acanthus leaves and scrolling, which kind of decoration was carried, in a modified form, on to the ceiling, there being used as a frame for paintings, in the Renaissance style; mythological scenes, pastorals, apotheses, figures disporting themselves amidst clouds. The modelling of the plaster, however, was far more restrained than contemporary work in Italy and France, and so more suited to its surroundings. This was even so when Italians were at work.

With Sir Christopher Wren, the Continental exuberance had more influence here, partly, no doubt, owing to the importation of craftsmen from Italy. But though the carved plaster was given excessive projection, human figures, war trophies, flowers, leaves and fruit introduced in profusion, extravagance was subdued, the work even of the foreigners being modified by native supervision, while a school of English plasterers sprang up whose work somewhat resembles that of the wood carvers with Grinling Gibbons at their head. This sculptured plaster was found mostly on the cornices and as surrounds to the painted ceilings, executed by such artists as Verrio, who was followed by Laguerre. Both painted mythological subjects, allegories, cupids and so on, often larger than life-size, and occasionally they introduced portraits surrounded by enriched, coloured and gilded plaster frames.

The Georgian period was one of transition. At first the heavy Renaissance style was to the fore, but this gradually gave way to the Rococo style, in which we see unbalanced foliated scrolls, shells and trellis-work carried out in medium relief. Often this was the sole decoration on the ceiling, but it is also associated with painting, which, however, gradually disappeared as the Chinese type of Rococo, so often associated with Chippendale, asserted itself. Much of this was left uncoloured.

This style prevailed until ousted by the work of the brothers Adam, who headed a revolt against the excesses of the Regency style. What is known as the Adam type of decoration is based on a study of the

Græco-Roman work as seen in Magna Grecia. The motifs included the partially draped human figure, mythological monsters, lion masks, tripods, urns, fan ornaments, the honeysuckle design, palms, acorns and the charming fuchsia drop (a real inspiration of Robert Adam). All these were transferred to ceilings, together with laurelling, small floral festoons and patteræ, traced on a small scale, in low relief and thin lines, widely but sparsely distributed. A good deal of this work was carried out in patent compo, much of it being moulded and then stuck on. It was painted, mostly in pale tertiary tints, on tinted (pinkish, milky blue or primrose-yellow) grounds, the compo work frequently forming frames for oblongs, ovals, as well as circular and other panels, painted with classical figures and scenery, executed by Zucchi, Cipriani, Angelica Kauffmann and many more, who usually painted with a full brush in strong colours.

The general lay-out of the ceiling decoration is regular, based on geometrical laws, and very carefully balanced: centre, four corners, or compensating compartments, with a border linking up with the cornice. Whether plain mouldings or laurelling is used, the patterns traced by the lines are perfectly symmetrical, and these, together with the figure work on a small scale and in low relief, are deliberately subordinated to any pictorial panels. In fact, the plaster work, although often quite elaborate, is subdued in breadth and projection, and in less competent hands was apt to degenerate to the level of piped sugar work on cakes. The low tone of the colouring also suggests this analogy. It lacks vigour, is rather finicking, yet did not always escape providing a rather provocative focal point, provided by the miniature paintings. Thanks, however, to the use of modelled plaster panels within strong mouldings on the walls, this does not prove so disturbing as it might otherwise do. All this goes to show that Adam work is difficult to imitate, as it requires exceptional powers of designing and execution to overcome the faults inherent in it. Exact copy is the safest course when this style is in question.

Robert Adam was followed by the architect George Richardson, who employed Joseph Rose (who had large workshops and a numerous staff of craftsmen) to carry out his plaster work. It was rather an emasculated type of the Adam style, very formal; figures, flowers and mouldings in low relief, painted in low-toned colours.

After this, decorative plaster work on ceilings fell into abeyance, until partially revived about the seventies of last century, when a band of clever plasterers arose, who carried out much valuable restoration work, and produced some original designs, in which the realistic representation of flowers and leaves, often in high relief, was a marked feature.

Types of Ceilings

In the foregoing rapid survey we have dealt with the subject rather as though the ceiling was always flat, above a more or less well-developed entablature, with possibly a coved frieze and upper cornice. As a matter of fact, however, from quite early days, the form of ceilings has varied within a wide range, and these have all had their influence on the design of ornament and the work of the plasterer.

Flat Ceilings

If we derive civilisation from Egypt and Greece, the flat ceiling was the earliest type, being formed of wood planks or stone slabs placed on joists with their ends resting on the walls. To make such covering weather proof, the underside was daubed with clay or lime mortar. From this type the panelled and ribbed types evolved.

As regards the flat ceiling, providing a smooth, uninterrupted surface, this must be obtained by plastering boards or laths nailed to joists. Boards were soon discarded for the lighter, cheaper reeds, and then wood laths. These last are now frequently replaced by metal lathing or expanded metal. The success and durability of the work largely depend upon the quality and fixing of the joists and laths. These must be of sufficient strength, stability and enduring nature to sustain the weight of the plaster. It is generally estimated that an average well-made ceiling weighs, including joists, 1,200 lb. per square of 100 superficial feet. In the best kind of work, flooring joists and ceiling joists will be separate timbers, to prevent any vibration from the flooring to be transmitted to the ceiling. In the cheaper work, flooring is fixed to the upper part of joists and ceiling laths nailed to the under part. When such is the case, special care must be taken that the joists are sound and properly fixed.

The Scottish brandered or strapped ceiling is an excellent type, particularly to be recommended where vibration is to be feared. In this kind fillets of wood, called straps, are nailed to the underside of the joists at about 12″ to 14″ from centre to centre, and to these the laths are nailed. The straps should be 1½″ square for light laths, and 2″ square for heavy laths and plastering.

Joists should not be placed farther apart than 12″ from centre to centre for single laths, or 14″ for double laths.

Flat ceilings carried out by plaster applied to laths must be carefully done. If the plaster is too thin (especially in one- or two-coat work) the positions of the laths soon become apparent, the plaster being stained; then cracks appear, probably leading to falling down in patches. Even in better class work, the surface is often disfigured by dark, broad bands, corresponding to the interspace *between* the joists. These joist lines are generally the result of infiltration of water or dust from above, and the nuisance can be obviated by back plastering the laths, which, however, adds greatly to the weight and cost, or by pugging, which we shall describe presently.

Cornices.—The sightliness of a flat ceiling depends largely upon the character of the cornice. In a large and lofty room the cornice should be prominent, comprising several mouldings of different proportions, advancing well on to the ceiling and well down into the frieze, which should have an architrave or lower cornice. Cornices and frieze should be treated decoratively, so as to link up the walls with the ceiling, and all three parts will require the attention of the plasterer, though the frieze itself must be quite plain, for painting or papering, though in an important room it should bear modelled plaster ornament. Instructions for running cornices of different descriptions will be found in another chapter. It will be understood, of course, that cornices are associated with every other type of ceiling, though insistence on its importance is emphasised here, because its assistance is more needed with this kind.

Provision for Lighting.—A modern demand on the cornice relates to present-day lighting practice. It is not only necessary to find get-at-able channels for electric wires, but also provide a screen for tubular or other concealed electric lamps used in the indirect-lighting system. For these reasons the plasterer is frequently called upon to run a cornice which will project well forward and upwards, so that a concealed channel of the required width and depth is formed. Pre-cast work meets the demand with the least cost and trouble, enabling the work to be carried out with the smallest amount of material possible, and so keeping down weight. While this treatment of the cornice may make balance in design difficult (and therefore, strictly speaking, is only suitable for large, high rooms), it need not interfere with the character of mouldings or other ornament.

Panelled Ceilings

Panelled ceilings are those whose surface is divided into panels by beams, girders or screens (arches or beams supported by pilasters or pillars). These divisions may be natural or artificial. In the old houses where timber played so important a part, heavy and small beams were often carried horizontally across a room from wall to wall, or in halls, churches, etc., one end resting on the wall plate, and the beam was carried up at an angle and the upper end housed in a ridge pole. Such beams were often left uncovered, and the spaces between either boarded or bridged by joists carrying laths and plaster. Beams of this kind were either left square, ornamented with rebating, or moulding, or carved, and the space between treated decoratively by pargeting or moulded plaster. Many old panelled ceilings of this kind were beautifully decorated, and the fashion has now been revived, chiefly for country houses. This treatment of the ceiling affords splendid opportunities for the skilled plaster craftsman. Such ceilings do not require elaborate cornices; a plain moulding or two will suffice, though a decorative frieze is often indicated. In some ceilings these beams are merely make-believes, and even may be formed of plaster, usually a box beam being formed with planks and then plastered. Such box beams look well if ornamented with stamped or modelled plaster. It sometimes happens that steel girders have to be used in place of wood beams. These should be cased in with wire mesh or expanded metal and plastered. Steel girders should be cleaned and painted with anti-rusting paint before being cased in, otherwise staining may occur.

In the case of screens, the arch or beam, if plastered, should be treated in character with the ceiling and cornice.

Another type of panelled ceiling is known as the "ribbed." With these the panels may be simply formed by running mouldings. When this is done care should be taken to adjust the size and shape of the panels to the size, shape and character of the room, in accordance with the rules laid down in our paragraph on the designing of decorative ceilings. The mouldings used for forming the bays may be quite plain, or broad and decorative, they may suffice for ornament, but often constitute framing for modelled plaster or for painting. In the Tudor period "ribbed" ceilings were often elaborate and extremely beautiful, some having floral ornament, others stamped diaper patterns, while yet others were treated architecturally, like the ribs in Gothic groining.

Mouldings and ribs are run on brackets, those of small diameter on rope, dipped in glue and spiked; larger mouldings and ribs on one or more laths spiked in position. Before spiking, the floating on the ceiling should be cut down to the lath, and proper care should be taken that where joists are not available, quartering is fixed to receive the spikes. The bracket is then roughed out with rough stuff to receive the fine stucco for running.

Coffered Ceilings

Coffered ceilings are an elaborate form of the panelled class. Instead of the ceiling being divided into bays, it is covered with a series of small compartments, square, oblong, octagonal, circular or oval. These may cover the whole surface or be arranged in bays, and are found both on flat and vaulted ceilings. The mouldings are generally of high projection, and are run on brackets, either of spiked rope or shaped wood, though spiked wire mesh is also used on big work. The mouldings may be plain or elaborate. Screeds are formed on the top of the brackets to support parallel running rules, by means of which half the mould is run on one side and then completed on the other. The coffers may be left plain, contain painting or plaster ornament.

Coved Ceilings

Coved ceilings are those in which the junction between the wall and ceiling is formed by a concave curve; it is, therefore, an incipient form of vaulting. The coving may be very slight or very deep, and spring either direct from the frieze or from a cornice. It may be plain or

BARRELL VAULTED CEILING IN PLASTER, GUILDHALL, ABINGDON.
(Harry Redfern, Architect.)

PLASTER CEILING, BANK BOARD-ROOM.
(Evrard, Son & Pink, Architects.)

decorated, and is generally in the form of a quadrant of a circle or of an elipse. The cove is generally formed on a cradle or bracket cut to the desired size and shape, fastened to the wall and ceiling by means of plugging and spiking, and subsequently lathed and plastered. A jack template is usually employed for running the internal and external angle mouldings.

Vaulted Ceilings

Vaulted ceilings are usually in the form of cupolas, which can be made highly decorative. The commonest type met with is the hemispherical dome, covering a circular chamber, or at the intersection of two aisles or wings of a chamber or building, and having a smooth, ribless under or concave part, the top being truly spherical or flat. But as three or more wings of a chamber or building may meet, the Roman form of barrel vaulting is sometimes adopted, the three or more semicircular barrels intersecting each other. While the Roman, the Renaissance or classical vault was hemispherical or semicircular, in Gothic architecture it was generally pointed, often consisting of three or more sections intersecting each other.

A vaulted roof or dome may be perfectly spherical and smooth, or composed of a number of diminishing panels, generally divided by mouldings. Thus we may have a hexagonal or octagonal dome, each panel being smooth or divided up into coffers or other designs, while the ribs may be plain mouldings or broad bands carrying modelled ornament. For plastering purposes, special methods have to be adopted, which are explained in detail in a later chapter.

Sometimes false vaulting is adopted, and then brackets or cradles of wood have to be built up, to which lathing is fixed to receive and hold the plaster.

In Gothic work the vaulting is usually decorated with ornamental ribbing, which may take the elaborate form of fan vaulting, in which rib mouldings spring from the cornice or wall-plate level and spread over the surface. In early classic work the panels are often crowded with coffers of different shapes, but at other times the panels are left free for painting, and the plasterer's task is confined to the floating of the ground and the running of the main mouldings or ribs, springing from the cornice and rising to the apex, with possible secondary mouldings running vertically, horizontally or diagonally. For these mouldings screeds must be formed for the main mouldings, but in small work the mouldings and ornaments are generally cast and then planted, the floating on the cupola being cut back to the foundation for the purpose of secure fixing. The secondary mouldings may be in the nature of framing for painting, and then they are often composed of raised, modelled or moulded ornament. Circular or oval mouldings may be required when bays or openings for lighting are pierced in the panels. In Saracenic and Byzantine work the piercings for lighting are often comparatively small, running slanting upwards from the outer covering to the inner shell, and may be left open or filled with plain or coloured glass, the openings being partially concealed by means of plaster mouldings or ornament.

Barrel Vaulting.—Barrel vaulting was a Roman form of construction which again came into favour with the Renaissance. It is semicircular vaulting, tunnelling, mostly applied to long halls. It presents no great difficulties to the plasterer, though as a rule such vaults are really carried out on an extended principle of coving, and then suitable brackets and lathing will be necessary. This type of vaulting is occasionally an afterthought, adopted to diminish the exaggerated height of a hall, or to improve its acoustical purposes, and in such cases it will be necessary to run a prominently projecting cornice from which to spring the cradling for the vault. Vaulting of this kind is also known as wagon vaulting, and is found in Early English Gothic, as seen in Canterbury Cathedral.

Curved Surfaces

The floating of curved surfaces, whether it be a smooth, spherical dome or semicircular vault, or the segment of an octagonal or other polygonal cupola, is not easy, and when work on a small scale is in question, sections are generally cast and then fixed in position. For this purpose fibrous plaster is mostly used, as it permits of freer handling of light castings, which are more easy to fix securely. As we shall see in due course, the use of pre-cast sections and fibrous plaster slabs in no way precludes the introduction of hand finish and modelled ornament.

Groined Ceilings

We have referred to ribbed work in Gothic ceilings, which may develop into fan tracery vaulting, characteristic of the Perpendicular, Florid or Tudor style of Gothic. In this the vault is covered with main ribs and secondary ribs or veins of tracery, of which all the principal lines diverge from a point, the datum line, corresponding to the cornice level in a classic building. This kind of ribbing is seen at its best in Henry the Seventh's Chapel, Westminster Abbey, and King's College, Cambridge. It is very beautiful, and usually carried out in carved stone, but the Tudor plasterers introduced a modification of it in domestic architecture with superb effect.

It sprang naturally from the groined type of ceiling which began with the earliest style of Gothic vaulting. Groins are the lines or ridges which the surface of vaults form by their intersections, called an edge on the external mitre and a ridge on the internal mitre. These groins are covered with ribs or mouldings, which may be plain semicircular in section or elaborately decorated. In Early English Gothic these ribs were simple and in parallel lines; later they intersected each other, and at these intersections roundels, rosettes and pendants were placed, adding to the elegance and interest of the ceilings, for these ornaments were either purely decorative (floral and architectural) or symbolical, including figure-work, heraldic devices, etc.

Ceilings of these descriptions require careful planning. The datum, or cornice level, line is first fixed, and the ribs made to spring from this, and all intersections are calculated therefrom. Vertical and diagonal screeds are formed and the ribs run by means of appropriate templates or running moulds. When the ribs are run, pressed screeds are formed and the various bays floated. If the groined ceiling is to be decorated, knobs are roughed out at the intersections of the ribs, and duly carved or modelled, or pre-cast rosettes, pendants, etc., can be planted, first cutting down the ribs to the foundation. Heavy ribs are best run on the spike and rope brackets already referred to in a previous paragraph. If this is done, when cutting down in order to plant ornamental bosses, care must be taken not to interfere with the brackets, which must be left intact, otherwise the stability of the whole length of rib would be endangered. Where metal lathing is used, it is easy to form a projecting metal rib to form a foundation for the plastering.

Colouring

It is possible to use coloured plasters, pigmentary material being added when gauging. But this is not advisable for ceiling work, as it compels the artist to be satisfied with a very limited palette, mostly neutral tints or dark, luminous colours, if the setting is not to be interfered with (as only earth colours and a few oxides are available). Unless a yellow or coffee-coloured ground is desired, it is far better to use white or natural plaster, and colour it as desired after setting. This is the method generally adopted by the old masters of stucco on the Continent, and by our own pargeters and plasterers at home. It is to be noted that some "patent" cements require to be painted immediately after floating.

The whole subject of painting plaster and cement is dealt with at length in a special chapter.

Pugging

In ceiling work pugging serves many useful purposes. Primarily it is intended to deaden sound, and to prevent the spread of odours either upwards or downwards. Incidentally pugging does away with the disfiguring joist lines by arresting the fall of dust and wet on the back of the ceiling. If properly laid down with suitable materials, it is a real safeguard against the spread of vermin and fire, while it provides an effectual absorbent mat which will prevent the percolation of moisture under ordinary conditions. Pugging is placed so as to form a fixed insulating plate between the flooring and ceiling. For this purpose fillets are nailed to the joists about one-third up, and laths are nailed to the fillets. These laths are plastered and back plastered commonly with coarse stuff; or with lime mixed with chopped hay, bracken or sawdust, the fibrous materials assisting in muffling sound waves.

If coarse plaster is mixed with fine blast slag, engine ashes, or brick-dust, a fire-resisting mat is formed, always provided that the laths are well covered back and front, or that expanded metal is used. Whatever the mortar employed, it should be seen that the plaster covers the whole surface, without cracks, and is carried right up against the joists, and is laid on sufficiently thick. The joists themselves should not be in direct contact with the ceiling plaster or the flooring boards, strips of asbestos cloth being placed between them to act as washers or insulators.

In some cases the space between flooring and ceiling is packed with ashes or sawdust. But while such practice is effectual for deadening sound, it is not at all advisable, for such packing will harbour vermin, permit wet dripping through (and then stains are intensified), and should any accident occur proves a great nuisance, making repairs difficult. Such packing is also too heavy and, should there be the slightest settling, is sure to shift, becoming too compact at the lower level and too thin at the upper or higher end, so that the weight and insulation become unevenly distributed.

For one reason or another the old-fashioned pugging of "coarse stuff" mixed with chopped hay or bracken fern is going out of favour as a sound-deadening packing for floors and partitions, though, as stated, it has its merits. It errs, perhaps, in being rather heavy, and that is why sawdust in thin layers and coarse ungreased wool is being substituted. Both of these are not as serviceable for the purpose desired as seems generally to be supposed; the first because it is too compact, the second because it becomes expensive if handled so as to be effective. But the most vital objection is that they are both dangerous in case of fire, and also are apt to harbour vermin. If ordinary pugging is too heavy, slag wool should be used. This, when carefully packed, is a quite efficient deadener of sound; it is practically fireproof, or at all events fire-resistant, and does not harbour vermin. While slightly heavier than ordinary wool, it is lighter than sawdust and far easier to handle.

Where space is limited, insulating slabs may be used. These are made with coarse plaster mixed with fine slag and a little asbestos fibre, applied in two coats to both sides of wire mesh or expanded metal. The slabs are fixed to the fillets nailed to the joists, and should be sealed at the edges with a little of the slag mortar.

Pugging, when not of a fire-resisting nature, should not be carried under fire-places or too close to chimney flues without taking proper precautions as regards insulation.

CHAPTER V

PLASTERING MATERIALS

Limes—Varieties of Lime : Hydraulic Limes, Lias Lime, Poor Lime, Stone Lime or Grey Chalk Lime, Hydrated Lime—Slaking Lime—Gypsum—Gypsum Cements—Plaster of Paris—Hard-baked Plaster—Gypsum Plaster and Wall Boards—" Patent " Plasters and Cements : Keene's, Parian, Martin's, Thistle Hardwall, Granite Silicon, Victorite, Siparite, Selenetic—Plastic Magnesia—Lime Plasters and Gypsum Plasters Compared—Natural Cements : Roman, Medina—Portland Cement—Sand.

Limes

Limes for building uses are prepared by the heating of chalk, limestone and other minerals and rocks of a calcareous nature, the effect of which is to dispose of the moisture and carbonic acid, the resulting material being known as quicklime or caustic lime.

The " burning " or, as it should more accurately be called, " heating," of the limestone is carried out in some form of limekiln, the average weight of pure lime obtained being in the region of half the limestone placed in the kiln, this amount being slightly increased by the inert matter present in all commercial lime and amounting to roughly one-third ton for each four tons of limestone consumed.

In other words, from four tons of limestone approximately two and one-third tons of commercial lime are produced.

A knowledge of the types of kilns used in the production of lime being needed as an aid in the selection of limes from competing plants, to obtain the best results in the finished work, the following notes are of value.

Of the older methods of burning lime, the more economical is by the use of the " tunnel," also known as the " perpetual," " continuous running " or " draw " kiln, the internal shape being cylindrical, an inverted cone or two inverted cones base to base. At the lower end of the cone or cylinder are placed loose fire-bars, with a space beneath connected with the outside by a tunnel, through which, as the burnt lime is withdrawn through the bars, it is taken from the kiln.

The kiln may be constructed of brick, stone or concrete—the inner lining being of fire-brick with an air space between this and the main structure—and may be built either above or sunk into the ground. In either case the tunnel or draw-hole is usually protected from the weather by a shed to preserve the freshly drawn lime.

When lighting the kiln, a layer of brushwood is laid on the fire-bars, and then alternate layers of coal and moistened limestone—the latter being usually six times the thickness of the layer of coal—are built up to the top of the kiln. Smaller pieces of stone are placed at the bottom of the kiln, the larger pieces of limestone (which are never larger than a man can carry) are placed in the middle, where they obtain more heat. After the kiln is well started, and as the burnt lime is withdrawn through the bars at the bottom, the space at the top is refilled with layers of coal and stone. The kiln should be built high enough to ensure that the limestone shall be sufficiently long in contact with the hot fuel and attain a temperature which will drive off the carbon dioxide, but anything beyond this is complete waste. The temperature of complete decomposition is only about 900° C.

Though this type of kiln requires more management than the flare type, it is more economical in the use of fuel—only about one-fifth of the weight of lime produced being required—and is cheaper in maintenance and gives a continuous production, but the lime is not equally burnt, and through being in contact with the fuel, is not clean. Though perhaps passable for use as mortar for brickwork, its use for plaster would be doubtful policy, partly due to difficulty in proper slaking with consequent blow-holes and cracking.

The "flare" kiln is generally circular in shape, with a fire-hole, or draw-hole, at the bottom, and is constructed with rough bricks or stone.

In working, an arch is first formed of the larger pieces of limestone inside the kiln round the fire-hole, the rest of the stone being piled on top of this arch, under which a fire, often of wood, is built or maintained for about three or four days, until the carbon dioxide is driven off, the lime being then withdrawn through the fire-hole.

By this method of burning the flames and heat only come in contact with the lime, cleanliness is obtained, together with evenness of burning, combined with ease of management, but fuel for this type of kiln is about one and one-third times the amount required with the "tunnel" type, and there is the inconvenience of having to let the fire out after each charge—with consequent loss of production—and also the cost of lining the kiln at least once in twelve months.

In the modern type of kiln, the burning fuel is kept away from the limestone, the process is continuous, and a purer and more evenly burnt lime is obtained.

Varieties of Limes

Fat lime, or rich lime, is made from practically pure carbonate of lime coming from the upper chalk, marble or other beds containing 98 to 100 per cent. carbonate of lime. This lime slakes fiercely, swelling to two or three times its former bulk, throwing out great heat and steam, and becomes a bulky powder which has little setting power when made into mortar. It is very soluble, and used chiefly for plastering and white-washing. Because it is nearly pure carbonate of lime, rich lime will carry the most sand, the various impurities in the others replacing sand required in mixing ; in fact, with this lime it is necessary to use a greater volume of sand to prevent shrinking.

This lime should show an analysis of 95 to 98 per cent. of quicklime, and should slake rapidly and well. It should be white in colour and should be in lumps, much powder showing that it has been exposed for a long time.

Fat lime or, as it is also called, "chalk," or "pure" lime, for the following reasons is used for plastering:

It slakes easily and thoroughly, and, due to the slaking process causing it to become about three times its original bulk, it is economical, added to which its original cost is low.

It works easily, and though deficient in strength, this is not a deterrent, as strength is of no importance for this purpose.

It has the property (due to its porosity) of temporarily absorbing moisture which condenses on a wall, and also it improves by being mixed some considerable time before being used.

Hydraulic Limes.—Hydraulic limes are those produced from carbonate of lime with an admixture of clay (silicate of alumina), which gives the property of setting in water—hence the name. The hydraulic property depends largely upon the amount of the silicate of alumina (clay) present, as:

Feebly hydraulic limes contain up to 10 per cent.

Ordinarily hydraulic limes contain from 10 to 20 per cent.

Strongly hydraulic limes contain from 20 to 30 per cent.

Hydraulic limes slake almost without heat, and do not fall into powder; and to obtain thorough hydration it is advisable to grind them finely; even in damp situations the set is vigorous and access to air is very necessary. The colour of these limes is pale grey.

Lias Lime, or Blue Lias lime, is a natural hydraulic lime and, according to the locality from which it comes, contains generally 20 to 30 per cent. of clay; slaking is very difficult, commencing after long and uncertain periods, with heat only sensible to touch, and then without any cracking. This lime sets firm under water within twenty-four hours; in two to four days it is hard; within a month it is very hard; and in six months it can be worked like hard limestone and has similar attributes.

This lime is not suitable for plaster, and is practically never used for the purpose; its main uses being for retaining walls and for foundations in damp soils.

Poor Lime.—Poor lime, containing 60 to 90 per cent. of carbonate of lime combined with inert impurities, slakes imperfectly and sluggishly without much increase of bulk. Owing to the presence of clay in small quantities it sets harder than fat limes, but is deficient in strength.

Stone Lime or Grey Chalk Lime.—Stone lime or grey chalk lime is very free slaking. When water is added it pauses for a short time and then slakes, breaking up into powder, becomes hot and gives off vapour. Owing to having 5 to 12 per cent. of clay in its composition, it will set in sitll water, and in fifteen to twenty days it is firm, gradually becoming harder. It is feebly hydraulic, is of a light buff colour, and is suitable for building ordinary stone or brick walls.

This lime in plasterers' work is used at times for rough-cast in place of blue lias limes, which, owing to difficulty in thorough slaking, often cause scaling. It is also used for " scratch " or first coat for interior plastering.

Hydrated Lime.—Hydrated lime is heated and slaked by special processes, and produces a fine dry powder, which can be safely stored for a long time. If 5 to 10 per cent. is added to Portland cement, or to Portland cement concrete, a more plastic material is obtained, which will dry with slight retardation, hard and water-tight.

Slaking Lime

Slaking is the chemical combination of lime with water—when quicklime (CaO, calcium oxide) has water added, the whole mass gets hot, falls to powder and becomes dry ; the weight will be found to have increased, owing to the combination of calcium oxide with the water to form slaked lime ($Ca(OH)_2$, calcium hydroxide or calcium hydrate).

The amount of water required to form slaked lime from pure lime is 32 per cent. of the quicklime by weight, and the increase in bulk during the operation is from two to three times. Completely slaking lime is not an easy operation, for too much water results in the weakening of the lime, and if too little water is used, incomplete slaking results.

Also it must be expected that some portions of the lime will be over-burnt and some under-burnt, and that parts will contain silica. These portions are very slow in slaking, and cause blow-holes and cracking after the mortar is made up and used ; though for ordinary mortar the rough slaking usually carried out is probably good enough ; for lime plaster greater care is required, and the lime should be slaked and made into a cream some weeks before use—such material is usually known as plasterers' putty.

The purchase of lime already slaked (hydrated lime) from a manufacturer—a practice of long standing in America—is a far better plan, for the lime is slaked under proper conditions. It has also been carefully sifted, and, being ready for use as lime plaster, besides giving certainty of results, saves a considerable amount of inconvenience.

With hydraulic limes slaking is much slower, the heating up and increase of bulk less, and though the lumps may crack all over they do not fall to powder—clay, carbonate of magnesia, alkalies, metallic oxides and sulphate of lime all retard the slaking action, though they quicken the setting or hardening of the mass after being made into mortar.

The slaking of lime is usually carried out in a mortar box or on a platform, either with a ring of sand on the base, or on a bed of sand with a ring of the same material to contain the lime ; water (usually far in excess of that actually necessary) is poured on to the lime, and, slaking commencing, the top is covered in either by a layer of sand or, preferably, with a tarpaulin, in order to prevent the gases escaping. If the material is required for mortar, when slaking finishes more sand (if required) is added, the whole mixed well together and left to " cool " (finish slaking) until required for use.

For plaster work more care is required in the preparation of the lime, and slaking should be carried out preferably in a receptacle without sand, the same care being taken to conserve the vapours which come off. Then, when the lime is broken down, either in the same container or some other, water should be added, the whole stirred into a cream and allowed to stand for at least two weeks to " cool " ; this period should be the minimum, most specifications calling for the "running," as it is called, to take place at least one month before being required for use.

In this state the lime is known as "plasterers' putty." When required, the additional water is poured off, and the lime is passed through a sieve to remove all over- or under-burnt particles and other lumps, and either the required amount of sand is added to form the plaster, or it is allowed to dry until of sufficient consistency for use.

Milk of lime is water made turbid by dissolving in it slaked lime. It is often used for slaking purposes. Whitewash is milk of lime brought to the consistency of liquid cream by the addition of more lime, and usually a small quantity of size or sugar, to secure better adherence.

Lime requires to be carefully and thoroughly burnt, or it is apt to slake badly and then to " blow," owing to the presence of quicklime.

The slaking is also important, because if imperfectly carried out it will slake of itself, and if used in mortars will cause blisters. That is why care is necessary, and that no lime should be used until at least three weeks after slaking.

Gypsum

Rock gypsum, which is found in many places in England and is abundant in the outskirts of Paris, answers to the chemical formula $CaSO_4 2H_2O$. In a pure state it contains: lime 32·6 per cent., sulphuric acid 46·5 per cent. and water 20·9 per cent. Before use it is crushed and heated in special kilns to a temperature between 170° and 140° C. By this means three-fourths of the water of crystallisation is driven off and the material becomes calcined gypsum, or hemi-hydrate, commonly called plaster of Paris, or plaster in accordance with the following reaction :

$$2(CaSO_4 2H_2O) + \text{heat} = 2CaSO_{4\frac{1}{2}}H_2O + 3H_2O.$$

Should the heating be continued until a temperature of 200°C. is reached the remaining water is driven off and the material obtained is a soluble anhydrite, thus :

$$2CaSO_4\ H_2O + \text{heat} = 2CaSO_4 + 3H_2O.$$

Both products have their uses, the second setting quicker than the former.

Gypsum, besides giving common plaster of Paris, is the basis of numerous special cements. For this purpose it must be pure white, free from red or yellow ferruginous stains.

In its fibrous state it is known as Satin Spar, and the pure coloured granular rock is Alabaster.

Characteristics of Plaster

Good plaster is in the form of a very fine powder, which when gauged with water produces a smooth, slow-setting cream. When dry it should be quite hard. Quick-setting plaster is less durable than slow-setting, and is more difficult to work. Plaster prepared in lumps when gauged will set in five minutes or under; the powdered will take over ten minutes to set, and if gauged with "strong water" (glue water), it will take fifteen to sixteen minutes to set. If gauged with lime water, it dries much harder.

In selecting plaster, fineness should be a guiding rule. The colour should be pure white or pale pink. Coarse, dark plasters are usually sandy, containing silica, and are generally weak. Tests should be made with pats of plaster gauged both with plain and glue water, to ascertain setting properties and strength. Good plaster will have a compressive strength of 120 lb. to the square inch if gauged with plain water and 160 lb. if gauged with lime water.

Gypsum is heavier than lime. In the natural state it weighs 140 lb. to the cubic foot, as compared with 125 for chalk and 53 for limestone. Plaster of Paris at 80 lb. per cubic foot compares with 44 lb. for lime and grey chalk.

Gypsum Cements

Under the title of Keene's cement and Parian cement there are available on the market many cements of varying merit, having for their main constituent calcium sulphate or dehydrated gypsum. Most of these cements are acid in character, some are markedly alkaline, and a few, very few, are neutral.

Neutral Keene's or Parian Cements.—Generally speaking, these consist, essentially, of anhydrous calcium sulphate; that is, dehydrated gypsum ($CaSO_4$). If properly prepared they should quickly come to a finish under the trowel, and subsequently continue to hydrate and harden for a period of about 8 to 10 days, becoming practically gypsum ($CaSO_4 2H_2O$).

For all mural, cornice and ceiling work, and most decorative work, they are eminently the most suitable of all cements of the Keene's or Parian group. They do not injure, nor are they injured by, the large majority of pigments used in paints, distempers and wallpapers.

For the manufacture of artificial marble one may say that they are essential, an acid or alkaline Keene's or Parian so restricting the range of pigments that may be used with them as to render them inadmissible.

Gypsum cements for mural, decorative and architectural work should conform to the following standards:

(1) They should be neutral to litmus in powder form, or rapidly become so on using.

(2) On gauging, work easily and smoothly, giving sufficient time for manipulation. For mural work, good behaviour under the trowel is essential.

(3) Work well with and become hard with large amounts of suitable sand.

(4) After the first set, which is reasonably quick, the subsequent completion of hydration and hardening should be slow and progressive giving any lath- or other wood-work to which it is applied time to swell without cracking the cement.

(5) Expansion (or contraction) must not be such as to cause any disturbance whatever of the surface.

Acid Cements.—Acid cements, if properly made, have their uses. They hydrate and harden after setting with greater rapidity than a neutral cement. They give density and strength even when gauged generously with water and used on non-absorbent surfaces. Their main uses are for casting, in gelatine or other moulds, tile slabbing and flooring.

In marine architecture they are used for decking.

Under certain circumstances they may be used, mixed with sand, as a backing for a neutral cement face. They should not, however, be so mixed if the surface is to be ultimately papered or distempered with delicate colours. For such work as ceilings, this objection usually disappears.

An acid cement may be easily detected. Add a drop or two of water (preferably distilled) to a pinch of the cement and apply a blue litmus paper to the paste formed. The colour of the litmus is at once changed to red if the cement is acid.

Alkaline Cements.—A markedly alkaline cement for general plastering purposes has little to recommend it. Long waits are necessary to allow the calcium hydrate, the cause of the alkalinity, to carbonate and become neutral. To test for alkalinity, moisten a pinch of the material with water (preferably distilled) and apply a red litmus paper. A decided change from red to blue indicates a cement of the above character. A faint indication of blue, or an odd blue spot or two, may be disregarded.

Plaster of Paris

Plaster of Paris is made by the heating of gypsum, a soft stone of crystalline texture consisting of hydrated sulphate of lime.

Gypsum mines and quarries are worked in this country in Nottinghamshire, Sussex, Derbyshire and Cumberland among other places, the principal deposits being those found in the Nottingham district.

In the manufacture of common plaster of Paris, the gypsum is first ground and then heated, at the same time being constantly stirred, in a temperature of approximately 200° C. On heating, the material loses the water of combination, the amount of this loss being proportionate to the heat to which it is subjected.

As the temperature rises, the powdered gypsum bubbles like boiling water, due to the disengagement of the water in the form of steam, the heating being carried out until the correct percentage of water is driven off, when the resulting powder is known as plaster of Paris.

When again mixed with water, plaster of Paris absorbs a great quantity without any noticeable heating up, sets hard in a short time, and, at the same time, expansion takes place.

The reason for the change which takes place is that the calcination (heating) drove off the water of combination and other matters from the original gypsum and left it " thirsty " (i.e. with a strong chemical affinity for moisture), and in " slaking its thirst " the water recombines with the calcium sulphate to form the original gypsum, and during the recombination sets into a hard, compact mass. The plaster of Paris takes about three to four weeks to arrive at its full strength after having set.

Owing to the fact that calcium sulphate is slightly soluble in water, plaster of Paris is considered in this country as being only suitable for interior work, though there are examples of the material being used as mortar in mediæval buildings and having stood the test of time, and it has been also used for external construction in certain districts on the Continent.

In this country plaster of Paris is used chiefly in connection with plastering, for the making of " gauged stuff " and for the moulding of interior decorations for centre-pieces, cornices and other ceiling ornaments, console brackets, etc., where the material is made into a paste and cast before fixing.

It is also used for filling cracks and holes in walls and ceilings, etc., on account of its quick-setting properties and the fact that it expands in setting.

For " gauged stuff," plaster of Paris is used in the proportion of 1 to 4 and 1 to 20 of the ordinary stuff, the amount varying according to circumstances, the speed with which the work has to be finished, the state of the weather, etc., etc.

On the Continent, and more particularly in America, plaster of Paris—in America it is called gypsum—is used to a very considerable extent as an interior plaster. For this purpose it is, in manufacture, mixed with " retarders," such as glue, sawdust, blood and dried ground cactus, which allow time for working before setting takes place.

Such plasters have proved very successful, and aid in the rapid finish of the building. The usual time of setting of plasters is two hours, and for variations of climate, temperature, etc., " accelerators " and " retarders " can be added on the job, for the purpose either of quickening the time, as would be required in cold or wet weather, or for slowing up the time of set, as would be required in very hot weather and under certain other conditions.

In America "gypsum" is also supplied ready mixed with hair and wood or other fibre for mixing with sand in the ordinary way, or with sawdust, paper pulp and other materials to be made into plaster without sand by having only water added. These latter are usually sold under trade names.

In this country, as also abroad, plaster of Paris is used for manufacturing "plaster boards," used in place of lath, etc., as a base for plaster in ceilings, partitions and other constructions, or for boards to replace plastering. For this purpose the plaster is mixed with sawdust, or other material, in place of sand, and is run by special machine between two layers of paper or "pulp boards," the thickness of the finished board being about $\frac{1}{4}''$ to $\frac{3}{8}''$, and it is claimed that they are damp proof, fire resisting and vermin proof.

Plaster of Paris is also used for the manufacture of partition slabs, and though the brittleness of the material has led to tardy recognition in this country of the bearing power of such slabs, it is a fact that a 3″ partition made with slabs of plaster with coke breeze and reeds (to facilitate handling) 8″ high, only collapsed under a load of 1 ton 16 cwt. per foot super placed on top of the wall. Reeds weaken the strength of the slabs; and, from the above, it is evident that 3″ partition slabs without reeds can carry with safety considerable loads. In France partition slabs are made of plaster of Paris and coke breeze in the thickness of 2″ to 4″, and in Canada $3\frac{1}{2}''$ slabs are used and support the ordinary loads of cottage construction.

Official tests made in this country (by the Munitions Inventions Department) show that ordinary pink plaster of Paris breaks down under an ultimate load of 0·94 ton per square inch, and this figure closely agrees with results obtained in America.

Hard-baked Plaster

By heating gypsum to a higher temperature (700°–900° C.), the water is completely driven off, and the material loses its power of recombining with water, and is therefore valueless for constructional purposes. The plaster so formed sets quite slowly, but forms a very hard and durable surface, and is much used on the Continent for flooring plasters, linseed oil often being used for treating the finished surface to prevent dusting.

Such plasters were formerly used in England for flooring. In fact, on the examination of old floors in the neighbourhood of Nottingham, they were found in good condition and with a thickness of about $2\frac{1}{2}''$ to be spanning $16\frac{1}{2}''$, reeds having been placed over the beams to carry the plaster.

Gypsum Plaster and Wall Boards

According to a *Bulletin* (1926) issued by the United States Bureau of Standards, the first structural gypsum products used to any great extent in America were gypsum plaster board and wall board. These products were placed on the market about 1908. Gypsum plaster board is used as a lath for the reception of plaster. The nominal size of the sheets is 32″ × ⅜″. Wall is similar in construction and composition with a slight difference to be noted later. It is not intended to be plastered, but serves as a complete wall or ceiling finish. The nominal width of the board is 48″ and the thickness ⅜″. Its length is between 6′ and 12′.

Gypsum plaster and wall board are essentially of the same composition, consisting of one or more layers of hydrated calcined gypsum between sheets of a fibrous material. They are both erected like ordinary wall boards by nailing directly to the wooden studding or by clipping with metal clips to metal channels or angles. Inasmuch as the surface of the wall board may be the exposed surface of a finished wall or ceiling, it is usual to have at least one side of the wall board sized or given other special finish. After erection of the wall board the joints are either pointed up, papered over or covered with wooden strips or mouldings.

There are a number of features obtainable in wall- or plaster-board construction, which account for its popularity, several of which include: Low cost, relatively less combustibility, ease of erection, economy in plaster and lightness. Besides its use as a lath, plaster board has found use as a protection against fire for wooden beams and columns, and as a sound deadener for concrete floors.

The processes of manufacture of gypsum wall and plaster board are identical. They are made upon the same machine; the only difference being that with wall board a little more attention is paid to details, so as to obtain a more uniform product. Hemi-hydrate, wooden chips, an accelerator and the water necessary are mixed together in a container. The accelerator is usually ground gypsum. Sufficient water is used to give a mix of easy pouring consistency, which is run out of the mixing hopper upon a belt carrying a strip of surfacing material. The wet gypsum mix is distributed uniformly over the width of the surfacing layer of paper or other material, after which another layer of surfacing material is brought into contact with the top surface of the wet mix and the whole is passed between steel rolls to give a uniform thickness. The board then travels along the belt for a distance of about 200′ at a speed of 10′ per minute. This allows the gypsum to set, so that when the end of the belt is reached the board is hard and rigid and easily cut into the desired lengths. After cutting, the board is dried and is ready for the market.

"Patent" Plasters and Cements

By the heating of plaster of Paris with small percentages of any of the following or a mixture of them—cream of tartar, petroleum, borax, alumina and potash salts—plasters are obtained which, by reason of their setting with a much harder surface, are useful as finishing plasters. Several plasters other than those named below are on the markets under various names, but their manufacture and properties are based upon these facts.

Tests show that their crushing strength is not equal to that of common plaster of Paris, and they should not, therefore, be used for constructing slabs, etc., to carry loads.

Keene's Cement.—Keene's cement, contrary to most other cements, has the advantage that when partly set it may be re-tempered or knocked up again with water without the final set being interfered with, or without loss of either tensile or compressive strength. It is made by steeping calcined gypsum in a solution of borax and cream of tartar. It is very fine, perfectly white when of good quality and takes on a brilliant polish. It is applied, usually on a backing of Portland or Selenetic cement, for covering columns, pilasters, architraves, arrises, reveals, splays, etc. It is also largely used in the preparation of scagliola and other artificial marbles. Keene's cement can be painted immediately after it is applied, but care must be observed, and we would refer the reader to the chapter on " Surface Treatment of Plaster and Cement," in the section on painting.

Parian.—Parian cement is composed of an intimate mixture of gypsum and borax, which is calcined and ground into a fine powder. It makes an excellent stucco when mixed with equal volume of fine worked white sand, eight bushels of the stucco covering ten superficial yards ½" thick. It can be used neat for the same purposes as Keene's, and like this can be applied to newly built damp walls, and can be painted or papered soon after.

Martin's.—Martin's cement is made of pure gypsum slaked in a solution of carbonate of potash. It dries hard and, like the two just mentioned, can be painted or papered soon after application. It is fairly economical, 1 cwt. with an equal quantity of sand will cover from six to seven yards superficial ½" thick. It is excellent for walls, especially for dadoes. For this purpose an undercoat of 1 part Martin's to 1½ parts dry sand is laid ½" thick to take a finishing coat of pure cement ⅛" thick. The same method is suitable for flooring.

Thistle Hardwall Plaster.—" Thistle " plaster is another gypsum product, which possesses considerable sand-carrying capacity, mixes easily, works freely, sets promptly and uniformly, and possesses great adhesive powers and strength. It can be painted as soon as it is dry. This plaster is utilised in manufacturing building boards.

Granite Silicon Plaster.—This is a quick-setting material, remarkable for its elasticity and hard-drying qualities. It is not apt to crack. It is marketed in four different types: (1) Render, containing hair, and is used, with 2 parts of sand, for applying to lath-work. (2) Brick, without hair, gauged with 3 parts of sand for use on walls. (3) Finish. (4) Superfine, for cornices, mouldings, etc. It is usually applied in two coats of ½" thick. It is extensively used for fire-resisting partitions, 2" of plaster being rendered on expanded metal or similar lathing, strengthened with steel rods.

Victorite.—This is a superfine plastering cement which will carry 2 to 3 parts of washed sand to 1 of cement. It hardens with age, is largely fireproof and sound deadening.

Sirapite.—Sirapite, another fire-resisting cement, is made with gypsum slaked in petroleum. It can be used alone, when it dries with a hard, white surface, or can be mixed with ordinary lime plaster acting as a strengthener. For lath-work, coarse sirapite is mixed with an equal quantity of pit sand; for brickwork, with 2 to 3 parts of sand. The finishing coat is pure, fine sirapite.

Selenetic Cement.—Is a preparation of gypsum and lime, which is hydraulic, has high sand-carrying capacity and dries quickly and hard. It makes an excellent backing for Keene's, Martin's, Parian and similar cements. It will receive the finishing coat more promptly and with greater safety than Portland cement.

Most of these cements are finely ground and fairly stable, not deteriorating by storage. But very choice work—for instance, when running intricate mouldings or producing artificial marbles—they should be carefully sieved. The coarser particles retained on the mesh need not be wasted, they can be mixed with the general bulk for preparing undercoats or ordinary plastering. Most of these cements do best when gauged with plain water, though Keene's may be gauged with weak size water to secure retardation and a higher degree of polish.

Plastic Magnesia

The United States Bureau of Mines devotes *Bulletin* 236 (1924) to a study of plastic magnesia, which is extensively utilised in America and Germany in the manufacture of stucco and composition flooring. It is obtained from magnesite and dolomite deposits.

It appears that magnesia stucco has gained headway in competition with cement stucco, chiefly because it makes a stronger, harder and more elastic cement, which adheres more firmly to wood surfaces and takes and holds dashes of ornamental coloured sand and pebbles as well as other stuccos. This stucco gives a house a more finished appearance, which is probably due to the coloured pebbles, but it has one disadvantage—it is not as weatherproof as cement stucco. Weatherproofing compounds are, therefore, usually added. The structural firmness and density of the stucco are the qualities that best ensure protection, and these qualities largely depend on the quality of the plastic magnesia used, which in turn depends mainly on selecting suitable raw material and accurately controlling the proper calcining temperature. Judged from the cumulative experience of the industry in America, properly made magnesia stucco is resistant enough to weathering to be considered entirely satisfactory, and, of course, conditions are far more severe there than they are in England.

Interior plaster made with plaster magnesia offers a field as yet little exploited.

Composition floors, if carefully laid, are quite satisfactory. They are permanent, warm to the touch, elastic, fireproof and fairly ornamental, and, moreover, can be laid on new or old wood flooring. Bathroom floors and walls are now quite frequently covered with the composition, and even bath tubs encased in it.

Slabs of the same material, often appropriately tinted, are used for draining boards over kitchen sinks and for many other similar purposes.

Usually manufacturers of stucco and composition flooring buy plastic magnesia and prepare their mixtures ready to be moistened with magnesia chloride solution at the place of application.

It appears that recarbonisation of free lime in burned dolomite has proved beneficial, but not in the magnesites.

Lime Plasters and Gypsum Plasters Compared

Lime plaster is prepared by calcining limestone (calcium carbonate) to a high temperature, and then slaking the calcined stone in water. It should then be left under water for a long period to hydrate thoroughly. After this soaking in water, it is ready for use as a plastering material.

To carry out lime plastering so as to make a wall that is really strong and hard right through is a very lengthy process. It must be applied in several thin coats, and each coat must be allowed several weeks' exposure to the atmosphere for the purpose of absorbing carbon dioxide. Until this absorption has taken place, the plaster remains alkaline, and without strength.

If an attempt is made to plaster a wall with one thick backing coat of lime and sand, only a very thin crust on the exterior of this backing coat attains any strength, the remainder, not being allowed free contact with the atmosphere, cannot absorb the necessary carbon dioxide which alone gives the strength to a lime-plaster wall. The addition of a skimming coat still further prevents the backing from attaining any strength.

Any good gypsum plaster, such as Parian cement, Keene's cement, Victorite or Murite, when used with sand as a backing and applied in one thick coat, will set quite hard right through in one or two days, and at the end of that time will attain greater hardness and strength than could be obtained with a lime and sand backing spread on in several thin coats with many weeks interval between the application of each.

It will thus be seen that gypsum plaster is a much quicker and stronger material in use than lime plaster.

Murite is the most rapid of the gypsum plasters in attaining this full strength ; it is also the cheapest in first cost.

Corners and Angles.—From the foregoing it can be readily seen that where lime plastering is being used it is very necessary that all corners and angles should be made of some material that is sufficiently strong and hard to withstand the inevitable rough usage that must occur at times.

It is usual in such cases to specify that corners or angles shall be finished either in Parian Cement, metal corner beads, wooden corners or angle staffs.

Wooden corners are, perhaps, the most widely used in the cheaper forms of building. When a good gypsum plaster is being used instead of lime plaster, the need for any of the above types of corner entirely disappears. If walls are being plastered with a good gypsum plaster, this of itself is sufficiently hard and strong to make a perfectly satisfactory corner.

Hitherto the entire plastering of walls and ceilings with a good gypsum plaster has not been attempted very often, owing to the high cost of such plasters, but with the introduction of Murite, the disadvantage of high cost disappears. Wooden angles and corners are, in fact, very undesirable for use with gypsum plasters. In the first place, when the plaster is wet, the wood, being in contact with the plaster, swells, and is liable to push the finish coat of plaster away from the backing coat for a space of about one to two inches from the corner. The blame for this is often laid against the plaster, whereas the real truth is that it is due entirely to the expansion of the wood. In the second place, when the wall has thoroughly dried the wooden corner shrinks and is always visible under the wallpaper or distemper, and looks very ugly.

Setting Time and Working Time.—There is a great difference between the working time and the setting time of a gypsum plaster. The difference is not always fully appreciated by the plasterer.

The setting time of gypsum plaster is always about the same; but the working time when used on a wall depends almost entirely on the dryness or otherwise of the wall. A very dry wall will give a short working time. The reason for this is that the dry wall absorbs the moisture out of the plaster quickly, long before the actual set takes place, and the working time depends upon the retention of the water in the plaster.

It follows, therefore, that to obtain a long working time with gypsum plaster on walls, the walls should be thoroughly wet before the plaster is applied.

To test the actual setting time of a gypsum plaster, a gauging should be made to the desired consistency, and put on a non-absorbent surface, such as glass.

There is always a certain amount of danger, when patching is being done on a wall, that the moisture will be pulled out of the plaster too quickly, and consequently the plaster will not set properly. It is, therefore, always advisable for patching work, to soak thoroughly with water the part of the wall to be patched. Such a plaster as Murite, on account of its quick-setting properties, may be regarded as a safe material to use for patching.

Natural Cements

Natural cements are combinations of lime and clay, the typical example being hydraulic limes, like the blue lias. There are many varieties, according to the character of the stone found in different countries and districts.

Roman Cement.—Roman cement is obtained by calcining and pulverising lime nodules found in London clay. It contains about 45 per cent. of lime and 30 per cent. of clay; is of a rich brown colour and setting commences within a few minutes after gauging. It must, therefore, be used in small quantities at a time. It is useful for work in damp places and those subject to flooding. For rendering, it will carry an equal quantity of sand. Its hydraulic character makes it a useful under-coat for ordinary plastering in exceptional positions. It possesses about half the strength of Portland cement. The best quality comes from the Isle of Sheppey.

Medina Cement.—Medina cement is much the same as Roman, but rather weaker. It is best suited for under-coats in interior decoration, when the plaster finish and ornament is not too heavy. It is not largely used, as other types—for instance, Roman, Selenite or Portland—give better, more reliable results.

Portland Cement

The manufacture and properties of this material have been dealt with elsewhere, but attention is called to the importance of this being properly cooled for plastering purposes, otherwise slaking will take place after the work is completed and cause great trouble.

In modern Portland cement plants great care is taken that the materials are neither under- nor over-burnt, and that the finished material is thoroughly hydrated by scientific means. Therefore, English-made cements may generally be used as received without fear. At times, however, cement has to be used of which the source is not known, and should this prove to be not properly hydrated, it is necessary to air slake by emptying the cement from the containers on to the dry and damp-proof floor of a shed, protected from wet and damp, to a depth of not more than 2′, by leaving it there for not less than fourteen days, and turning it over from time to time with a dry shovel.

Rough tests for heat in cement are:

The bare arm should be plunged into the dry, fresh cement, which should feel cool, but not cold or dead. When the cement is hot, it is likely to be under-burnt and to contain an excess of lime; it must then be cooled by air slaking.

Gauge (mix with water) stiffly the sample of cement to be tested, and fill into a thin glass test-tube; if the tube cracks the cement has not been sufficiently matured (air slaked). If the cement becomes loose in the tube during setting, due to shrinkage, the cement has been under-burnt. In both cases air slaking is required.

Sand

This material has been fully dealt with elsewhere, but note should be taken of the great importance in plaster work particularly of faults in this material. Salt in sand is of particular importance as regards interior plaster, on account of the efflorescence which, apart from being unsightly, will not allow any decoration of the walls, etc., to be carried out with success, and is of particular importance in connection with lime plaster, as lime hardens mainly by the water drying out, and as salt has the power of attracting moisture, its presence may retard or prevent setting of the plaster—a far more serious matter than would be the case if used in ordinary mortar.

CHAPTER VI

PORTLAND CEMENT

Clay and Lime—Roman Cement—Materials—Mixing—Burning—The Rotary Kiln—Cooling—Grinding—Uses of Portland Cement—Testing: Purity and Homogeneity; Specific Gravity; Fineness; Soundness; Setting; Tensile Strength.

It was not until the middle of the eighteenth century that the principle of what we now call Portland cement was discovered, and this we owe to that eminent engineer, John Smeaton, who carried out a number of experiments with a view to producing a cementing material which he could confidently use in the erection of the Eddystone Lighthouse.

Clay and Lime

He wondered why certain kinds of lime when burnt would set under water while others would not. After considerable research, he found that those limes which contained an admixture of clay possessed, when burnt, this valuable property, which was lacking in the purer varieties. This was an important discovery, because it established the principle that by burning a mixture of lime and clay a substance was produced which possessed the property of hydraulicity; that is to say, a substance which, on the addition of water, became hard and solid.

Here, then, was the nucleus, and although this was not Portland cement, it was, as we have said, a valuable discovery, and the fact that Smeaton's lighthouse stood for 120 years on the Eddystone rock is evidence of the success of his experiments. As a matter of fact, when it was eventually removed, it was not on account of any weakness in the structure, but because the rock itself was being undermined.

As often happens, very little notice was taken of Smeaton's work on cements. It was not until fifty years later a distinguished chemist, Vicat, went a step further and produced a cement by burning together finely pulverised chalk and clay, after having mixed them into a paste. His product again received but little attention.

Roman Cement

Just previous to this, however—that is to say, at the end of the eighteenth century—another cement was invented—Roman cement—which marked a distinct advance in the industry. This was produced by burning what were described as "noddles" (really nodules) of clay containing veins of chalky matter, and it was on these lines that the early inventors worked.

But Portland cement was not yet.

These investigators were on the right track, but by a curious trick of fortune the very portion of the burnt material which would have given them Portland cement, they carefully picked out and discarded.

In 1824, Joseph Aspdin, a bricklayer of Leeds, found that by mixing in specified proportions finely pulverised chalk and clay, burning them at a high temperature and grinding the resulting clinker, he was able to produce a hydraulic binding material far superior to any product known up to that time.

As the colour of this material bore a real or fancied resemblance to that of Portland stone, Aspdin called his product "Portland" cement, the name under which it was patented, and the name by which this class of cement has been known ever since.

Of course, the Portland cement of that day, produced by crude methods, and with comparatively little chemical knowledge, was not comparable with the Portland cement of to-day, the manufacture of which is a highly scientific process, in which highly specialised machinery is employed, carried out by skilled operators and checked by expert chemists.

In fact, in an up-to-date cement works the manufacture is controlled at every stage, and tests are carried out in the laboratory every hour and often every half-hour day and night. Only in this way is he reliable article we know as Portland cement produced.

In order to state exactly what Portland cement is, we cannot do better than refer to the official description. The definition of the British Standard Specification for Portland Cement is as follows:

"The cement shall be manufactured by intimately mixing together calcareous and argillaceous or other silica- and alumina-bearing materials, burning them at a clinkering temperature, and grinding the resulting clinker, so as to produce a cement capable of complying with the specification."

That specification,[1] which is too long to reproduce here, is more exacting in its requirements than any other national specification in the world. This accounts for the high reputation British Portland cement has achieved, and the confidence with which it is used.

An essential part of this specification relates to setting properties. Thus it is stipulated that a mortar shall be prepared with 1 part of Portland cement and 2 parts of dry standard sand, and that when tested these mortars should show three grades: the *quick*, setting in not less than ten minutes or more than thirty minutes; the *medium*, in not less than thirty minutes or more than two hours; the *slow*, in not less than two hours or more than five. In practice, setting can be regulated to a certain degree, hastened or slackened, by adjusting the dosage of sand and working, or the addition of hydrated lime.

[1] *British Standard Specification No.* 12 *for Portland Cement.* Crosby, Lockwood & Co., London. 5*s.* 2*d.* post free.

Materials

Portland cement, then, can be produced from any suitable materials containing the necessary constituents, which are lime, silica and alumina. The raw materials are usually carbonate of lime, in the form of chalk or limestone, and silica and alumina, in the form of clay or shale.

The two materials for the necessary combination are not always found upon the same site. In the Thames and Medway area, for example, a chalk deposit is usually found on the site where the works are situated, and the clay is taken from pits near the river banks.

On the other hand, the Cambridgeshire and Bedfordshire works are able to obtain from one deposit all the necessary raw materials for the manufacture, in the form of marl. In the Rugby and South Wales districts, limestone and shale from the lias deposits are used, and in Northern Ireland a hard chalk with clay dredged from Lough Larne are the raw materials.

Formerly, the chalk and the clay were dug by hand, but, as the demand for cement grew and a greater output became necessary, machinery was introduced. Steam navvies are now used for the chalk, and the crane and grab for the clay. These machines seize mighty handfuls of from one to two tons of chalk or clay, and deposit it in the waiting trucks, which carry it off to the works.

It can easily be imagined that in cement-making districts, the face of the countryside is often entirely changed. Where once was a hill of chalk is now level ground, and where once were green fields are now huge pits which, when the supply of clay is exhausted, will gradually fill with water and become lakes.

Mixing

Since Portland cement is a *chemical* product, all the operations must be carried out with mathematical precision. The first operation is that of mixing the raw materials, and it is essential that the utmost care be exercised in obtaining the correct proportions of the chalk and the clay, or whatever other raw materials may be used, so that the resultant mixture may contain the exact chemical constituents necessary for the production of a sound cement.

When it is remembered that a variation of even 1 per cent. in the proportion of, say, the carbonate of lime would result in an inferior cement, the great care and skill required at this stage of the manufacture will be appreciated. This is why, as long as the manufacture is proceeding, the raw materials are checked day and night by the works' chemists in order that corrections may be made when necessary.

The chalk and clay are weighed and tipped into a wash-mill, which is a circular pit ; here a considerable quantity of water is added.

The whole is agitated by revolving tines or harrows, which break up the lumps and convert the mixture into a liquid of thick, creamy consistency, which can only leave the mill when fine enough to pass the screens which cover the outlets. This creamy liquid is called " slurry."

From here the slurry passes to a storage tank, where it is again kept in constant agitation in order to prevent the solid particles sinking and rendering the mixture uneven. The fineness to which the solid material has been reduced is shown by the fact that all but 5 per cent. will pass a 180 × 180 sieve—that is to say, a sieve with 32,400 meshes to the square inch. This is a gauze finer than a silk handkerchief, and one that will almost hold water. At this stage, again, the composition of the mixture is ascertained by analyses and tests before it passes on to the next stage of manufacture, which is that of burning.

Burning

A century ago, in Aspdin's time, the raw materials were burned in a lime-kiln, or kiln of that type, and here it may be interesting to read an extract from his specification of 1824:

" I take a specific quantity of limestone, such as that generally used for making or repairing roads, and I take it from the roads after it is reduced to a puddle or powder; but if I cannot procure a sufficient quantity of the above from the roads, I obtain the limestone itself, and I cause the puddle or powder, or the limestone as the case may be, to be calcined. I then take a specific quantity of argillaceous earth or clay and mix them with water to a state approaching impalpability, either by manual labour or machinery.

" After this proceeding, I put the above mixture into a slip pan for evaporation either by the heat of the sun or by submitting it to the action of fire or steam conveyed in flues or pipes under or near the pan till the water is entirely evaporated. Then I break the said mixture into suitable lumps and calcine them in a furnace similar to a lime-kiln till the carbonic acid is entirely expelled. The mixture so calcined is to be ground, beat or rolled to a fine powder, and is then in a fit state for making cement or artificial stone."

After the lime-kiln type, special kilns were devised. At first their working was intermittent—that is to say, they were loaded and fired, and when the calcination was completed, the kiln was allowed to cool when the burnt material was drawn. Such were the " bottle " and " chamber " kilns, now nearly obsolete. An improvement on this was the *continuous* kiln, in which the burning takes place without a break and the drawing of the clinker is carried out at the same time without interruption.

The Rotary Kiln

The most modern form of continuous kiln, and that most widely adopted, is the *rotary* kiln. One writer says: " The manufacture of cement by the rotary kiln may be said to have revolutionised the industry during the past twenty years, and this method of burning the raw materials is, perhaps, the most scientific and practical invention that has been introduced into the manufacture since Portland cement was first known."

The first rotary kilns were erected in this country about 1887, and were between 25′ to 30′ in length, and about 5′ in diameter.

The rotary kiln of to-day is a slightly inclined steel or wrought-iron cylinder from 100′ to 250′ in length, and 9′ to 11′ in diameter, being inclined to the horizontal at about 1 in 30.

This kiln is lined with fire-bricks, and is slowly rotated while burning is proceeding, the speed of rotation being 1 to 2½ revolutions per minute.

The slurry is pumped from the storage tank, and is fed continuously into the kiln through a pipe at the upper end, moving by gravity from one end of the kiln to the other.

The fuel consists of finely ground and thoroughly dried coal, which is introduced into the lower or outlet end of the kiln by a jet of air issuing from a blast fan. In the lower third of the kiln an intense heat is generated by the perfect combustion of the finely ground coal—a heat of between 2,500° and 3,000° F. The raw material gradually moves towards this burning zone and parts first with any water it may contain, then with its carbonic acid, and, when passing through the fire zone, is converted into grains of clinker about the size of a large pea.

The heat is carefully regulated so that actual fusion does not take place, since this would produce a slag instead of clinker, and also to prevent underburning which would produce an unsound cement.

Cooling

The hot clinker is cooled by passing through rotary cooling drums similar in form to the kiln, but much smaller in size, and through which a current of cold air is constantly passing. Inside is an arrangement—a series of cascading channels—for lifting and dropping the clinker as the coolers rotate, which they do more rapidly than the kilns.

Although the clinker is formed at practically white heat, when it emerges from the cooling drum it can be handled without discomfort.

Grinding

The next step is the grinding of the clinker, and this is important, because if all the previous operations have been properly carried out, the fineness to which the cement is ground has a great influence upon its quality.

During the last few years there has been a steady increase in the fineness of grinding, since this is essential to the production of great strength in the concrete. It is economical, too, because a finely ground Portland cement will go further than one that is coarsely ground.

A quarter of a century or so ago a grinding plant consisted of mill-stones. As requirements became more stringent, more efficient means had to be provided, and this has resulted in the designing of new machinery.

The clinker is first ground in what is termed a "ball-mill," which consists of a cylindrical grinding drum lined with overlapping steel plates, and containing a number of steel balls of various sizes, between which, as the drum revolves, the clinker is crushed and pulverised. When the material is reduced to a certain size, it passes through slots or holes in the lining plates to a sieve surrounding the drum, which retains the coarser portion and returns it to the drum, while the finer portion passes out to another mill—the "tube-mill"—for finer grinding.

The tube-mill consists of a wrought-iron revolving cylinder about half full of rounded flint stones or other grinding medium. The coarsely ground material is fed from the ball-mill into one end of the tube-mill, and as it passes through the falling flint stones, the cement—for it is now Portland cement—is so finely pulverised, that, in the ordinary way, the residue left on a sieve with 32,400 meshes to the square inch is often as low as 4 per cent. or a fraction over, and certain cements, such as "Ferrocrete," are even more finely ground than this. How fine this is will be realised when it is remembered that the British Standard Specification allows a residue of 10 per cent. on the same sieve.

The immense strength of Portland cement has rendered possible many great engineering enterprises, which without it would never have come into being. Its plasticity allows it to be moulded into articles and structures of any desired shape at a fraction of the cost of the same things in stone; its durability reduces the cost of the upkeep practically to nil; its impermeability to moisture makes for comfort and cleanliness; and the artistic possibilities of which it is capable add to the beauty of the home and garden.

Uses of Portland Cement

In plastering Portland cement is only used to a limited extent for interior work, chiefly for skirtings and arrises. But it is also used under exceptional circumstances, either in coarse mortars or concretes, for certain kinds of decorative moulded work and modelling and carving. Also for flooring, and to some degree for ceilings.

Externally, it is largely utilised both for plain and decorative plastering. Indeed, in the form of a mortar, composed of 1 part Portland cement to 3 or 4 parts of sand, laid on in two coats, each $\frac{3}{4}''$ thick, is rapidly displacing common, rough, bastard and trowelled plaster. For this purpose, it is now often applied with the cement gun. But this appliance, though extremely useful for heavy work on industrial buildings and the like, does not permit of the pleasing variety obtainable in trowelled work, variations which are now very much sought after. Character in texture and surface is more esteemed than the perfect smoothness once so insisted upon. This, indeed, applies equally to interior and exterior plastering with lime plaster or cement mortar.

For plain interior and exterior work, quick-drying cements are to be preferred. This is particularly the case in exterior work exposed to moisture. When cement mortars are to be moulded and shaped by means of templates, a medium cement is indicated. If it is to be modelled, carved or otherwise elaborately decorated, then a slow grade should be used, with the admixture of hydrated lime where prolonged work is needed, as will be explained more at length in a subsequent chapter.

Although hydrated lime is sometimes mixed in Portland cement mortars, both in order to retard setting, and so make it easier to handle, and to increase hardness, fast lime should never be used. While fast lime undoubtedly makes cement mortars and concretes more malleable, it weakens them and renders them liable to crack, owing to the difference of the materials and their failure to form a true amalgamation.

Portland cement should never be used "hot"; if necessary, it should be air-dried. Nor must it be exposed in such a way that it will absorb moisture, which would cause it to swell, in bad cases to cake, and lose strength. It must also be kept perfectly clean.

Testing

For plastering purposes, the tests usually applied are quite simple. They aim at ascertaining purity, perfect homogeneity, soundness, setting properties and tensile strength.

Purity and Homogeneity.—These qualities are revealed by delicate chemical tests to ascertain (1) loss of weight by calcination, (2) amount of insoluble residue, (3) specific gravity. There must be no excess of lime—that is to say, there shall be no more than is sufficient to saturate the silica and alumina present; the percentage of insoluble residue must not exceed 1·5 per cent.; that of magnesia not more than 3 per cent., and that of sulphuric anhydride not more than 2·5 per cent.

Specific Gravity.—The specific gravity should not be less than 3·15 at the works, nor less than 3·10 when delivered.

Fineness.—The fineness tests are carried out by continuously sifting 4 oz. of cement on two different sieves for fifteen minutes. The residue on the sieve with 32,400 meshes to the square inch must not be more than 14 per cent.; and on the sieve with 5,776 meshes per square inch not more than 1 per cent. This, of course, is for standard cement, but coarser grades are suitable for much of the work carried out by plasterers.

Soundness.—Soundness is tested by gauging pats of pure cement, 3″ in diameter, ½″ in the centre and tapering towards the edges. They are placed on smooth, clean sheets of glass to mature in a damp atmosphere; are then steeped in water (or exposed to steam or damp air in an oven). When mature, the pats should be flat and free from cracks. If they curl up, or the surface is curved with fissures, even if there are no deep cracks, the cement should be rejected.

Setting.—This test is made by subjecting the pats to pressure from needles; when they penetrate half-way through, we have the initial set; when they will not penetrate at all, we have the final set. The penetration is timed with a minute watch.

Tensile Strength.—Finally there is the tensile strength (compression, impact and abrasion not being needed in connection with plastering). The tensile strength is ascertained by moulding flat briquettes of hour-glass shape and placing them in automatically revolving machines which tear them asunder without crushing them. The Standard Specification demands that briquettes at seven days after gauging shall not break under a strain of less than 450 lb. per square inch of section. This has to be considerably increased at twenty-eight days.

CHAPTER VII

HOW TO MAKE CONCRETE

Reinforced Concrete—Materials: Cement; Sand; Coarse Aggregate—Suitable Aggregate: Coke-breeze; Clinker; Broken Brick; "Granite" Concrete; Size of Coarse Aggregate—Proportioning—Mixing: Mixing Board; Measuring Boxes—Placing—Protection and Curing—Concreting in Frosty Weather—Impermeability—Surface Dressing—Mistakes to be Avoided—Materials Required: Tools; Timber; Cement, Sand and Coarse Material—Tables—Summary.

PORTLAND cement is so inseparably connected with the manufacture of concrete in its various forms, that it is necessary to set forth clearly how concrete can be made, and what are the chief mistakes to be avoided if the work is to meet with success.

Concrete, then, is a substitute for stone and timber, but more adaptable than either, and possessing great strength. It is made by binding together, by means of Portland cement and water, particles of sand and gravel, broken stone or other substances known as the aggregate.

The term "aggregate" is somewhat loosely used. Sometimes it refers to the coarser particles of the material as distinct from the sand, but, strictly speaking, it includes the sand, which may be described as the finer aggregate, the whole of which is less than ¼" in diameter.

REINFORCED CONCRETE

In certain kinds of concrete work it is necessary to strengthen the material by embedding steel in the concrete in the form of rods, wire-netting or some other form of metal mesh. When thus strengthened, it is known as reinforced concrete. For any important work involving the calculation of strains and stresses a specialist should be consulted, since what might appear to the casual observer to be a very simple matter is not really so, requiring as it often does expert knowledge; but in the smaller uses of concrete no special knowledge is necessary to enable the user to manipulate the reinforcement successfully. The fundamental principle underlying the various applications of reinforced concrete is that whereas steel is elastic and extremely strong in tension, concrete is strong in compression, but comparatively weak in tension—that is to say, concrete will resist an enormous crushing load, but it will not resist strains tending to bend or stretch it. Thus, what the concrete lacks the steel supplies, and in reinforced concrete, therefore, an ideal constructive material is obtained.

Materials

1. *Cement.*—Portland cement is a chemical product, and the selection of this material need entail no difficulty, if a brand of good repute, which is certified to comply with the requirements of the British Standard Specification, be chosen. The cement is delivered in bags or barrels, and the user should see that the maker's name is legibly marked upon them. The weight of a sack of cement is, approximately, 200 lb., and that of a barrel 400 lb.

Storage.—Portland cement readily absorbs moisture from the atmosphere and soon becomes lumpy, or even a solid mass, if exposed to damp. It must, therefore, be stored in a dry place. Should there be any danger of damp rising from the floor, the cement should stand on a raised platform of boards.

2. *Sand.*—The essentials of a good sand for the purpose of making concrete are that it shall be clean, coarse, graded, pure and, according to many, sharp.

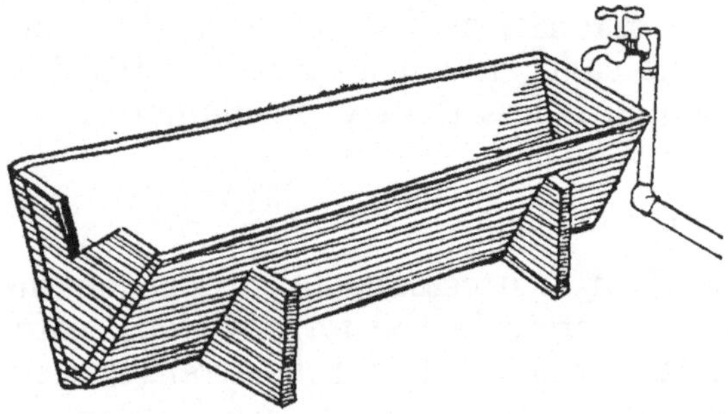

Fig. 1.

It must be clean, because if coated or mixed with particles of clay, earth or other foreign matter the cement will not properly adhere to it, and thus a perfect bond is not effected between itself, the cement and the coarser aggregate, a weaker product being the result.

To test whether the sand is clean or not, rub a little between the hands. If the hands are considerably soiled the sand is not clean enough; or drop a handful into a glass of clean water; if the water is much discoloured the sand is dirty. When this is found to be the case, the sand must be washed. A simple way of doing this for a small job is as follows: Procure or make out of plain boards a simple trough, 8′ or 10′ long, such as that shown in Fig. 1. Put the sand in the trough, and while gently stirring the material let a stream of water run through it, either from a tap or a garden hose. The water will overflow at the lower end, carrying away the particles of dirt, and the sand will remain.

If the sand is fine, the particles will present a much greater total surface than those of a coarse kind, and will, in consequence, require considerably more cement to coat them, and also more care in mixing. A coarse sand is therefore necessary, in order that a minimum of cement may attain a maximum of strength.

The sand should consist of particles varying from the coarsest to the finest, the coarse predominating. To understand the reason for this, consideration should be given to the following. If gravel or broken stone, say an inch in diameter, are piled together, there will be spaces, or, as they are technically termed, "voids," between them. If now smaller stones are mixed with them the voids will be partially filled, and if coarse sand be added, it will practically fill the voids between the stones, but there will be small voids between the particles of sand. If these, again, are filled with finer sand the voids will be reduced to a minimum, and thus the amount of cement, which not only has to coat all the particles of aggregate, both coarse and fine, but must also fill the voids, is also reduced to a minimum, while the resulting mass is compact and strong. This, it will be seen, makes for economy.

The sand must be pure—that is, it must actually be sand, and not merely have the form or appearance of such.

It is often specified that the sand should be "sharp"—i.e. that the particles shall be angular. Recent tests, however, have shown that as good a concrete can be made from sand consisting of rounded grains as from that whose grains are angular. Indeed, it is very seldom that the latter are obtained, since practically all sand has at some time or other been water-worn, and therefore will consist of rounded grains. If river, lake or sea sand can be obtained conveniently it may be used, provided it is coarse and graded. The objection to this kind of sand is that it is often too fine and the grains too uniform in size. The salt contained in sea sand retards the setting and hardening of the concrete, although it has no ultimate effect on its quality. Salt, too, is apt to cause efflorescence on the finished surface. This, however, may be removed in extreme cases by washing with a weak solution of hydrochloric acid, popularly known as spirits of salt. If sea sand is used it should be taken from above or below the tide-water, otherwise it is likely to contain organic and other deleterious matter.

3. *The Coarse Aggregate.*—The selection of the aggregate is important, because upon its strength and hardness depend the strength and hardness of the concrete. The cement, in itself, is not intended to add hardness to the product, its function being that of binding only. Hence the saying, which is true as far as it goes, that "Concrete is no stronger than its aggregate." Sometimes the cement is unjustly blamed when the work is found to be defective. The truth is that inferior concrete is almost invariably the result of unsuitable aggregate or bad workmanship.

A good aggregate must be clean, graded, hard and, when occasion demands it, fire-resisting. Clean—i.e. free from clay, loam or vegetable matter—for the same reason that sand must be clean and free from organic impurities; graded, so as to fulfil, on a larger scale, the same functions as graded sand; and hard, to impart strength and to resist abrasion. In buildings, or wherever the material is liable to great heat, all aggregate that expands or splinters under the agency of heat should be broken into small pieces, such, for example, as limestone, dolomite, etc.

Suitable Aggregates

The aggregates most suitable for making good concrete are granite, gravel (such as Thames ballast) and crushed stone; but other materials, such as coke-breeze, clinker and broken brick may be used for certain purposes with satisfactory results, provided care be taken that they contain no deleterious matter.

Coke-breeze.—Coke-breeze—i.e. small pieces of coke—is often used because of its cheapness, but the chief objection to it is that it is porous and may allow moisture to penetrate the concrete, which would be fatal if the concrete contained any embedded steel-work. Coke-breeze, however, is often confused with "breeze," which includes pan-breeze and ashes, which are entirely unsuitable, containing, as they do, pieces of unburnt or partially-burnt coal and coke, and other deleterious substances, such as sulphur; they are, moreover, often full of dust.

Clinker.—This, like coke-breeze, is light and porous, and produces a concrete which is not so strong as that made with heavier and denser coarse materials.

Broken Brick.—This is often quite a good material, but great care should be exercised in choosing it, as some bricks contain sulphur and unslaked lime, and others have been found by practice to possess qualities that render them unsuitable for concrete; where they are to be employed, local experience should be consulted, and only a hard-burned variety selected. Any plaster remaining on the bricks should be removed, and excessive dust should be avoided.

Where available, then, granite, broken stone or gravel are the best materials to use as coarse aggregate. The last-named is widely distributed, and can therefore often be obtained more conveniently and at less cost than the others; and if it be clean and well graded, gravel will result in as good a concrete as stone.

"Granite" Concrete.—A handsome variety, known as "granite" concrete, is a mixture of white Portland cement, fine white sand and such aggregates as quartz, felspar and other coloured stone, broken up rather small. The aggregate will vary according to taste or to what is available locally. Usual proportions: 1 part cement, 1 part sand, 2 parts aggregate. Aggregate should vary from $\frac{1}{16}''$ to $\frac{1}{4}''$ in diameter. For walls under 6" thick this concrete is generally used in mass, but when 6" thick or more casing is more common in order to save expense. Concrete of this kind can be used to cover the lower part of walls (say, up to first-floor level), panels or to case pillars. It must be so laid that the aggregate shows on the surface, and it is often found necessary to wash down the surface with a moderately stiff brush as soon as the initial set has commenced. A rich-coloured sand may replace the white sand, and the aggregate can be broken more or less in the form of cubes or in thin flakes to vary effects.

Size of Coarse Aggregate.—The size of the coarse material should be

SECTION OF EARLY GEORGIAN CEILING.

Painted on the flat, or the ornament in plaster, medium relief, the panels usually painted by well-known artists.

varied according to the purpose for which the concrete is to be used, but must in all cases be graded, for the reason already given. For mass concrete, such as foundations or large structures, pieces up to 2½″ diameter may be used. For floors and for walls up to 12″ in thickness, and for reinforced concrete, the largest size should be ¾″.

Proportioning

Having selected the materials for making the concrete, the question arises as to the proportions in which they should be mixed.

The ideal of good and economical concrete is to make it strong enough for the particular purpose in view, but not to make it of any better quality than is absolutely required. In order, therefore, that the concrete may afford the necessary strength with the greatest economy, the proportions will vary according to the materials used and the purpose for which the concrete is to be employed. If the materials used are of good quality and of average character, this aspect of the question can be neglected in small work, and the purpose for which the concrete is to be made alone taken into consideration. Where a very good concrete is required—such, for example, as in important reinforced work and for water-tanks—the proportions usually adopted are 1 part by volume of Portland cement, 1½ parts sand and 3 parts coarse material. This is generally referred to as a 1 : 1½ : 3 mixture.

Only in exceptional circumstances is a richer concrete than this used. For ordinary purposes the proportions most commonly adopted are 1 : 2 : 4, while for a medium quality concrete, such as is used for foundations, floors and steps, the proportions should be 1 : 2½ : 5.

For heavy walls and foundations, piers and other structures where actual strength is not so important as weight and stability, the proportions may be 1 : 3 : 6.

In surfacing factory floors, public footpaths and similar work where there is heavy usage, a richer mixture of concrete is employed, such as 1 part of cement and 2 parts of fine granite chippings. A surface of 1″ in thickness generally suffices.

Bank Gravel.—One caution must be given. Gravel as it comes from the bank or pit contains a varying quantity of sand, and, therefore, cannot be used as though it were all coarse aggregate. The only safe way of treating such material is to "screen" or sieve it on ¼″ mesh, when the sand will pass through and the coarser material will be retained on the screen. The two materials can afterwards be mixed in the proper proportions.

Mixing

Consider, now, for a moment what is the object to be aimed at in mixing. It is that in order to produce a dense, strong concrete, every void shall be filled and every particle of sand and stone coated with the cement. If this end is to be accomplished, the mixing and watering must be very thorough, as it is obviously highly important.

On large jobs the concrete is usually mixed by a machine, but in small jobs which do not justify such an outlay this part of the work is done by hand. The materials are first mixed dry, but before this can be done the following preparations must be made.

Mixing Board.—If a clean, level floor is available, the materials may be mixed on this, but if not, build a simple platform or stage, such as is shown in Fig. 2, about 10′ long and 8′ or 9′ wide, or, if large batches are to be mixed, 12′ by 10′. One-inch boards may be used, but they must be free from knot holes and planed on one side and two edges. If they are tongued and grooved, so much the better. Cross-pieces of 2″ × 4″ timber must be provided, and to these the boards are nailed, smooth side outermost. When nailing, take care that the boards fit closely together so that none of the cement can pass through. A border can also be made to the platform by nailing boards round the edges. If much mixing is to be done, it is advisable to cover the stage with thin sheet-iron or zinc. Before commencing the operation of mixing, place the mixing-stage as near to the materials and to the work as is convenient, and see that it is quite level.

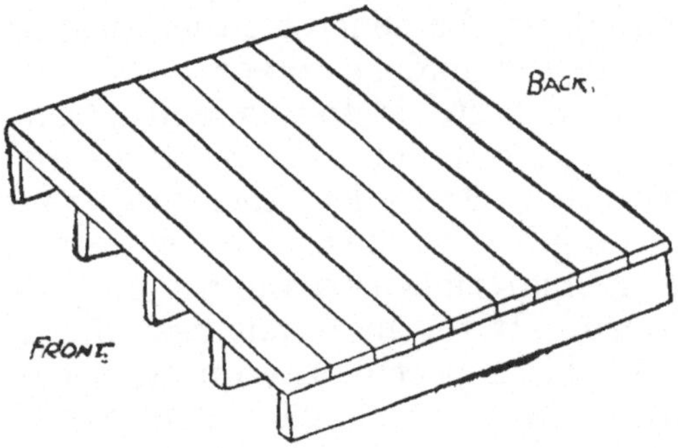

Fig. 2.

Measuring Boxes.—It will be remembered that the materials for making concrete are proportioned by volume and not by weight. A convenient way, therefore, of effecting this is by the use of measuring boxes; that is to say, boxes without top or bottom (see Fig. 3). These may be any convenient size, but each should contain an exact number of cubic feet, and for very small jobs a bucket or a box whose capacity is 1 cubic foot will be found useful.

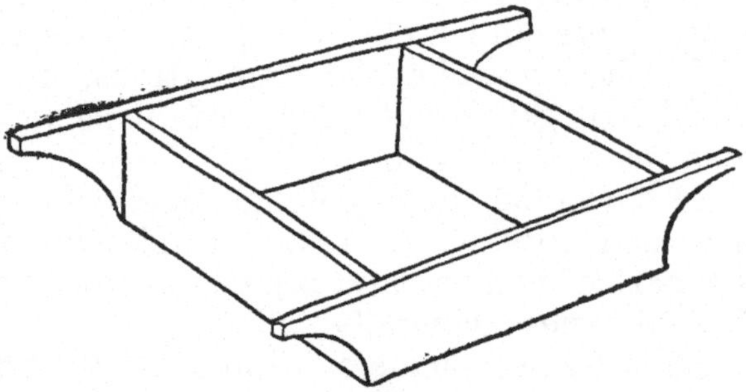

Fig. 3.

Suppose it is desired to make a measuring box 2′ square and 1′ deep, inside measurement.

Cut two pieces of board 1″ × 12″ × 4′, and two pieces of 1″ × 12″ × 2′. The two longer pieces will form opposite sides of the box. Mark off on these 11″ from each end; the portions so marked off may be shaped to form handles (see Fig. 3). The two shorter pieces form the other two sides, and must be fastened in position so as to leave exactly 2′ between them (see Fig. 3). The box thus made will contain exactly 4 cubic feet of material. If a greater quantity of concrete has to be mixed in each batch, a larger box may be constructed; say, 3′ square. This will contain 9 cubic feet or ⅓ cubic yard.

Variations of this kind are possible, but it is inadvisable to have boxes of too large a size, as they become unwieldy to handle and rather encourage carelessness. When two or more measures are required the workers are apt to scamp the detail of levelling in order to obtain a full measure, and may lift it prematurely, so that either too little or too much is provided. This may not be of great consequence when sand or aggregate is in question, but is of great importance in the case of cement. These remarks apply with equal force whether making concrete or mortar.

Another plan is to use a wheelbarrow of known capacity as a unit of measure for the aggregate, care being taken that the material is levelled off each time flush with the edges of the barrow.

Let us now imagine we are going to mix a batch of 1 : 2 : 4 concrete, consisting of 2 cubic feet cement, 4 cubic feet of sand and 8 cubic feet of coarse material.

Follow these instructions closely.

Place the measuring box near one end of the mixing-board as at X (Fig. 4), and exactly fill it with sand. Lift off the box and spread out the sand in a layer about 4″ deep. Then lay the cement, after measurement, as evenly as possible over the sand. These should now be mixed while in their dry state by two men in the following manner: The men, provided with square-ended shovels, stand on opposite sides of the pile (A and B, Fig. 4), and each, starting at the outside of the material and shovelling away from him, takes a shovel-load and turns it over some little distance towards the other end of the board, as at C and D (Fig. 4), thus beginning a fresh pile. In throwing the material from the shovel, it should be flicked or shaken from the end and sides, so that the sand and cement are mixed as they fall. When the whole of the material has been turned over and shovelled from one end of the board to the other, it should be shovelled back again in the same manner into its original position, and again spread out. The material

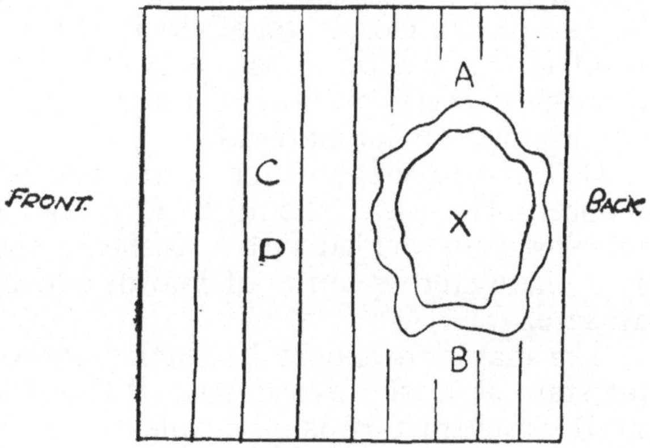

Fig. 4.

will now have been turned over twice, and if the work has been done thoroughly this should be sufficient, but if the colour of the material is not uniform, indicating that the mixing is not complete, it must be turned over again. A good rule is to give the material an extra turn whenever there is any doubt as to the thoroughness of the mixing.

The coarse material should now be measured to the requisite quantity, laid on the top of the cement and sand mixture, and the whole turned over three times. If the coarse material is of a porous nature, it should be well wetted before being added to the cement and sand mixture.

The dry material, having now been thoroughly mixed, is ready for the water. This should be clean, and free from acid, alkali or organic impurities.

No definite rule can be given as to the amount of water to be used, as this will differ according to the nature of the aggregate, the purpose for which the concrete is to be used and the temperature at the time of mixing; the necessary quantity must be determined by experiment; it will be such, however, that when the concrete is placed and tamped—i.e. beaten down—a little water will just flush to the surface. The correct quantity having once been determined, the same proportion, as a rule, can be used for the other batches.

It should be noted, however, that in reinforced concrete, particularly in such portions as may contain a large amount of reinforcing bars or the like placed closely together, it is essential that the concrete should be sufficiently plastic to pass between the reinforcement, and to surround thoroughly every portion of the steel, though not so wet as to allow any dripping of the material.

Do not throw the water on from a bucket, and do not put it all on at once. The water should be applied a little at a time by means of a watering-can provided with a rosehead, and filled from measured buckets, or, if the watering-can is of known capacity, it may itself be used as a measure.

The materials should be further mixed by being turned over in the wet state at least three times; if the workmen employed are inexperienced, an extra turn is advisable.

Placing

As soon as the materials are properly mixed, the concrete should be placed without delay. The rapidity with which this can be done should govern the quantity of material mixed at one time, since concrete which has begun to harden should under no circumstances be used, or failure will be the result.

Wherever it is possible, it is better that the concreting should be completed in one operation; but when this cannot be done, the surface of the concrete, where the junction or "bond" has to take place, should be left as rough as possible and, before the work is resumed, should be washed, and a $\frac{3}{4}''$ layer of 1 : 1 cement and sand mortar applied, and the fresh material tamped into it.

The concrete in its plastic state is deposited in moulds or "forms" previously arranged according to the work to be done, and these are constructed to meet the requirements of each particular job.

Where the concrete has to be laid on a foundation, the latter should be thoroughly wetted before the concrete is placed.

Except in the case of floors, lay the concrete in layers of 6″ and tamp in order thoroughly to work the material into place, to expel the air and produce a compact mass. The concrete next the forms should be well worked with a spade or a spading tool or a wedged stick (see Fig. 5), so as to force back the coarse aggregate and allow the mortar to flush in. This gives an even surface to the finished work, and the danger of the formation of air-pockets is obviated.

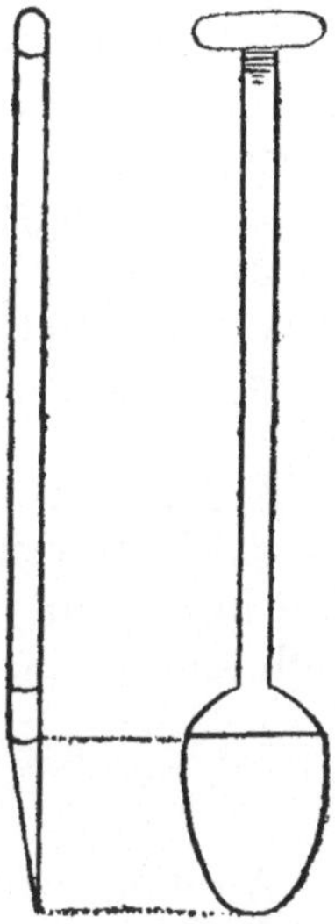

Fig. 5.

Protection and Curing

Newly-laid concrete should be prevented from drying too quickly, and must be protected from the sun and wind. Drying must not be confused with setting and hardening. The latter, which are chemical processes, go on quite as readily under water. It is in order that the setting and hardening may be thorough that the concrete must be kept damp for a time. The surface should be sprayed with water as soon as it has sufficiently hardened to withstand pitting; and as soon afterwards as it can be done without damaging the surface, the concrete should be covered with not less than 2″ of wet sand or other material which will afford equally good protection, or in the case of vertical work by hanging cloths kept continuously wet for ten days. If this is not convenient, the concrete should be sprayed with water daily for the same period.

Concreting in Frosty Weather

The execution of concrete in frosty weather should be avoided as far as possible, as frost prevents the bonding of the different layers, and often causes a thin scale to peel off the surface, and sometimes a disrupture of the work through the expansion of the water.

When the work is urgent and cannot be delayed, an expedient that may be adopted is to add to the water 1 per cent. by weight of salt for every degree F. below freezing-point. As mentioned above, the salt delays the hardening and retards the development of strength, but the ultimate result is not affected. This expedient, however, is rarely resorted to, as it is preferable not to concrete in frosty weather.

Impermeability

For many purposes to which concrete is applied, as, for example, for water-tanks, it is absolutely essential that the material shall be impermeable. This will be secured when:

(*a*) The aggregate is impermeable and well graded.

(*b*) The materials are well-proportioned and well mixed.

(*c*) Sufficient water is used to make a fairly wet mixture.

(*d*) The concrete is kept thoroughly wet for ten days or more after being placed.

A number of waterproof compounds in the form of paint or washes are sold for rendering concrete impermeable, but given a mixture fairly rich in cement, and the above-mentioned conditions thoroughly fulfilled, such should be unnecessary.

Surface Dressing

A defect commonly seen in concrete work is that known as "crazing," or fine superficial cracking. These "hair" cracks, as they are sometimes called, do not affect the strength of the concrete in any way, but they are, nevertheless, very unsightly. They are due to the unequal shrinkage of the surface concrete and the bulk portion, and generally result from the use of a mortar for surface dressing too rich in cement, or from over-trowelling. This defect can be entirely avoided by using cement well mixed with sand or other very fine aggregate in the dressing of concrete work, and trowelling no more than is absolutely necessary.

Mistakes to be Avoided

The ways in which concrete may be spoilt by the user are many, and it may be useful at this point to summarise the most common mistakes. They are:

(*a*) The use of too much or too little water during mixing, or water carelessly applied.

(*b*) The incomplete incorporation of the aggregate with the cement.

(*c*) Using concrete which has already begun to set.

(*d*) The placing of concrete on dry foundations, and without thoroughly saturating their surface with water.

(*e*) The use of dirty aggregate or water containing earthy matter, clay or loam.

(*f*) The repeated "floating" or trowelling of concrete surfaces.

Materials Required

Tools

2 square-ended shovels.
2 tampers (Figs. 7, 8 and 9).
2 spading-tools (Fig. 6) or garden spades.
Measuring boxes.
Wheelbarrow.
Watering-can with rose.
Water buckets.

Timber

For mixing-board 10′ × 9 ′.

Nine lengths of 1″ × 12″ × 10′ = 90′ for the platform.

Five lengths of 2″ × 4″ × 9′ = 45′ for the cross-pieces.

38½′ of 1″ × 4″ for the border.

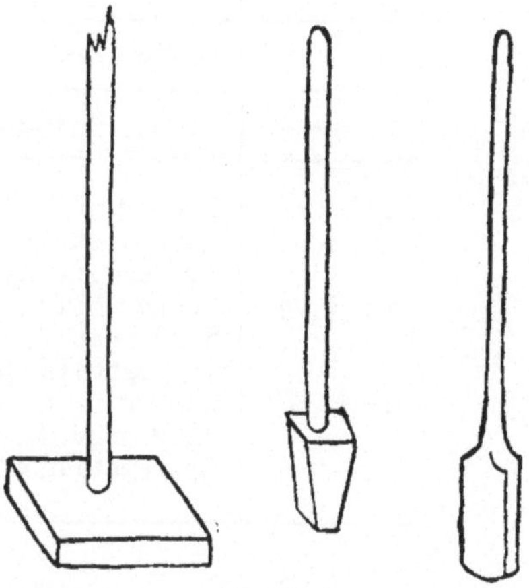

Fig. 6.

Cement, Sand and Coarse Material

To ascertain the quantities of cement and aggregate required, the number of cubic feet or cubic yards of finished concrete must first be worked out.

It must not be supposed that a mixture of 1 part cement, 2 parts sand, and 4 parts coarse material will produce 7 parts of concrete. This will be clear when it is remembered that the sand and cement go to fill up the voids. As a matter of fact, the volume of the finished concrete is very little greater than that of the coarse aggregate, and for very small jobs a convenient method, and one that will not be very far wrong, is to make this the basis; that is to say, for every cubic foot of concrete required provide a cubic foot of coarse aggregate, and the sand and cement in proportion.

As the cement is bought by weight, it is more convenient to apportion it in the same manner, but no difficulty will be experienced if 1 cubic foot of loose cement be taken to weigh 90 lb. Where more exact quantities are necessary, they may be calculated from the table on page 88, which gives the materials required for each cubic yard of finished concrete.

In connection with this table, it is necessary to point out that if the aggregate is large, of nearly uniform size and very irregular in shape, the mixture will bulk less than if the aggregate was not too irregular in form and varying in size, because in the former case there would be more cavities to be filled by the sand.

MATERIALS FOR 1 CUBIC YARD OF CONCRETE

Based on loose cement weighing 90 lb. per cubic foot with an average specific gravity of 3·12 and a cubic foot of loose moist coarse sand weighing 89 lb. when dried.

Proportions.	Kind of Coarse Material.	Lb. Portland Cement in 1 cu. yd.	Sand, cu. yd. in 1 cu. yd.	Coarse Material cu. yd. in 1 cu. yd.
1 : 1½ : 3	Shingle (40% voids) . .	666	0·41	0·82
Do.	Broken stone (45% voids) .	697	0·43	0·86
1 : 1⅔ : 3⅓	Shingle	610	0·42	0·84
Do.	Broken stone . . .	640	0·44	0·88
1 : 2 : 4	Shingle	520	0·43	0·86
Do.	Broken stone . . .	548	0·45	0·90
1 : 2½ : 5	Shingle	430	0·44	0·88
Do.	Broken stone . . .	450	0·46	0·92
1 : 3 : 6	Shingle	364	0·45	0·90
Do.	Broken stone . . .	383	0·47	0·94
1 : 4 : 8	Shingle	280	0·46	0·92
Do.	Broken stone . . .	294	0·48	0·97

AMOUNT OF MATERIALS PER SQUARE YARD

Proportions.	Materials.	4" thick.	8" thick.	12" thick.
1 : 2 : 4	Cement in lb.	59·3	118·6	177·9
	Sand in cu. ft.	1·32	2·64	3·96
	Aggregate in cu. ft. . .	2·62	5·24	7·86
1 : 2½ : 5	Cement in lb	48·9	97·8	146·7
	Sand in cu. ft.	1·35	2·70	4·65
	Aggregate in cu. ft. . .	2·7	5·4	8·1
1 : 3 : 6	Cement in lb.	41·5	83·6	124·5
	Sand in cu. ft.	1·43	2·86	4·29
	Aggregate in cu. ft. . .	2·86	5·72	8·58

SUMMARY

1. Calculate exactly how much finished concrete will be required for your job.
2. Ascertain how much cement, sand, coarse aggregate and timber you will require.
3. Choose good materials. Wash the sand and coarse aggregate if necessary.
4. Build your mixing-board and make your measuring-boxes if the size of the job warrants it.
5. Construct the " forms " and place them in position.
6. Do not guess at quantities. Measure them.
7. Do not mix too large a batch at one time.
8. Mix your materials thoroughly and a little more.
9. Place the concrete in position without 20 or 30 minutes of wetting.
10. Tamp it well and spade the edges.
11. Protect the finished concrete from sun and air.
12. Do not remove the forms until the concrete is hard.

CHAPTER VIII

SELECTION OF MATERIALS

Selections of Plaster—Lime Plaster—Impervious Plaster—Cement Plaster—Exterior Work—On Metal Laths—Finish for Interior Work—Use of Hair in Plastering—Hair for Cement—Gypsum Plasters—Nature of Hair Used—Even Distribution—Amount of Hair Required—Hair for Portland Cement—Patent Plasters.

Selections of Plaster

Until a few years ago, even if such is not the case to-day, the most common material for plastering used in this country was lime plaster. In spite of its many disadvantages in the way of delaying the work on account of slow setting and the time taken finally to dry out, it is doubtful whether this, or at least some other absorbent plaster, is not the best material for interior finishes, and particularly for dwelling-house purposes; as by its use condensation trouble, so often met with where impervious plasters are applied, is avoided.

Amongst the impervious plasters, apart from Portland cement, may be counted those made up with a gypsum base and the combination of various other materials, such as Keene's, Parian plasters, and others sold under various trade names, which, though setting quickly and readily working up to a good surface, at the same time do not absorb from the air any moisture whether from condensation or other reason. This quality on scientific grounds may be good, but, nevertheless, the result is unsightly, and it is doubtful whether this humidity is not less dangerous distributed through a considerable thickness of, say, lime plaster, and again taken out to the atmosphere of a room when this is sufficiently dry, or lying on the surface of the walls and thence being absorbed into furniture, etc.

Lime Plaster.—Lime plaster, there is no doubt, was assisted to its bad repute by the time and trouble required in preparing the lime, but as it becomes more common to obtain lime properly and scientifically hydrated (slaked) at the kilns, and supplied in powder form ready for mixing with the sand and water, this material will doubtless regain its popularity for interior finishing. To overcome the delay in being able to apply the different coats, the setting may be quickened by mixing with plaster ordinary plaster of Paris in the two last coats. It is difficult definitely to state a proportion of the latter which should be used, as this will vary according to weather conditions as regards the working time, and again, if too much is used the absorbent nature of the plaster will be unduly interfered with ; but the proportion of 1 part to 10 of the lime can safely be relied upon, and this can often be increased. The plaster of Paris can, of course, only be mixed with the lime plaster in small quantities as required for use, and if the quantity of this material is too great there will be combined cracking and hollow work.

For the " scratch " or rendering coat in lime plaster work grey stone lime properly slaked should be used, combined with about 1 lb. of hair to 3 cubic feet of plaster. This is best used without any addition of plaster of Paris, as this will set much more quickly than " fat " or " chalk " lime plaster.

Impervious Plasters.—Where walls are to be painted, or covered with non-absorbent material, there will be no object in not using an impervious plaster, for in any case the pores would be filled up and, under these circumstances, advantage should certainly be taken of the quicker setting of the hard gypsum plasters, which give the added benefits that they may usually be brought to a good surface with two-coat work, either on a Portland cement back or with two coats of the same material, thus showing a considerable saving in labour.

With all these hard plasters, however, there is the danger of resonance, this being particularly noticeable in rooms having stud partitions and those of small size ; therefore, even though it is desired to use such plasters for the walls, for the ceilings it is advisable to use an absorbent plaster such as lime, which will be found to have a neutralising effect as regards this failing.

Where damp or waterproofing is required, Portland cement mortar should be used, either with or without the addition of a waterproofing compound ; but if this material is resorted to, it is preferable to select one which is added to the plaster in a liquid form—in the mixing water if possible—in order to be certain that it is carried to all parts of the mix, as with powder compounds it is difficult to get, and retain, these evenly distributed through the mix.

Cement Plasters.—Cement plaster should be applied in two coats each of not less than $\frac{3}{8}''$ thick, the first coat being in the proportion of 1 of cement to 3 of sand, and the finishing coat of 1 of cement and 3 of sand.

In using cement plaster, the addition of 10 per cent. (of the cement) of properly hydrated lime not only makes the working easier, but also assists in the water-resisting qualities of the plaster, and, contrary to the old belief, where this percentage is not exceeded, actually increases the mortar in tensile strength—the quality which is required for plastering. Tests have demonstrated that a 1 to 3 mix with the addition of this proportion of lime added was at twenty-eight days 242 lb. as against 201 lb. for the ordinary mortar.

When using a waterproofing compound, however, it is advisable, before adding hydrated lime, to consult the makers in order to ascertain whether this addition will interfere with their composition.

In the use of cement plaster there is a point to which sufficient consideration and care are often not given—that is, the protection of the finished work against cracking. This is of particular importance with waterproofing work, for however impervious in itself the plaster may be, should cracks occur the water will find its way through. The main causes of cracking are too much cement, too much trowelling (particularly with a steel float instead of a wooden one), and too rapid drying out of the water, either by suction of the base or of the first coat.

Cement in setting first expands and then contracts, chemical action which requires water going on for some considerable time, and unless this is present, particularly during the first ten days, the plaster loses its tensile strength, and, having nothing to resist the considerable strain caused by contraction, cracking occurs.

Hydrated Lime.—Hydrated lime assists both by increasing the tensile strength and by retaining water more readily than cement, which is actually using it by its action in setting, and in interior work it is fairly easy to prevent too rapid drying by closing all doors and window openings, and sprinkling the surface, once it has hardened, frequently with water.

Exterior Work.—On outside work, however, it is a more difficult matter, for in hot weather the water will be dried off the surface practically as it is applied, and probably the only satisfactory way of meeting the proposition is to hang sacking or other absorbent material close to the finished work and keep this constantly wet. This is a method which is in general use in America, where the matter has received much attention.

Another point which is of much importance in waterproofing coats is that of joints between the work from day to day. If it is at all possible, cement plastering should be continuous without any break of more than an hour, particularly in the final coat and during hot weather; but this is not, as a rule, possible, and then the edge of the previous work should be chiselled off at an angle of 45 degrees, be scrubbed down with a 10 per cent. solution of hydrochloric acid, which should afterwards be well washed off with clean water. Cement mixed with water to the consistency of cream can then be brushed on, to be immediately followed by the plaster.

On Metal Laths.—For plastering on metal lath bases, there is no doubt that cement plaster should be used, at least for the first coat, for though metal lath coated with lime and gypsum plasters, applied years ago, is in existence in perfect state, cement gives both the best protection to the metal and the strongest base for applying the finished coats.

Cement plaster for applying to metal lath should be used stiffer than for ordinary work, and should, particularly for ceiling or horizontal work, contain more hair than for ordinary work, 1½ lb. per 3 cubic feet of plaster being considered a good proportion. Sometimes a rich mix of cement and sand is proposed for this purpose, but 1 part of cement to 3 parts of sand with 10 per cent. of hydrated lime has proved quite satisfactory.

In applying the plaster to metal lathing the trowel should be taken across the lath as far as possible, all in one direction at an angle of about 45 degrees, and only sufficient pressure to push the material through to a height of ⅜″ at the back should be used.

Finish for Interior Work.—The finish for interior work may be any desired plaster; but, for the reasons already given, a lime or other absorbent plaster is preferable. For exterior work, however, the complete work should be carried out in cement plaster, the finish ½″ or at least ⅜″ being of 1 of cement and 2 of sand, and, of course, without any hair, which would tend to carry water through the plaster.

Where using lime or soft plaster, all external angles, narrow reveals, splays and returns under 12″ wide should be finished in a hard gypsum plaster, such as Keene's cement, either on a cement backing or on a first coat of the same material, according to the type of patent plaster used.

Lime plaster is too soft to form any sharp exterior angle or to resist any rough usage, and previously, it was customary to have these corners and angles made of wood, but these should on no account be used, as when they are in contact with the wet plaster, the wood swells, tending to push the plaster away, and afterwards the wood contracts, leaving a crack which can never be hidden. Gypsum plasters are the most satisfactory for forming such corners, for they do not contract in hardening, and therefore make a good bond with the other plasters.

Metal corners are sometimes used, but, though these save labour to a great extent, it is not possible to obtain the same finish as with a hard plaster, and if made of iron, even if galvanised, there is always the danger of rusting at a later date, with at least destruction of decoration.

The Use of Hair in Plastering

Hair is required in plastering mainly to strengthen the mortar used; it therefore acts in very much the same capacity as steel in reinforced concrete.

Reason for Using Hair in Lime Plaster

Lime plaster requires a considerable time to harden, due to the fact that "setting" is brought about by the drying out, by the air, of the water used in mixing the mortar—thus being dependent greatly on the season of the year and the weather—and the gradual absorbing of carbon dioxide (the gas driven off when limestone is heated or, as it is termed, "burned," to form commercial lime), thus recarbonising and, in fact, forming artificial limestone. Due to the above facts, lime plaster develops no actual tensile strength, or cohesion of one particle of sand or lime to another, for a considerable period of undetermined duration. In fact, some authorities maintain that lime mortar in brickwork can never become really hardened except on the face of the joints, and this would also be true, to a certain extent, in the case of plaster work.

The incorporation of hair in lime mortar used for plastering is of more importance than in the case where Portland cement is used, for the latter material quickly develops tensile strength, and the separate components of sand and cement are bound together, commencing from the time of initial set—usually about one hour after tempering the mortar with water by a colloidal or jelly-like substance.

Patent plasters, and those under trade names, having a gypsum (the material from which plaster of Paris is made) base also, have less need for reinforcement than lime, as water added to such plasters actually enters into chemical combination with the calcium sulphate—the main constituent—to form the original hard crystalline substance, the full strength being developed in a few weeks only.

Hair for Cement and Gypsum Plasters

In spite of the greater strength of cement and gypsum plasters, it is customary to add hair or some other " binder " to mortars containing these materials for plastering, but where it may be possible to carry out satisfactory plaster work with such mortars without any addition this is done. In the case of lime plaster, however, hair or other material to act in the same capacity is absolutely necessary, and this is so in greater quantities for ceilings or horizontal positions.

Another important function of hair in plaster is that it prevents cracking, and this particularly so in the case of lime, which shrinks to a greater extent in hardening, so much so, in fact, that were a considerable proportion of sand not used with the lime it would be unusable on account of this tendency.

Portland cement mortar also contracts in setting, though to a much less extent, but as expansion takes place during the initial set and previous to the contraction, hair in this plaster also prevents cracking.

On the other hand, most plasters prepared from gypsum expand in hardening, and the hair serves mainly to strengthen the plaster and to overcome the brittleness of the substance.

Nature of Hair Used

Ox or cow hair is the material generally used in this country for mixing with plasters, and owing to its strength and its ready binding with the mortar, it is the best material that can be readily obtained for the purpose. As hair is required for binding the plaster together, it is necessary that it should be long enough for the mortar to obtain sufficient grip, or key, to prevent any sliding of the mortar from the hair where the plaster tends to break. Therefore, within easy working possibilities, the longer the hair that is used the better for the strength of the plaster. For this same reason of grip or key, the hair should be of the curled variety.

The above conditions are usually met in specifications by inserting the words "The hair to be long curled ox (cow) hair, of strong quality, to be used unbroken."

Where it is impossible or difficult to obtain ox hair, that of long coated horses, donkeys, camels or other strong-coated animals may be used, but care should be taken as regards length and curl as in the case of the material generally used. Where hair is not obtainable, tow or other equal fibre may be used.

It is necessary that the hair, or other substance, be free from dirt, and particularly from grease, which, apart from having a repellent action on the water used for mixing the mortar, thereby preventing a good and even mixing of the hair with the plaster, may also have chemical action on the materials forming the mortar, thus preventing setting of the plaster and causing saponification (soap formation).

These precautions are usually met in specifications by the proviso: "Shall be free from grease, dirt or other impurities."

Even Distribution

As it is necessary that the hair, or other binder used, should be evenly distributed through the plaster, care should be taken that the hairs are not clogged together; and if there appears the slightest tendency to form into lumps or to mat, the hair should, before being mixed with the mortar, be switched with laths, or be immersed in water until it is seen that the hairs are well separated.

The above requisite is usually met in specifications by the proviso: "and shall be well separated before being mixed with the mortar, either by switching with a lath or by being immersed in water."

In mixing the hair with the mortar it is customary to place this on top of the previously slaked lime in the basin of sand and then thoroughly to mix the whole together, seeing that the hair is evenly distributed in the mass; as it is necessary that the hair should remain unbroken in the mortar, care should be used in the mixing, and this is best done with a three-pronged rake, the shovel being used as little as possible on the mass.

Generally, after mixing in the hair, the lime mortar is allowed to remain to "cool," i.e. to complete slaking, but this would seem doubtful practice if slaking is actually still required, for in that case quicklime would be present, which has a burning effect on the hair, and would therefore weaken the plaster. It would, under such circumstances, appear far better to be perfectly certain that the lime was entirely slaked, either by purchasing this hydrated from reputable manufacturers, or by keeping the lime in a fluid state with water for a longer period, stirring from time to time and passing through a sieve, to remove any lumps, etc., before mixing with the hair.

Amount of Hair Required

The amount of hair to be used for the " scratch " or " pricking up " coat should be 1 lb. per 3 cubic feet of lime mortar (" coarse stuff ") for vertical surfaces on lathing. The same proportion is often used for ceilings, though for this purpose it is better to allow 1 lb. of hair to 2 cubic feet of plaster, and this proportion should certainly be used for better-class work or where any vibration is to be expected.

This necessity for increasing the hair in plaster for ceilings can be readily understood when it is realised that the weight of the ceiling is carried on that portion of the plaster entering the space between the laths (usually $\frac{1}{4}''$ to $\frac{3}{8}''$ wide).

Where the same " coarse stuff " is not used for the second or " floating " coat, the proportion of hair used for plaster prepared for this layer may be half of the above.

As a general practice, no hair is used in the finishing coat, though in two-coat work on walls a small proportion (about 1 lb. to $4\frac{1}{2}$ to 6 cubic feet) may be added, when the hair should be white.

When rendering rough brick walls with well raked-out joints the " rough stuff " may have the amount of hair reduced to 1 lb. to 4 cubic feet of mortar, and the " floating " coat contain half of this.

Due to the increased tensile strength and the quicker setting of Portland cement plaster, the amount of hair mixed may be reduced; and for scratch coat on ceilings 1 lb. of hair to 4 cubic feet, and for walls 1 lb. to 5 cubic feet of plaster should be sufficient.

Hair for Portland Cement

In fact, cement plaster is often used without the addition of any hair or other binder, but this is doubtful practice, and the addition of a little hair pays well in the class of work obtained.

When mixing hair, or other binder, with Portland cement this can, of course, only be done with small amounts of plaster, such as can be used within half an hour of being wetted, as cement plaster should never be re-tempered after set has taken place.

For outside plastering, when with cement or hydraulic lime plaster, care should be taken that no hairs used in any of the coats work to the finished surface, as these would have capillary properties, and would conduct water to the inside, thereby counteracting the purpose of the coating.

Patent Plasters

Patent and other prepared plasters should only have hair added under the instructions and in the proportions advised by the makers; as these, in most cases, contain the proper amount of hair, or other fibre, for strength and other purposes.

CHAPTER IX

PREPARING THE BASE FOR PLASTERING

Ceilings: Counter Lathing; Counter Lathing for Steel Beams; Encased Beams; Wooden Nailing Strips; Process of Lathing; Brackets—Partitions: Studding; Bricknogged Partitions; Metal Lathing—Errors in Preparing Bases: Cracks; Faulty Spacing of Laths—Insecure Framing; Settlement Cracks; Plaster Board—Fixing Partitions: Framing Doors and Windows; Plastering on Breeze Concrete Slabs; Plaster of Paris Slabs; Partition Details; Terra-cotta; Brick Walls: Concrete.

THE state of the base upon which plaster is applied is perhaps the most important item in the obtaining of satisfactorily completed plastering. This is particularly so in the case of lathing in lath-and-plaster work, and of this type especially ceilings, due to the fact that they are generally of comparatively large area and that the weight is suspended in a horizontal position, that this part of the work is usually not covered with any material which will tend to conceal cracks or other defects which occur, and that the appearance of the ceiling, attracting the eye as it does, will have a large part to play in making or marring the complete finish of the room. For these reasons this class of work will have first consideration.

CEILINGS

In ceilings, it is useless to expect the completed plaster work to be free from cracks unless the joists to which they are attached are sufficiently stiff to carry their load; the deflection of joists should never exceed 1/360 of their span if they are to carry plaster ceilings, and for important work, and large unbroken areas, their stiffness should be greater than this.

When commencing the lathing for a ceiling, as it is necessary that the plaster shall be as uniform in thickness as possible, the first care should be to see that the undersides of the floor joists, or other supports to which the ceiling is to be attached, are level, or otherwise in accordance with the drawings, and that the supports are level and in line with each other. Should there be any unevenness in the lining of the supports, this should be made good before starting to fix the laths proper, by tacking on laths or other wooden levelling strips (running in the direction of the joist) to bring the supports into correct line. Strips used for this purpose should be as narrow as will allow for secure nailing of the laths which follow, and it is preferable that they be not more than 1" wide.

PANELLED CEILING IN FIBROUS PLASTER.
(Naylor & Sale, Architects.)

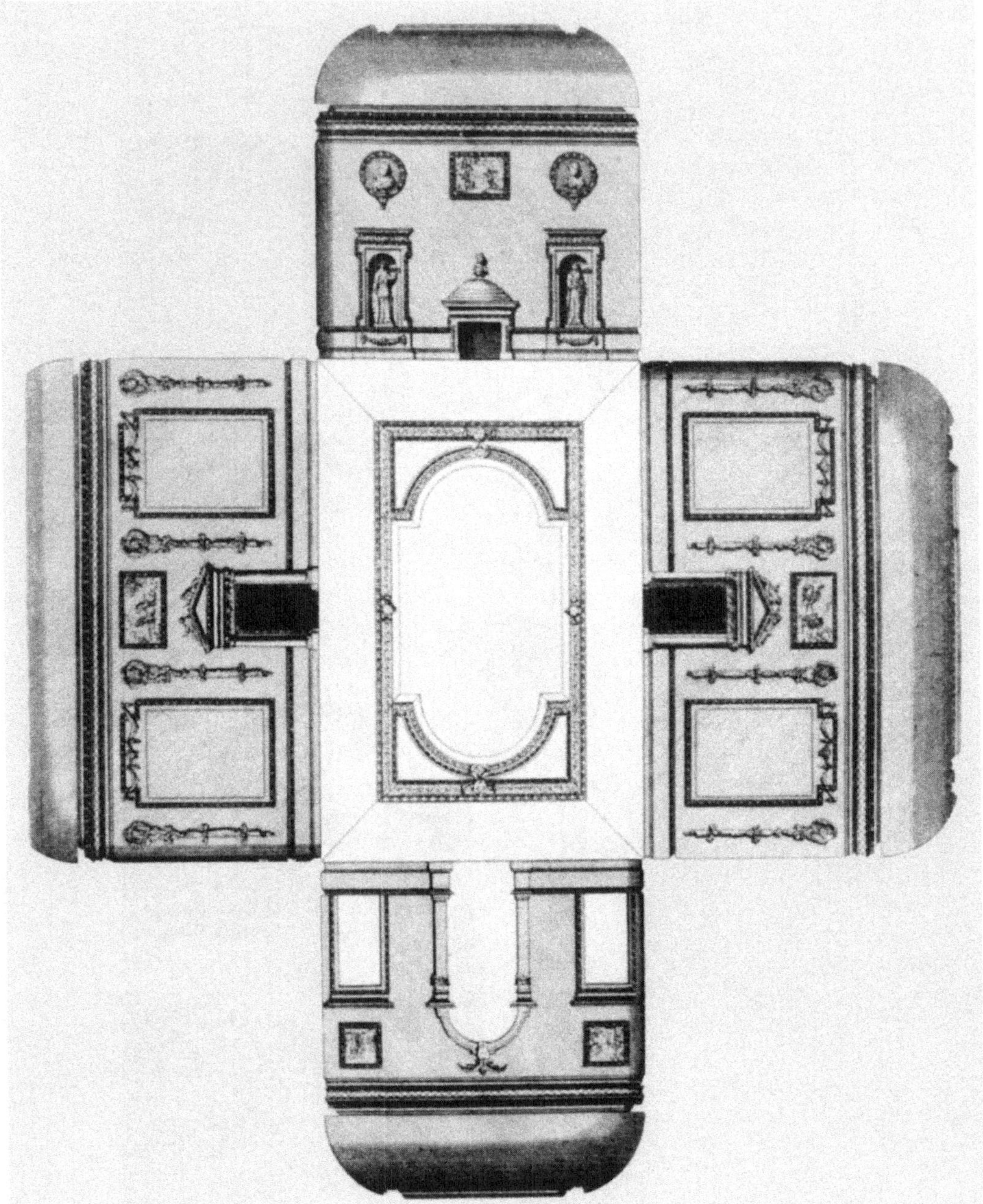

LAY-OUT FOR PLASTER ENRICHMENT TO CEILINGS AND WALLS.
(By George Dance, the Elder, Mid 18th Century.)

The importance of obtaining a good and even key for the plaster, particularly in ceiling work, has already been pointed out in discussing laths, and in this connection it is well to consider the effect that nailing lathing direct to supports may have on causing bad plaster work. The width of a joist may at first sight appear to be of little importance; but with joists of 2½", and often more,, in thickness, spaced at, say, 12" centre to centre, the ratio of support surface to the whole area is considerable. Moreover, it should be remembered that, apart from a small percentage from bonding, or adhesion, to the face of the lath—depending greatly on the roughness of the wood—the weight of the plaster is secured by that part of the mortar which turns over on top of the spaces between the laths. Owing to the shrinking of the lime and the swelling and subsequent shrinking of the lath, no bonding can, of course, be looked for on the edges of the laths on either side of the spaces. Furthermore, where laths are nailed direct, as is often the case, to such plain surfaces, the plaster is merely pushed up against the joists without obtaining any key over these comparatively large areas.

Counter Lathing.—For the above reason joists or supports which are more than 2" wide should be "counter lathed," that is, laths of not less than lath and half thickness should be tacked to the centre of the support running in the same direction as these; such laths will only require sufficient nails to keep them in place temporarily, as in due course they will be firmly secured by the nails used in fixing the lathing proper.

In connection with counter lathing, it should be noted that a longer nail will be required than that specified for the thickness of lath for the work, i.e. if "lath and half" is called for—and nothing less should be used for ceilings—the length of nail usually called for with "double lath" thickness should be used.

Wooden beams, or other construction, running parallel with the direction of the lathing will also require counter lathing in order that the main laths may run all in the same way. In this case, the counter laths will run across the width of the beam at twelve-inch centres, the same thickness being used as for running along the centre of the joists.

In regard to beams into the lower parts of which joists are framed in order to obtain a ceiling without breaks, as counter lathing will be required on these beams, a little foresight in the framing may save work and labour, particularly where the joists used are narrow enough to permit of nailing the main lathing direct to the joists; if the joists do not require counter lathing, their lower edges should come about ¼" below the beam; and, if the joists are wider than 2", their lower edge should be not higher than flush with the lower face of the beam. Unless this precaution is taken in order to obtain reliable work, it would be necessary to either counter lath joists which—though this would lead to better keying—could do without it, or to counter lath the joists with a greater thickness of wood than would otherwise be required.

Counter Lathing for Steel Beams.—Steel beams are often used in modern long span and heavy load construction. Where a ceiling on wooden joists, resting on the steel beam, has to be formed to cover in the lower flange without a break, unless such beams are more than 12″ in width, their counter lathing may be carried out by "bridging," i.e. by carrying the lath from the joist on one side to the joist on the other side of the steel beam and double nailing on either joist. Where this method is used, as it is, of course, impossible to drive the lathing nails into the steel, it will be necessary to pack between the counter lath and the beam with wood to give sufficient thickness to take the lathing nails. This thickness should never be less than ¾″. To permit of this, the framing of the joists will have to be carried out so that the relationship of their faces with that of the beam will give the required thickness.

Encased Beams.—Where steel beams are used in such construction, either through being obligatory for fireproofing purposes or in order to protect them from corrosion, they are often encased in solid concrete or in expanded metal and plastering with cement mortar to the requisite thickness. In these circumstances sometimes the wooden lathing is brought up to and level with the lower encased edges of the beam, and the roughened surface of the concrete or cement mortar is depended upon for keying the ceiling plaster in this portion. This method, however, is practically certain to cause cracking of the finished ceiling at the junction of the wood lathing and concrete on account of differences in the coefficients of expansion in the various materials and also due to the effect of damp on the wood without corresponding action upon the concrete.

A better method is to arrange the framing of the joists in such a way that, presuming the beam is not more than 15″ wide, expanded metal of heavy gauge not more than ⅜″ mesh may be carried over it with a space of about ¼″ to ⅜″ between the back of the metal and the face of the concrete casing. To carry out this method satisfactorily, the expanded metal should be carried over the lathing—which will be brought close to the edge of the concrete—for a distance of not less than 6″ on either side, and be firmly fixed by stapling at both the edge of the metal and adjacent to the beam, these staples being at a distance of not more than 4″ in line and staggered one with the other. To give sufficient stiffness for stapling the metal, at the lines of fixing, and running parallel with the beam, there should be firmly fixed in between the joists, level with the lower face, 2″ × 1″ timber, on edge. The expanded metal should be well stretched when fitting, the sheets at joints should be overlapped for at least 2″, and be firmly wired together at both edges of the lap at distances not exceeding 4″, while the ties at each edge should be staggered.

Wooden Nailing Strips.—There is, however, no doubt that it is far better to keep the base for plaster work all of the same material in order to obviate cracks caused by unequal expansion and contraction; and it is, therefore, by far the best practice to arrange for nailing strips to be fixed in, or to, the beam casing when this work is being carried out, in order that the wooden lathing may be taken right across. These strips should be placed as the counter lathing for wooden beams, should be not less than 1″ in thickness and not more than this in width on the lower face, and if cast into the concrete—care will be required in this case that no regulations as regards thickness of concrete in connection with fireproofing, etc., are interfered with—the sides should be tapered, the narrow face being placed downward, there being a space of not less than ¼″ between the lower surface of the strip and the concrete casing, in order that a key for the plaster may be obtained at the back of the lathing.

Where it is not practical to embed these strips in the concrete, or where they come in cement plaster laid in more than one coat, they may be fixed at the required level either by metal clips on to the steel beam before the casing is carried out, or, by having strong wire nails driven through them—these being hooked over at the back to as great a length as possible—turned in different directions and embedded in the concrete or plaster. In placing these strips care will be required that they are level on the underface, with each other, and with the joists, or with the longitudinal counter lathing, which will be fixed on these.

Process of Lathing.—The lathing proper may now be considered—specifications generally call for the laths to be fixed with spaces of ¼″ to ⅜″ between, but it would appear that the greater of these dimensions should be maintained more particularly for ceiling work. This will be readily appreciated when it is realised that the weight of the whole thickness of plaster over the area, centre to centre of the spaces, is carried mainly by the neck of plaster between the key formed by the mortar falling over at the back of the lath, and the face plaster itself, that the material in this neck is in tension, depending chiefly on the adhesion of one particle to another of the constituents of the mortar, and the length of time required for the plaster to set, particularly where lime is used, during which period of softness these necks or " hangers " are carrying quite considerable loads.

The ends of the laths should form butt joints over a support, and in no case should a lapped joint—that is, the end of one lath laid on top of the lath it adjoins in the same line—be allowed, as this would give unequal thickness of plaster (amounting to only about ¼″ cover over such joints where the usual general thickness of ½″ is used), and therefore leading to cracking of the finished work. It is advisable that a space should be left between the ends of two laths the same, or perhaps a little less, as at the sides, though this is often neglected and rarely called for in specifications. This end space, coming as it does against a plain surface and leaving no void for mortar to fall over on the back, can add nothing to the actual key, but, probably owing to the roughness of the ends of the laths, does give some added support to the body of plaster, its main use being that it allows free expansion of the lath.

The expansion of wood lath when fixed dry and then wetted is very considerable; and where the ends of, say, 4′ long laths are butted tight up against each other, this expansion must go somewhere, causing buckling either upward or downward, and twisting of the laths. Moreover, as the wood dries out with the plaster there is a tendency for laths to straighten out again, causing breaking down of the key, with consequent loose and cracked plaster.

Nails of the proper length for the thickness of lath used, and allowing for whether the joints, or other supports, have been counter lathed, should be used, one for each support the lath crosses, and the ends of the laths should be double nailed. All nails should be driven well home, though without sufficient force to bruise or break the lath at the point of support.

It is important that the joints of adjacent laths should not come over the same support, but that they should break joint, as otherwise, from various causes, cracking will take place on this line; and though it is common practice to nail the laths in bays, say 3′ or 4′ deep, it is far better if they break joint entirely. This can be done by using 3′ and 4′ laths together alternatively in one line and reversing the order in the next, or, if laths of equal length are used, by commencing the lines of lathing on alternate joists; this method of fixing the lath will give a far stronger ceiling, particularly if the space of two joists comes between end joints in adjacent lines of lath.

Brackets.—Where a cornice in lath and plaster has to be run, brackets—that is, pieces of wood cut closely to the outline of the finished work—will have to be provided upon which to nail the lathing. Such brackets should be cut from wood no thicker than will allow end butt joints to be securely nailed, should be firmly fixed to previously prepared and levelled grounds, and should be spaced about 12″ centre to centre.

However close brackets may come to the corner of a room, special angle brackets should be provided, so that the laths from the two sides of the room may be firmly nailed where the mitre is to be formed. The same care should be taken in the fixing of laths, breaking of joint and spacing as on the flat portion of the ceiling; the cornice base will, of course, be completely lathed and finished before any of the ceiling plaster is commenced.

Wetting of the lath before plastering is of considerable importance in the prevention of cracks and loosening of the plaster, besides retaining the nature in the plaster, this latter being of perhaps more importance where Portland cement or plasters with a gypsum base, such as Keene's or Parian cement, etc., are used, than is the case with lime mortar. But dry wood lath will absorb very considerable amounts of water, and even with this latter so much wetness will be taken away that the plaster, instead of obtaining any degree of hardness, will be "crumbly" and without tenacity or strength, and will perhaps cause the ceiling to fall away even before the finishing coats are applied.

Wooden lath exerts tremendous pressure in swelling, with consequent buckling, warping and twisting, and unless such bases are swollen to their limit before any plaster is applied, the keys will be sheared off, leaving the plaster with practically nothing to retain it in place; for this reason wood lathing should be thoroughly wetted the day before, and again an hour or so before plastering. It is practically as bad to have the lathing unwet as to have the swelling going on whilst plaster is being applied and is drying.

Partitions

The next most important lath-and-plaster work is in connection with *Partitions*. The same care should be taken in the selection of the laths for this class of work as is the case with those used for ceilings, though perhaps a little more latitude may be allowed as far as the question of imperfections which would cause staining of the plaster. Walls being generally covered, either by papering, distempering or painting, this becomes of minor importance as compared with ceilings, where stains are practically certain to show through. This latitude is only suggested where great difficulty may be encountered in obtaining good lath, which is not likely to be the case in this country; and it will be readily appreciated that it is far better not to allow two qualities of lath on the job, on account of the supervision required to ensure that none of the partition material is used for ceiling work.

It is generally considered that a lighter lath can be used for partitions than for horizontal surfaces; and looking at the case from the point of view of the dead load (that is, the weight of the plaster) and the manner in which it has to be carried, this is doubtless correct. But it must be remembered that the forces which a partition has to resist, particularly from blows in the lower part of the wall, are very difficult to determine, and therefore, taking into account the little extra saving by using a thinner lath—more so where it is a question as between " lath " or " lath and half"—it is doubtful practice to use anything less than "lath and half" for partitions, this being particularly so where sawn laths are used, owing to the possibility of cutting through the fibres of the wood when sawing.

Studding.—It may appear unnecessary to point out that it is useless to expect well-finished work if the studding is not stiff enough to resist, without bending or vibrating, any strain likely to come upon the partition. This question of stiffness, however, is not merely one of the strength at right angles to the surface of the wall, but also one of the width of the stud. During the wetting of the lath previous to, and during, plastering, as also during the time until the plaster has dried out, this studding is in contact with considerable amounts of water and moisture, and if narrow timber is used there is the possibility of warping in the direction of the least thickness, with consequent buckling and twisting of the lathing.

The size of timber to be used for lathing in partitions will, of course, vary according to the length of span between supports, as also the type of usage likely to come upon the finished work, but generally speaking it is not advisable to use anything less than 3″ by 2″ for this purpose. Should the partition be of a considerable height it is advisable to bridge between the studs and not depend upon the lathing for lateral stiffness, for these laths, when wet, will have a tendency to movement and buckling, and, in place of stiffening, the main timbers will be more likely to assist distortion, with consequent trouble when the plaster and timber dry out. It must be remembered that plaster, and particularly lime plaster, takes a considerable time to become really dry, and it may be readily realised what may happen to the timber base if distortion occurs during the early stages of plastering, and then, when drying out takes place, the wood studs and lathing tries to get back to its original shape. Needless to say, the timber used for studding should be of good quality and free from imperfections which would assist any expansion or contraction to take place unequally.

Before commencing to lath, the studding should be tested for plumb and trueness and should be tuned up if necessary. In partitions the extra thickness of plaster caused by the lathing not being true is not of the same importance as in the case of ceilings, for there is not the same danger of the added weight breaking away the key; but it must be remembered that there is always danger of cracking from shrinkage strains when one part of the plaster is thicker than that adjoining.

Counter lathing will be required as for ceilings; and though, for partitions, it is often considered that this is only necessary where the timbers do not exceed 3″ in width, for first-class work it is advisable to limit the width which may be crossed without this preparation, to 2″.

The length of nails to be used should be the same as for ceilings with equal thickness of lath. The number of nails to be used should be as previously noted, and the spacing between, and at the ends of, the laths should be the same, namely, about $\frac{3}{8}$″. It is important, also, that the end joints of laths should be staggered, and though it is quite common for these to be "snatched" (or "matched")—that is, with the joints in line—for 2′ to 3′, for good work it is advisable that the lath should be applied in groups of not more than 5 to 7 laths deep, the joint of the next group being moved to the stud to right or left of the previous one.

Before any plaster is applied, the same care should be taken as regards wetting the lath, but care should be taken to keep water away from the studding as far as is possible.

Bricknogged Partitions.—Bricknogged partitions are often plastered (or rendered) without any previous preparation, the unstruck joints of the brickwork, or the raking out of these, being depended upon for keying the plaster, but this is a course not to be commended. In the first place, the expansion of timber and brick is not equal, even when dry; and when the base is wetted previous to applying the plaster the wood will swell without corresponding increase in the brick. On the contrary, there will be a tendency to push the bricks and mortar away—the swelling force of wood is very considerable, so much so, that in the East the granite is quarried by inserting plugs of wood, soaking with water, and allowing the force of expansion to cleave the rock—and, upon subsequent drying out, there will probably be cracking of the plaster at the point of contact of brick and timber. This tendency will be assisted by the lack of bond over the face of the timber, and should cracks occur on both sides of the timber stud there will be this area of plaster without any bond to the base, and with the liability of dropping away upon the least provocation.

Metal Lathing.—For first-class work, bricknogged partitions should be counter lathed and lathed; but, where such expense is not permissible, an alternative is to bring the brickwork slightly in front of the timber, or to run a ½″ iron rod down the centre of the timber and place lightweight expanded metal lath, wide enough to extend 3″ on either side of the stud, over the wood. This metal lath should be firmly attached to the brickwork on either side, and should be without buckles where it meets the brick. The idea of slightly recessing the main timber, or of running up the rod, is to form a key for the plaster at the back of the lath.

Instead of plastering, or rendering, direct on to the brick wall it has been the practice, more in the past than in latter years, in order to prevent the moisture from damp outside walls entering the interior of a building, to place battens and to lath and plaster, leaving an air space between the brick and the interior plaster surface. These battens, or furring strips, as they are sometimes called, are fixed at about twelve-inch centres, to plugs set into the walls; and, after truing up, the lath is attached in the usual way.

This practice would appear to have nothing to recommend it, for if the walls are damp the plugs will in time carry the moisture through to the battens, lath and plaster, and the work will be useless for the purpose for which it is intended. Moreover, this system may well be regarded as an actual danger to the general woodwork of the building, for sooner or later dry-rot must be expected to take place in this part of the construction, and there is every probability that this fungus would spread to any wood in any way in contact.

If for any reason such work be carried out, the only hope of obtaining a good job will be by thoroughly creosoting the plugs after they are prepared, and also the battens, or furring strips, and laths before fixing, and to again touch over the nail holes or points where the lath has been cut before plastering.

This furring is sometimes carried out with expanded metal lath, and is satisfactory; but this will be dealt with in another part.

Errors in Preparing Bases

Before closing on the question of lath and plaster, it may be well to mention the faults upon finished work which are generally taken to indicate errors in the preparation of the base for this construction.

Cracks.—Cracks running in the direction of lath are practically always caused by lath conditions and the plaster is often bulged out along the cracks. These cracks may be in groups, following the lines of several laths, then a crack-free space occurs, and again a further group of straight-line cracks; in this instance the groups of cracks are often joined by a crack at approximately right angles, following the end joints of a group of laths. This condition of straight cracks running parallel with the line of laths may extend for long or short distances, for the laths of different lengths may vary in texture, with consequent variations in swelling, buckling and twisting.

The most common causes of such cracks are: Insufficient wetting of the laths before plastering, or wetting only just before applying the plaster—not allowing spacing at ends of laths.

Faulty Spacing of Laths.—When the plaster sounds hollow and loose upon being tapped with the hand, faulty spacing of the lath may be looked for. The same condition may be encountered where the plaster has been oversanded, thus causing the keys to drop off through lack of strength; and such lack of strength of the key would be assisted or brought about also by not mixing or adding insufficient quantities of cow-hair to the mortar.

This hollow sounding of the plaster may also be caused by the various coats of plaster not bonding properly together, due either to the surfaces being too smooth or to being too dry before the succeeding coats are applied. Oversanding of the scratch or floating coats may also cause this hollowness on sounding on account of absorbing the water from the following coat. This hollow sound may also be due to lack of strength in one coat, allowing the later coat upon setting to draw away the material of that to which it is fixed. That the fault on sounding is due to these latter conditions, and not to the faults in the true base of wood lath, may usually be ascertained by applying pressure to the plaster. The finish coat, if the fault lies between this and the floating coat, being thin, will readily, and without the application of much force, bend back and, sensibly to the feel of the hand, come to rest on the next hard coat. Looseness of the floating coat away from the scratch coat may often also be ascertained in this way, though, of course, greater pressure will be required. Should there be any doubt as to the cause of the hollow sounding, when cutting away to make good—the only cure—it is advisable to hack out gently, so that should the fault be due to the floating coat leaving the scratch or "pricking-up" coat, there will be no need to remove the latter, thus saving delay in the drying out of more coats than are absolutely necessary—this being particularly so in the case of lime plaster—and also minimising the risk of breaking away the key of surrounding plaster, previously sound, by vibration on the lath.

Insecure Framing.—In partitions, particularly, though occasionally the same indications are found in ceilings, cracks running in the direction of the lath, that is, horizontally in partitions, or at right angles to this, or vertically in the same construction, will, where no bulging of plaster takes place, point to movement of the framing, either due to wetting and drying during plastering, the using of unseasoned wood, or to the normal expansion and contraction of the material used for the purpose. The vertical cracks usually follow end joints of the laths, though in extreme cases they will continue across the laths. Such cracking in ceilings, where there is no bulging or loosening of the plaster, is generally due to weakness of the joists and consequent stretching of the plaster due to deflection of the joists—1/360th of their span being all the stretching that any plaster can stand.

Settlement Cracks.—Cracks due to settlement of the building as a whole, or of the floor upon which a partition is built, should not be confused with those due to faults in the lath base. Settlement cracks may usually be recognised by their running diagonally across the construction, either alone or in combination with cracks at the junction of wall with ceiling or wall with wall, i.e. in the corners. In lath-and-plaster construction, such diagonal cracks generally follow the lath joints. In partitions, if upon examination both sides are found to be cracked in approximately the same position, making due allowance for variation in the plan of staggering the joints on each side, it is a positive indication that the cracks are due to settlement.

"Map cracks," that is, cracks of a wavy or closely zigzagged nature running in all directions and usually found in groups, are not due to faults in the lath base. These cracks do not usually extend beneath the finish coat, and are caused by shrinkage of the material of the ultimate coat; in the case of a lime putty finish, this being generally due to not using a sufficiently high percentage of gauging plaster.

Plaster Board.—Plaster board used for the same purposes as lath has not in this country become the common practice its advantages over the material it displaces would appear to warrant, though it is steadily making headway, and will certainly lead to better plaster work both in ceilings and partitions. The use of this material does away with the three great causes of cracks in plaster which are present when wood laths are used. When of reputable make—which can be counted upon as a general thing, for it is factory made by expensive machinery which means specialising—it is uniform in quality, and not variable as is the case with wood laths. Plaster board does not swell, warp or buckle, and therefore cracks are not caused from this most common fault with wood lath. Properly applied there is no danger of water getting to the joists and studding, as the joints between the boards are filled in previous to putting on the main plaster, and also there is no need to wet the base before plastering; thus swelling of this part of the construction, with its consequent effects, is avoided. Again, unlike wood, which gives little or no permanent bond between the lath and the plaster—the grip being almost entirely dependent on the key at the back of the lath spacing—there is an excellent bond between the plaster board and the plaster, this being particularly so where plaster of Paris or plasters formed with a gypsum base are used.

Plaster board, if not fireproof, is strongly fire-resisting; it is sound-deadening, and also adds to the heat and cold proofing of the plaster. It is also claimed that such boards can be fixed in half the time required for an equal area of wood lathing, and that there is a saving of 50 per cent. of the nails usually required. It is also indestructible and vermin proof, and has the added advantage that there is a solid base for plastering upon.

In fixing there should be a space of not less than ¼" between boards —which in this country are usually just under 1 sq. yard in area —and these should break joint preferably in both directions, but at least every other board vertically on walls, and at right angles with the joists on ceilings; this can easily be done by commencing every other tier with a half board—the cutting being readily done by scoring with a knife, and then breaking, or by using a saw. The breaking of joints in the other direction is done by moving the joints to alternate supports. The nails used for fixing are generally flat-headed "wire" nails, 1¼" long. Should the fixing be to oak, these nails would require to be galvanised or preferably copper; and the usual allowance is 22 nails per square yard. The nailing should be commenced close to the edges of the boards, and all nails should be driven well home.

The joints between boards are filled, previous to the main plastering, with gypsum-plaster mortar usually in the proportion of 1 of plaster to 2 of sand; but manufacturers of these boards have special jointing materials which should be used as directed by them.

These boards should not be wetted or even sprinkled in preparation for plastering, for the moisture in the plaster is sufficient for perfect adhesion.

Cracking in plaster applied over plaster board which can be attributed to the wrong preparation of the base is that caused by insecure nailing unless there has been distortion of the supports either by using unseasoned timber or by insufficient stiffness in the joists. In the latter case, long cracks follow usually the line of the joints, horizontally in the case of partitions on account of unseasoned timber, and at right angles to the direction of the joists where caused by lack of stiffness in these, and, in the former case, by cracks extending round the joint of the board. However, before attributing the cause of such cracks to improper nailing, the thickness of the plaster covering the boards should be ascertained, for this is the most usual cause of cracking, particularly in ceilings. The grounds for plaster on these boards should be the same as for wood lath—that is, ½″ over the face of the board.

Cracks due to settlement will be indicated in the same way as in lath-and-plaster work.

Breeze concrete slabs are to-day the most common plaster base for partitions in this country; they have little to recommend them besides their cheapness, comparative lightness for thickness and the advantage of being able to drive nails into them for fixing joinery or other purpose. Due chiefly to their porosity, and variation in this quality, it is difficult with their use to obtain a first-class plaster job.

Fixing Partitions.—In fixing partitions of breeze concrete slabs care should be taken that the bottom slabs are well bedded to the floor, and they should be bonded to the main walls, and at corners, by metal ties between at least every other horizontal joint between the slabs, or by other means. The slabs should, of course, break joint vertically, and the mortar used in building up should be Portland cement and should contain little sand, otherwise cracking from settlement of the joints will result. Previous to laying, the blocks should be well soaked in water in a receptacle large enough to cover them, and should be laid before any of the water can dry out, care being taken that all joints are well bedded and filled, the top of the completed partition being wedged to the ceiling before being plastered. As it is advisable, in order that any contraction which is to take place, particularly in the horizontal joints, shall do so before plaster is applied—Portland cement in setting expands at first and later contracts—the partition should be left for some days before plastering commences, and the wedging should be tightened up before proceeding with this part of the work. Failure to properly wedge slab partitions not only is the cause of cracking at the junction between the ceiling and wall, but will cause cracks in other parts of the structure through slight vibration, and will also mean a weak partition and added danger of collapse in case of fire.

Framing Doors and Windows.—The wooden framing of doorways, etc., is one of the most frequent causes of cracks in slab partitions, for wood, when wet, swells with great force. If the timber uprights are taken to the ceiling, for purposes of strength, in close contact with the slabs or mortar joints, cracking of the plaster, and even of the partition itself, is more than likely to occur. Where partitions are not above 8′ in height and not long, with doors of ordinary width and using slabs not less than 2½″ thick, generally a satisfactory job can be made by bridging over the top of the frame if care is taken that no vertical joint between the slabs cuts through an angle of 45 degrees from the corner of the frame in either direction. Even under these conditions it is better to leave a small space, unfilled with mortar, between the slabs and the wood, the frame being spiked to the breeze blocks or fixed by means which will allow of expansion and contraction, and the architraves, if packed with plaster, can be relied upon to prevent movement of the jambs in the direction at right angles to the faces of the partition.

However, there is no doubt that framing doorways, and other openings of any size, right through from floor to ceilings, materially strengthens the partition, and prevents cracking due to vibration, and tendency to twist in opening and closing the door or from other reasons. In this construction, spacings should be left between the timber and the slabs, with the exception, of course, of over the lintel. The uprights over lintel, which will be reduced to a thickness not greater than the wall, on both sides of the partition, should be covered with light expanded metal lath, extending on to the slabs for a least 3″ on either side and securely fastened to them either by stapling or nailing. This metal lath gives a continuous bond, or key, for the plaster and prevents cracks on either side of the wood. For important partitions and where high-grade finish is required, it is far better to frame such openings in structural steel, such as light channel sections, to which the wood framing is affixed before building up the slabs. The slabs should be tapered at the edges to fit into the section, so that the metal face does not protrude beyond the faces of the slabs, and these joints can be filled with mortar. The uprights above the opening should preferably be formed of channel sections back to back. After fixing the slabs, expanded metal lath should be taken over them—as in the case of wood framing—in order to form a key for the plaster over the steel face; this method of construction is more expensive than wood framing, but the added cost is well worth while where certainty of finish and strength is required.

Plastering on Breeze Concrete Slabs.—As already pointed out, the difficulty of obtaining first-class plastering on breeze slabs is on account of the porous nature of the substance, which gives advantages in other ways. Even though the slabs are saturated with water when erected, as it is advisable to leave the base for several days in order that the horizontal joints shall have thoroughly contracted, by the time the plastering commences, much of the water will have dried out and it will be found almost impossible to prevent excessive suction, with consequent dry-out trouble with the plaster, and variations in the colour of the surface due to differences of density and wetness of various parts of the wall. A method which has been found satisfactory to overcome this difficulty is that of skimming over each slab as it is laid and still saturated with water with Portland cement mortar, rich in cement, and mixed with an integral waterproofing of a colloidal nature; this skimming coat should be "broomed" or otherwise roughened before it becomes hard, and should be wetted before applying other coats of plaster. This method is particularly useful where outside walls have to be rendered with cement, as in this case, apart from the defects already mentioned, discoloration from smoke, etc., will not be even but will depend on the porosity of the surface, and will give a mottled effect in patches of various sizes and shapes, and varying shades.

Where the above method is not adopted the partition should be thoroughly wetted the night before, and again immediately before plastering, and care should be exercised that each coat is well wetted before the succeeding one is applied, for the suction of the slabs themselves will tend to be carried through the covering coats due to an excessive amount of water being absorbed.

Plaster of Paris Slabs.—Plaster of Paris slabs, generally reinforced with reeds and other materials to allow of transportation and handling, are used for the same purposes as breeze concrete slabs, and it is claimed that such construction is stronger for equal thickness, and, particularly when machine made, are more even in density and do away with plastering troubles, especially where plasters with a gypsum base are used, and these are ideal, for the coefficients of expansion of the slab and plaster are the same. Due to the fact that plaster of Paris is slightly soluble in water, such slabs should not be used for external work.

Plaster of Paris slabs may be laid dry to form partitions, or they may be bedded, the former method being satisfactory for small partitions, and where they are to be rendered and set, but bedding is required always for one-coat finish.

The method of dry erecting is to cut off the tongue (these slabs are usually tongued on one edge and grooved on the other) and bed down preferably with a gypsum-base cement, to take up inequalities of the floor, and securely fasten down by skew nailing to base through one corner of each slab. Subsequent tiers of slabs will break joint and will be nailed, on the skew, at one bottom corner to the slab below, and at the top corners to the slabs on either side. Nails should be driven though door and other frames, into the slabs, a good plan being to drive one nail into the centre of one tier of slabs, and two nails into the next tier, each nail coming close to the top and bottom edge of the slab; secure nailing should also take place through the lintel into the slabs running over. Fixing to the supports at either end of the wall will also be by means of skew nailing as a general rule, but when such partitions come in a position on the supporting wall where they are close to a corner or position where, when the construction is completed, there will be little area of plaster on either side, wooden angle fillets should be securely fixed to the wall. The ends of the slabs coming against this should be chopped out to a corresponding angle, bedded with mortar, and skew nailed to the wall by nailing through faces of slab.

Where bedding of the slabs is required, the edges should be wetted well to prevent absorption of moisture from the bedding plaster, and to guard against the plaster drying before the slabs are in place. This wetting is best done in a bath, or other watertight receptacle long enough to hold the slabs, and wide enough to contain twelve of them; the depth of water should be maintained at about 4", and, as one edge is wet enough, the slab should be reversed to place the other edge in the water; and as one slab is used, it should be immediately replaced in the bath by a fresh one.

The plaster to be used for bedding should be of a gypsum base; most manufacturers of slabs supply or specify special materials for this purpose.

In other particulars the placing and nailing of the slabs is the same as for dry erection of partitions.

Partition Details.—Where partitions run at right angles to joists they should be wedged between ceiling and the top edge, and the space filled with plaster; if the partition runs in the same direction as joists and directly under one, the top slab should be spiked or nailed to this; if in the same direction as, but between the joists, light timber may be fixed between, at right angles to the joists, at intervals of 3′ to 4′ to which the top slab of the partition is attached by a wood screw. Such wood should not be less than 1″ thick.

In erecting slab partitions, accurate lining up and speed can be obtained by fixing temporary timber profiles to the walls at either end, and one or two in between if the partition is long. These profiles act as guides, and are taken away as soon as the partition is erected.

Partitions coming to an angle are fixed at the corner by nailing the slabs through the face of one into the end of the other in the same tier of the opposing wall. These joints should be dovetailed—that is, where the nails are driven through the face of the slab in one tier of the one wall, in the next tier the nails will be driven through the face of the slab in the opposing wall. After completely nailing, the corner is chopped off to form a "bull nose."

For plastering on plaster of Paris slabs, gypsum plasters are best, and with the use of these the suction of the slab may be stopped by going over the face with a wash of the plaster.

Where lime plaster has to be used, the base should be well wetted before plastering commences; the slabs are usually supplied roughed in order to form a key.

Terra-cotta and other special tile and block partitions require the same care in erection as regards wedging, etc., as those formed with slabs, and generally the chief care before plastering commences is to see that the base is well moistened in order to prevent "dry-outs" in the plaster.

Manufacturers of special terra-cotta and other building blocks and tiles usually call for special treatment of their products, and it is best to ascertain these.

Brick Walls.—Brick walls should be left with the joints unstruck and with the mortar protruding, or have the joints raked out, if they are to be "rendered," as plastering on such surfaces is termed. Of the two, it would appear better to "rake out" the joints, as this gives one even thickness of plaster, and will tend to prevent cracking, as, particularly if the unstruck joint has had some time to dry before plastering, there will be no combination between the mortar of the wall and the plaster.

Walls to be rendered should be formed of rough brick, for if the surface is too smooth and dense, hacking, or roughing, will be required, and should concrete lintels, etc., or stone come to the face of the wall, these should be hacked. Old walls should be well cleaned by hard brushing with a wire brush, and if these are exterior and exposed to much smoke and fumes, they should also be roughed by hacking. Generally the exterior wall of a building which is old, and has not been previously rendered, will require hacking, as such a wall will have been built with dense brick in order to resist the elements. Any cornices or other projections should be "dubbed out" with brick or tile previous to plastering, in order that the plaster may not be suddenly considerably increased in thickness.

There is no need, as a rule, to wet brick walls when rendering with lime plaster, but the first coat, or rough rendering, should be applied in a moister state than would be the case for laths, as this moisture assists it to adhere to the brick. Should the brick be very dry, or the work be taking place in very hot weather, it will be an advantage to wet the walls thoroughly the night before plastering commences. The rough rendering with lime plaster need not be roughened, but it should fill up all cracks, holes, etc., in the wall.

Where cement plaster is used, the wall itself should be well wetted in order to prevent the absorption of the moisture in the plaster too quickly. The first coat of cement should be roughened to form a key for the succeeding coat and should be well wetted before this is applied; in this connection the sprinkling of the surface whilst plastering is going on is likely to cause trouble by loosening of the plaster, for though the surface should be damp, there should be no free water which will break down the key.

Any efflorescence on the walls to be rendered must be brushed off with wire brushes, and the walls wetted down before plastering takes place.

Concrete.—Concrete surfaces to be rendered should be left rough in moulding, but for ceilings or for horizontal surfaces it will be necessary to hack the surfaces.

If oil has been used on the forms it will be necessary to hack the surface and wash down with a 5 per cent. solution of muriatic acid, which should again be washed thoroughly away with clear water.

Concrete surfaces should be well brushed down to remove all dust, and the surface well wetted before plastering.

CHAPTER X

LATHING

Wood Laths: Qualities Requisite; Oak and Fir; Riven or Sawn; Sizes and Quantities—Metal Laths: Comparison with Wooden—Types: Diamond Mesh Expanded Metal; General Points; Hollow Ribbed; Partitions; Solid Ribs; Metal Plate Bases; For Ceilings; For Partitions—Fixing of Metal Lathing—Protection of Metal Lathing: Painting; Galvanising—Weight—The Process of Fixing—Corners and Cornices—Suspended Ceilings—Further Points in Fixing—Miscellaneous Types of Metal Lathing: Hy-Rib; Trussit; Self-Sentering; B.R.C. Fabric; Bricanion.

WOOD LATHS

THE material of which they are formed and the finish of the laths, as also the manner of fixing, have a considerable bearing on the quality and lasting power of plaster work in which they are used. Moreover, as ceilings of most buildings are constructed of lath and plaster, faults in which are easily apparent to the eye, and are not cheaply remedied, the consideration of quality in this part of the construction is important.

Qualities Requisite.—The qualities chiefly required in laths are that they shall be strong, not subject to undue stretching or warping, free from knots and sap, free from crookedness and not too smooth.

It is essential to avoid a thick lath, particularly in a horizontal position such as in ceilings. Apart from the added weight of the structure and extra cost in timber and plaster required, undue thickness causes difficulty in getting the mortar in between the laths in such a way as to form an efficient key.

If, however, the lath is not sufficiently strong without being unduly thick, there is not sufficient firmness to work upon, especially when applying the "scratch" or "pricking up" coat, which, apart from making the labour of applying the plaster costly, causes a breaking away of the "key," and, if this fault is not noticed and remedied upon completion of the first coat, a breaking away of the heavier plaster during or at the completion of the later coats.

In ceiling work when laths unduly thick are used, even if an efficient key to the plaster is obtained, downward bulging between the points of fixing to the joists will take place, leading to expensive "dubbing out" (bringing to a level by filling up hollows) before applying subsequent coats of plaster, and by giving uneven thickness of material in different parts, tend to cracking of the finished work and breaking down of the key and bonding.

There is also the added weight of the construction, which is of importance, on the underside of floors. This consideration of strength is, of course, quite apart from the purely mechanical requirements caused by varying loads which may come upon partitions and walling into the composition of which lathing enters.

Freedom from knots is also chiefly a question of the strength of the lath, though the presence of knot holes, or knots, is considered by some to cause buckling of the lath when in contact with the wet plaster, with consequent breaking away of the key; but it is doubtful whether this consideration is of any great moment unless the knot is of large size and causes breaking of the grain on one side of the lath and extends well across the width.

It is, however, easy to understand that with laths of 1″ in width and about ¼″ in thickness, the presence of large or many knots causes a weakness of the lath, and even if one such is adjacent to others that are perfectly sound and strong, subsequent bulging and the danger of breaking away of the key of the plaster is involved. The question of knots, though one of degree, portends a fault which should be guarded against, particularly as knotty timber is generally from badly grown trees, and, apart from these defects, is not likely to be as strong as clean material.

Sap in laths leads to decay of the wood with consequent short life of the structure. The term "sap" covers pitch, resin, gum, oil or tar, all of which faults are generally visible almost at the first glance, and even if this is not the case can usually be discovered in the handling. Any of these sap conditions are likely to cause staining of the finished plaster, a matter of great importance where the structure is not to be covered by paper, such as in ceilings.

Crookedness of the lath does not allow of even laying on of the plaster, and consequently involves liability to cracking when the work is dried out. Furthermore, when this fault is in the direction of the width it is impossible to get even spacing lengthways between the laths, thus causing faulty keying.

Visible crookedness is not found in sawn laths (though unevenness of grain in this case will lead to twisting, which is a parallel fault), but if these are cut from timber which is crooked, the grain will be cut across, thus giving a lath lacking in strength.

A lath which is too smooth will give a faulty bond between the plaster and the wood, thereby placing additional strain on the keys formed on top of the space between the laths, and tending to cause loosening of the finished plaster. For this reason "riven" laths could come from wood which will permit of the grain being torn apart and should be made in such a way that this can be done. In the case of sawn laths, a fine saw should not be used, the laths being left rough on all surfaces.

Oak and Fir.—Oak, on account of its great strength and freedom from decay and other qualities, would appear to be an excellent material rom which to form plastering laths ; but, though much used in the past, oak is little used now because it has the fault of warping badly, thus tending to give bulging and the breaking away of the key.

Baltic fir is the timber from which laths are generally specified to be made in this country, and they should come from the heart of the tree and be of first-class material ; that is, they should not be cut from the sapwood, which is that part of the tree nearest the bark, consisting of immature wood, soft in nature, which readily decays. Sapwood may usually be detected by the absence of grain, its greenish tinge and its power of absorbing water readily. Bark on laths can generally be taken to indicate that they are cut from sapwood, and they should be rejected.

Riven or Sawn.—Riven laths—that is, those split from the log along its fibres—are far preferable to those made by sawing, as, apart from being stronger (the grain in sawn laths is often cut through) any crookedness or twist of the grain immediately shows itself. But sawn laths are cheaper, and have to a great extent superseded riven laths, but where the latter can be readily obtained they should be used, and particularly for first-class work.

Sizes and Quantities.—Laths are sold in three sizes, namely: " Single " (average $\frac{1}{8}''$ to $\frac{3}{16}''$ thick), " lath and half " (average $\frac{1}{4}''$ thick) and " double " ($\frac{3}{8}''$ to $\frac{1}{2}''$ thick), and are about 1" wide. They are in bundles of 360 to 500 lineal feet, it being usually reckoned that one bundle of laths with 500 nails is sufficient to cover five super yards with the space of $\frac{3}{8}''$ usually allowed for key between laths.

The length of laths varies from 2' to 5', but 3' to 4' are the usual lengths.

In good-class work the lengths of nails used should be $\frac{3}{4}''$ for " single " laths, 1" for " lath and half " laths, and $1\frac{1}{4}''$ for " double " laths ; these nails should preferably be at least galvanised.

In very good-class work zinc nails are used, but these are expensive. Should oak laths, for any reason, be used, or ordinary laths have to be fastened to oak beams or joists, on account of the corrosive action of this timber on iron neither from reasons of strength, nor because of staining, would it be safe to use iron nails, and these should be either of zinc or copper, preferably the latter.

" Single " lath is sometimes specified for partition work where the studding is set closely and where no particular load is likely to come upon it, but the lath generally used is "lath and half," and it is not considered practicable to use anything lighter than this for ceiling or horizontal work. In partitions likely to be subjected to heavy usage, or where for any reason the studding is more widely spaced than usual, " double " lath should be used.

Metal Lathing

Metal lathing, though making rapid progress in recent years, has not reached the same stage as a base for plaster work in this country as is the case in other parts of the world, more particularly in the United States of America and in Canada.

Metal lathing of proper design, correctly applied, offers so many advantages over other bases for plaster work that it would appear a better knowledge of using it can only lead to its becoming a matter of general practice for ceilings, partitions and other constructions which have to be attached to rafters or studding.

Comparison with Wooden.—Compared with wooden laths, the advantages of metal lathing for the same purpose are: there is no warping or twisting caused by expansion and contraction due to wetting and drying, or by imperfect or improper timber, neither is there any question of dry-rot; the key obtained for the plaster is definite and does not mainly depend, as in the case of wood lathing, upon a neck of more or less friable material pushed through narrow spaces of about $\frac{1}{4}''$ in depth placed at 1″ on centres, and depending for its holding power on the material bulging over at the back (as previously pointed out, the face of wood lath can be depended upon for little or no bonding value); the key on a good type of metal lathing extends over the whole surface; in fact, where the work is properly carried out, the metal is practically buried in the plaster and forms part of the slab of material.

Another obvious advantage of metal lathing is that it is fire-proof, and this, combined with the perfect key, is a protection which cannot be obtained with any other base, for there is practically no fear of scaling due to expansion by heat. In the case of other bases, such as brick, tile, breeze slabs, etc., the plaster will become hot before the base, and will expand with great force, causing the material to bulge and scale off, either coat by coat as it was applied, or in one thickness by breaking the bond with the base. In the case of wood lath bases, this falling away of the plaster will be even more rapid, leaving exposed a highly combustible material, which will simply add to the fire.

Where properly designed metal (steel) lathing is used with cement plaster—and it is advisable that cement should be used at least for the "scratch" or base coat—as the coefficient of expansion of cement and steel are for all practical purposes identical, little or no cracking can take place by heat, and even where plasters are used whose coefficients by expansion are different from that of the lathing, the metal is so interwoven with the plaster that any cracking can only be purely local and minute. To fully realise this point one must visualise a great number of small cells of plaster absolutely in contact and combination, and yet with each cell surrounded by a practically unbreakable boundary; under these conditions it can be appreciated that expansion or contraction with consequent cracking can only be small and not sufficient to break away the key of any portion, and that plaster work on metal lathing, properly applied, can only cease being fire-resisting when the heat is great enough to disintegrate the plaster itself.

Metal lathing is, of course, vermin proof, and, by giving a good and reliable plaster surface with only two-coat work, it saves both in material and labour.

Metal lathing has a field of its own for suspended ceilings; that is, ceilings hung on tension members clear of the floor and floor rafters above, which are more and more coming into use, either with the object of using the air space between the ceiling and the floor as additional protection against fire, or to provide space for necessary services such as gas, water and electricity; as also, in modern practice, for heating systems, and also for sound deadening.

Suspended ceilings also offer a ready and economical method of forming barrelled, vaulted or other interiors which do not conform to the actual structural part of the building, and also for false arches and false work generally.

The main reasons that metal lathing holds the field for such purposes are that there is no need for heavy framework to carry the plaster, the members to which the lathing is usually attached consisting of either flat iron on edge or rods with lighter rods or bars hung down from the structure above to carry the weight. This lack of necessity for heavy framing, combined with the fact that the lathing can readily be made to conform to the exact shape required, and that deep grounds for the plaster are not necessary, means a very considerable saving in the weight which the main structure has to carry.

A further advantage of this lathing is that for cornices or other structures of curved or intricate design, the base can economically be made to conform to the finished design; but there is little danger of cracking on account of variations in the thickness of the plaster, and also there is a certainty of key in all parts which will permit of intricate mouldings, etc., being carried out.

With all these advantages, it does appear strange that metal lathing has not come into more general use, and the writer would suggest that one of the great causes of this is that there are many plasterers who have not sufficient experience in the fixing of the metal and in subsequent plastering. The former of these difficulties is overcome in America and other countries by employing men specially for this work. There is nothing particularly intricate or skilled in the fixing of lathing, but, as in most other things, knack and constant practice are responsible for rapid work; and whilst in some of the large districts it is possible to find men who claim to specialise in this work, it would appear that, generally speaking, there are too few of them, with a consequent tendency to keep up the cost of fixing metal lathing by yardage.

In regard to the latter difficulty, men who are used to working in other bases attempt to plaster on lath in the way they are accustomed to, with consequent springing of the metal and falling off of the mortar, or " cutting " of the plaster on the face of the lath by excess of pressure, thereby leaving no thickness of material on the underside to form a surface for the subsequent coat or coats to bond to, and insufficient material to stiffen the lath for later working. With expanded metal laths there is no need to use pressure on the plaster in order to make it adhere and bond, as is the case with brick, tile or solid bases, nor the amount of force that is required to drive the mortar through the narrow openings between wood lathing. On the contrary, using a mortar stiffer than is usual for other bases, the application of the first coat particularly is more a question of gentle pressure than of hard work.

It may be well, now, to discuss the different types of expanded metal lath designed as bases for plaster work, in order to arrive at their adaptability for different classes of work.

Types

Diamond Mesh Expanded Metal.—The most common type of metal lath met with, and the one which can usually be obtained from stock from any builders' merchant, is what may be termed the diamond mesh expanded metal. This material is manufactured from mild steel sheets by machines which slit and expand the metal into diamond-shaped meshwork without welds, joins, separate strands or attachments, but all the strands or members are connected rigidly, and the meshes are continuous.

The size of mesh usual for plaster work is ⅜″ (measured the short way of the mesh), though, should it be necessary for any reason to use plaster without hair or other reinforcing fibre, a ¼″ mesh lath should be used. But this is rarely necessary, and hair or fibre should always be used if at all possible, particularly for ceiling or horizontal work. The former size of opening is the best for the purpose of plastering; it allows the plaster to pass easily through between the strands to form the key, and at the same time the back cover to the metal, without any undue pressure which may spring the lath—a most probable cause of falling off of the wet mortar and looseness of the finished plaster. Furthermore, treating the area of one diamond mesh as a unit carrying a weight of plaster suspended on the area of material contained in a single mesh—a case which might occur if the key to the plaster in and on the surrounding meshes broke down—the cross section of the material is sufficient to carry some weight in tension, this being the quality in which plaster is most deficient, particularly until the material has fully set.

Another point which should not be lost sight of is that the mesh of a metal lath should be large enough to allow mortar containing fairly coarse sand to readily pass through without separating it from the more fluid and finer constituents of the plaster.

A strong mortar cannot be made with sand that is not graded from coarse to fine, and this strength is of importance, particularly in the "scratch" or "pricking up" coat; for the lath and plaster really becomes a reinforced beam, or slab in miniature, dependence having to be placed upon the strength of the plaster in compression to resist the greater force which is required to make a bond between the subsequent coats of material after the first one.

Should the mortar be not quite evenly mixed, it can readily be seen that a patch of plaster may be obtained which contains an excess of large particles, which wedge and block small meshes, thus preventing a proper key being obtained on the lath. This fault may not be visible from the outer face, as the plaster may have enough bond to remain in place, but trouble will occur by springing when applying the next coat, for even if the first coat does not come down, not only on that area where the keying trouble occurred, but also the breaking away of a larger area of good work by loosening from the lath through vibration and movement of the base may happen. It should be remembered that the " scratch coat " can practically never be left until the plaster has properly set and developed its full strength before the next coat is applied.

Another advantage of diamond mesh expanded metal lath, when well made, is that, in manufacture, the strands of metal are not formed vertical from the face of the lath, but slope all in one direction, which, while aiding the application of the plaster and the formation of a perfect key, even if the mortar is not turned over at the back of the metal on part of the area, will tend to hold the plaster in place. Contrary to the effect in wood lathing, where, if there is no bulging on top of the neck through the spaces between the laths, the material pulls away in a vertical direction, the plaster in this case would have to move first in a horizontal direction, which the surrounding plaster prevents it from doing once any setting has taken place.

For the foregoing reasons, and because expansion and contraction take place equally in all directions, the metal of the lath being evenly distributed over the whole area, and because the face of the lath offers a plain surface without any projections to plaster upon, therefore allowing a minimum thickness of material to be used, the ordinary diamond shape expanded metal undoubtedly makes a good base for plaster work.

General Points.—This type of expanded metal lath can also be used for solid or hollow partitions combined with either rods or studding, or may be obtained with curved corrugations for solid partitions, this being with the object of obtaining the maximum strength compatible with the thickness of structure and equality of resistance to thrust from either side.

There are also on the market metal laths which are as light, or lighter, in weight than the diamond mesh. These, while having no hollow or solid ribs to obtain the effect, it is claimed are equally as stiff, or even more so than the foregoing type, such claims being based usually upon a distorted mesh and key running at approximately right angles to straight strands, following in the direction of the length of sheet—such strands being lightly grooved, or corrugated, to obtain stiffness in their direction.

The key for the plaster in such laths is often mechanically very good; but, in deciding upon their adoption for plaster work, there are several points to be taken into consideration, such as whether expansion or contraction is likely to be equal in all parts of the sheet. In some types the strands across the width are far greater in area than those running in the opposite direction; and, moreover, these are often "kinked up" or distorted from the face of the lath in order to obtain stiffness and key. Consequently there would appear to be a tendency for increased movement in one direction of the lath, with consequent cracks running in the direction of the sheet, which faults, as such lath is laid lengthways across supports, would be increased by any lack of stiffness in the latter, particularly in the case of floor joists. Another point to consider is whether the size of opening in the lath is such as to allow free passage of the mortar, especially in case by accident one portion should contain an excess of coarse material. A further point is the probability of the plaster applied on the face falling over at the back to form a protection to the metal, for in many cases these special laths do not allow of this taking place. A point often made about such laths is the lack of, or the smallness of, lap required. But, as with small diamond mesh lath used for plaster work this overlap generally requires to be only one mesh, or $\frac{3}{8}''$, in a sheet of, say, 2′ wide, it appears to be not of great moment, but may be considered in conjunction with yardage prices of various laths.

There are other laths of expanded metal designed to span greater distances between supports with the object of economising, under the same conditions, in the preparation of the framing to which such lathing has to be fixed, and also in the cost of fixing the lath in place. This greater rigidity, or strength, is gained by forming in the sheet, either before expanding or during the carrying out of this operation, either hollow ribs, varying from about $\frac{1}{4}''$ to more than $\frac{3}{4}''$ in depth, according to the stiffness required, or solid ribs, formed from the steel sheet and set more or less vertically to the face of the mesh. The distance apart of these ribs across the width of the sheet varies, but is usually between 3″ and 5″ for the hollow, and 1″ and $1\frac{1}{2}''$ for the solid ribs; the space in between these ribs is filled with an expanded mesh, either of the diamond or some other design, as is the case with the larger weight and short-span laths.

Hollow Ribbed Types.—In regard to the hollow ribbed type of long span lath, though in many cases there is a great advantage in being able to span, without supports, long distances, say even up to 8′, this type of lath would appear to offer difficulties against obtaining really first-class finished plaster work, particularly in the deeper ribbed type.

The material is formed with the base of the hollow rib level with the mesh, so that unless the lath is to be used for forming solid partitions by plastering on each side, the base for working upon will consist of narrow strips of mesh alternating with long hollow air-tight pockets—

the ribs, upon which the rigidity of the sheet depends—and unless care is taken in evenly filling the rib when applying the first coat, there will be a danger of air cushions being formed, which will tend to force the plaster away from this and also, to a certain extent, from the surrounding mesh. On the other hand, when the rib is filled with wet plaster, especially where the work being carried out is a ceiling or other horizontal, or even a sloping, surface, this weight will have to be carried by the unset " scratch " or " pricking up " coat. In this part there is no mesh to form a bond, the inside of the rib being smooth and generally painted, and in fact, as these ribs usually slope inwards from the base to the top, there is everything to assist the weight in sliding downwards, and as the distance from mesh to mesh on either side of the ribs is rarely less than 1″ wide, over which the plaster has no bond whatever, combined with a considerable change in thickness of plaster, there is every tendency for cracking to take place, even though the plaster remains in place.

Another point worthy of consideration in the use of such ribbed laths, is whether there is not a great tendency for expansion and contraction to concentrate on these ribs; as before pointed out, the mesh is usually level with the bottom of the rib, the two sides of this being connected together at the top, say ¾″ from the base, and therefore having a levered hinge effect and acting as an expansion joint, which will tend to concentrate all movement at that point. This would doubtless be an excellent thing if the joint had not to be covered with plaster, but such being the case, there will be every probability of cracking taking place. In practice it is usually found that cracks follow the lines of such hollow ribs; and, whether such defect is caused by the weight of plaster in the ribs, by cushioning of air when plastering, or on account of expansion and contraction being concentrated at this point, it would appear that, to form a good base for plaster work where such hollow ribs are used to increase the strength and rigidity of the lath, it would be preferable that the mesh extended completely over the opening of the rib, thus giving a continuous key and equal expansion and contraction in all directions.

Partitions.—In regard to the use of this type of lath for plastering to form solid partitions, though it certainly offers advantages in stiffness and the lack of necessity of providing either permanent or temporary supports over fairly large spans, it would appear that the ribs are not calculated to crack-free work unless the partitions are more than 2″ thick—and it is doubtful whether it is economical to construct such walls thicker than this under ordinary circumstances. A partition must take pressure from either side, and, to obtain this, the structure needs to be central about the centre of gravity of the reinforcement, but in these ribbed lathings this point will be found close to the base of the rib and consequently on one side of the plaster work. In a properly designed partition, the top of the rib will come fairly close to the face and, combined with unevenness of thickness of plaster, will tend to cracking.

Solid Ribs.—The other types of laths having small solid ribs are not intended to span the same distances as those with the deep hollow rib, and, whilst they offer advantages over the latter in the matter of bond, etc., about the rib, are not yet all they might be as a base for plaster. Most of such laths are formed so that in cross section they are zigzag, or stepped, the top of one stiffening rib, running lengthways of the sheet, being connected to the bottom of the next one by means of strands or mesh made by slitting the sheet and then expanding. Though these ribs are of no great depth, there is not the equality of thickness of plaster over the face of the lath which is so much to be desired if cracking is to be prevented.

Apart from the points raised above concerning the stiffening ribs, before deciding on the use of any particular type of long-span lathing, it will be advisable to take into consideration the question of size of mesh which was mentioned in connection with ordinary expanded metal lath.

Solid Sheets.—Further types of expanded metal which, while not coming under the heading of metal lathing, are of interest, as they are intended for plaster work in forming partitions and walls. These are in the form of sheets consisting of a series of V-shaped troughs of expanded metal, or of main strands connected together either by a mesh or thinner strands. Such sheets should offer a maximum of stiffness for plastering against, in comparison to their depth and weight, would bring the centre of gravity of the reinforcement in the middle of the wall and, by having the main reinforcing strands the apexes of the Vs, would give a stronger partition as regards resisting thrust from either side than is the case with most other types, but the writer is of opinion that the steeply sloping sides of the troughs may cause loosening of the plaster on one side of the base from that on the other, particularly if the mesh on the sloping sides is not of ample size to give a strong key through, and this is a condition which is not generally met with.

In order to give the strength required for these partitions, cement plaster is required at least for the main portion of the plaster work. Cement, in setting, first expands and then contracts, and these two movements acting against sloping sides inclined at, roughly, 45 degrees, presuming the key through is not strong, would appear to have the tendency of separating the two layers of plaster from the metal. The resulting structure could not, therefore, act as a reinforced concrete beam, which, of course, is the condition desired. Moreover, if any loosening of the plaster takes place through vibration, sound-proofing of the partition —one of the claims made for a thin reinforced cement plaster partition —breaks down, and in case of fire there would be great danger of the comparatively thin layer of plaster expanding and scaling off the metal —it has to be remembered that in a 2″ partition the plaster at its minimum thickness will only be about ½″ thick—and the resistance to this danger will be slight.

Metal Plate Bases.—Another metal base for plaster, which does not really come under the name of metal lath, is one consisting of metal plates perforated with specially shaped holes to act as keying space for the plaster, and either plain, grooved or ribbed, according to distances between supports which it is intended to span ; the materials are designed for use for such structures as ceilings, partitions or for furring.

For Ceilings.—In considering the use of this type of base for ceilings, it must be borne in mind that the holes through which the plaster passes are generally large in area and spaced some distance apart, so that the material which falls over at the back in the forming of the key for the plaster covers only a small part of the steel plate of which the base is formed. Unless back plastering is carried out, practically no protection from corrosion can be expected from the face plaster, and should back plastering be resorted to, apart from the added expense, the weight of the structure will be very materially increased.

The actual keying value of this base will, of course, depend very largely upon the size and number of openings in the steel, but in any circumstances it is impossible for the key to be the same as it would be by using an expanded metal lath, where the base becomes practically buried in the plaster; and it is very doubtful whether, when set, the plaster is in intimate contact or bond with the steel. Therefore the finished structure could hardly be looked upon as acting as a reinforced beam, but rather as a more or less rigid sheet steel plate supporting a slab of plaster. Though there would appear in practice to be a saving in plaster required for keying purposes, it certainly does not seem that the use of such bases has much to commend it for ceilings, for though economy in plaster may be achieved, the key is not the same. In fact, such a base approaches the faults of wood lath as regards lack of adhesion or dependence of a neck of plaster, though of course much shorter in this case, to carry a suspended weight.

For Partitions.—In regard to the use of such a base for solid partitions, though the key for the one side may appear sufficient, it is doubtful whether there would be enough bond to the plaster coming through the plate to form the key on the opposite side to resist considerable shock or expansion and buckling under heat. For hollow partitions such a base would appear to be good, with the exception of lack of protection against corrosion due to there being little back plaster.

Fixing of Metal Lathing

The fixing of metal lathing has a considerable bearing both on the finished structure and upon the ease with which the work is carried out, for though the actual finished strength will not be decreased by slackness of the lath before plastering—presuming that good plaster work is obtained and that the fixing is strong enough—if there is constant dropping of the wet scratch coat through movement of the base, the labour costs will be high. And if there is breaking away of the key by movement being transmitted from a portion of the lath being worked upon to a part where the plaster is partly set, the least trouble that can be expected is considerable cracking.

The fixing of metal lathing requires little in the way of actual skill, and should be carried out with the exercise of a little care, though experience, as in other things, leads to better and quicker work, and therefore lower costs.

Though the particulars of fixing which will be given apply in general to all types of metal lathing, they are intended specifically for that expanded metal which is more commonly used, and which can be obtained from stock in most parts of the world, namely, the diamond mesh, and it is advisable when using other special types to obtain the maker's specific instructions.

As with other bases in plaster, it is necessary that the rafters or studding to which the lath is to be fixed shall be true and plumb, as otherwise to obtain truly finished surfaces there will be uneven thickness of plaster, which is liable to cause cracking. Though metal lathing will give a stronger slab of plaster than is the case, say, with wood lath, it is important, as with other bases, that wood or metal framing shall be strong enough to resist, without bending, any strain likely to come upon it ; and, particularly as regards joists, it is well to see that deflection under full load will not exceed 1/360 the part of the span.

Protection of Metal Lathing

Painting.—Metal lathing for plaster work should always be well painted, preferably with bitumastic paint, unless it has to be used in a very damp situation without back plastering, when it should be galvanised. This process should be carried out after the lath has been manufactured; that is to say, the lathing should not merely be made from galvanised steel sheets, as this would leave the cut edges without any protection should the plaster not fall over sufficiently to entirely cover them.

Galvanising.—Galvanising adds considerably to the cost of the finished work, and it will be well to consider whether it is not cheaper to back plaster, if it is possible to do so on account of the position in which the work has to be carried out. Should the latter course be decided upon, an allowance for the added weight may be required in the structural members in the case of ceilings, etc., to prevent cracking in the back plaster—which might allow the dampness to get through to the lath—the thickness should be not less than $\frac{3}{8}''$, and preferably $\frac{1}{2}''$ from the back of the lath ; and, on account of the key coming through the mesh, an allowance of three to four pounds per square foot above the weight taken for the ceiling should be sufficient.

Where it is impossible, from a practical point of view, to get sand without traces of salt—a condition which scarcely ever applies in this country, though it does in some parts of the world—only galvanised lathing should be used, as salt attracts any moisture from the air, and will carry it through the plaster.

Weight

Generally speaking, the minimum weights of lathing per square yard that should be used for different spans between supports are as follows: 12″, $2\frac{1}{4}$ lb.; 14″, 3 lb.; 16″, $3\frac{3}{4}$ lb.; and 18″, 5 lb. For ceilings, lathing made from sheets of less than 24 gauge should not be used; for suspended ceilings it is advisable to use the weight of lathing which is given for the next greater distance, centre to centre, between the supports to which the lathing is to be attached.

Before fixing, it is advisable to flatten any sheets of lathing which have become bent or distorted during transport, or in handling the bundles, as this can be more readily and efficiently carried out on a bench or plank than when fixing in position.

The Process of Fixing

Dealing firstly with the fixing to wooden supports, the sheets of lath should always be carried lengthways across the bearers; that is, the long way of the diamonds will span across the space between the two supports, and the strands of the various sheets should all slope in one direction. This sloping of the strands will be made clear upon an examination of a piece of expanded metal, when it will be noted that in manufacture the strands are not made vertical to the face of the sheet, but are inclined towards one edge in the direction of the width of the sheet. In fixing on a vertical face, such as a partition, these strands should slope inwards and downwards, as this prevents any tendency of the wet plaster of the "scratch" or "pricking up" coat to slide away from the lath.

The fixing should be with galvanised slice-cut staples—unless for very first-class work, when staples of copper are used, and this should always be the case if fixing is being done to oak, which decays iron very rapidly—the length being not less than 1″ for vertical work, and $1\frac{1}{4}$″ for ceiling or horizontal work.

As in the case of wood laths, counter lathing also applies to a metal lath base, and for good work it is advisable that where the supports are more than 2″ in width, the lath should be "stood off" in order to obtain continuity of key. In fact, as this counter lathing in the case of expanded metal can be obtained with little expense of labour or material, it would appear worth while to carry this out on faces of less than 2″ wide; for, apart from getting a better key, it prevents wet plaster being forced against the wood with the possibility of twisting and warping; also the absorption of water from the mortar combined with change of thickness of the plaster is likely to lead to cracking on either side of the supports. This "standing off" or counter lathing may be either $\frac{1}{4}''$ or $\frac{3}{16}''$ diameter down the centre of the support; this being stapled to the wood sufficiently to hold it in place, and when fixing the lath the staples should be driven through the strands and over the rod at the same time. When fixing in this way, the diameter of the rod should be added to the length of staple required, as given above. This counter lathing could, of course, be carried out with wooden strips, but it will generally be found cheaper and quicker to use the iron rods, and there will be no danger of warping or drying out, with consequent marking over this narrow face.

In commencing the fixing of the lathing, one end of a sheet should be fixed to the end support with staples at 4″ centres for lighter weight meshes, say up to and including 3 lb. per square yard; and at 3″ centres for the heavier weights, or for all weights where excessive vibration is likely to be met with, or where particularly heavy plaster, ornamentation or tiling, etc., has to be carried on the plastered face. But the sheet should not be fixed nearer than 3″ from the edge where the join in width will come, this being for the purpose of easy wiring together of the two sheets of lath, and in order that a staple may afterwards be driven over the joint of the two sheets at each support.

The sheet should now be stretched across the next support, this being easily carried out by a tapered iron spike, small enough to fit through the mesh up to 4″ from its end, which is used as a lever against the side of the wooden, or other, support, to tauten the lathing and hold it in place whilst driving in the staple. It must be remembered that the stretching required is not extreme, but merely sufficient to form a firm ground for plastering upon, and one that will not sag under the weight of wet, or partly dry, plaster.

A method of stretching expanded metal lathing over supports, and one which offers advantages as compared with the single spike, in that it does not so readily lead to distortion of one part through an excess of force and that it straightens a considerable area at one time, is by means of an implement shaped like a small garden fork of steel having four prongs, or spikes, spaced to alternate meshes of the lath, the whole being about 12″ in length including the handle. This is used in the same way as the single spike, but stretches at four points instead of one; and where, by the ends of the diamond meshes not coming true with the support, there would be unevenness of stretch, this can be allowed for by inclining the handle, so that the unevenness is compensated for by variation in the leverage of the spikes.

The staples should be staggered in position; that is, instead of being in one line across all the supports, in numbers one and three they will be in the same position, but in numbers two and four they will come in the middle of the spacing, and so on across all the supports. This method of fixing gives more uniformity of tautness, and does away with any tendency for the lathing to form into slight furrows, or grooves, in between staples, on account of excess stretching.

Should the end of a sheet to which a further length of lathing has to be added across the bearers come on a support, but beyond the stand of rod or strip, this end will not be stapled until the next sheet has been joined by wiring. If it can be arranged, either by predetermination of the spacing of supports or by using a short piece to fill in between the first support and, say, the wall, or without undue waste, it is better that they should so come. The end lap should be at least one mesh in length, and where lathing is being used for the longer spans, and particularly where the joint comes about the centre between bearers, the lap should be two meshes long, the connection being made with not less than 18-gauge soft iron galvanised wire spaced at the same distances as the staples, or closer should the joint between the ties tend to open. As, however, ungalvanised wire can, with proper plastering, be counted upon to be covered with plaster, both top and bottom, and therefore protected, there would appear to be no reason why this should not be used, but the galvanising is an increased precaution which adds practically nothing to the cost of the job.

The tightness of the ties at joints of sheets will have much to do with the obtaining of well-finished work, for loose joints will not only lead to falling plaster on the scratch coat, but are likely to cause cracking later. To obtain a good joint the tie should come as close as possible to the end of the lap which is on the working face, the mesh of the lower sheet matching as closely as possible that over it, in order that there shall be no cutting of the key when plastering, and this can be readily done by the use of the tool used for straining the lathing.

Where the end lap is of one mesh length, good work may generally be obtained by tying at the one end of the lap on the working face, but a tie at the other end, say, at double the spacing given for the main ties, is worth while, particularly where there is a tendency for the other or inner end of the lap not to lie flat. In any case, this leads to greater rigidity, but where the end lap is two meshes long, the ties should come at both ends of the lap, should never be at more than 4″ centres, and should be "staggered," one end of the lap with the other. At the junction of the end lap with the side lap of the sheets, particular care should be taken in the tying, so that the three sheets shall be brought closely together.

A practice which is used in making joints in expanded metal lathing is that of lacing; that is, taking a length of wire and either passing it through the meshes and over the end of the lap, at the same time traversing across the width of the sheets, as in "over-sewing" of tailoring (though, of course, with the stitches much farther apart), or passing the wire in and out of the meshes across the sheets, as in ordinary sewing. But by this method it is practically impossible to keep the sheets in close contact or get tight tieing, and it should on no account be allowed, but each tie should be an entirely separate short length of wire.

A convenient way of making these ties is to have the wire cut into short lengths and bent hair-pin shape, so that they may be dropped over the two strands, to be tied, from the back of the sheet with one hand, whilst with the other, either by means of the fingers or pliers, the two ends are given a half-turn, to hold the sheet temporarily until all the ties are in place. By this means the meshes of the two sheets can be made to correspond, and, combined with tightly tacking the rest of the unfixed sheet, in two or three places, to the supports, the fixing is greatly facilitated. The ties should then be tightened and secured by placing the nose of the pliers close to the mesh and giving a complete turn—or more if this is not sufficient to bring the sheets in close contact—to the two wires, and then bending back the untwisted parts of the ends on the face of the lath. There should be no need for such ends to exceed, without cutting, 1″ long.

A fault to be guarded against in tieing expanded metal lathing, whether to metal supports or in joining sheets together, is that of commencing the twisting too far from the lath; for even though, after twisting several times to take up the slack, a tight joint is obtained, it is a costly proceeding, and there is every probability that before this aim is attained, the wire will have sheared, likely enough without being noticed until the plastering has commenced.

Another incorrect practice is that of taking a turn in the wire round the support or strand to which metal lath is being fixed, before passing through the sheet and tying. This will not, as is commonly supposed, give a stronger tie, for the strain will still fall on the two straight pieces of wire which come through the lath and are twisted. This method is likely to lead to bad fixing, as it is difficult to tighten up the loop without much twisting of the tie, even if then; and unless this loop is fitting tightly on the strand or support, the effect will be that of attaching the lath to a ring fixed on a bar smaller than its internal diameter.

Before passing to the question of side laps—that is, the side joins at right angles to the supports to which they are fixed—it should be pointed out that where an end lap comes over a bearer, there is of course no need to wire the lap together, as stapling will be all that is required. But if rods or strips of wood are used for standing off, it will be necessary for the lap to centre on these, and it would not be good practice to staple through on one side.

In fixing the sheets which build up the width in the direction of the supports or bearers, they should be so arranged that the end laps of each line of sheets break joints with each other. That is, that no end lap in one line of sheets comes nearer to an end lap in an adjacent line 2′ distant, if possible, but under no circumstances nearer than 1′ in the direction of spanning across the supports. This can be arranged for by commencing each alternate row of sheets with a length of, say, 2′, and then continuing with the full-length sheets; or, of course, the same result may be obtained by shortening the first sheet in each alternate row or line by the same amount. This method of breaking joint prevents a weak spot being formed across the base, where cracks may follow in the finished work.

The side joints need only lap one mesh, the tieing being on the edge of the lap on the face, and should be at the distances apart previously stated; though if there is any standing away of the inner edge of the sheet on the back of the lath, a few ties should be put on the other edge of the lap, in order to prevent breaking of the key of the plaster. For the same reason, care should be taken to keep the mesh in the two sheets corresponding as closely as possible.

The laps, both end and side, should all follow the same planning; that is, if number one sheet lies with the edge lapping over number two on the plastering face, number two sheet will be over number three on the same face, and on no account should one sheet lie at the back of two others in such a way that it forms a sunk panel.

Corners and Cornices

Where corners have to be formed, or where a vertical face connects to a horizontal surface, both in expanded metal, the lathing should not be cut at these angles, but the bend should be formed in the metal. This operation can readily be carried out over the edge of a bench or plank, and such method gives a far better result than could be obtained by the closest and most careful tying at such junctions.

This same precaution follows in forming cornices and like structures; but where these are curved it will not be possible to make the lathing continuous at the angle, and where such joint has to be made, the ordinary spacing of the wire ties will not be sufficient—they should be not more than 1½″ apart, and preferably 1″ only in centres, in order to make a perfectly firm and strong joint. It is in such angles that one often sees "lacing" carried out; and, needless to say, it is the worst position that such a practice, which is in any case bad, could be used.

In forming cornices, etc., if these are of any size, brackets will be required, as for wood lath constructions; but where these are small they can often be fixed without other support than attachment at the wall and ceiling, by spanning the metal the long way of the mesh between the two points of fixing, and forming the metal to required shape on the ground. Where the lathing has to be attached to brackets, either of wood or iron—and the use of round rods for such supports will often prove an economy—the lathing should run in the opposite direction; that is, the long way of the mesh across the supports.

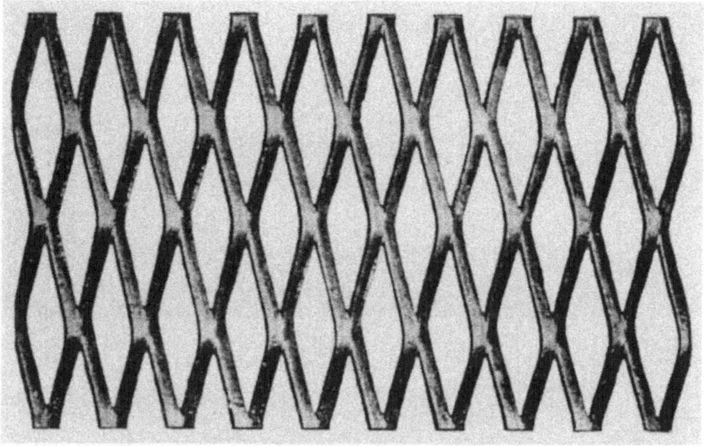

DIAMOND-MESHED EXPANDED METAL LATHING.

EXPANDED METAL LATHING USED IN CEILING AND PARTITION PLASTERING.

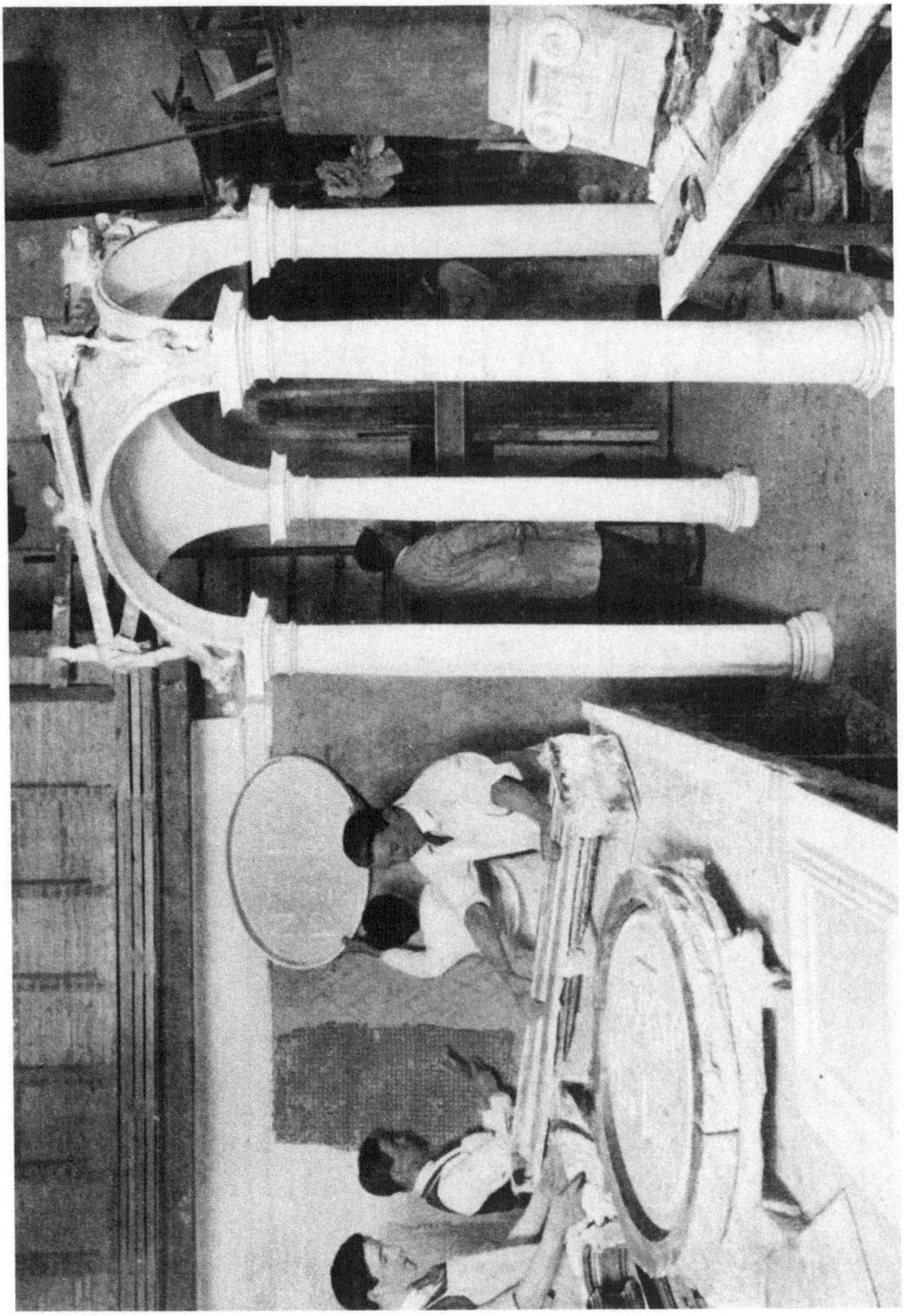

MOULDING A COLUMN, WITH COVED CEILING SUPPORTED ON PLASTER COLUMNS.

Further Points in Fixing

Before leaving the question of fixing to wooden supports, it should be noted that for walls or partitions, whether hollow or solid, where the sheets of lathing run horizontally across the studs in the long way of the mesh, the slope of the strands already referred to should be inwards and downwards from the face to be first plastered upon.

With the exception of using metal lathing having ribs, of one kind or another, intended for long spans (that is, over 18″ or 20″ in plain structural work), there is little fixing which can be called "direct" to other materials than wood, owing to the other types of construction having their supports, whether of steel or concrete, spaced at too wide spaces. Therefore, the majority of such buildings using metal lathing for plaster will have such bases on the "suspended" principle, but it will perhaps be well to have a few notes on fixing of long-span types.

Ceilings under reinforced concrete floors, where the beams or joists which practically form part of the floor slab proper, and are therefore spaced about 2′ apart, form the most common type of such support for fixing to; though there are other constructions where the concrete beams are spaced farther apart, yet closely enough to use long-span metal lathing directly fixed.

As a means of fixing such lathing to the beams or joists, wooden strips have been cast in the bottom of the beam when pouring the concrete, and to these the lathing has been fixed either with staples or nails. This method, however, would appear far from ideal, for the wood will not have the same expansion and contraction as the reinforced concrete or the lath and plaster. The wood, therefore, will have a tendency to warp and buckle whilst in contact with the wet concrete, with consequent movement when drying out, and unless it is carefully chosen, being in a confined space, it will be liable to "dry-rot," and therefore prove an unsafe means of fixing.

A far better method is to arrange for hangers, of $\frac{3}{16}$″ round, hooked at one end to fit over the reinforcement of the beam, and the straight end long enough to bend over a $\frac{3}{8}$″ rod, to be cast in the beam when pouring at 2′ centres along the beam. The hangers are, at least as regards the straight position, annealed before fixing, and the portion protruding from the beam is easily hammered over the rod, leaving a space of about $\frac{1}{8}$″ between the rod and face of the beam—this can be readily gauged by placing a piece of steel of required thickness between the two—for the tying wire to pass over.

Should the bottom faces of the beams be not level, differences can be allowed for when fixing the rod by increasing the $\frac{1}{8}$″ space by this method. If the short spacing pieces are to be removed, the rod should be wired to the hanger—to prevent any movement when plastering—and the rod and hangers given a good coat of bitumastic paint before the lathing is fixed, which may be carried out in the usual way by means of galvanised iron wire (soft) of not less than 18 gauge.

Should the beams be already cast without any arrangement for fixing the lathing—$\frac{3}{4}''$ round holes may be punched through the width of the beam, not less than $1\frac{1}{2}''$ above—the reinforcement should be at 4′ centres along the length, and into these a piece of $\frac{5}{8}''$ round rod, 2″ longer than the width of the beam, should be grouted with the 2″ protruding on the side where the hanger is to be fixed. The bar to which the lathing will be fixed, and which for this spacing of pegs should be 1″ by $\frac{3}{16}''$ flat on edge, should be hung at the correct level by means of $\frac{1}{4}''$ round rods which are first hooked and wired to the pegs and then bent up at the bottom into a hook at least the depth of and holding the flat, which is tied by wire to prevent any upward movement. The whole of this steelwork should receive a good coat of bitumastic paint—not forgetting the end of the peg on the opposite side to that carrying the hanger—before fixing the lathing by wiring in the usual way.

The above method may be considered expensive, and it is possible to devise a hanger of flat iron which will entail cutting into the side of the beam 2″ deep; but this would have to be hot forged and would probably work out more expensive. In any case, it is better to make some firm fixing in the side of the beam than to cut into the bottom face and grout in a hanger; for, in the first place, it is probable that the reinforcement will be struck before sufficient depth has been obtained, and then it is not safe to rely upon the bond of the new cement to the old concrete to hold a weight in tension. Moreover, there would be every danger of such fixing becoming loose as soon as it is attempted to bend the hanger for holding the flat.

For fixing long-span lathing to steel beams, the same method of running a rod along the centre of the lower flange of the beam, to which the lath is wired, may be used; but in this case the clips may be made of $\frac{1}{4}''$ round rod bent at one end to clip on to one side of the lower flange, and long enough to go across the bottom of the flange and be hammered over on to the upper face of the flange on the other side, in order to make it secure; in the centre of the part going across the flange a downward depression should be formed through which is passed, before fixing, a piece of annealed $\frac{3}{16}''$ round, bent in the form of a staple, both legs being of sufficient length to be bent to form a hook for the rod; the method of fixing this latter is to hammer the two legs of the $\frac{3}{16}''$ staple round the latter. As the rod will be fixed against the bottom of the main clip, presuming the link is long enough to come well up in the direction of the flange, there should be no need for wiring to keep the rod from moving, and the thickness of the clip will act as a spacer to allow room to pass the wire for tying the lathing.

There are on the market special clips made for attaching either the rod to the flange of the beam for ordinary wire tying, or for fixing on the lower flange of the beam and for bending round each rib of the lath

—the spacing at which attaching is generally required on this long-span material. Such clips are usually manufactured from flat iron about $\frac{1}{8}''$ thick, with one end bent to fit one flange, and of varying lengths to fit different widths of steel beams. They can, as a rule, be purchased at a price which does not make it economical to form one's own clips for such fixing.

Suspended Ceilings

Suspended ceiling is a term covering a quantity of differing types of construction, for it may mean the ceiling attached to bars bearing on two walls a few feet apart, an area of 100′ square, a vaulted or barrelled ceiling below a roof or floor of quite a different shape, or any other form of ceiling or like construction suspended below the part of the building over it.

The metal lathing is fixed to iron bars placed apart at distances to suit the type of lath used. Generally speaking, the most economical section to use is a "flat" on edge; though for vaulted or other work, which requires curving or shaping, round bars are better. These, of course, give a greater weight of metal for an equal area covered, but this is more than compensated for by the saving in labour and the rapidity of fixing.

Taking ordinary diamond mesh expanded lathing and ordinary conditions and thicknesses of plaster, the following table gives safe general practice for the size of section required for different spans, bearing in mind that these are intended for use with heavier weight lathings spanning 16″ to 18″, as it will be cheaper to use this than a lighter weight with the bars at, say, 12″ apart.

For Spans not greater than	4′,	section	$\frac{3}{4} \times \frac{3}{16}''$
" " "	5′	"	$\frac{7}{8} \times \frac{3}{16}''$
" " "	6′	"	$1 \times \frac{1}{4}''$
" " "	7′	"	$1\frac{1}{4} \times \frac{1}{4}''$
" " "	8′	"	$1\frac{1}{2} \times \frac{1}{4}''$

Generally, when the span exceeds 6′, it will be more economical to suspend rolled steel sections from the main structure at about 5′ or 6′ centres, and to attach to these, running at right angles to them, flats on edge to carry the lathing. Angle irons or T-irons may be used for this purpose, or rolled iron joists of the lighter sections and narrow flange widths—there is no need to have a wide flange, as the flats will stiffen the joists across the narrow direction and so prevent any tendency to buckle in that direction. In selecting these it must be borne in mind that it is not merely a question of weight carrying but of deflection, for beams used for this purpose would usually carry loads considerably in excess of those which would be allowable on account of bending. In working from the tables supplied by makers of these sections, it should be remembered that with flats fixed to the lower flange of the joist at distances of, say, 16″ apart, the load is not a distributed one in the true meaning of the term, and if it is not desired to work out the actual bending moment, it is advisable to use only three-quarters of the safe deflection load given for the section, if first-class work is required.

The simplest type of suspended ceiling is a narrow one, such as a corridor, not more than 8′ wide with walls on either side into which the supporting bars can be pinned, not less than 2″, at each end, at the required distances apart. Should such a corridor be long, however, it is wise to have a bar at the half span on top of, and at right angles to, those on edge, in order to keep them in a vertical position, as should these twist over in tying the metal lathing in position, or from any other cause, the strength of the bar will be diminished.

In cases where the span is too long to be carried on walls at either side without any intermediate support, the bars will be carried either directly by hangers fixed into the structure above or attached to rolled sections, which are hung from the main structure by rods at distances of 4′ and upwards. Which of these systems shall be used will depend on various conditions, though if a first-class construction is required or where excessive vibration is expected, the second method should be adopted, for by this a far stiffer construction is obtained. Again, if the hangers have to be fixed into a reinforced concrete floor which is already constructed, or in any cases where it is difficult to fix the hangers, it will prove, in spite of the extra weight of metal entailed, far more economical to use the method of fixing rolled sections at greater centres, with the "flats" on edge attached to these.

Dealing with the light "flats" suspended direct, it will usually be found best to use for the hangers a round rod of $\frac{3}{16}$″ diameter if the distance between the main structure and the bars is less than 12″ and $\frac{1}{4}$″ up to 18″—the length of the rod, of course, makes no difference to the strength, but by using a longer rod where there is any length, greater resistance is obtained to movement when applying the plaster—these being fixed in rows spaced in one direction at the distance apart determined for the type of lathing to be used, and in the other direction as centres, according to the size of flat employed. This distance will, in most cases, prove to be most economical if fixed at 4′. These hangers, if to be fixed to wood at one end, should be formed with an eye through which a screw or coach-bolt passes, according to the weight supported by each hanger, for attachment, but if to be cast in concrete beams or floors, this end should be at right angles to rest on the reinforcement, if for a beam, and again bent into a curve with a radius of about 1″ in order to get a grip in the concrete, and to be fixed for length either by resting on the steel or form work.

The other end of the hanger should be left unbent for this class of work, but should be annealed for some distance up so that it may be readily bent in place at the required point. The bending of the end to hold the bar before fixing in place, such as one sometimes sees done, is a practice which is likely to prove expensive, for it is most improbable that the hangers will be fixed exactly either in length or line before the concrete is poured. Even if this is so, there is always settling of the form work to contend with, and, once the hook for holding the bar is made, it is difficult either to lengthen or shorten the distance, to bring the bars all level.

In calculating the length of hanger required, it is advisable to leave an inch or so in excess of the amount actually needed for the length of suspension plus the formation of the hook—and the leg of this should be at least as long as the depth of the "flat"—so that should there be any variation in the height of fixing at the one end, there will be sufficient to give an ample hook at the bottom.

Flats of, say, $\frac{1}{2}''$ by $\frac{1}{8}''$ or $\frac{3}{16}''$ are sometimes used for these hangers, but whilst they weigh more than the material proposed, if they are not fixed truly as regards position, they are difficult to rectify in the direction of width, particularly if they are short.

In fixing the ceiling, presuming the hangers to be already in place, the proper level for the position of the bars is taken across in both directions and marked on the hangers by chalk, or other means, after which these are bent up at this point, a convenient way of doing this being by means of a tube with a bore slightly bigger than the diameter of the rod used for the hanger, of about the same thickness as the "flat," and of sufficient length to give the required leverage, which is slipped over the end of the hanger to the required point, and then twined upwards, thus forming a hook into which the "flat" is placed. The hook is closed over against the side, if necessary, after which a wire tie should be taken round the hook and bar to prevent any upward movement of the latter when plastering takes place.

The whole of the bars, at least in a section, should be in place before any lathing commences, and the bars and hangers should be given a good coat of paint (bitumastic, of slow-drying quality), not forgetting to work well round the eye and screw if the upper end is to timber.

In planning out ceilings on directly supported "runners" (bars), a supporting line of bars should come parallel with the walls, or other continuing parts of the structure, at about 3″ from them in the direction of the "run" of the bars. The ends of the runners should be pinned into the walls, if of brick, for a distance of 2″; or, if they come up against steel or concrete, they should butt closely and be provided with a hanger not more than 3″ from the end of each bar.

Joists at the end of bars should always be made in a hanger, and should lap at least 2″. Greater stiffness will be obtained by wiring the bars together at each end of the lap.

In fixing runners to rolled sections, the attachment should be made by clips either such as are on the market made from punched strip, as already described, or other types which are obtainable. Or a clip can readily be made from $\frac{3}{16}''$ round rod. This clip for the lower flange of joist sections is made by doubling a piece of rod so that the two diameters lie side by side, and then bending at this point in the direction of least thickness to form a hook to go $\frac{1}{2}''$ on to the lower flange of the joist. The one leg is of sufficient length to span the flange, and to be hammered over when fixing to form a hook of not less than $\frac{1}{2}''$ on the other side of the flange; the other leg being carried to the centre of the flange and bent down at right angles, and of sufficient length when fixed to form a hook to hold the runner (bar). These clips should be made away from the job, and should, of course, be well annealed in order to assist the fixing.

Where the rolled sections have to be suspended, particularly where the hangers have to come some considerable distance apart, some consideration will have to be given to the construction of these, as the weight they carry may be considerable. If a hanger of greater diameter than can be readily bent in position has to be used, probably a turn-buckle screwing on to a hanger, already fixed, at one end and carrying a clip for attaching to the upper flange of the joist at the other, is the most satisfactory means.

However, such cases are not frequent, and usually the ordinary hanger is sufficient. This may either be bent right round the section and wired on top, or a clip made, to fix on the upper flange, from flat iron, having at the top a hook round which the hanger is bent and then wired.

The fixing of the lathing is the same as the case of wiring to rods previously described, but in this case, as a considerable amount of the tying wire is exposed above the plaster when tying to "flats" on edge, it is advisable to give the wire a good coat of paint where it comes round the flat. This can readily be done as each sheet is fixed, or at least before too great a breadth is fixed to allow of reaching over.

This painting of the wire is in the nature of an extra precaution, for it is usually considered that rusting does not take place in an enclosed space without any current of moist air, and as such constructions are not usually in a damp situation, but are quite dry, ceilings would appear not to require it. Moreover ceilings constructed without any such protection more than twenty-five years ago have been found to be in perfect preservation.

If difficulty is found in getting the lathing taut on account of the flexibility of the runners, a piece of timber notched to fit over two bars further back where the lathing is fixed than that being worked upon and bearing against this, close to where the leverage is being taken, must be used. Such piece of timber will be carried along as the fixing takes place.

Miscellaneous Types of Metal Lathing

"*Hy-rib.*"—Hy-rib also consists of steel sheets—28, 26 and 24 gauge, slit in a peculiar way which produces a kind of herring-bone pattern in three strips, with plain rib between them. They are made 12′ long by 12″ broad, to interlock at the sides and ends, so that large, practically continuous, surfaces are obtainable. The slots are so arranged that they provide a perfect key, and in plastering automatically ensure back-plastering. Hy-rib is used for suspended ceilings, partitions (which need not be more than 2″ thick), walls, tanks, etc. It is sufficiently flexible to form columns, barrel ceilings, etc., and is very light for its strength.

"*Trussit.*"—Trussit is yet another type of slit and expanded metal sheet, supplied in corrugated form. It is made in sheets 12′ long by 19″ wide, and in 28, 26 and 24 gauge. It is claimed that with Trussit plastered walls 1½″ to 2½″ thick can be quickly finished, and will prove sound-proof and fire-resisting.

"*Self-Sentering.*"—Self-Sentering, which is manufactured by the same people who produce Trussit, is a ribbed type of expanded metal, the diamond mesh between the ribs being thin strips like wires. It is very easily handled and proves useful in ceiling work.

"*B.R.C. Fabric.*"—B.R.C. Fabric is made up of wire mesh, the wires at their points of contact being electrically welded.

"*Bricanion.*"—Bricanion is a new type of metal lathing, consisting of a square-meshed iron wire netting encased in brick earth; in other words, it is a peculiar form of reinforced brick fabric. The brick earth is not applied in continuous sheets or strips, but as small crosses, with rebated sides, at the intersections of the wire meshes. Thus each cross is a separate unit, and the sheets remain wonderfully flexible after they have been kilned under careful heat control. This reinforced brick lathing can be used as a base for any kind of plastering or cement-rendering on partitions, ceilings, arched and curved surfaces, for casing columns, beams, etc.

It is made in long sheets of fair width. Joining together is easy, it only being necessary to break off the brick crosses at the edges and twist the exposed wires of the two sections tightly together. The mesh is cut with specially designed slips. In this way adjustment to long, broad or narrow spans is quickly carried out.

For ceiling work the sheets are cut to length, so that they should be stretched at right angles to the joists to which they are to be fastened. Each sheet should be longer than the span by four rows of crosses, as two rows of these are broken off at each end and the exposed wires wound round and fastened to two $\frac{5}{16}$″ iron stretching rods, which are inserted in large cramps placed at intervals of 12″. The sheets are lifted, loosely cramped to the joists. They are then carefully and uniformly stretched by means of a special screw spanner. It is essential that the stretching should be uniform, for sagging would mean defective plastering. On large ceilings, as an additional support, galvanised wires of 14 S.W.G. should be tightly stretched across the joists under the sheets at intervals of 8″, stapled to each joist and then fastened to the

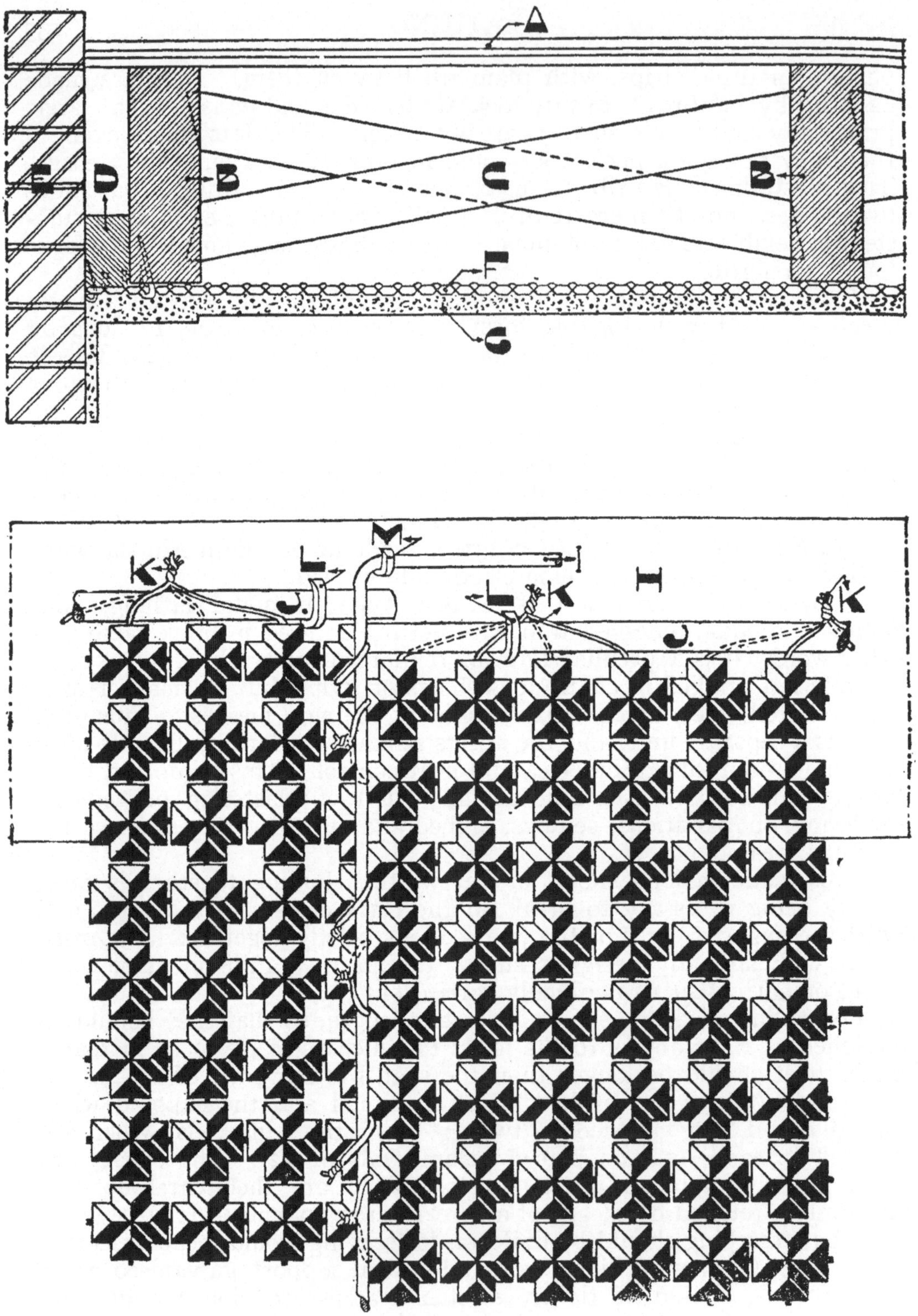

Fig. 7.—Ceilings under joists with Bricanion lathing.

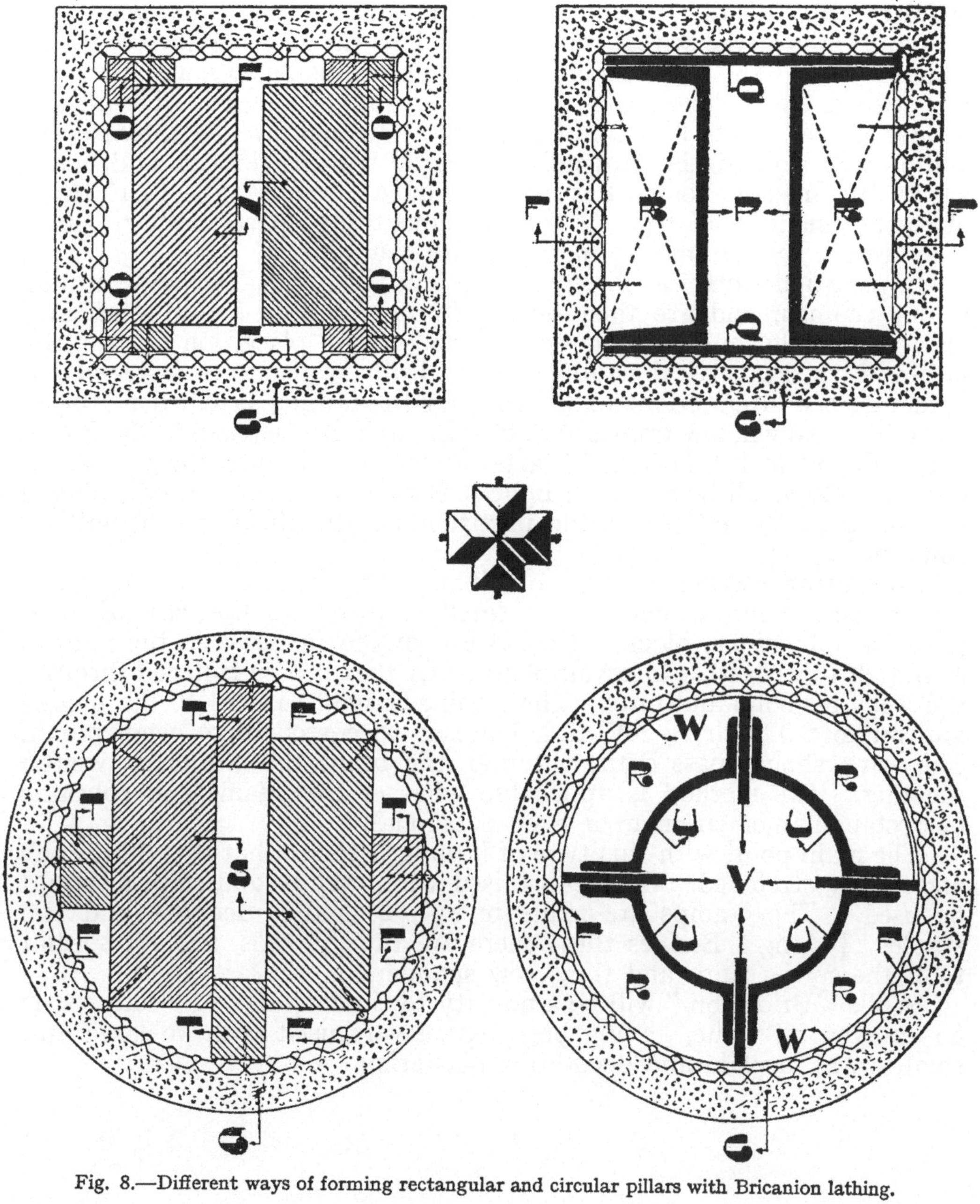

Fig. 8.—Different ways of forming rectangular and circular pillars with Bricanion lathing.

lathing by wiring once between each pair of joists. On small ceilings straining rods are dispensed with, and instead 14 S.W.G. wires are spanned across and immediately below the joists, at intervals of about 20". These wires are fixed to the joists by means of long staples, slightly driven in. The lathing is cut in length to span across the space between a pair of joists, reaching to the centre line of its joists. It is fixed below the joists and above the wires, being fastened to the latter with stout galvanised wire. The staples may now be driven home to make all taut. For ceilings under steel joists, the straining rods are supported from the flanges of the joists by strong S hooks. Ceilings are back-plastered, and if well worked the protruding material will provide an additional key to the finishing coat. For heavy ceilings a rendering $\frac{7}{8}$″ to 1″ thick of mortar, made up of 1 part Portland cement to $3\frac{1}{2}$ parts of sharp sand, is advised. But ordinary practice is quite suitable.

In partition work the sheets are stretched between floor and ceilings as for ceilings, and are then reinforced by $\frac{5}{16}$″ rods placed vertically and horizontally 20″ apart, horizontal rods being placed on side of the lathing and the vertical on the other. Both systems are fastened together at the intersections of wiring, and the lathing fastened to these. Over door and window frames the straining rods are fastened to the frame and wall, while the lathing is fastened by cramps into the plastering grooves. In small houses with light partitions a single iron bar, placed horizontal in the middle of the partition for its whole length, will be sufficient.

For casing wooden beams or columns, the lathing is merely bent round tightly and fastened with small staples. In the case of steel joists or stanchions, blocks of wood are driven in between the flanges until tightly wedged at intervals of 20″. Or they can be fastened through the webs by bolts and nuts. The lathing is next secured to the blocks and reinforced with 14 S.W.G. galvanised wire secured to every block. The wire should pass entirely round the beam and the ends twisted together. The method is applicable whether the beams and columns be rectangular or circular, or even octagonal.

The main point with this type of lathing is to see that it is quite taut and evenly stretched. Only galvanised wire, cramps and staples should be used. The cramps are made in $1\frac{1}{4}$″, 2″ and 3″ lengths, and the staples $1\frac{1}{4}$″ long. Besides these there are the $\frac{5}{16}$″ rods, and for special tools the spring snips and the screw spanner.

While "Bricanion" will be a novelty to the majority of plasterers, it has proved its value, having been extensively used both on large and small work, in plain and decorative plastering.

CHAPTER XI

PLASTERERS' TOOLS

Compasses—Level—Plumb Bob—Line and Reel—Jointer—Trowels—Drags—Scratch Tools—Small Tools—Knife—Modelling Tools—Tack Hammer—Pad Saw—Bradawl—Gimlet—Planes—Hawks—Skimming Float—Cross-grained Float—Wire Scratcher or Comb—Margin Trowel—Angle Trowel—Plasterers' Hammer—Steel Joint—Rules—Stock or Water Brush—Tool Brushes—Muslin Bag—Fine Sieve—Snips and Cutters—Files—Steel Triangular Square and Wood Square—Mason's Square—Saws—Wooden Tool Stick—Splash Brushes—Felt or Cork Faced Floats.

A GOOD craftsman will always have a fine and varied assortment of tools, whether shop made or of his own contriving. For it is a fact that the man who takes interest in his work, is anxious to possess the right sort of tools and to keep them in perfect order, is the successful man, for these help towards successful results.

In plastering many kinds and odd-shaped tools are needed which are not stocked in tool shops. It rests with the enterprising plasterer to gather together an assortment in wood, iron or steel—to suit his requirements.

Recent discoveries in Egypt disclosed plaster work that had been completed perhaps five thousand or more years ago and yet is in perfect condition. It was also found that the tools used by those ancient craftsmen were for all practical purposes identical with those used to-day. The fact is it matters not how crude the tools are so long as they answer the purposes for which they are intended. Tools do not make the craftsmen, though fine and appropriate ones will help him in his task.

The following is a list of those things necessary for solid plaster work on the job or when working in shop.

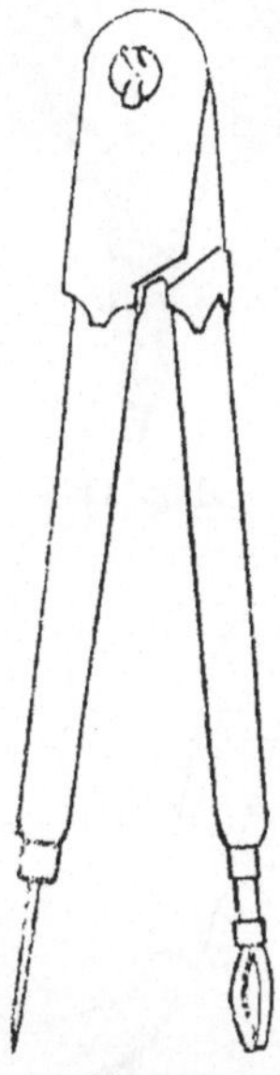

Fig. 9.

WOODEN COMPASSES

1. Wooden compasses are utilised for setting out blocks in cornices, and also for setting out geometrical panels to ceilings; besides being serviceable for numerous other setting-out jobs (Fig. 9).

LEVEL

2. Spirit level, with upright tube. This is useful for plumb-upright faces to girders, etc., when putting in dots (Fig. 10).

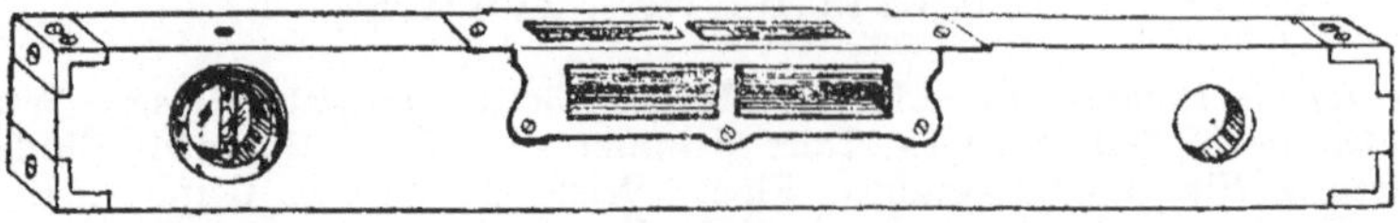

Fig. 10.

Plumb Bob

3. Centre plumb bob is required for plumbing dots for floating, either with the two gauges or plumb-rule (Fig. 11).

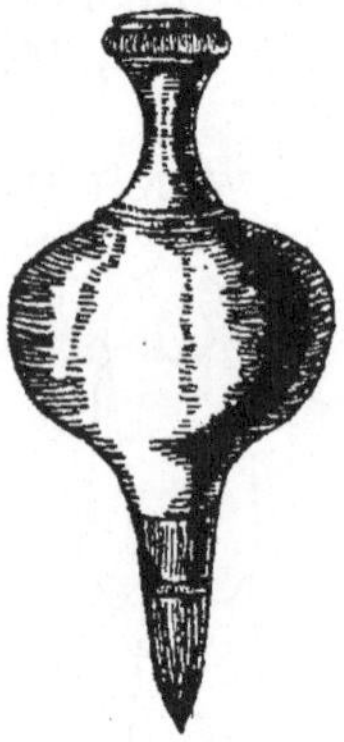

Fig. 11.

Line and Reel

4. The line and reel is used for lining in dots to floating, also for snapping the line when fixing running rules for screeds, etc. (Fig. 12).

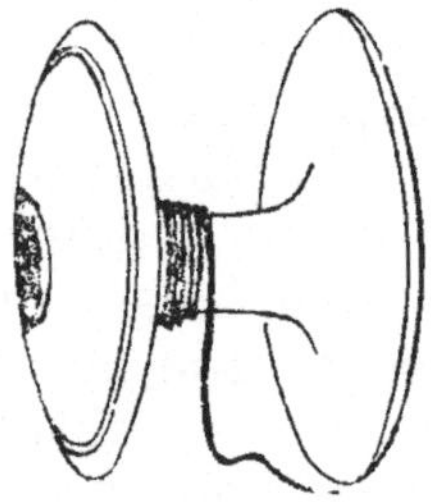

Fig. 12.

Jointer

5. The jointer is required for jointing soft lime stucco or dinging old walls to imitate the joints of stone and brickwork (Fig. 13).

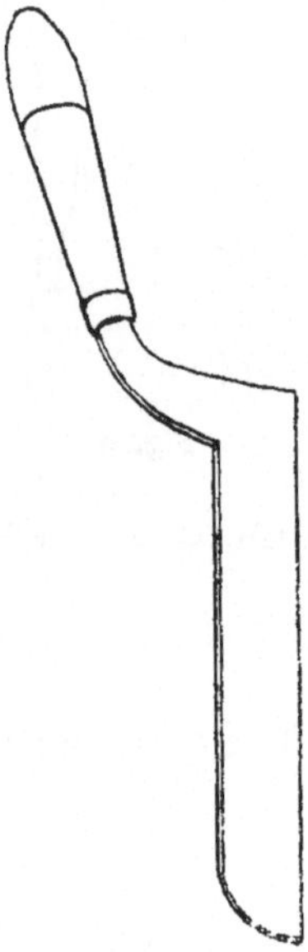

Fig. 13.

Trowels

6. There are at least two special trowels. The plasterer's laying-on trowel is the principal tool in solid plastering; the experienced plasterer using this for the greater part of his work, even to laying on stuff to cornices.

7. The gauging trowel is needed in gauging small quantities of materials, also in plastering small places in repair work (Fig. 14).

Fig. 14.

Drags

8. The steel drag, which is required for roughening hard cement when more has to be added; also for dragging off plaster when fixing fibrous plaster (Fig. 15).

Scrapers

9. Thin pieces of shaped steel less than $\frac{1}{16}''$ thick are used when planting fibrous plaster moulding, etc., for cleaning off purposes (Fig. 16).

10. Another piece of shaped steel which will be found very useful for all kinds of scraping purposes (Fig. 17).

11. This is the most useful tool for plain face fibrous plaster; it is called a piece of bust. The thinner the blade is the better, as it should bend to fit circular or plain work (Fig. 18).

Fig. 15.

Fig. 16.

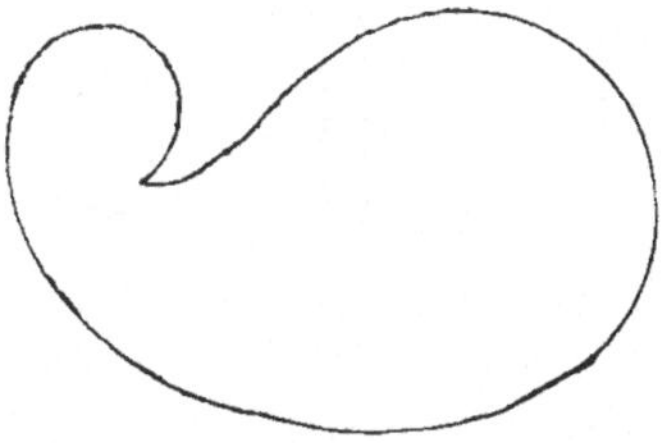

Fig. 17.

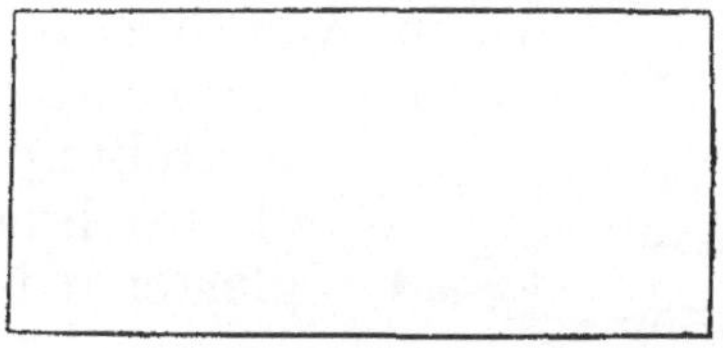

Fig. 18.

Scratch Tools

12. Various kinds of scratch tools, and some with smooth edges which are required when touching up or making good ornament in fibrous plaster work (Fig. 19).

Small Tools

13. Plasterer's small tools. Years ago these were called spoons. These are required when forming the mitres to cornices or mouldings. Various shapes can be

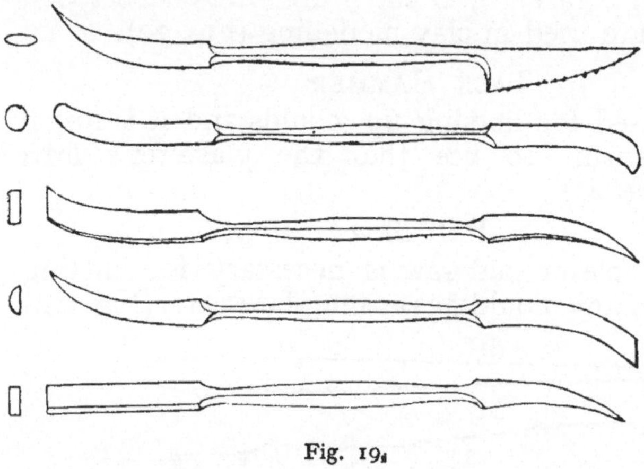

Fig. 19.

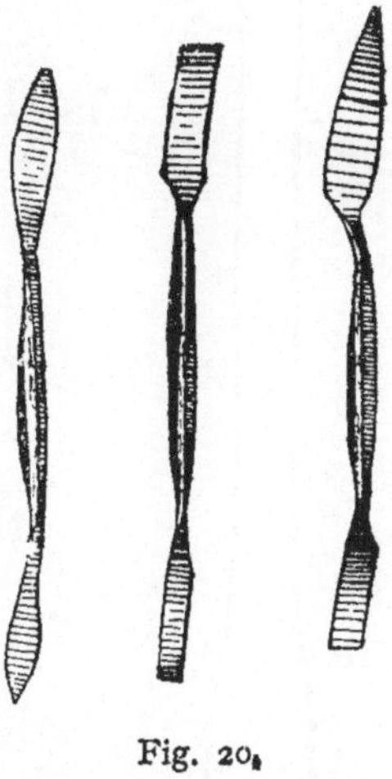

Fig. 20.

bought at tool shops. A good assortment of these tools is necessary if the plasterer works on ornament (Fig. 20).

Fig. 21.

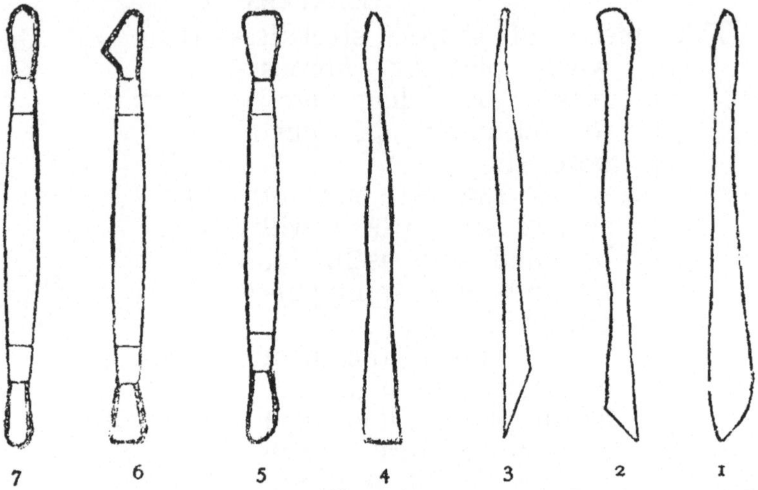

Fig. 22.

KNIFE

14. The canvas or trimming knife which is required when working in the plasterer's shop, used for cutting canvas for casting purposes. Also for trimming up and scratching the back of cast-work, such as short lengths of enrichment, and for general work when on fibrous plaster it is most useful (Fig. 21).

MODELLING TOOLS

15. 1 to 4 of these are boxwood tools, which are necessary when modelling in clay, though these are very useful when the plasterer has a small model to clay over when making a case. 5, 6 and 7 are wire-ended tools. These also are used in clay modelling (Fig. 22).

TACK HAMMER

16. Is used for horsing up moulds and it is much more convenient to use than the plasterer's lath hammer (Fig. 23).

PAD SAW

17. Keyhole or pad saw is necessary for cutting the profile of any mouldings or templates, etc. (Fig. 24).

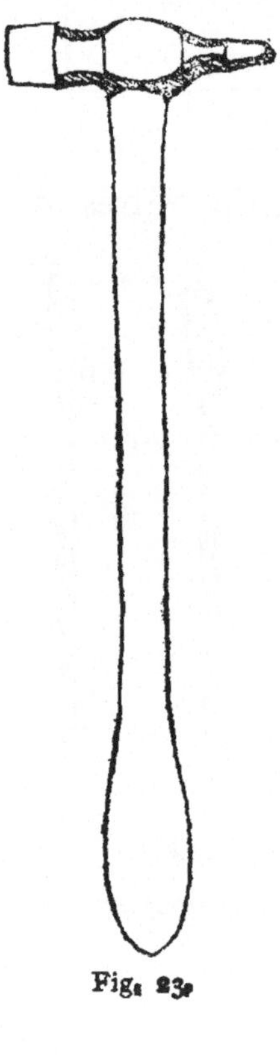

Fig. 23.

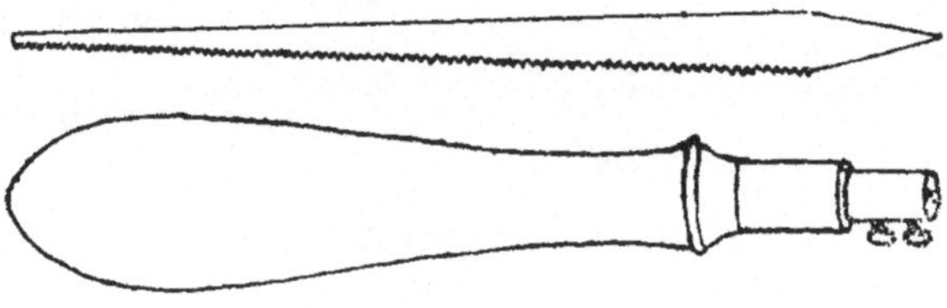

Fig. 24.

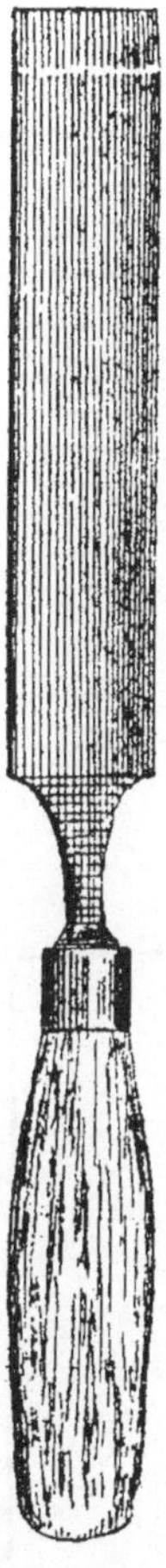

Fig. 25.

Fig. 26.

Wood Chisel and Gouge

18 and 19. These are necessary in the horsing up of running moulds, of which various sizes are required. The gouge may also be used for carving in plaster and the chisel in repair work, for trimming off the old edge of plaster to enable the plasterer to make a neat joint (Figs. 25 and 26).

Gimlet

20. Required for boring holes in the trammel rod for running ovals, also boring holes in fibrous plaster mouldings, etc., when fixing with screws (Fig. 27).

Bradawl

21. Used for punching holes in the zinc or iron for mould and shoes, and very useful to drive in temporarily to find the correct position for the screw in running arches (Fig. 28).

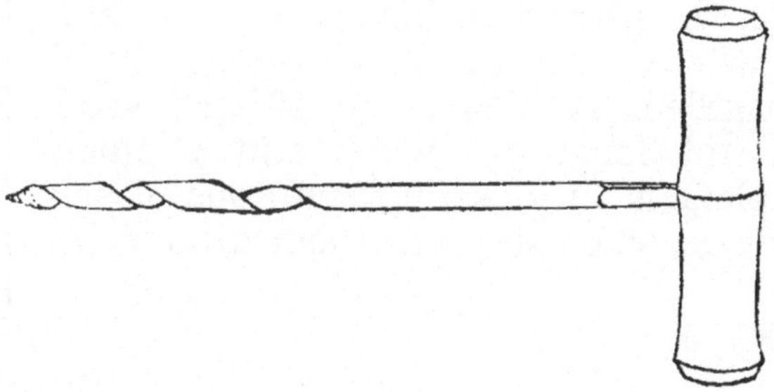

Fig. 27.

Plane

22. For smoothing up the wood for running mould, template, running rules and general use (Fig. 29).

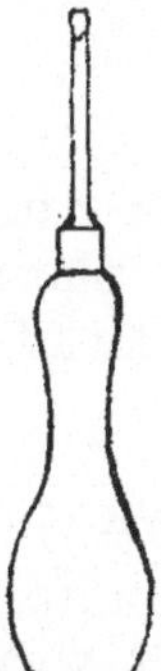

Fig. 28.

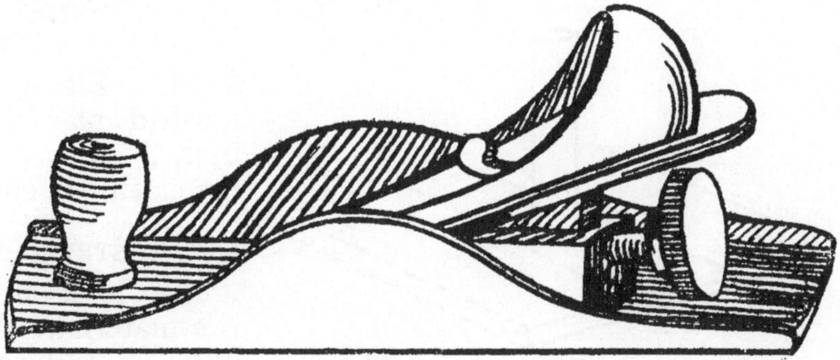

Fig. 29.

Plasterer's Planes

23. These are made in various widths, from about ½″ to 1½″. The plasterer usually makes this sort of plane himself, to suit his own taste

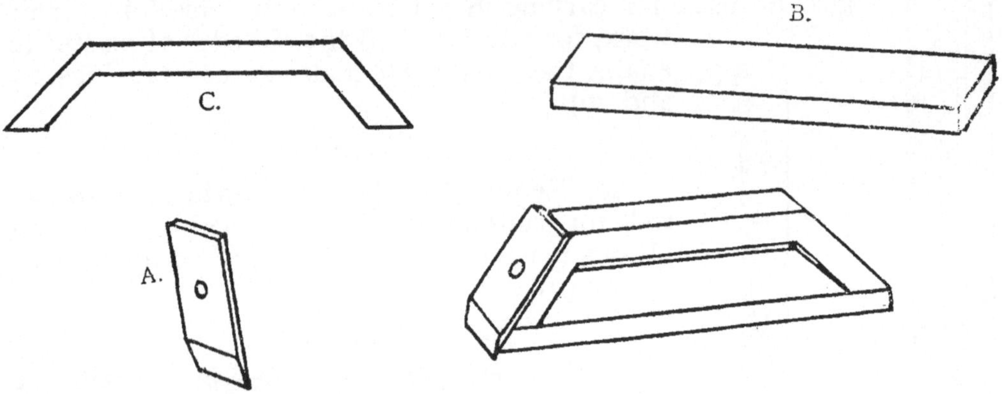

Fig. 30.

and requirements. It is made by fixing a small plane iron on the front part of a sawn handle, the plane iron having a screw hole to allow of its being fastened with a screw. The handle is fixed on a piece of hard wood, about 10″ long, 1½″ wide, and ½″ thick. This plane is very useful for fibrous plaster mouldings, when stopping them after being fixed. Fig. 30 shows the finished plane, A its plain iron, B its hard wood and C the sawn-out handle.

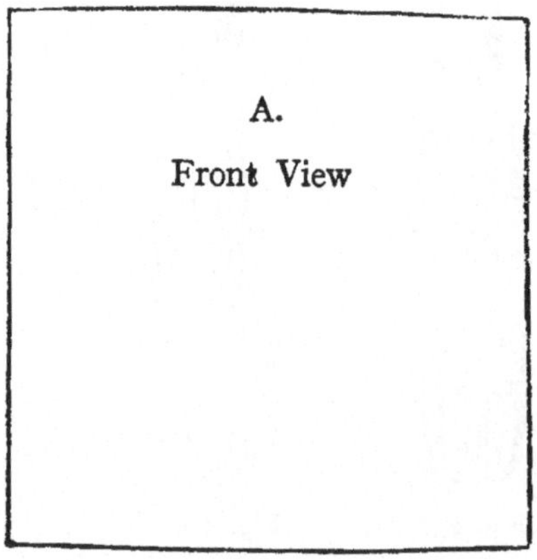

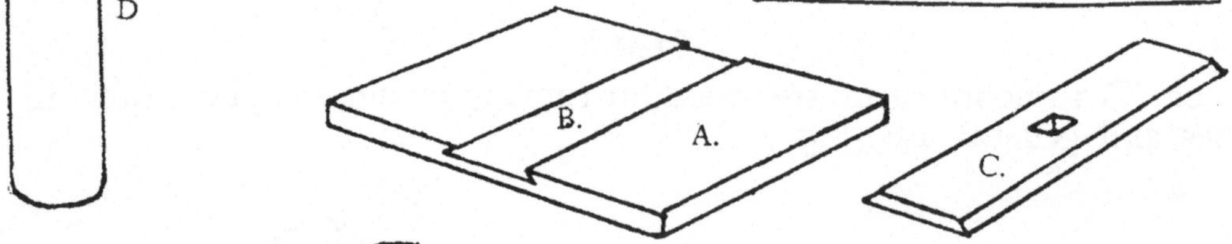

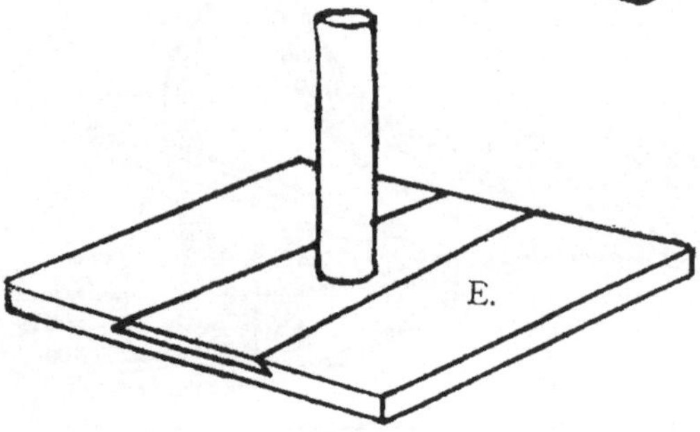

Fig. 31.

Plasterer's Hawk

24. Is used for holding the material, whatever sort of work is being executed.

SECTION OF CEILING IN THE STYLE OF GEORGE RICHARDSON, LATE EIGHTEENTH CENTURY, SHOWING THE INFLUENCE OF ROBERT ADAM.

Colours in low tones, on the flat or on slightly raised moulded composition ornaments.

Formerly this was made of 1″ pine wood, but to-day all kinds of wood are used. It is 12″ × 12″, $\frac{3}{4}$″ to 1″ thick. If a dovetailed piece for handle is used, a sinking is cut in the thickness piece as with the cross-grained float, and the handle fastened to the piece, which is 12″ in length, 3″ wide, tapered to 2″, and cut to fit in the sinking. The handle is about 2″ in diameter, 6″ long and cut to a shoulder to fit the dovetail piece. Fig. 31, A hawk face, B sinking, D dovetail, D handle with shoulder, E finished hawk. The simplest hawk to make is a piece of wood of $\frac{3}{4}$″, 12″ × 12″, and a cleat nailed or wood pegged to the face with a long screw through the centre to hold the handle.

Skimming Floats

25. These floats are used for various purposes, and are made of soft wood, pine if procurable. The panels of old doors make very good float faces, and being old they are well seasoned. The average size is about $4\frac{1}{2}$″ wide and 13″ long and $\frac{5}{8}$″ thick. Smaller sizes are made to suit the requirements of the various kinds of work. Narrow floats from $\frac{1}{2}$″ in width and upwards and hollow and bead face are very useful for mouldings in cement work. In Fig. 32 A the handle, B the finished float, which is made complete by nailing through the face with fine pins in the handle.

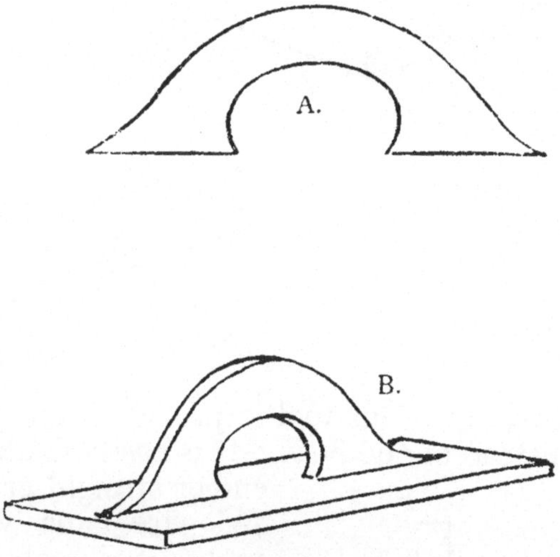

Fig. 32.

Cross-grained Floats

26. These are floats needed to rub up and down the angles either in setting or in finishing the Keene's angles or rubbing up Keene's in the finishing, or arrises in interior work or external cement work. These are made from a cross-grained piece of wood (not like a skimming float face, which is straight grained), slightly shorter than the skimming face, being about 12″ long and $4\frac{1}{2}$″ wide, 1″ to $1\frac{1}{2}$″ thick. The handle is fastened on to a dovetail piece 12″ long and $1\frac{1}{2}$″ × $1\frac{1}{4}$″, $\frac{1}{2}$″ thick, and a dovetail piece each side. A sinking is cut in the back part of the float face, and the piece with the handle attached is let in. The handle which is made from a piece of soft wood, $1\frac{1}{2}$″ × 9″ × $2\frac{1}{2}$″, planed up; part of the centre is cut out for the hand, about 4″ long in centre by $1\frac{3}{8}$″, and rounded off to suit the plasterer's hand for using. Fig. 33, A the handle, B the dovetail piece, C end section, D back face of float, E finished float.

Wire Scratcher or Comb

27. This scratcher is required when rendering walls with coarse **stuff,** or coring out mouldings in cement, or covering wire lathing. It also can be used for combing cement or other kinds of material used **in**

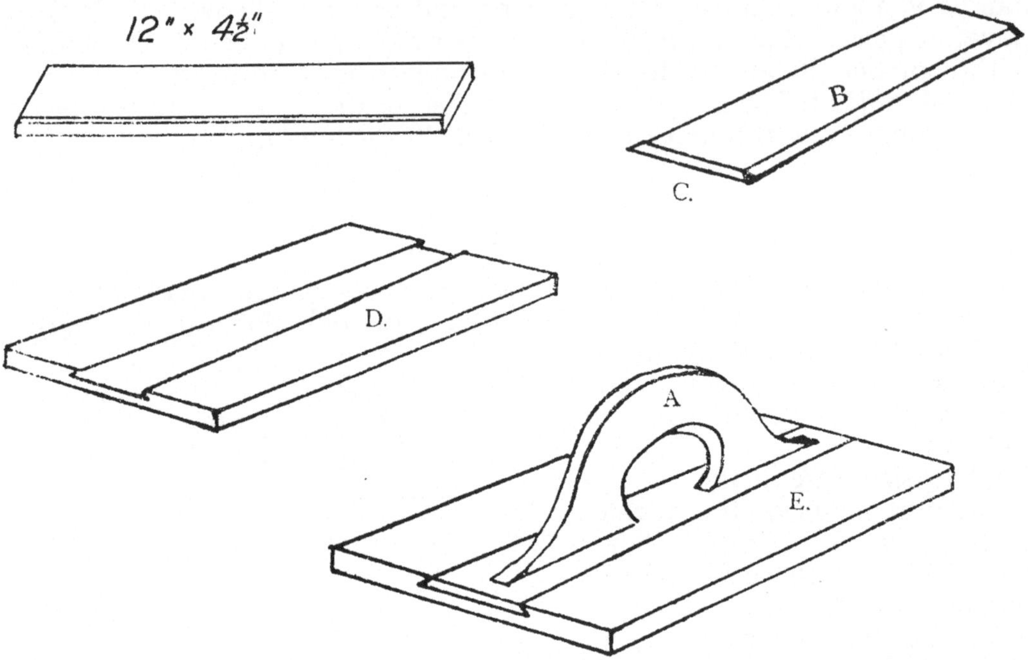

Fig. 33.

floating. This tool is preferred sometimes for the develling of floating, instead of the float. It is easily made by cutting a piece of $\frac{3}{4}$" board, the end or straight grain, about 4" or 5" long, and 4" wide, $1\frac{1}{2}$" wire nails with heads cut off, or umbrella ribs about 3" long, are driven in about $\frac{5}{8}$" apart and about $1\frac{1}{2}$" out of the wood. A short piece of 1" × 1" wood, 4" long, can be tacked on as a handle (Fig. 34).

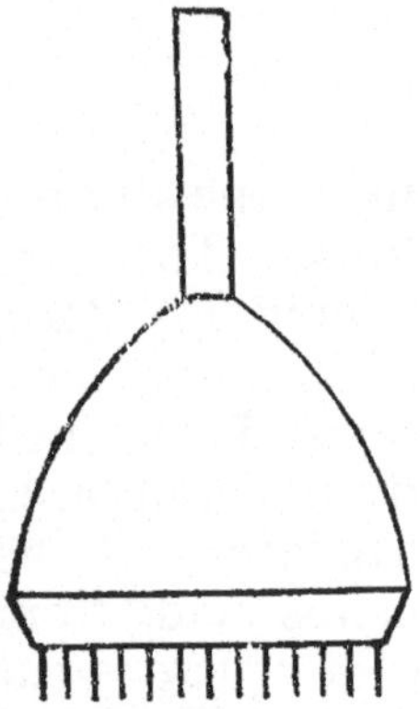

Fig. 34.

Margin Trowel

28. This is a small narrow trowel, about 2" wide and 4" in length, and is very useful for narrow margins between doors or window frames, or any other narrow places where the laying or gauging trowel is too big to fit in. Apart from small places, it can be used for other purposes, such as in repairing or for scaling off old setting coats (Fig. 35).

Fig. 35.

Angle Trowel

29. This trowel is useful for drawing up internal angles to rooms when finishing, for, being the shape of a margin trowel with the two sides turned up, it presses both sides of the finishing coat in the angle. It is made of thin steel, about 4″ long with turned up sides about 1″. Before using the trowel, the angle is rubbed up by the cross-grained float and then the trowel is used. It is very useful in cupboards and other small places, where it is awkward to use the trowel in the angle (Fig. 36).

Fig. 36.

Plasterer's Hammer

30. Also called a lath hammer, though it is used for all kinds of work, having a rough steel head for driving nails and lathing, etc., and a hatchet blade at the other end. This is very useful for chopping off old plaster work when repairing. The blade also has a small slot for extracting nails (Fig. 37).

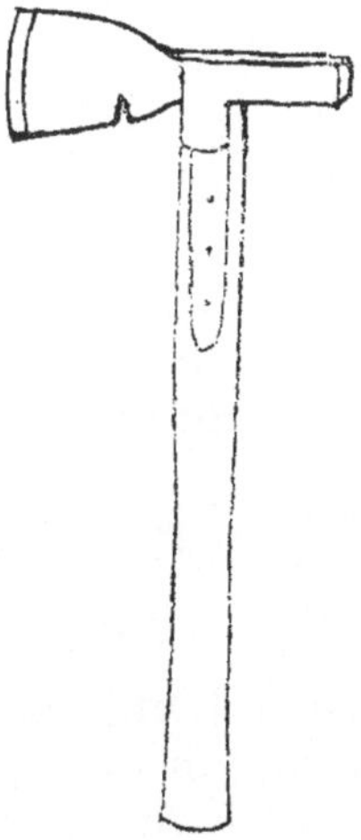
Fig. 37.

Steel Joint Rules

31. Joint rules are required when working in mitres or returns, both in interior and exterior work, and also when repairing small places in walls or ceiling. It will be found useful to keep the joint flush by drawing the joint rule over the place under repair, bearing the rule on the old existing work. These rules are made of sheet steel, in sizes from $1\frac{1}{2}$″ long up to 18″, the most useful length for use is 14″. Some plasterers have a thin piece of wood about $\frac{5}{8}$″ thick fastened on the end for holding, which gives a better grip for the hands, as Fig. 38.

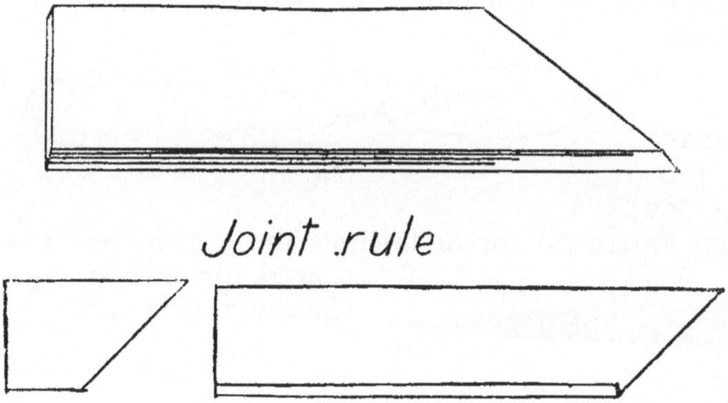

Fig. 38.

Stock or Water Brush

32. Without the water brush the

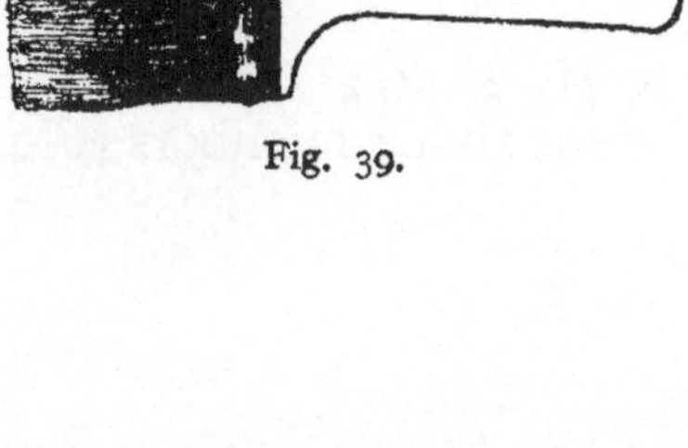
Fig. 39.

plasterer would not be able to proceed far with his work. In plastering, it is essential that the background to be plastered be well damped. When finishing the walls or ceilings the water brush is needed to assist in the trowelling in of the work, also when repairing old plaster work the places to be repaired require wetting before being filled in, and in addition for use in plastering the brush is necessary to wash the floats and trowels, etc. (Fig. 39).

TOOL BRUSHES

33. Small brushes, such as used by painters for painting sashes, are very useful to the plasterer; as when he is mitreing in interior work with Keene's or plaster, the stuff is gauged very soft in the gauge pot and put on with brush for the finishing off part with the joint rule. Small brushes of this description are also required when doing shop work, when making up models with ornaments, or preparing or oiling jelly or plaster moulds (Fig. 40).

Fig. 40.

MUSLIN BAG

34. Made with fine muslin 10″ × 8″ is a very useful size. This will be needed to sift Keene's cement or plaster for the finishing off of mitres.

FINE SIEVE

35. For exterior work a useful sieve for the plasterer on the scaffolding is made with a piece of fine perforated zinc nailed on to four pieces of ½″ × 2″ wood of a convenient size.

SNIPS OR CUTTERS

36. Snips are necessary when cutting zinc or iron profiles for moulds or strips for slippers; also when fixing expanded metal lathing the cutters are required for piecing out the metal (Fig. 41).

Fig. 41.

Fig. 42.

FILES

37. After cutting out the profiles of moulding, etc., in zinc or iron, it is necessary to have a varied assortment, such as ½ and ¼ round, three corners, ¾″ × ⅛″, ¼″ × ⅜″ flat and various others, to be able to file up the different shape members (Fig. 42).

Fig. 42.

STEEL TRIANGLES AND WOOD SQUARES

38. Steel squares are much better for use by the plasterer, as they do not buckle or crack like a wood square, being used so much in the wet material. A useful size is a 10″, and being at the angle of 45 degrees they are serviceable in marking off mitres when cutting fibrous plaster cornices for fixing. Also when cutting up short lengths of mouldings on the bench the triangle square is useful to mark off on the bench the 45 degrees for the mitre cut. The 6″ square also is a very useful size (Fig. 43).

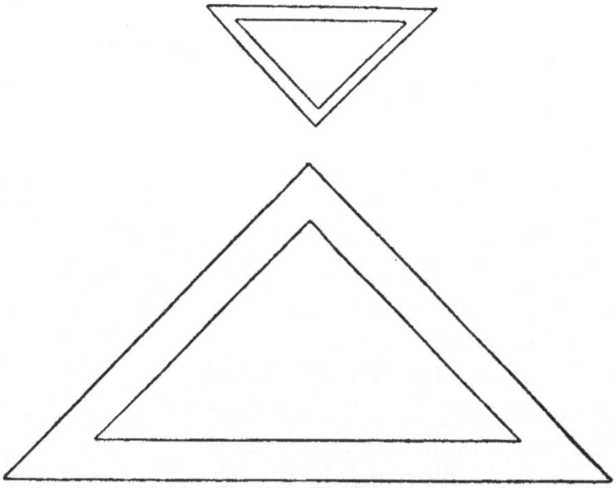

Fig. 43.

39. Wooden squares of the triangle shape are handy for filling out small breaks when floating, as being solid they save the stuff from falling on the floor when being drawn up (Fig. 44).

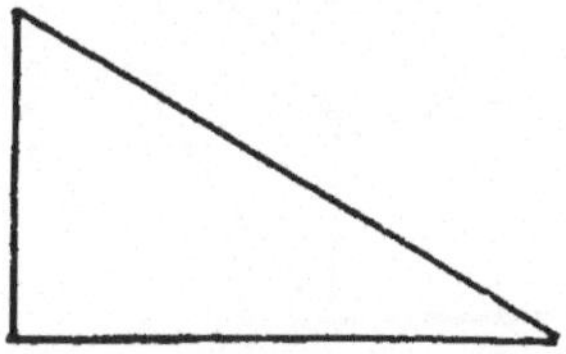

Fig. 44.

L-SHAPED OR MASON'S SQUARE

40. This steel square is about 12″ long and $1\frac{1}{4}$″ wide. It is useful when working in cement returns, for squaring the return members from the front face (Fig. 45).

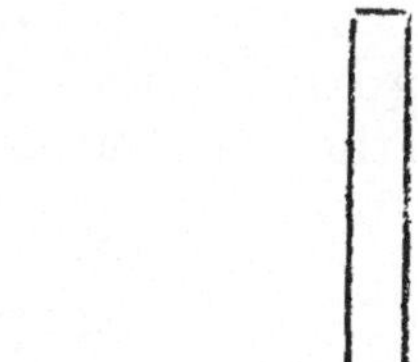

Fig. 45.

SAWS

41. The plasterer's saw is of the same kind a the carpenter's, though it has rougher work at times It is usually about 18″ long, so it can be carried in his box or bag. It is necessary to have a saw, as the plasterer usually horses up his own moulds. Then when he is running, there are the running rules to cut to fit. If he is fixing fibrous plaster moulding, the mitres have to be cut (Fig. 46).

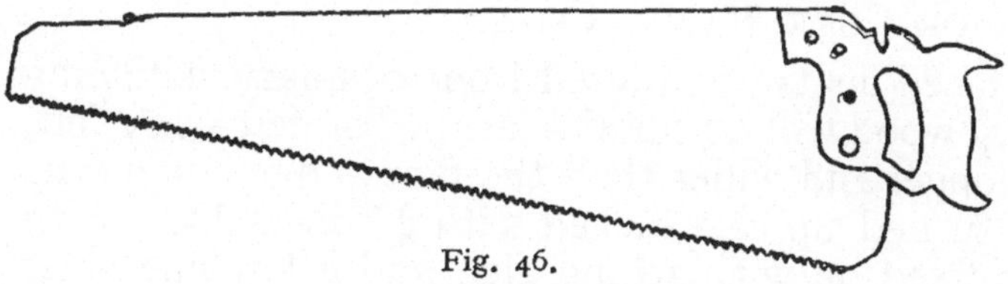

Fig. 46.

WOODEN TOOL STICK

42. The tool stick

is cut out of an ordinary piece of $\frac{3}{4}''$ wood, is $3''$ to $4''$ long and shaped to a handle end, the various tools being rubbed over on the spot board with the addition of Portland cement and water (Fig. 47).

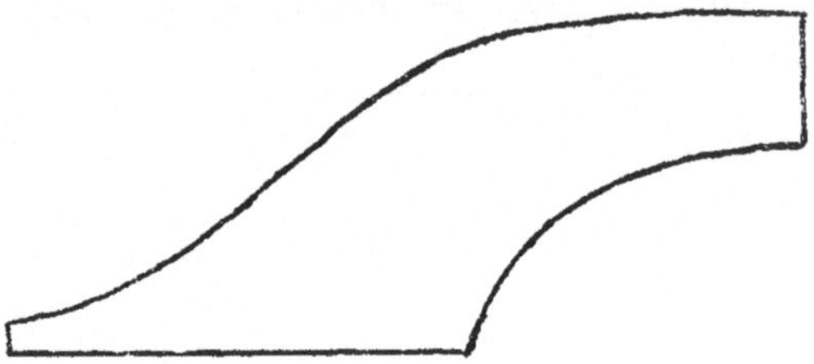

Fig. 47.

Plasterer's Tool Box

43. The box, for convenience of carrying from job to job, is most suitable about $2' \times 14'' \times 9''$. This is capable of carrying the hawk if necessary (though, as a rule, the plasterer tucks the hawk and floats up in his apron). A tray can be fitted up at one side to put in small tools, such as pincers, gimlets, bradawls, etc. For small tools it is best to have a tin box, as it keeps them together, and when not in use they can be greased, as they very quickly rust when not in use. A small chain or piece of cord is necessary to prevent the lid falling back, and a handle fastened on the front for carrying (Fig. 48).

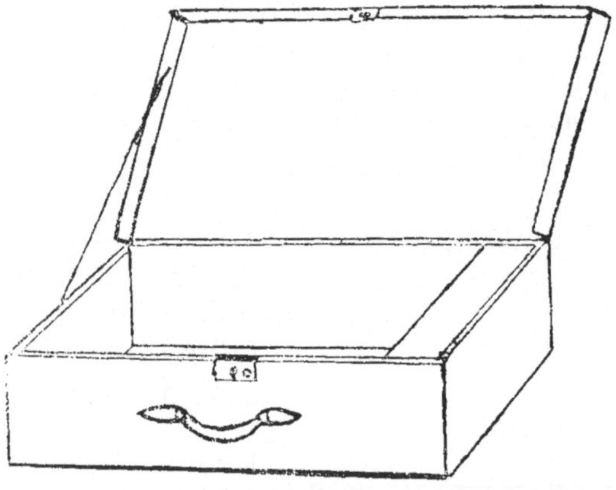

Fig. 48.

Splash Brushes
(For Shop Work)

44. Splash brushes or filling-in brushes are used when working on shop work, and chiefly when filling in plaster or jelly moulds. They are made by Messrs. Burton, Balls Pond Road, London, who have supplied these brushes for years past. This is a special make of brush mounted up in one knot, bound with copper wire round a flat handle about $8''$ long and $2'' \times \frac{1}{2}''$ thick, the bristles being a coarse stubby hair about $4''$ long (Fig. 49).

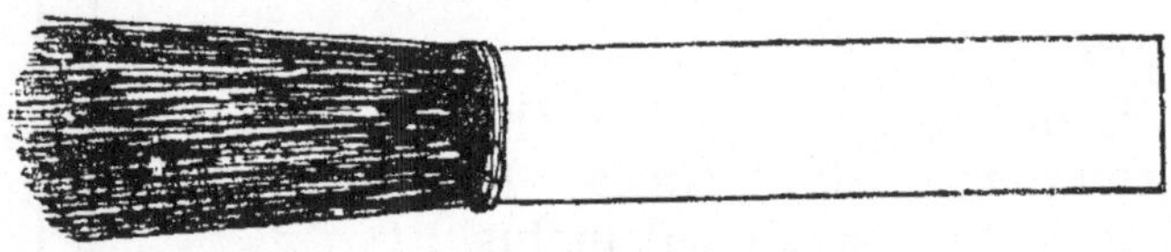

Fig. 49.

Felt or Cork Faced Floats

45. Felt faced or cork faced floats are altered from ordinary skimming floats, though any ordinary wood will do for the face. The felt—any felt, such as an old hat—is cut longer and wider than the face and is laid on the wood, the ends and sides turned up and tacked with $\frac{3}{8}''$ tin-tacks. With the cork face, the cork is fixed on with strong glue and a few fine brass

pins driven in the face. Care must be taken that this float is not left in the water pail, otherwise it will soften the glue. These kinds of float are useful in stucco or "stuke" finishing, also in cement work. The cork varies in thickness from ½″ upwards. In some cases it is 1½″ thick, and the handle fastened on to the cork by brass screws.

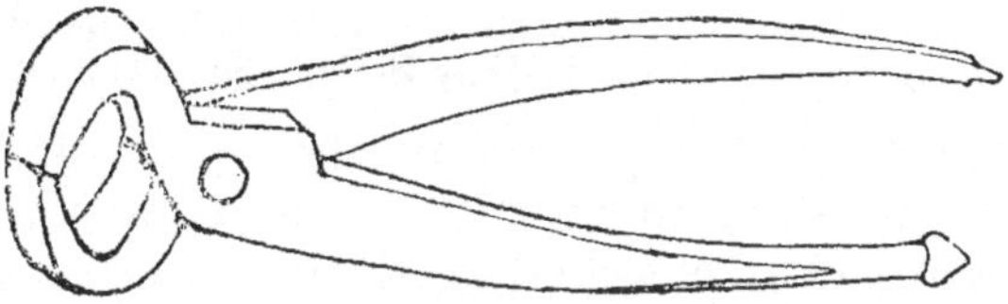

Fig. 50.

Pincers

46. The pincers are needed for drawing nails from running rules after completing running (as the nails have to be driven in the rules so as not to damage the plasterer's hands when passing the mould along the rules), and are useful for many other purposes (Fig. 50).

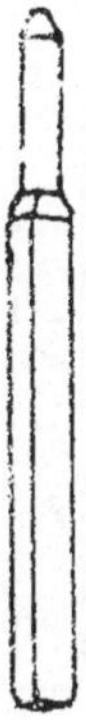

Fig. 51.

Steel Punch

47. This is required when nailing up fibrous plaster, as the nails require punching in clear of the surface, and by using the punch only the heads of the nails require stopping (Fig. 51).

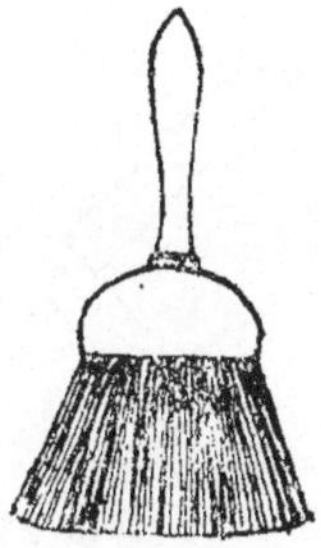

Fig. 52.

Dusting Brush

48. Very useful when working on ornament for cleaning off, or when on repair work for dusting out cracks, etc., before wetting (Fig. 52).

CHAPTER XII

SCAFFOLDING AND PLANT

Stands—Spot-boards—Split Heads—Scaffold Plaster Box—Scoop for Gauging—Larry—Plasterer's Hair Hook—Hair Beater—Sieves—Washed Sand Sieve—Tubs and Tanks—Lime Stirrer—Feather-edges—Parallel Rules—Running Rules—Trestles—Gauge Pot—Iron Pail—Wooden Steps—Bench Brooms—Broom—Double-handed Bowls—Baths—Screen—Clay Board for Shop Work—Funnels—Jelly Boiler—Rib Templates—Wheel Barrow.

FORMERLY the plasterer when at work had his hawk kept replenished with plaster, mortar or cement by the hawk-boy. Now the method is somewhat changed, and the craftsman works with a stand and spot-board.

STANDS

Folding stands (or spots, as they are sometimes called) or frames each have two legs 3′ long made with 3″ × 1″ battens, a top piece 1′ 6″ long and a lower piece 2′ long, nailed together. These two frames are bolted together, as in Fig. 53, with a nut and screw. Thus the frame Fig. 53A can open out as in Fig. 53. Fig. 54 shows the frame. A small link chain is fastened at each side to keep the stand to the required height. Most stands are made either with battens 2″ × 1″ or with battens for braces and stouter stuff for the legs. A light stand of two frames can be made with four pieces of batten 1′ 10″ long, with a cross brace as shown in Fig. 54, and four pieces 2′ 3″ long. These are nailed together to form the two frames, and these are connected by

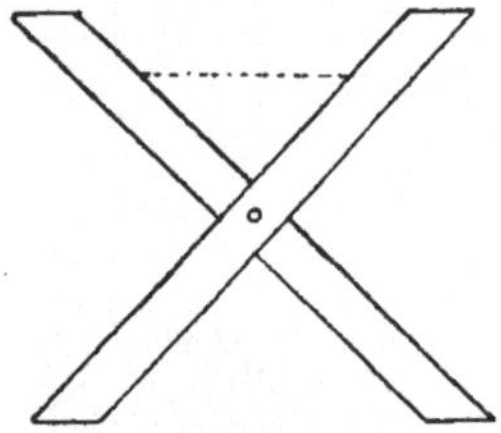

Fig. 53.

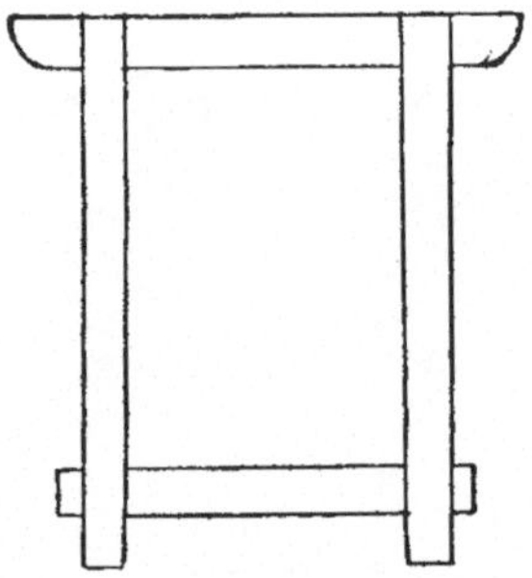

Fig. 53A.

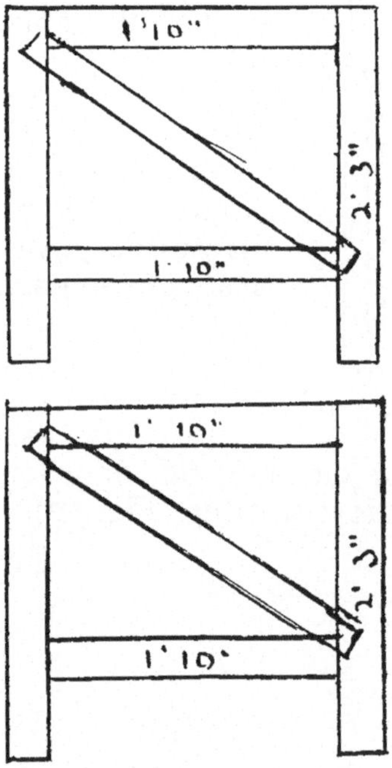

Fig. 54.

means of four pieces 2′ long and the braces; the completed stand being shown in Fig. 55.

Spot-boards

The board from which the plasterer takes his stuff, or on which he gauges it, is called a spot-board. Probably the most useful size is 3′ × 3′, though in big rooms, or on large scaffolds, where heavy cornices have to be run, the gauging boards would be of a larger size, because the materials for running must be gauged soft. In such cases a board 4′ square will be found more useful.

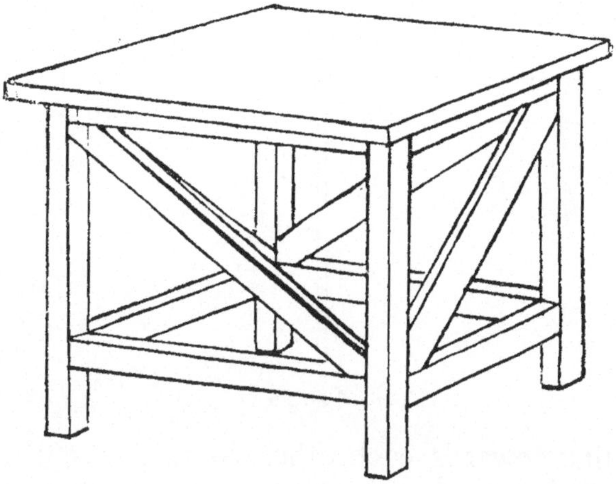

Fig. 55.

These gauging boards are made of any kind of $\frac{3}{4}''$ boarding, but floor boards make much the best, especially if required for gauging for running mouldings, etc. The best method to adopt with the spot-boards is to have rough ones for rendering and floating, and planed faced boards for finishing. When making these boards several lengths are cut to whatever size is required, with two pieces a little shorter for cleats underneath, and a cross one to stiffen it and to keep it straight. These boards are nailed through the face into the cleats and then clinched over to grip. Figs. 56 and 57 show the back and front of a finished board.

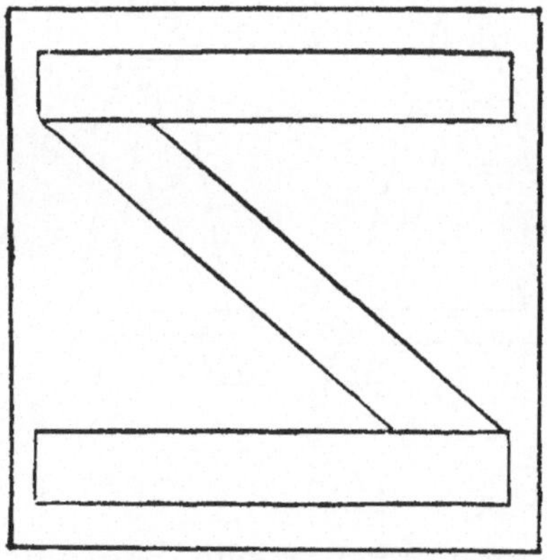

Fig. 56.

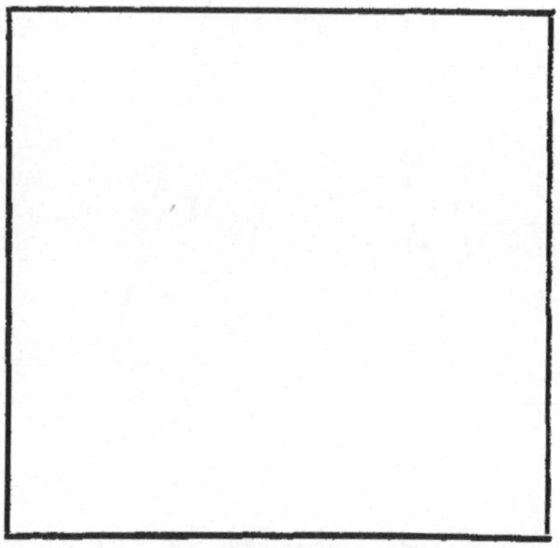

Fig. 57.

Split Heads

In the making of plasterers' scaffolds, a very different method is adopted to that for bricklayers, carpenters or other trades. When plastering interior work, walls and ceilings, in ordinary-sized rooms the scaffolding needed is of very light construction, so the split head method is very useful, up to about a 12′ pitch. After that a pole scaffold would be necessary. I have used split heads up to a length of 6′, allowing for a 5′ 10″ scaffold, which is considered the average height for a plasterer's scaffold. While these split heads are excellent, we also have iron uprights with a circular iron base, the uprights having clips with iron pegs to take a scaffold board at whatever height is required. The boards are laid on this transom to form the scaffold. Wood uprights and bases are also in use.

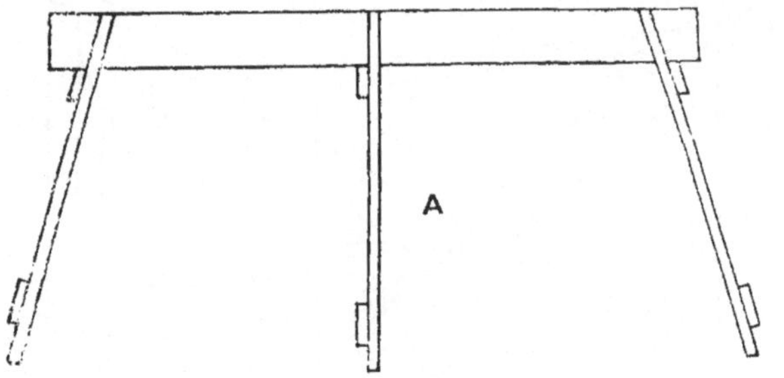

Fig. 58.

To return to the split heads, these are much the best, as they can be made from any rough boards from 18″ upwards, and are easily moved from place to place. When using these for scaffolds, the heads should incline towards each other, as in Fig. 58, which shows the two heads in position, and also the transom to receive the boards; the transoms being wedged in the split heads with wood wedges. Fig. 59 shows the two legs; B the two legs in position, C cleats nailed across the two legs to complete the split head. When the transom has long bearing, and it is necessary for a centre head, a shorter one is used, as shown at A, Fig. 58.

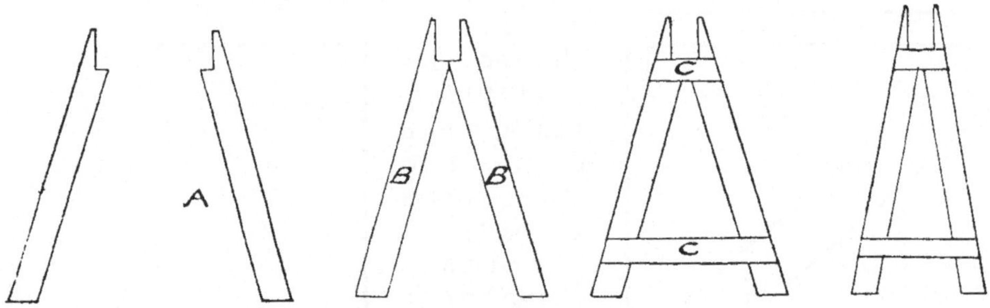

Fig. 59.

Scaffold Plaster Box

When plaster is required for use on the scaffold, a rough box will be found useful, though of late the practice of using these seems to be dying out. Now the plaster sack is turned upside down, and the bottom rolled up, then turned up again and the mouth of the sack turned over.

A rough box can be made and four legs nailed on it. In my opinion this is still the best method of obtaining the plaster for use, as the box can be put close up to the gauging board and a cleat nailed on the side, which is very useful to hold the gauging trowels in. Another advantage with the box is that it saves stooping to obtain the plaster when gauging. Fig. 60 shows the box and the cleat at the side for tools, the legs being made the same length as the plasterer's stand; Fig. 61 shows the finished box.

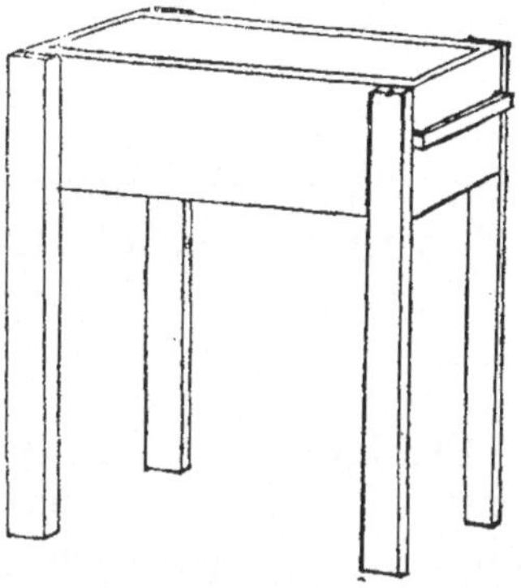

Fig. 61.

Scoop for Gauging

The scoop is required when gauging any kind of material. It is easily made by cutting a piece of wood in circular shape, and a piece of zinc slightly larger and bending it round and nailing it to a piece of wood. If required, a piece of wood, 1″ × 1″, can be nailed on the top edge to use as a handle. Fig. 62 shows the finished scoop.

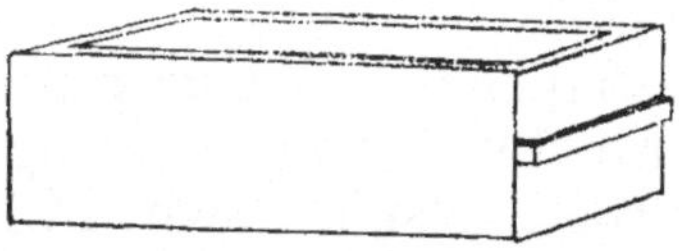

Fig. 60.

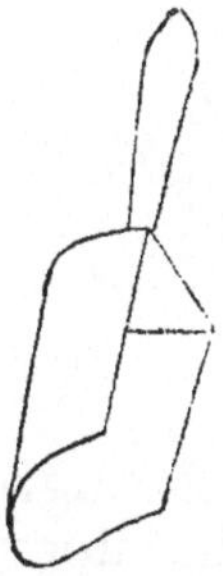

Fig. 62.

Larry

This tool is used by the plasterer's labourer for various purposes, either for turning out the coarse stuff, such as lime and hair, or gauging coarse stuff, as it mixes better and more easily than the shovel. It is fixed on a handle of lance wood 5′ or 6′ long, which allows it to bend when in use. Fig. 63 shows the iron larry without the handle.

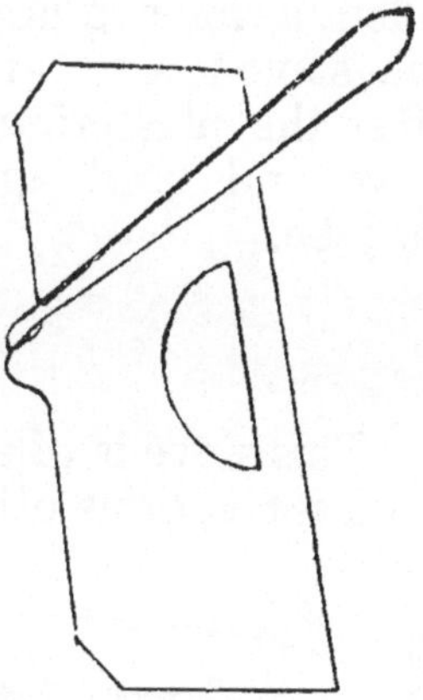

Fig. 63.

Plasterer's Hair Hook

The hair hook is required by the plasterer's labourer when he is mixing up the coarse stuff (called turning out), as the hair must be well beaten and soaked before being put in the lime and sand. When the coarse stuff is being made up, the hair is thrown in amongst it and larried about with the larry and hook. Fig. 64 shows the hook without the handle.

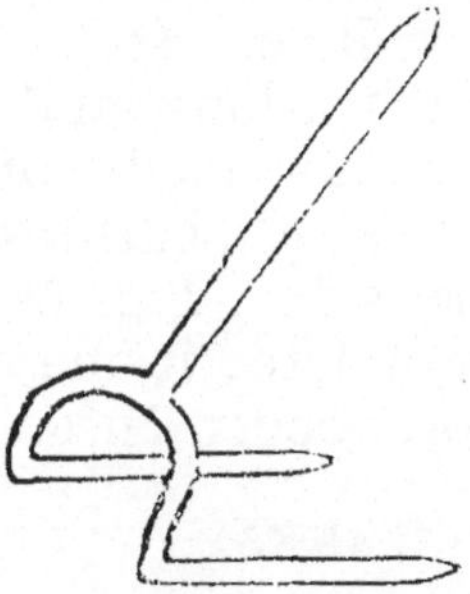

Fig. 64.

Hair Beater

The hair beater is used for beating the hair for various purposes. It is made with two long laths, about 2′ 6″ long, nailed flatways together at one end with a piece of lath nailed between the two, about 6″ or 7″ from the handle end. This makes the remaining part something like a clapper. When beating the hair, two of these are required, one being held in each hand. Fig. 65 shows the two laths nailed together, the ends being tied with a piece of string.

Fig. 65.

Sieves

Plasterers require sieves of different sizes and shapes for various purposes, apart from ordinary sifting operations. For instance, when beating hair in small quantities, a sieve of $\frac{1}{4}$″ mesh is turned upside down and the hair laid on top and then beaten with the lath beaters. By beating the hair on a sieve or screen the dirt and the old lime dust in which the hair is pickled falls through. After the hair is beaten, it needs soaking in a pail before using. Fig. 66 shows a strong sieve suitable for this kind of work. A very useful sieve can be made with a piece of expanded metal lathing nailed on four pieces of 2″ × $\frac{3}{4}$″ batten, which is quite suitable for beating hair on. The $\frac{1}{4}$″ sieve is a useful size, as when a finer quantity of sand is required it can be sifted through the sieve; also any material which is collected up from the floor after the plastering has been completed can be passed through the sieve and used again by adding more lime or cement as the case may be.

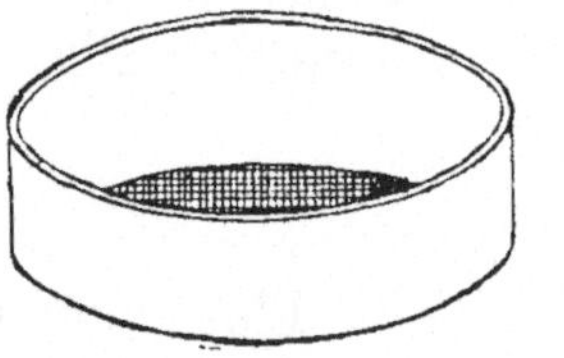

Fig. 66.

Washed-sand Sieve

This sieve is of a fine mesh. It is called a washed-sand sieve, but it is used for various other purposes, such as sifting cement or lime. It has two handles for use in washing sand. The handles are usually made of three pieces of 2″ × $\frac{1}{2}$″ battens and nailed on the side. Fig. 67 shows the sieve complete; it may have iron handles or wooden ones.

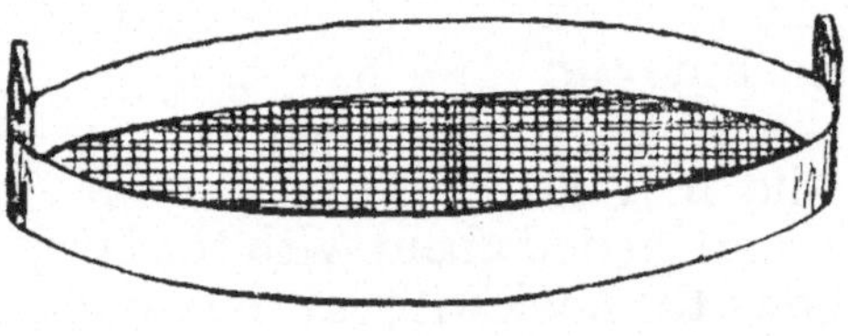

Fig. 67.

Tubs and Tanks

Tubs and tanks are necessary for the plasterer; the tubs being used for running lime and the tanks for holding water. The tub for running lime is usually a large wine cask cut in half, as a large tub is needed for this purpose. The smaller tubs are used for washing sand or holding water, and can be made by cutting a paraffin or oil cask in half. A tub of this size is useful to have on a large scaffold to contain water. Fig. 68.

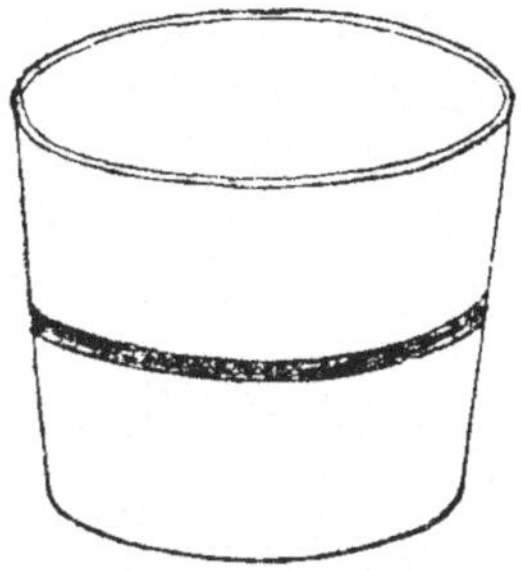

Fig. 68.

Lime Stirrer

The stirrer for use in running lime is a piece of 7″ × 1″ floor board, about 3′ 6″ long, shaped off to a bat end. When the lime is put in the tank and begins to boil, it is stirred up with the stirrers. Fig. 69.

Fig. 69.

Feather-edges

These are used for many purposes, though they are mostly required when setting or finishing. Also in external work for finishing reveals or laying in the finishing; while in interiors the feather-edge is necessary for the wiping out of the internal angles or laying in Keene's or setting stuff finish; also for laying in when floating. They are made in various lengths up to 6′, and are about 4½″ wide, and chamfered down to a fine edge one side

Parallel Rules

Though these are made any length, 10′ to 12′ is a useful size. These rules are mostly used for levelling purposes, either for levelling a line round the room or for levelling in dots when screeding or laying floors.

Running Rules

Running rules for moulding or cornices are of any length and 2″ wide by ½″ thick, planed up. They are also used for tacking up to angles when floating or finishing.

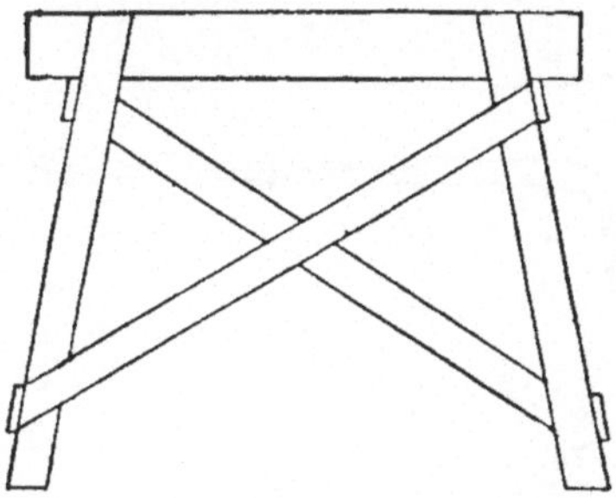

Fig. 70.

Trestles

Trestles are useful for forming a plasterer's scaffold, as in some cases the walls are required to be floated up to the ceilings before the scaffold for the ceiling is to be put up. Sometimes, also, the scaffold is taken down and the walls are finished from top to bottom without a joint. There again the trestles are needed. A trestle is made, like the split head, from about 3′ upward to whatever height is required, and wide enough to receive four scaffold boards; the legs and head being made of 3″ × 2″, and the cross braces ¾″, stuff. Fig. 70. Fig. 71 the end sections.

Gauge Pot

The gauge pot is greatly in request when soaking walls or when gauging material, for dipping the water out of the pail. It is also useful for measuring purposes; for instance, the proportion of cement and sand. The most common gauge pot used in the past was the quart beer-can, but any sort of preserved fruit tin answers the purpose.

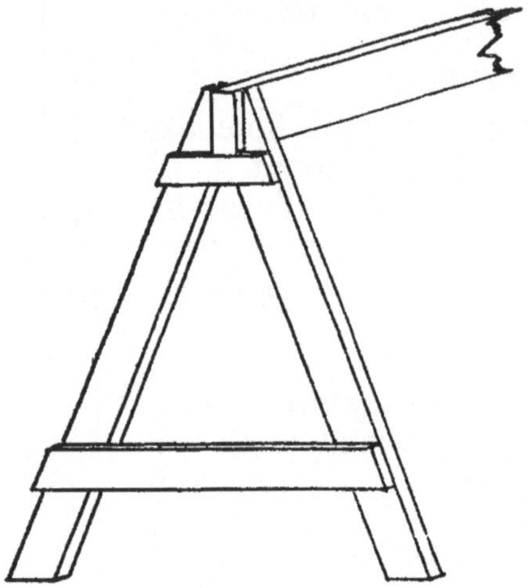

Fig. 71.

Iron Pail

A pail is needed when gauging materials and also for washing tools. The usual size pail contains about three gallons and requires to be strongly riveted at the sides. Fig. 72.

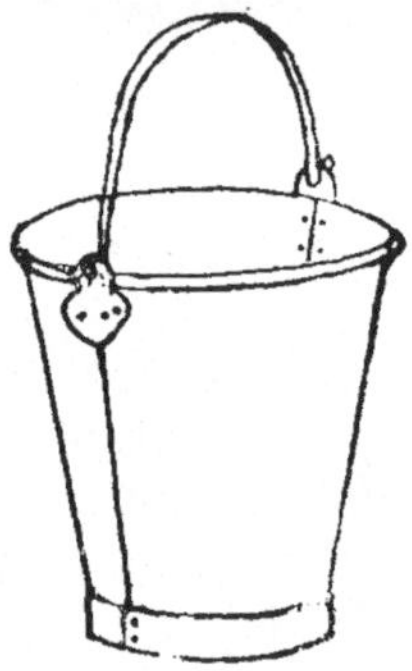

Fig. 72.

Wooden Steps

Steps are useful in repair work, as most of this is required on walls and ceilings in small places. Steps are also necessary to mount scaffolds when putting in dots. Fig. 73.

Fig. 73.

Bench Brooms

Bench brooms or hand brushes are made of stiff fibres, and are used in shop work when coring out mouldings and roughing them for running off. They also serve for brushing off the surplus material caused by the roughing of the core and roughing when running cement cornices. Fig. 74.

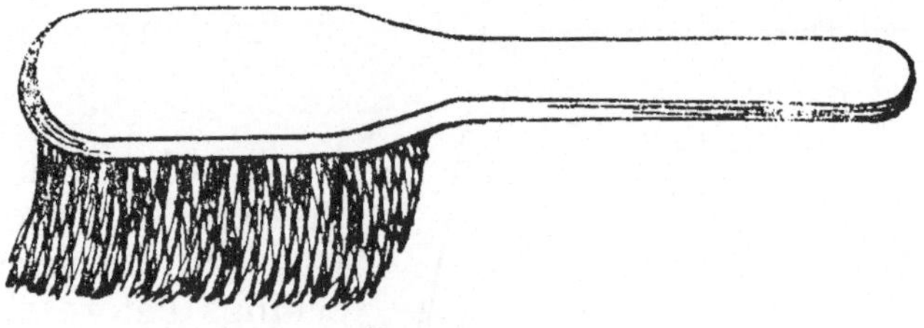

Fig. 74.

Broom

The broom is used for a great many things by the plasterer, apart from sweeping the scaffold or floor; for instance, for roughing the surface when rendering walls in cement, which is done by drawing it over. In plastering old walls it is needed for brushing the walls before soaking.

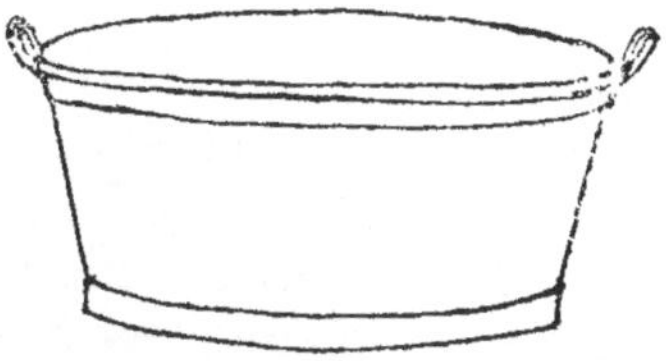

Fig. 75.

Double-handed Bowls

Double-handed bowls are used a great deal in shop work, and are obtainable in various sizes. As the quantity of material varies, the small bowl is used in preference to the pail or large bath, to economise in material.

Baths

When gauging for large mouldings, either for running or casting in fibrous plaster, it is necessary to have a large bath or iron tub. It is better to run off mouldings in the one gauging if possible, and this also applies when filling in moulds. Fig. 75.

Screen

Screens are required for sifting lime after slaking. These screens should have a stout wire bottom and sides. They are propped up at the head by a piece of wood 4″ × 2″, and the dry material of any sort thrown against it. This is a much quicker method than having to riddle or pass it through a sieve. Fig. 76.

Fig. 76.

Clay Board for Shop Work

When working on a model which requires a case for moulding, clay is used, and to regulate the thickness a clay board is needed. This is a piece of board about 11″ × 1′ 6″, having a piece of $\frac{5}{8}$″ fillet tacked on each side. After the clay is well beaten up, it is cut off in slices with a piece of wire, laid on the board, flattened down with the hand, and, when filled out to the fillet, it is dragged off with a short piece of rule.

Funnels

In the pouring of a jelly mould it is necessary to use funnels, which are fastened to the holes in the case with plaster. Funnels are made in various sizes to suit the models to be poured. These are made of zinc or block tin, about $1\frac{1}{2}''$ at the bottom to $4''$ upwards across the top. The average length is about $12''$. Fig. 77.

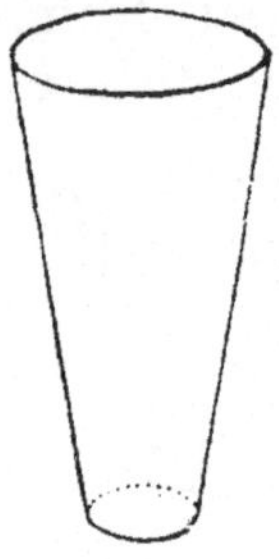

Fig. 77.

Jelly Boiler

This is a special kind of boiler for heating jelly used for moulds. It is a portable boiler jacketed; the jacket being filled with water, the inner lining is the container with jelly. As a temporary measure if only a small quantity of jelly is required, two galvanised containers, such as sanitary dust bins, can be utilised, the larger as a water boiler and the smaller one as the container. A small gas-ring can be used as the heater.

Rib Templates

Wood templates, or ribs of various radius, are kept in stock in shops, and are used to cast ribs for the setting up of drums for the running of circular mouldings, and for striking out the chord line on the bench for running purposes. The ribs are usually in lengths about 10′ long. Some fibrous shops have a large assortment of varying radius as Fig. 78.

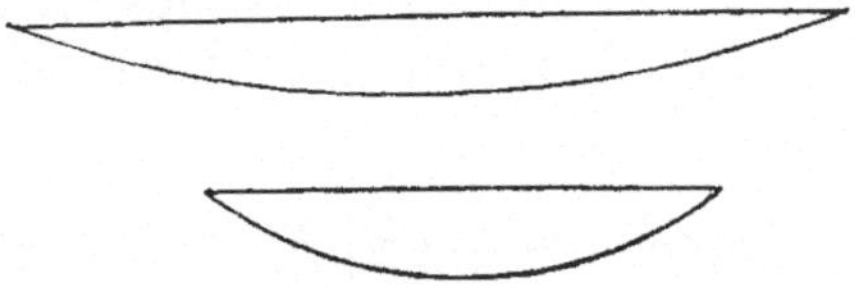

Fig. 78.

Wheel Barrow

A very necessary item for conveying materials in and about the various parts of the job.

CHAPTER XIII

INTERIOR PLASTERING

Fixing Laths—Rendering and Floating Walls—Plumbing Walls for Dots—Forming the Screeds—Floating—Three-coat Work for Ceilings—Levelling Dots for Ceilings—Setting Coat—Skimming—Trowelling—Alternative to Skimming—Finishing of Angles—Breaks to Chimney Breasts—Squaring in for Floating of Piers—Two-coat or Float and Set Work—Finishing with Patent Plasters—Render and Finish—Dinging—Parging—One-coat Base Work—Screeding—Patent Plastered Walls and Ceilings—Shallow Partitions—One-coat Lime-cement Plastering.

THAT part of interior plastering dealing with plain ceilings and walls comprises one-coat work, two-coat work and three-coat work. One-coat work is called a skimming, or setting coat, and sometimes dinging. Two-coat is known as render and set, or float and set. Three-coat work is known as render, float and set. These three processes will be described and followed on with an account of skimming, setting, dinging, coloured plasters and cement mortars, also the method of plumbing walls, etc., with gauges, and the mode of procedure for levelling in dots for ceiling screeds preparatory to floating.

For the plastering of plain ceilings a knowledge is required of the materials, so as to be certain which are best suited for the class of work in hand, whether it be lath work on concrete, plaster on Kliene tiles, or on reinforced work, such as hy-rib, expanded metal, or Briclath, or plaster slabs.

FIXING LATHS

For lath-work ceilings the ceiling joists to receive the laths are fixed to 1′ 3″ centres. That is usually the standard space apart to suit the length of laths. As both saw and render laths are made in lengths of from 2′ 6″ up to 4′, it will be seen that the ceiling joist being 1′ apart enables the laths to meet without being cut, with the exception, of course, where the finishing off parts end against the wall. Render laths are found in three varieties—the single lath, lath and half and the double, though the last has not been in use of late years. The render lath is obtained from fir props or logs, about 6′ long and 9″ in diameter, and are cut to the required length and split through with a special knife made for that purpose. This wood is easily rendered, being of a very straight grain and almost free from knots. Sawn laths are made the same length as the render lath, but of another kind of wood, something like a white deal. These are made about $1\frac{1}{8}$″ wide, and the single lath about a bare $\frac{1}{8}$″ thick, and the lath and half about a $\frac{1}{4}$″. These are cut by machine. An ordinary bundle contains roughly about 400′ run, and will cover an area of about $4\frac{1}{2}$ yards super, and require about $1\frac{1}{4}$ lb. of nails. The laths are nailed about $\frac{1}{4}''$ to $\frac{3}{8}''$ apart, the distance depending on the sort of material to be used in plastering. If the space is increased, the bundle would naturally cover more than the $4\frac{1}{2}$ yards; but the above gives a fair working average.

When fixing laths, each lath is held up to the centre of the joist at one end, and the nail is driven in. This is repeated at the next joist, and so

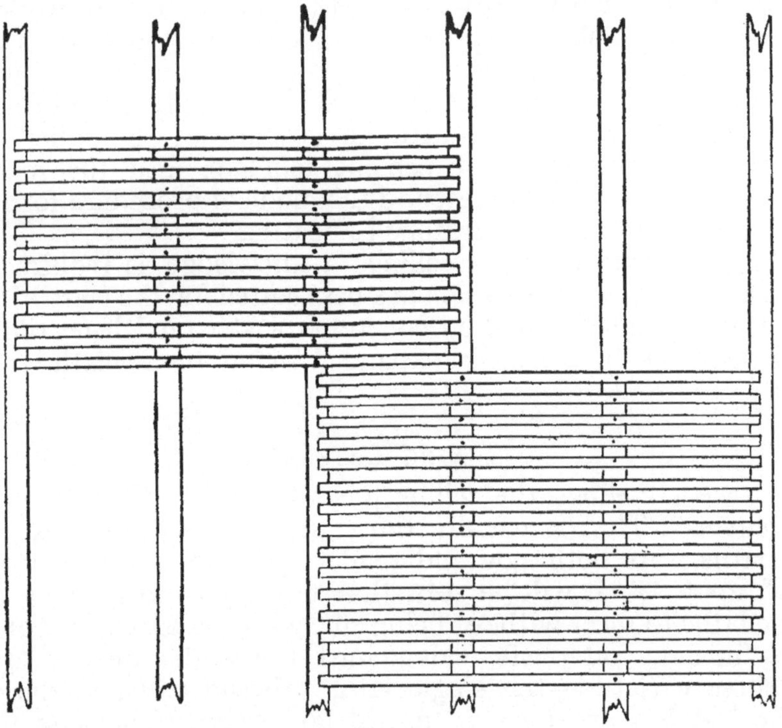

Fig. 79.

on to the end of the length of the lath. By starting at the end of the lath, it is possible to keep the lath straight. With practice it will be found that when putting the lath to the joist the tip of the forefinger will give a fair space between the lath when nailing, the space being required as a key for the plaster. One very important point in lathing is to break joint as shown in Fig. 79, which should be about 3′ square. The reason for so doing is that any vibration in the joist of the ceiling tends to crack the plaster work in a direct line, but by breaking joint as described the stresses are dispersed. Moreover, it is considered that by breaking the joint a much stronger job results. But joints should be carried out as in Fig. 80, otherwise there will be three thicknesses of laths to cover with material, and unless there is sufficient material over the laths, there would be a tendency for the plaster work to crack. In lap joints (when the joints lap over instead of butting) the ends are allowed to lap to save the time expended on cutting, while one nail fixes the two laths at the lap. This method is adopted in cheap jobs, such as speculative buildings. For lathing, the best nail to use is the 1″ wire, though with the single lath (which in a render lath is $\frac{1}{8}$″ to $\frac{3}{16}$″, or in a sawn about $\frac{1}{8}$″), $\frac{3}{4}$″ nails are quite long enough. This makes a difference to the amount of nails required per bundle, as 1 lb. would be sufficient for the fixing, while with the render lath the thickness of the lath and half would average $\frac{3}{16}$″ to a $\frac{1}{4}$″ and in the double from a $\frac{1}{4}$″ to $\frac{3}{8}$″.

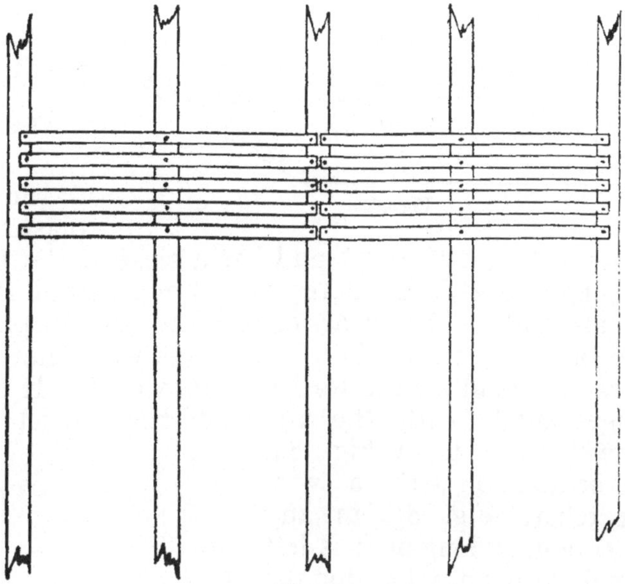

Fig. 80.

Old methods are often very interesting. A few years ago when repairing some ancient buildings in Holborn, London, I came across oak laths about 1$\frac{3}{4}$″ wide and over $\frac{3}{8}$″ thick, nailed with large clout-headed nails, while in the country I have found rushes instead of laths behind the plaster ceilings. This last method of fixing is shown in Fig. 81, where rushes held up at both ends (B), the laths nailed to the joists (A). The rushes being laid on the remaining part of the lath were then nailed up.

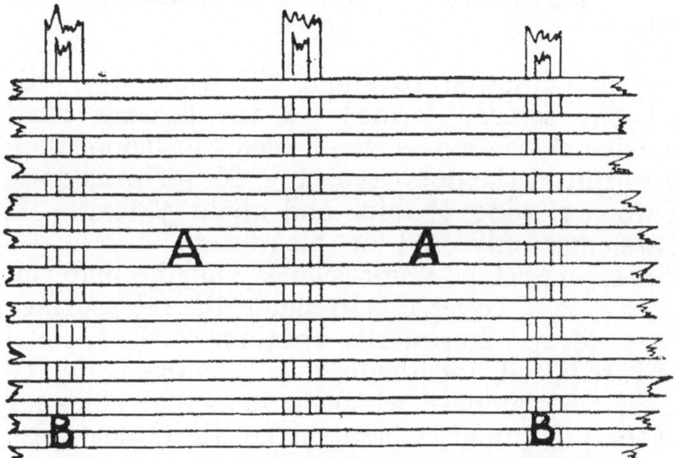

Fig. 81.

Yet the ceiling was sound and in good condition, apart from the small area needing repair. The same ways of breaking and butting joints, as shown in Fig. 79, for ceiling work, are applied in stud-partition work.

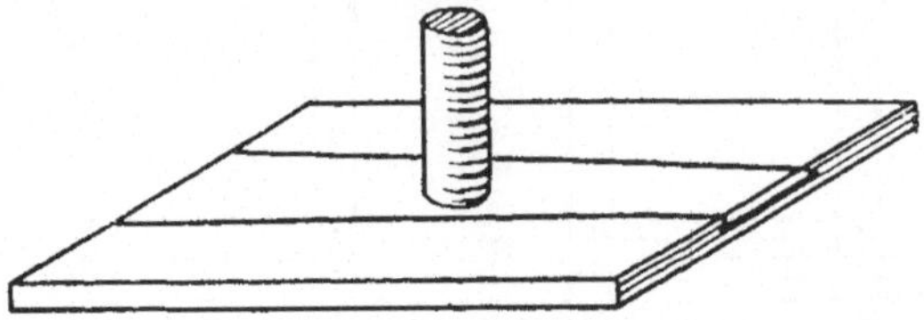

Fig. 82.

Rendering and Floating Walls

For this work a hawk, or plaster board, is used. It measures 12″ × 12″ × 1″, and has a cleat at the back, with a handle about 6″ long and 2″ in diameter, nailed or screwed through the centre of the square plate to the cleat, fastened to back of plate as in Fig. 82. On this the plasterer holds the plaster, and with the laying trowel, Fig. 83, the stuff is spread on the wall in a thin coat. It can either be left with the ridges made with the edge of the trowel or the front part of the trowel edge as in Fig. 84, or scratched with a wire scratcher, Fig. 85, in any direction, as long as it is left rough to form a key for the next coat, which is the floating coat. This process is called rendering, and is preparatory to floating. After all the walls have been rendered, and if possible allowed to stand till white dry, the plasterer proceeds to get the screeds formed for the floating coat. The nature and use of screeds will be seen as we proceed, but before they are formed it is necessary to put in the dots, as in Fig. 86. This is done by taking a thin piece of wood 1″ wide, ¼″ thick and about 5″ long. If laths are available, these are commonly used for this purpose.

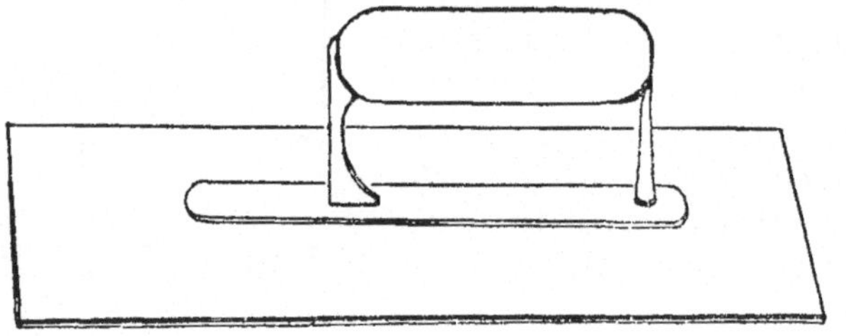

Fig. 83.

When plastering, the most important point is to lay on the material to an even depth, or if the base to be covered is irregular, then the surface must be brought to a perfect level. In order to obtain a perfectly level finish, some kind of gauge is required, and this in good practice is always provided by fixing dots at certain distances, which serve as the guides for the screeds. The whole of this preliminary work demands the utmost care.

Plumbing Walls for Dots

The following is the method of plumbing a wall with two wood gauges, when putting in dots for a flank wall or a high-pitch room, and also the way of using plumb rule and bob in plumbing dots in rooms on walls of ordinary height.

If it is a room of ordinary height, the dots are formed with the aid of a plumb bob and rule, as in Fig. 87. The rule is made of a parallel piece of ¾″ or 1″ seasoned wood, planed, and about 6′ long, with a centre line marked, and a hole cut about 6″ up from the end to allow of the bob to swing (Fig. 87). A is the centre line, B the

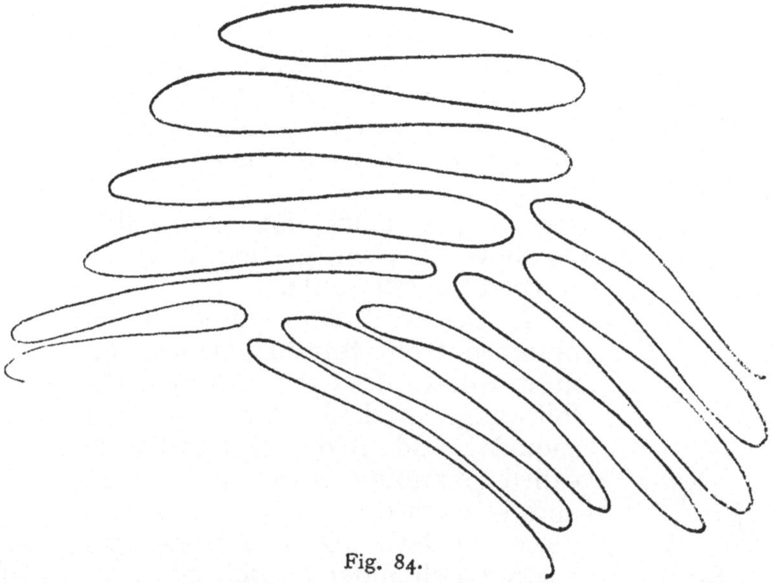

Fig. 84.

saw cuts in rule to pass the line through to hold up the plumb bob, C the hole to receive bob, and D plumb bob and line.

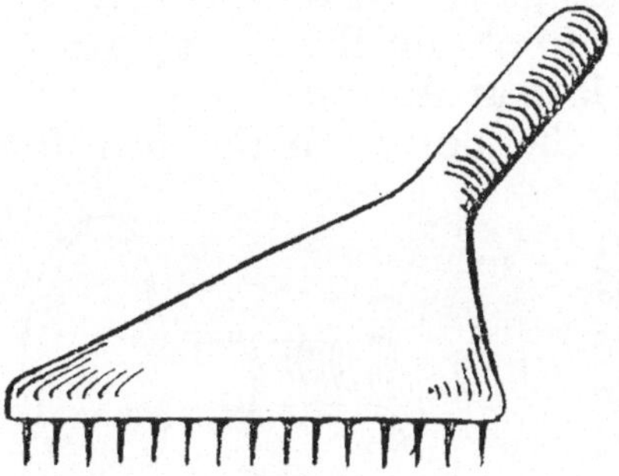

Fig. 85.

Take a small amount of plaster and lay it on the wall about 1′ from each angle and about the same distance from the floor. Then press in the lath, which is called the dot lath, to a fair thickness, about ⅜″. By the same process bed in another dot lath, at a convenient height, vertical with the first one. About 6′ will be found a very convenient height. In Fig. 86, A is the internal angle of the room, B the floor line, C the first dot, D the second. When the second dot has been bedded, the plumb rule is placed on the two dots, as in Fig. 88, and here D is pressed in till the plumb bob hangs vertical to the centre line. The same process is carried out at the other end of the wall. If the wall is long, a centre dot is needed and is put in and ranged with the line. The line, which is a thin piece of twisted

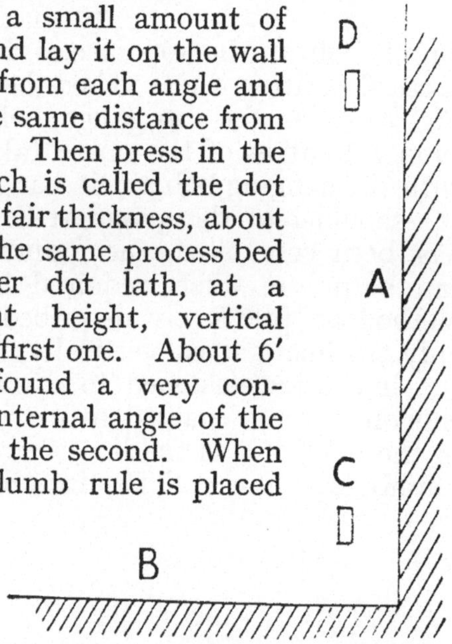

Fig. 86.

twine called a knot, is fastened on a large reel, as in Fig. 89. To range the dot, a nail is driven in the joint near dot at A and B, Fig. 90, and the line fastened on the nail and stretched to the other dot and fastened there. When this has been done, the centre dot at C, Fig. 90, is pressed in to clear the line, allowing the line to hang clear; allowance must also be made for the slight sagging of the line. The same procedure is carried out for the centre dot at D, Fig. 90.

It is necessary that as many dots as possible are put on to allow of their getting firm for working on with the traversing rule.

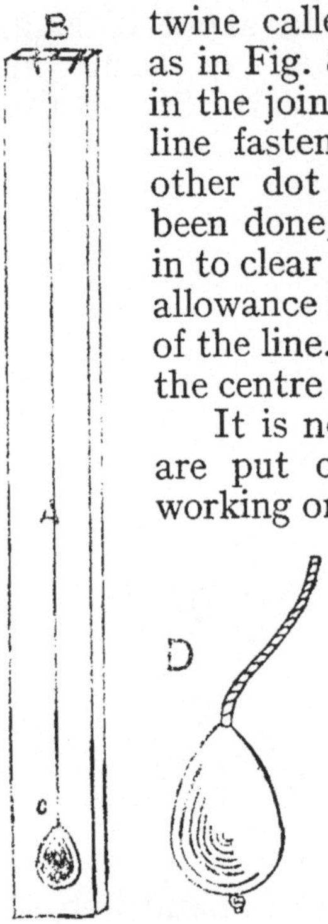

Fig. 87.

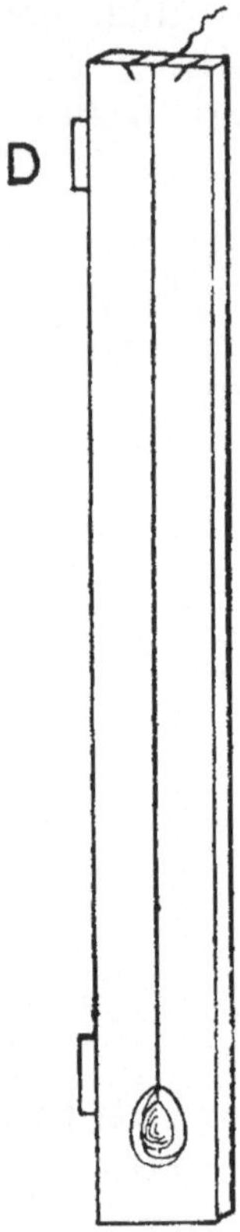

Fig. 88.

If the wall is required to be plumbed from top to bottom, and intermediate dots are required, the following methods for plumbing should be adopted. It is what is called plumbing with two gauges, and is carried out by taking two pieces of lath about 4″ long and cutting a shoulder in each, 2″ × ½″. When the dot has been bedded on the wall at C, in Fig. 91, one man holds the gauge and plumb bob at the top face of the wall, so as to obtain the vertical position where to bed the second dot lath at A. In Fig. 91, B is the top gauge, which presses the dot A in the bed far enough to allow of the gauge taking the line with the hanging bob till it hangs vertical in the shoulder of the gauge at C. After this has been completed, the intermediate dots can be put in where required by the same method as previously described in Fig. 90; only the line is fixed vertically, as in Fig. 92, the line being fastened to a nail (by doing this one man is enabled to carry on putting in the dots); B is the line, C and D the two plumb dots, E and F the intermediate dots. When the dots have been

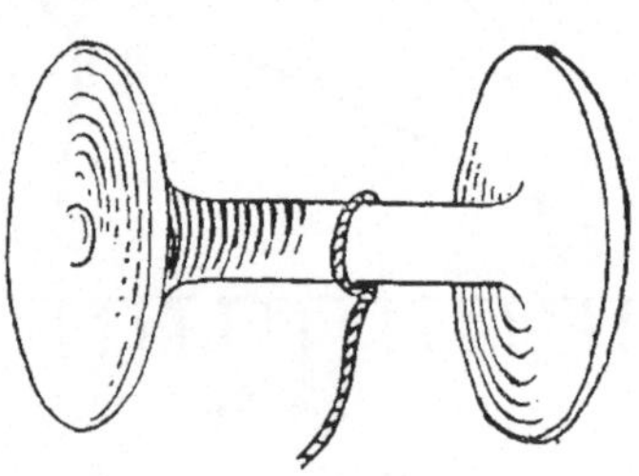
Fig. 89.

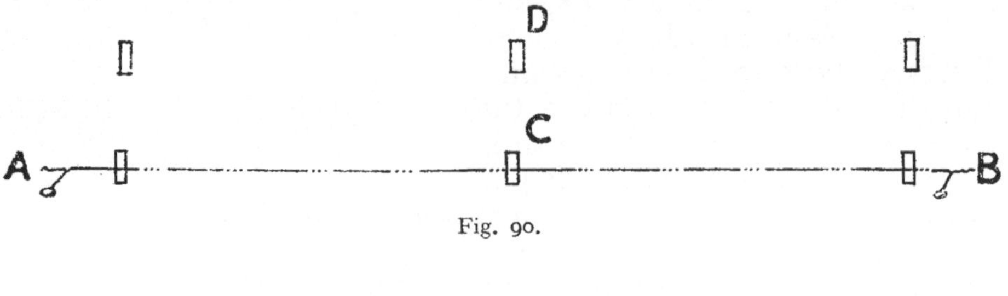

Fig. 90.

completed at one end, the same process is carried out at the other end of wall. Then the cross dots would be put in where required, as shown Fig. 90.

It will readily be seen that guidance of this kind is a great aid to accuracy, and can be applied to a wall of any dimensions.

This method of plumbing with gauges is usually adopted when fixing rules to float angles to piers and breaks to chimney breasts, etc., as in Fig. 93; where A is the rule nailed to angle at B, D the gauges as described in Fig. 91. The man with the bottom gauge at D holds the rule at E, and when the bob hangs vertically he nails the rule to the wall, when it is straightened and further nailed.

Another method adopted for plumbing the face of the walls, etc., in place of using dot laths, is to drive in nails, which are plumbed in the same manner. A very old method, now little used, is to put a little dab of plaster, about 1″ square. Quite recently when hacking off some old floating, I came across some of those plaster dots; they appeared as good as the day they were put there.

Forming the Screeds

Having bedded in the dots, the next process of filling in a line between these. These filled in lines are the screeds, used as guides in floating. The plaster is laid on between the dots about 6″ wide and outside the two ends, as in Fig. 94, with a traversing rule at each end. If a long rule, it is held by two men; but if of an ordinary length, it is held in the centre by one man. Care must be taken when traversing that the material which is to form the screed is kept down to the dot lath, otherwise the screed would not be straight. When the screed is finished, that is filled out to the face of the dots, the laths are taken out and the space filled in with the same sort of material as

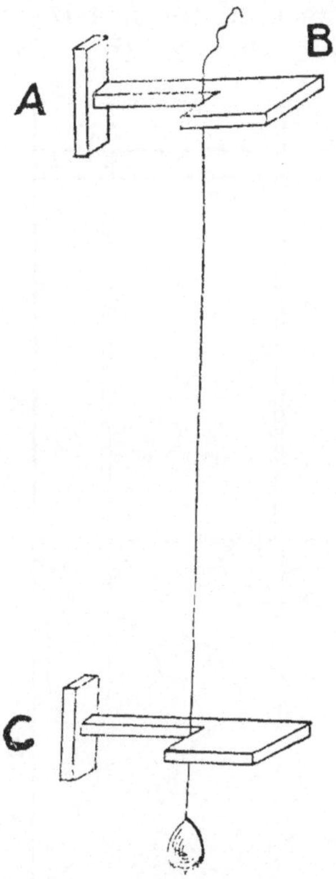

Fig. 91.

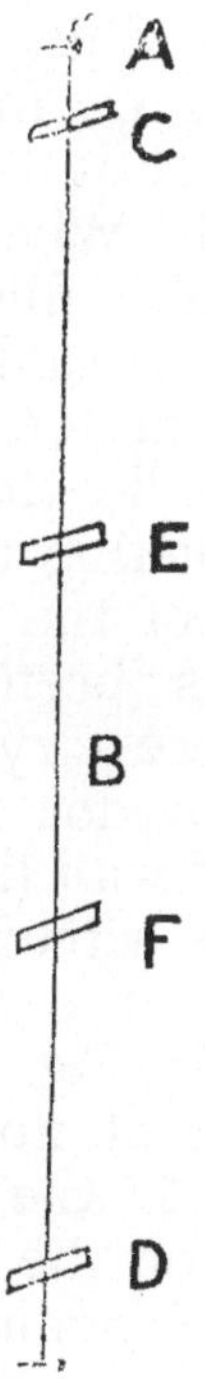

Fig. 92.

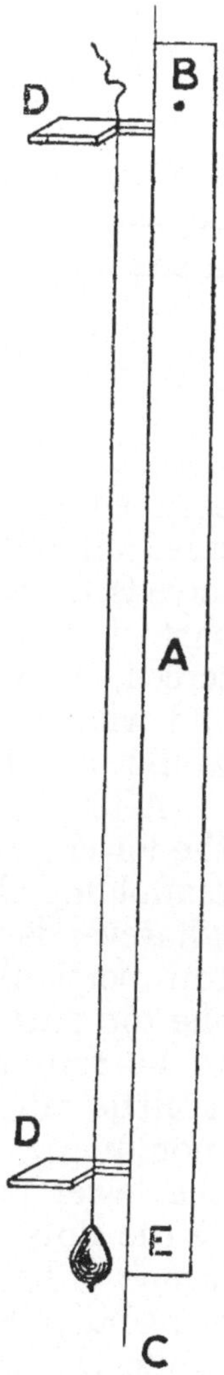

Fig. 93.

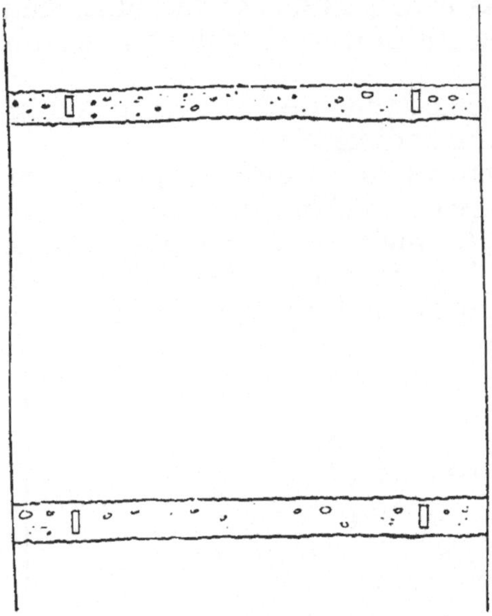

Fig. 94.

the screeds. This process of filling out to dots from D, Fig. 95, is repeated until all the screeds are filled in. This method is for dealing with the floating of walls in rooms of ordinary height.

Floating

When all the screeds have been put on, the next thing is to fill in the space between the screeds, as in Fig. 95, where A and B are the two internal angles of the room, or wall as the case may be; C and D the two screeds, C being at a convenient height to rule off with the traversing rule, which is one and the same rule as used for floating; E is the floating rule on the two screeds. When filling in the space between C and D, the material is ruled off until it is flush with the screeds C and D. When the bottom part of the wall is filled out to the screed, the wall is said to be floated and whatever, materials were used would be called floating stuff.

After completing the floating of the lower part of the walls and the scaffolding has been erected (or whatever is necessary to reach the top portion), a dot is bedded on the top part of wall (if a cornice is to be run, the screeds are formed a little above the depth of the moulding). In Fig. 95, A to B is the lower part of floating, C and D the dots, E is the floating rule, which is held on the finished floating and pressed against the dot at C and D; if a centre dot is required, it can be put in at F. The screed is then formed as previously described and the remaining part below filled in, using a shorter rule to reach

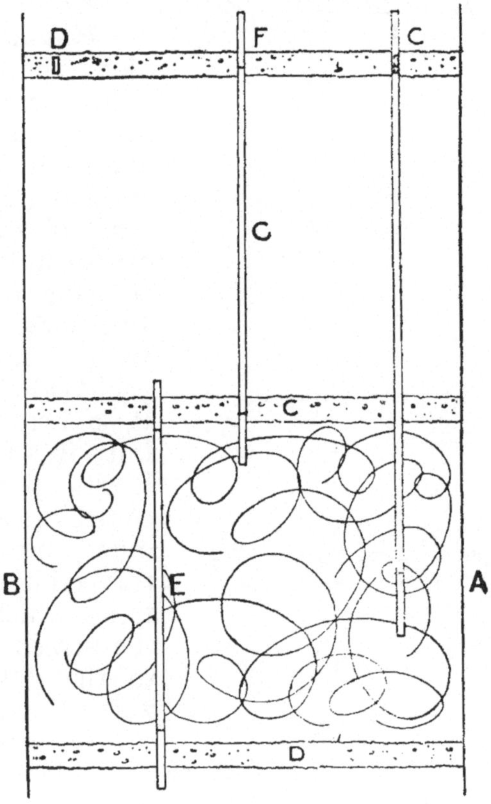

Fig. 95.

the lower screed, which is more convenient. G is the shorter rule. When the face is filled out to the screeds, a feather-edge rule will be found very useful to wipe out the angle. Fig. 96 shows the feather-edge rule. These are made in various lengths up to 6′ and $\frac{3}{4}''$ to 1″ thick and about $4\frac{1}{2}''$ wide, and planed back about $2\frac{1}{2}''$ from one edge, which is $\frac{1}{4}''$ thick.

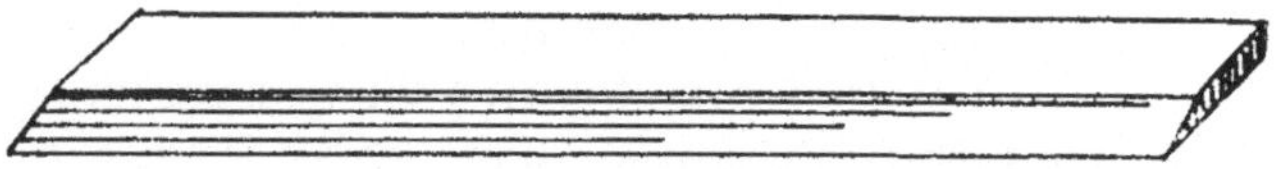

Fig. 96.

This method of screeding does away with a lot of upright screeds. If too many are formed, they have a tendency to become slightly full on the face of floating when filling in the bays. With upright screeds when ruling off the material, the floating rule catches much surplus stuff, which would otherwise fall on the floor, though when the plasterer is floating walls a "save board" is put against the wall on the floor. This is usually a scaffold board. I myself do not recommend upright screeds with inside work, as the fewer the screeds the better. If the material is laid on regularly with the laying trowel, there will not be such a great amount of surplus to rule off.

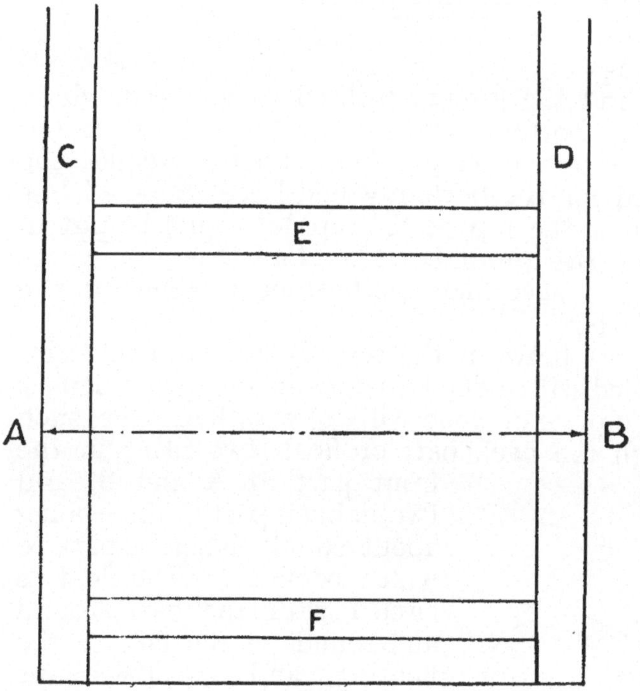

Fig. 97.

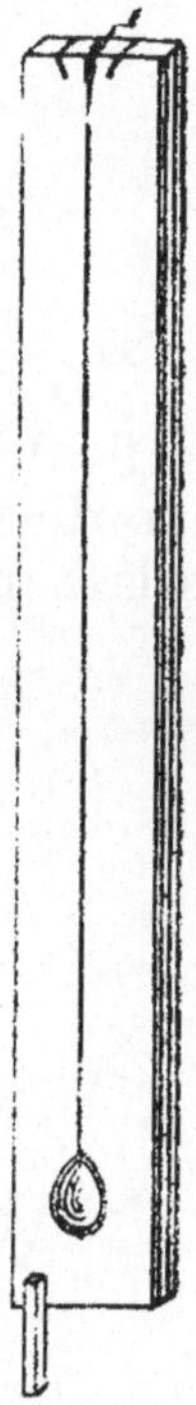

Fig. 98.

With the ordinary cottage work or speculating building, with rooms of ordinary size, the screeds are put right in the angle, as in Fig. 97, where A to B is the angle to angle, C and D upright screeds, which are about 6″ wide, and are "wiped out" by the long feather-edge. The top part of the screed is put on by standing on a short ladder or a pair of steps. These screeds are put on what is known as ordinary thickness, which is not more than $\frac{5}{8}''$ thick. After the screeds are wiped out, the plumb rule is tried and the screed adjusted to the plumb. The two cross screeds E and F are traversed across E; this being at a sufficient height from the floor to receive the top part of the skirting.

Fig. 99.

While I am dealing with screeds and floating, I must point out that in some cases the carpenters fix wood grounds, either for skirting or picture rails. These grounds are usually 2″ × $\frac{5}{8}$″ to $\frac{3}{4}$″ thick and nailed to plugs in the wall and straightened. If grounds are fixed, the first dot at C in Fig. 86 would be above the ground and the plumb rule used would require a short piece of wood about 6″ × 1″ nailed on at the bottom about $\frac{1}{8}$″ back, to allow of the finishing coat, as in Fig. 98. Then the screed is put on above the skirting ground, or a piece of wood may be tacked on the floating rule, as in Fig. 99, and the ground utilised as a screed, which does away with the bottom screed. If the dots are being plumbed with the bob and gauges from skirting ground, the gauge which is used for the bottom would have to be $\frac{1}{8}$″ less than the one used at the top, as the top dot would be put in and plumbed from the ground. Therefore, the top gauge would have to be $\frac{1}{8}$″ longer than the bottom, to allow for the finishing coat of plaster.

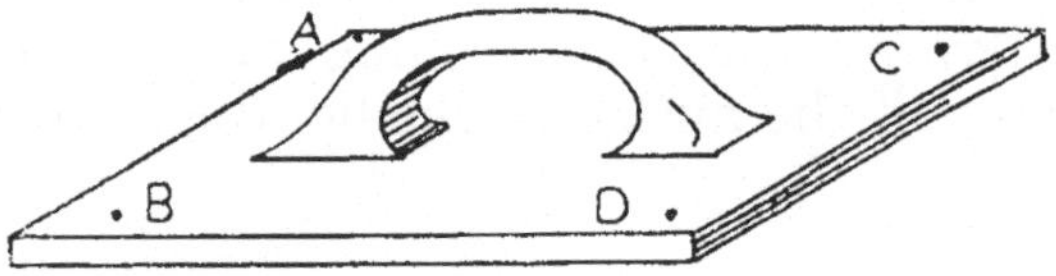

Fig. 100.

When all the area between the screeds has been filled in, the floated work is then devilled with a devil float, as in Fig. 100. This is an ordinary float (though any piece of wood will do) with four nails, such as lath nails, driven through the back part of float, two being in the front part at A and B, and two in back part C, these being about equal distance apart to width of float. The float is circled with the hand round and round on the face of the floating, and as the nails should be driven through to the front face of float about $\frac{1}{8}$″, it causes a slight circular sinking in the face of the floating, which provides a key for the finishing coat of plaster, Fig. 101. This I consider is better than the comb or wire scratcher mentioned in my notes on rendering walls, but it is chiefly used (when used at all) for combing floating work, especially cement floating.

Fig. 101.

THREE-COAT WORK FOR LATH CEILINGS OR STUD PARTITIONS

Three-coat work on lath is much the same as render, float and set on walls, with the exception that laths are added, so that in measuring up plaster work the items of supplying laths and lathing have to be added.

When the first coat is rendered on laths it is called pricking up, and the pricking up is the first coat. In covering lath work with the first coat of material it is much the best way to lay the stuff on the cross way of the laths, that is the reverse way to which the laths run. The reason for so doing is that the materials hang up better, for if the laths are covered the same way as they run (that is, the way they are nailed), the space between the laths is longer. This is shown in Fig. 102, where A to B, B to C, are the joists (the way the laths run), and D, the dotted line, is the cross way of the laths.

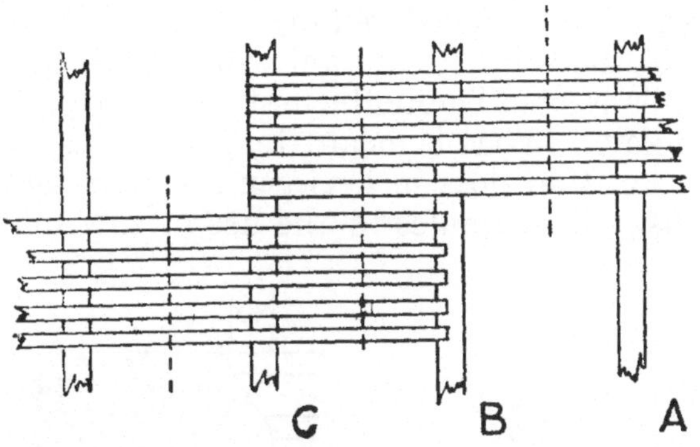

Fig. 102.

When all the lath work has been covered and allowed to stand sufficient time to get "steady," it will be firm enough to scratch. A scratcher is then made as follows: four or five laths about 14″ long are fixed, as in Fig. 103; A is a lath cut to a point about 3″ from the end and chamfered off both sides from B; the remaining laths, being cut off in the same manner and nailed diagonally together at C, keeping the points of the laths about $\frac{5}{8}$″ apart. If five laths are used, it would be necessary to nail a brace across to stiffen the scratcher at D. After this has been done the lower part of the laths below C are trimmed off with the knife,

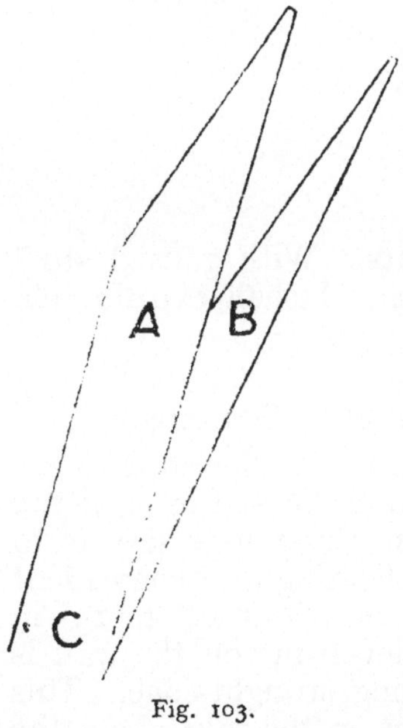

Fig. 103.

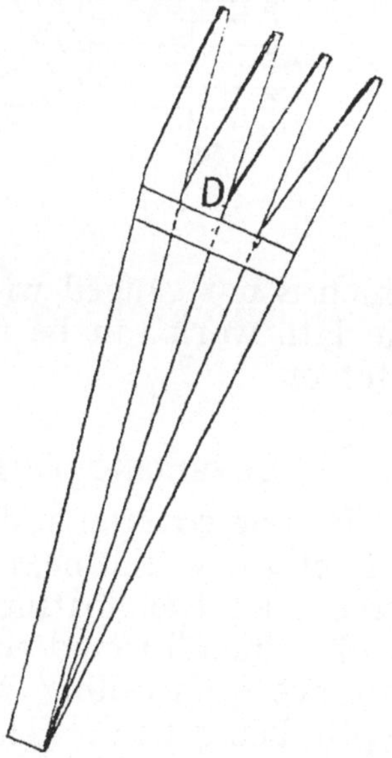

Fig. 104.

to make the scratcher handier to hold. Fig. 104 is the finished scratcher. One very important point in scratching lath work is to scratch diagonally across the laths as in Fig. 105, A being the lath work, B and C the cross scratching of the lath scratcher. This applies to the first render coat on laths, whether it be for ceiling work or stud partitions. Before proceeding with the floating of ceilings or any other lath work, it is always advisable to let the scratch coat get white dry if possible; otherwise, if the floating coat is started before the first coat is dry, there is the possibility of breaking the key. What is meant by the key is a certain amount of the pricking-up coat that goes in between the laths and laps over the back part of the laths. This applies to the use of raw material; that is to say, material

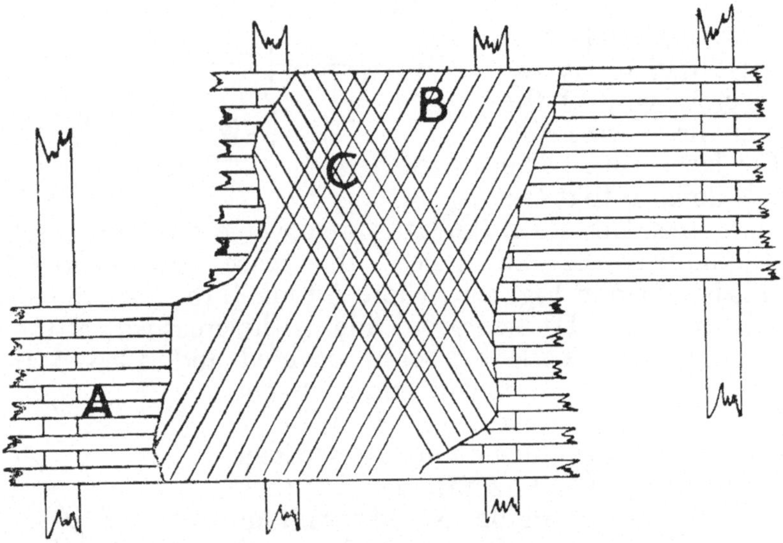

Fig. 105.

which is not gauged with plaster or other cements. With gauged stuff the lath work can be floated right from the lath. I shall explain this later on.

Levelling Dots for Ceilings and Forming Screeds

Having covered and scratched all the lath work and allowed it to get perfectly dry, the next proceeding is to bed in dots for the level of the screeds for the floating of the ceiling. The first thing necessary is to level a line all round the wall on the face of the floating at a convenient distance from ceiling, to allow of a wooden square, about 12″ to 15″ in depth, being used. The method of getting the level line on the wall is quite a simple matter, as one only requires a long straight-edge. This is placed against the wall. The spirit level is put on the rule, about the centre, and the rule adjusted till the spirit in the level is central. Then with the gauging trowel, or a sharp pencil, the wall is marked at each end of the rule and centre. When continuing the levelling, reverse the rule (that is, turn it end to end), and when the next length is levelled reverse it back again. This is done as a precaution, in case the levelling rule should vary a little in itself for straightness. When the plasterer has been all round the walls and arrives at the place where he first started levelling, should he find the two points do not meet, it will be necessary to check with the levelling, to find out where the variation occurs. When the level line has been corrected, a line is chalked and snapped at each end of the wall, by the following method. One man takes hold of the reel, as seen in Fig. 89, between his fingers and thumb, and with a piece of chalk or a small portion of dry plaster in his disengaged hand he allows the reel to revolve between his finger and thumb. While his mate is pulling the line, the one with the reel allows the line to run through the hand holding the chalk or plaster. When the line is chalked, both men hold the line at each end against the mark of the level line on wall as tight as possible; if a long length of line, another man is needed in the centre, who presses the line to the nearest level mark on the wall with his thumb, and with his other he snaps the line by catching hold of the line with his forefinger and thumb and pulling it slightly gently away from the wall and letting it go. He repeats this each side of the line he is holding with his thumb. This causes a white line to be made on the wall.

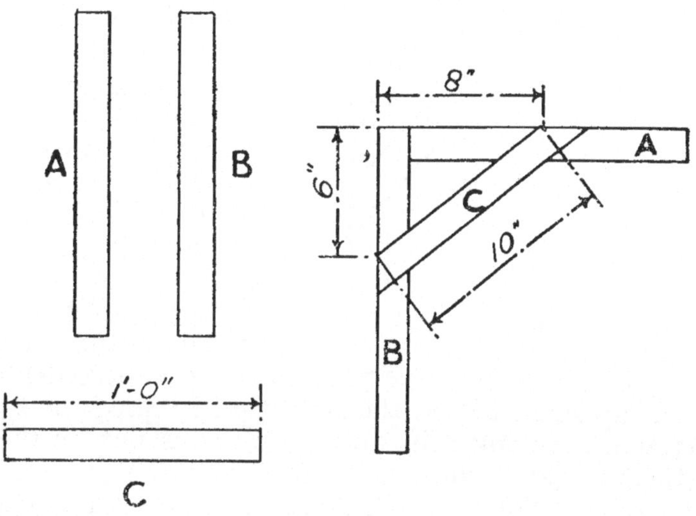

Fig. 106.

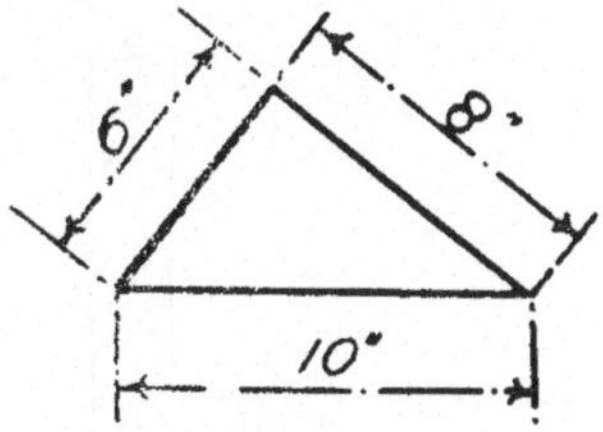

Fig. 107.

When the snapping of the line has been completed, a wooden square is needed, which can easily be made with three pieces of 15″ × 2″ × ½″. In Fig. 106, A and B, 15″ × 2″ × ½″, are nailed together temporarily at right angles to each other and lapped at the ends, as on dotted line C; this is 12″ long and is placed across the two diagonally and one end nailed; then with the rule of 6″, 8″, 10″, as in Fig. 107, the other is end nailed. When checked to see the measurement is correct, nails are driven in at A and B, and the cross piece and the ends of diagonal piece can be cut off flush with the edge of A and B. With the square complete the next process is to mark a point on the square at the level line, at the same time placing the square up to and against the scratch coat of the ceiling, as in Fig. 108, where A is the ceiling, B the square in position, C the level line and the mark on the square at D. The square is then tried along all parts of the level line to ceiling to find out which is the lowest line of ceiling. This is done because the dots for the floating screeds are put in from the lowest point. When this point is found, a very thin dot is put on, about 12″ from angle (this allows the plasterer to have room to work the traversing rule when forming the screeds), by pressing the square tight up against it, at the same time keeping the square firm on the wall. After completing this, another mark is put on the square at level line, which would be less on square than the previous mark, owing to the thickness of dot, Fig. 108, D, which shows the new mark. Continue putting the dots on the ceiling at convenient lengths for traversing with the rule to form the screed. It is not always necessary to have dot laths, as sometimes the plasterer gauges the stuff for the dot with a little plaster. This enables him to carry on filling out the screed, and the dots are hard enough to work the rule on. The reason laths are used for dots is that they allow the filling out of the screed without waiting for the material to set. When the screeds have been formed (round the ceiling line, as it is called, and I might add they are usually kept out from the angle far enough to receive the ceiling members of the cornice if any, though if there is no cornice the screeds

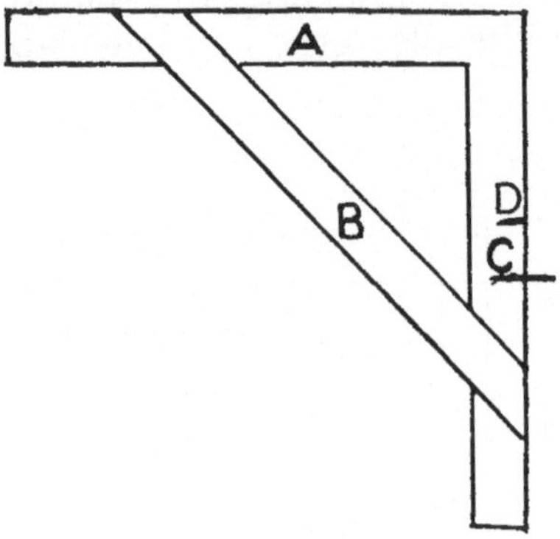

Fig. 108.

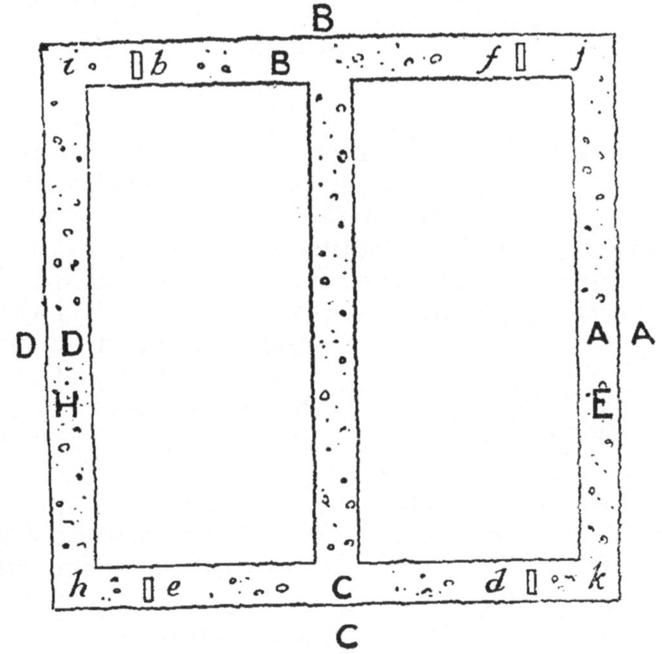

Fig. 109.

are kept out from the angle and the angle wiped out with the feather-edge after the flat is filled in), cross screeds are put in where necessary, as in Fig. 109, where A, B, C, D are the four angles of the room; likewise A, B, C, D the four screeds, for which it is only necessary to put in dots at the two ends or side; for if dots are put in at *d* and *e*, and *b f*, these two screeds being traversed first, the two remaining, at *h* and *c*, can be traversed off the ends of B and C at *h* and *i* and *jk*. As regards cross screeds, wherever possible these should be in the centre, the long side as at *l*, Fig. 110, so that the long feather-edge or floating rule reaches across, which allows the ceiling to be floated with the one cross screed.

The fewer the cross screeds the better the surface of the floating, as if the floating stuff is laid on regularly, which is easily done with a little practice, it is quite an easy matter to feather-edge the flat of the ceiling in evenly working the feather-edge, or rule in a zigzag manner on the screeds. If, however, the ceiling is large, it is necessary to have several cross screeds. When this is the case, the best method would be to traverse the cross screed the short way of the ceiling, as in

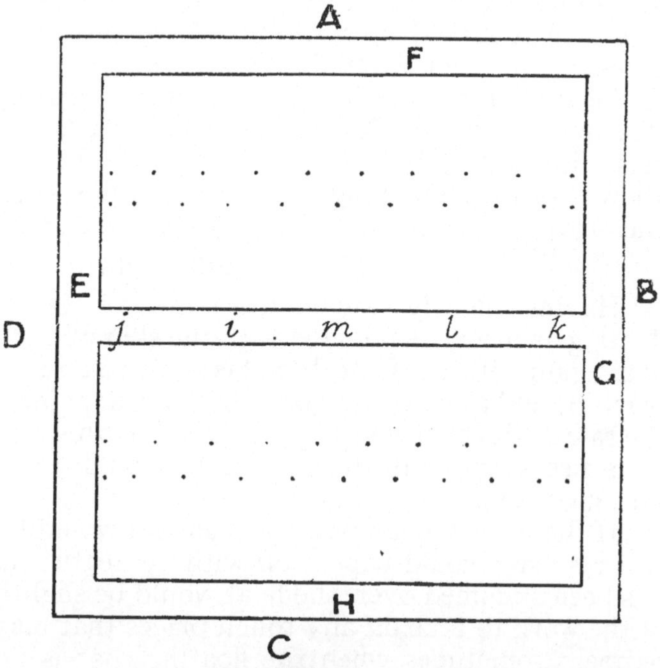

Fig. 110.

Fig. 110, along the dotted lines. Two plasterers are needed for this, and with the rule held at each end, the cross screed is ruled off as follows: One man holds his end up and against the angle screed, the other towards the centre of the room, working the rule backwards and forwards several times to a straight surface. This process is repeated from the other side of angle, as in Fig. 110, A, B, C, D are the four angles of the room; E, F, G, H the four screeds; *i*, *j* the first process of cross screeds; *k*, *l* the second; *m* the third. This is completed by filling in the space and traversing it, by each man bearing on to the outer screeds, which have been formed by the first process. Each cross screed is laid in the same method, and when completed the area between the screeds can be filled in and ruled off with a short rule or feather-edge.

When the whole of the surface has been filled in (not forgetting to take out the dot laths, if used), it can be said the ceiling is floated, and all it requires is to be devilled or combed, as I have previously described.

Setting Coat

The setting makes the finishing coat in three-coat work. The term setting applies where setting stuff is used for the finishing coat. Setting stuff is made of washed sand (placed in a very fine sieve) and lime putty. When the finishing coat of any plaster work is to be finished with other than setting stuff, the term given then is finishing.

One pail of setting stuff is put on the gauge board. It is always best to have it soft and pliable. A small quantity of plaster (about two handfuls) is mixed up with it. If the setting stuff is stiff, it would be necessary to open it out in a ring and add a little water before putting in the plaster for the first coat over the floating, as it helps to prevent what is called fire cracks; that is, small cracks hardly discernible after the plastering is dry, but not weakening the finished work. These are also known as map cracking; the cracks, like counties, varying in shapes and sizes.

Skimming

Having slightly gauged the materials, it is then skimmed over the floating as firmly as possible with the skimming float. It is then skimmed over again with the float, but this time with raw stuff (that is, without the plaster), and as plasterers usually work in pairs, one would follow on with the raw stuff, covering the gauge coat that his mate would be skimming on. It is necessary that the raw stuff should be skimmed over before the first coat sets.

If there are any angles, one plasterer would lay some stuff at the angle and his mate would wipe it off with the feather-edge. After the raw stuff has been skimmed over, the float would be slightly passed over the surface of the work to take off any rough places that may have been made. This happens sometimes when the floating coat is dry, which causes a lot of suction in the finishing coat.

Having gone over the ceiling or walls twice with the float, it is then laid down with the laying trowel. A very thin coat of stuff is laid over, starting from the angles, if any, or up to the edge of the cornice; if working on a ceiling, carrying on to the remaining part to be set.

Trowelling

The next process is what is called trowelling. This is done with a dry or clean laying trowel, passed backwards and forwards at arm's-length over the finishing coat. The plasterers are greatly assisted in this process by the use of the wet brush, for while he trowels over the work with one hand he passes the brush over with the other.

One very important thing in laying down and in the trowelling off in the finishing coat on lath work is that it is trowelled off the same way

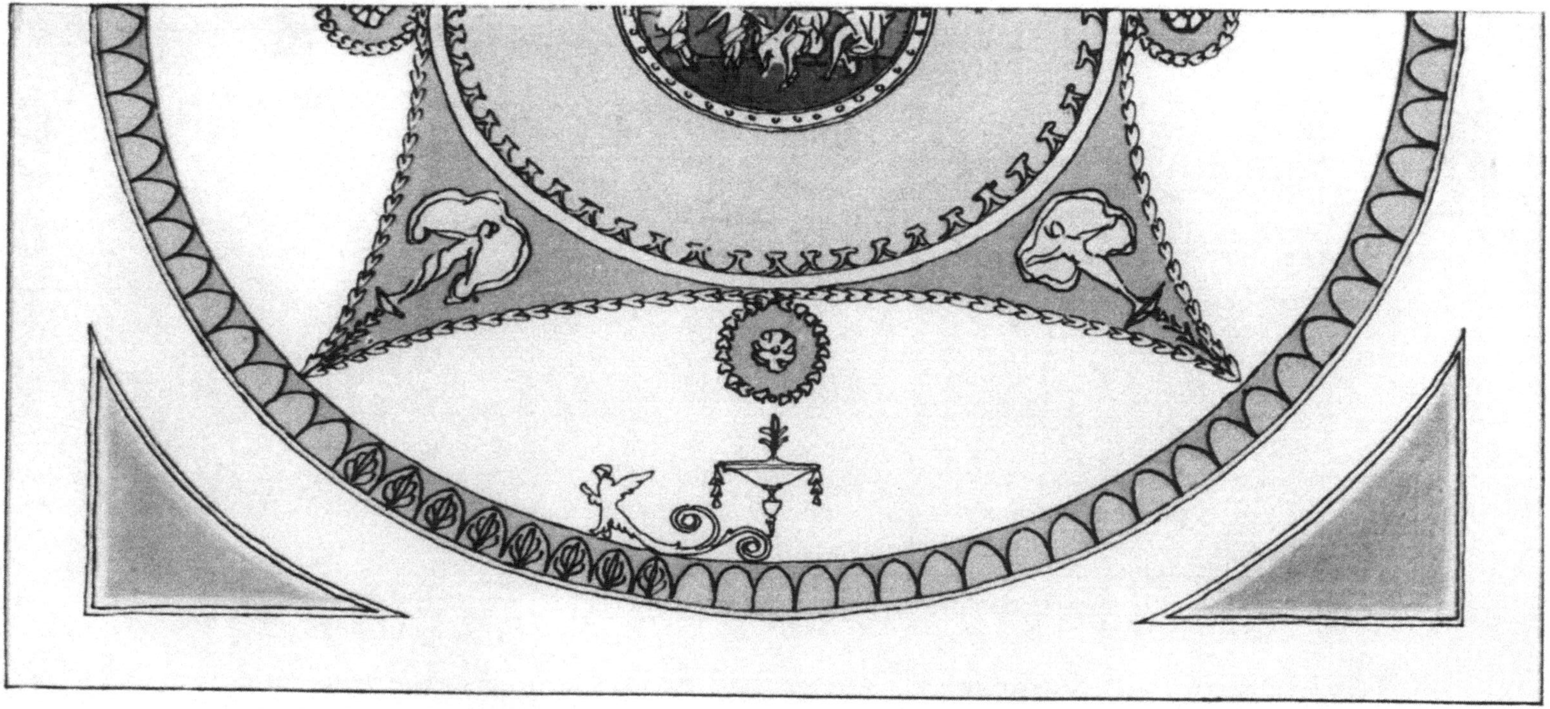

SECTION OF CEILING IN THE LATER STYLE OF ROBERT ADAM.

Colours in low tones ; ornament often in composition in low relief ; panels in raised composition or rich colouring.

as the laths run, and not cross ways over the laths, as it is not so liable to break the key of plaster by so doing. Another thing in trowelling off ceilings is the passing of the brush over the finished work to take off any trowel marks, or any fat which works up sometimes with the trowelling of the work. If this is done, always pass the brush over across the light; that is, if there are any windows, pass the brush over parallel to them. It is well to note this point, as when ceilings are distempered or painted, the painter always brushes off across the light, so this is an important item in finishing off ceilings.

Alternative to Skimming

It is also well to note that at the commencement of my describing the setting coat I mentioned skimming over. For this a skimming float is used. The term was more general in the North of England; yet a few years ago this method was adopted in London and the South, but of late years the skimming coat gradually died out, and the following process is the general mode of procedure, adopted in setting walls and ceiling with ordinary finishing materials, although this does not apply to the finishing with Keene's cement. Lay the first coat on (slightly gauged) with the laying trowel as firmly as possible, following on with another thin coat laid down with the float, and following on this with another thin coat, which is called laying it down with the trowel, and after this it is trowelled in with the laying trowel and water brush, and then finally is trowelled off.

The technical terms for these several stages are laying on with the trowel; laying down with the float; laying down with the trowel; trowelling it in; and trowelling off.

While I am dealing with the finishing coat it should be mentioned that in any kind of plaster or cement (with the exception of mastic), the first coat is laid on with the laying trowel, and I may add it is pressed into the key of the floating better this way than if it was skimmed on with the skimming float.

With patent plaster, such as granite, adamant, thistle, sirapite, xyolite, marbleite, asbestos, and other kinds, it is laid on and laid down with the trowel and trowelled off, as it is not necessary to go over the surface with the float. All these kinds of plaster are described as two-coat work.

Finishing of Angles

When finishing walls with the ordinary plaster finish (such as with setting stuff), it is usual that any exposed angles (such as the angles to reveals to windows or door openings, etc.), should be finished in Keene's cement. If this is the case, when floating the walls the angles are cut back about 2″ each side and filled in to the face of the floating with cement and sand, gauged 4 parts of sand to 1 part of cement. When the walls are being set, these angles are kept clear of the setting and cut off about 2″, and after the walls have been finished a small quantity of Keene's is gauged and laid on the angles and wiped off with the 12″ joint rule from the finished face, though it makes much the best job if the Keene's angles are put on first and traversed off straight, and then followed on with the setting of walls, and the setting stuff rubbed in with the angle. If the angles are left till afterwards and finished as before described, by wiping the Keene's off with the joint rule, the angle would have to be rubbed up with the cross-grained float and trowelled in as in the ordinary finish with Keene's in finishing angles to reveals of windows or breaks to chimney breasts or piers. When finishing reveals to windows or door openings, either with Keene's or other materials, the front face is put on and traversed off straight and a running rule tacked up on the face and straightened, and a small wooden square is drawn up the frame after the material is put on, care being taken that it is marginable from the side of the frame. If the frame is too narrow for the square to run on, a notched gauge can be made to run off the frame, as in Fig. 111, where R is the rule tacked on the face of work, F is the window frame, G is the gauge, which can easily be made with a piece of running rule or any thin piece of wood. This is notched out according to the margin required to the side of frame. The same method applies to the heads. The face is put on and the rule tacked across the opening. It can be levelled though. It is customary to take the margin from the frame head, and each side, with the wooden square. D the square, E reveal and gauge in position.

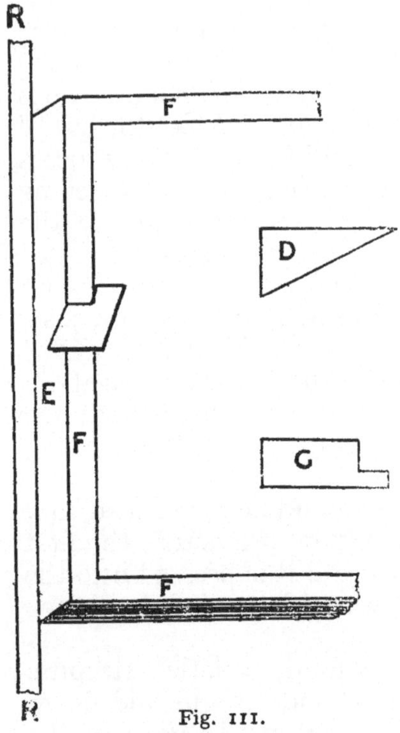

Fig. 111.

Breaks to Chimney Breasts

When finishing breaks, the face is put on and the rule tacked up and straightened, and the material laid on the break. Then the square is drawn up from off the back face of work, if the break is squared in from the back. When a break or pier is squared in from the front face, it is done in the following way, which also applies when floating: A short rule is put against the floated face of the breast or pier, as in Fig. 112, where A is the short rule, B the square, C the dot, D the second dot, which is a nail driven in and plumbed up the inside part of the break, or pier, and a dot lath bedded in with the square, touching the rule on face and the nail. Then, when the dots are steady, the break is filled out and wiped out with the floating rule or feather-edge.

Squaring in for Floating of Piers

With piers, the line of the first face is floated and each remaining face is squared in with nails by itself, from each face. Rules can be tacked up to these dots and filled out in between with a short rule across, as in Fig. 113, showing short rule with square and nails.

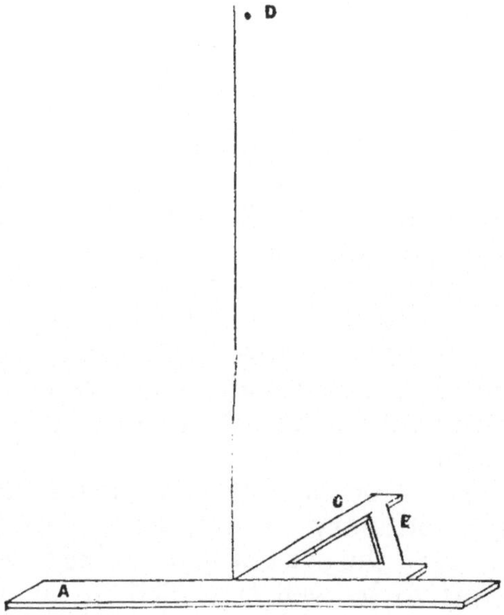

Fig. 112.

Two-coat or Float and Set Work

Two-coat work, also called float and set, is applied to concrete or other kinds of solid soffits of floors to form ceilings, or breeze blocks, brick, tile, pumice-stone blocks or any kinds of solid partitions, though in plastering concrete or other kinds of solid ceilings, it is best to render a picking-up coat all over, before getting the screeds on.

As I have previously described the method of levelling screeds for ceilings, I can pass on direct to the float and set on solid ceilings. The procedure is exactly the same as on the pricking-up coat on lath. The screeds are formed and the flat of ceiling filled in, though, if there is no cornice, and it is ordinary class work, the outside screeds would not be used, but a screed would be formed right along the angle and wiped out with the feather-edge, the cross screeds wiped off from these and the remaining part of the ceiling filled in. With partitions the screeds are formed as previously described when dealing with floating, with the exception that the screeds are built up direct from the material forming the partitions (this being only two-coat work), and the floating carried out in the same manner as before described. When using the patent plasters, more attention must be given to the floating, as it is not left so rough as in the case of ordinary materials, the finishing coat being laid on and laid down with the trowel.

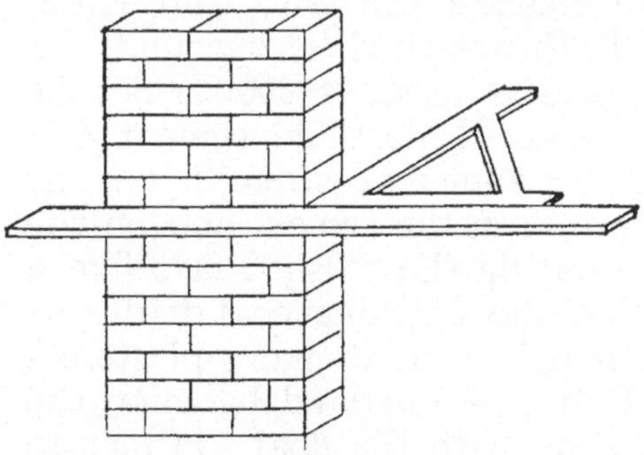
Fig. 113.

This I consider is much better, as the finishing coat is kept much thinner, and if the floating is of a good material it makes a better job.

Two-coat work on lath is really three-coat work. The screeds are formed with gauged stuff (if ordinary lime and hair, coarse stuff is used). With plaster this enables the plasterer to carry right on with the floating, as when the stuff sets the screeds are hard enough to work on. Then further stuff is gauged and the lathing covered, filled out and ruled off to the screeds.

Finishing with Patent Plasters

With patent plasters the screeds are formed, but as these materials are of a quick-setting nature, the plasterer is able to carry right on with the covering and floating of the lath work. Patent plasters are very good, providing the plasterer knows how to use them; if not, they very often are a failure, not through any fault of the material. The stuff must be used fresh and not killed. Ignoring this is very often the cause of it not being a success. My experience with all kinds of patent plasters (and that includes all that have been on the market for some years past) is that only sufficient should be gauged at one time as can be used before it sets.

The process of floating and setting, or finishing, is exactly the same as in ordinary plaster, but in finishing the plasterer must watch the setting of the stuff, which, in this case, means the hardening of the material. It has to be well trowelled in before it sets, because it is impossible to get a true and smooth trowelled surface after it has once set.

Two-coat work, which is the general run in plastering at the present time, is looked upon as $\frac{5}{8}''$ thick, but it is very often thicker, especially on concrete ceilings, owing sometimes to the faulty shuttering used in the formation of the concrete floors, though there has been a big improvement in concrete construction of floors. Much of the best work is done only in two coats, that is float and set, such as with Keene's cement or Parian, which are floated sometimes with Keene's and sand, and sometimes with Portland cement and sand. Both are good backgrounds for the Keene's finish, but if Keene's and sand is used, it should not be gauged more than 3 of sand to 1 of Keene's. If Portland cement and sand, 4 of sand and 1 of cement would be sufficient strength for the floating, though with Portland cement the acid in the cement acts on the mineral in the sand and causes it to come out on the finished face like a lot of hoar frost. With Keene's or Parian the finishing coat varies as to the method of the ordinary setting, though a great many plasterers lay the first coat on with the trowel. If it is a Portland backing, the floating (the first coat) can easily be laid on with the float; in fact to advantage, as the skimming float is larger than the trowel, and being a cement backing there would not be much suction. So, whichever way the first coat is put on, it is laid down with the float the second time, and laid in with the feather-edge; that is, the feather-edge is passed over to fill in any hollow places that may have occurred during the floating of the walls. In addition to this, the angles are wiped out.

After this has been done, the cross-grained float (a float with the face made of a cross-grained piece of wood, with a dovetail piece let in the centre to keep it straight, and also to fasten the handle to) is rubbed over the surface, assisted by the water brush if necessary. It is then laid down with the laying trowel and a little Keene's, and trowelled off in the usual way.

If it is to be a first-class job, the laying-down process with the trowel is omitted, and the use of the cross-grained float is continued until the surface is rubbed up to a silken face. It is then trowelled in, as the rubbing up brings a certain amount of fat to the surface, and the trowelling is continued until the Keene's has a perfectly smooth surface. This method is adopted for first-class work, but the general practice of finishing with Keene's is as I have previously described.

One very important thing when finishing Keene's for the painter is to apply a thin coat of sharp colour as soon as it is sufficiently hard for the brush to work on. This is necessary, as Keene's absorbs the colour if painted before it is hard (or actually set). This helps to keep back a lot of acid. In some cases the Keene's is given a coat of linseed oil with a small quantity of terebine to make it dry, but the general practice is a coat of sharp paint. In finishing a wall with Keene's, it makes much the best job if the whole of the wall is covered, so that there will be no joint. Besides, it can then be all feather-edged in at the same time. Apart from *Keene's*, where possible it is always an advantage to finish a wall right down, without a joint, as the angles can be wiped out all the way from top to bottom. Besides, the joining, which would occur if it was only finished half or part way down, would require rubbing up with the cross-grained float, and extra trowelling would be needed at the joint.

Render and Finish

Render and finish is applied to such work as scullery walls, w.c.'s lobbies, etc., in middle-class dwellings and cottages, etc.; also in warehouses. It is executed in Portland cement and sand, gauged about 3 of sand to 1 of cement. If the plasterer has a job of this description, and the sand is not too sharp, the first thing necessary is to damp the wall by throwing water on with a brush, or by having a small tin and a water pail, pouring the water on the wall from the top of the work to be completed. It is well worth while having plenty of water put on, as it stops a lot of the suction. Having damped the wall, a thin layer is rendered over as evenly as possible. Then the angles are wiped out with the feather-edge. After this has been done, another thin layer is laid on with the trowel, and followed up by rubbing the surface with the float.

This being only ordinary work, it is not necessary to rub it up with the cross-grain float, for if the ordinary skimming float is made of a soft pine wood it is quite suitable. When one gets adapted to this kind of work, it is quite an easy matter to make a good job of it. Some bricks have less suction than others; when this is the case, the water brush makes sufficient damping. If the sand is very sharp and full of grit, it is necessary to wash some of it through a fine sieve, and after rendering the first coat over with the sharp stuff, the coat for the finishing should be gauged with the washed sand and cement, as it would make an easier and quicker job. This kind of work is rubbed up and left with the float, but if a smooth face is required, the first coat is rendered over as before, and then fatter material is used to cover over (this fatter material is gauged more cement and less sand), and should be gauged $\frac{1}{2}$ part sand to 1 part of cement. A small portion of lime putty greatly assists in making it fat and better for working, as this coat is laid on with the trowel and trowelled on to a smooth surface. This is called trowelled cement, and is sometimes described as "Render walls with cement trowelled face," or as before, "Render walls and float face finish."

With manholes or cement tanks, or any description of work required to be finished in cement to hold water, great care must be used in floating, and when finishing neat cement must be used. On no consideration must lime be added. If the cement is of a fiery nature, a very small portion of washed sand can be used, and before trowelling off it would be necessary to rub the face up with the float. This would help to close in any pores that might arise in laying on the finishing coat. Strong rooms in banks, etc., are usually plastered out with cement and sand.

Dinging

Dinging is another kind of one-coat work. This work is generally done by bricklayers, as they consider this is part of their branch of trade; but seeing the wall has to be plastered with material of some kind, the plasterers claim the right to carry it out. Dinging is chiefly done to coal-hole walls, vaults, warehouses, stabling, rough old walls, and such like, and is very suitable for these requirements. When the plasterer is required to execute dinging as a finish, he proceeds as follows. Rendering a fair coat over the surface he wants to cover, he lays it evenly with the trowel. Sufficient quantity is laid on, and it must not be left too long to get dry. It is then rubbed over with a stock brush. A grass brush is very suitable for this kind of work, or an old piece of sacking can be used, being rubbed over as if cleaning windows. It is nearly always lime-washed. Air conduits and shafts are sometimes done this way, but in most cases air shafts are parged in the same manner as the bricklayer parges the flues to the chimney, doing this as he builds his way up.

Parging

Parging is the inside lining of a flue. It is, or should be, done with a special kind of mortar, which is spread over the brickwork inside the flue. It is a plaster mortar, mixed with cow dung. A similar kind of treatment is applied to old walls, sometimes adopted in cases where the brickwork, through long exposure to the weather, has greatly deteriorated. Instead of cutting out the bad bricks, the surface is covered with cement and sand, and jointed to imitate brickwork. This does not require much describing, as all that is needed is for the plasterer to cover the bad wall in the same way as the dinging method, but instead of using the brush a jointer (see tools) is used by holding a straight-edge horizontally on the face corresponding with the old joints of the brickwork, and drawing the jointer along the straight-edge into the soft stuff. The same proceeding is carried on vertically.

One-coat Finish Face

One-coat work is sometimes done on concrete floors, the underside forming the ceiling. For these the shuttering boards for the concrete must be carefully adjusted to procure a very smooth surface. Cement and washed sand, 2 parts of sand and 1 part of cement with $\frac{1}{4}$ part of Calime, will make a very good finish. It is laid on with the trowel as thin as possible, rubbed over with the skimming float, and when the whole of the surface has been treated with the float the flat of the laying trowel can be passed over to close in any rough places. This kind of work is very suitable for factories or warehouses, as with good finish concrete it can be finished less than $\frac{1}{4}''$ in thickness, as I have found by my experience of this kind of work.

One-coat Base Work

One-coat work does not always mean a finishing coat, as floating comes under this heading, such as when used for tiles on walls. But if for a tiled hearth or floor, it is generally called screeding. For tiles and on walls, the dots and screeds are formed exactly in the same manner as for ordinary floating, with the exception that the background is composed of cement and sand gauged not less than 4 parts of sand to 1 part of cement. This particular work is described as flanking when carried out with cement, otherwise floating. When flanking in for tiles, it is essential that no dead or killed stuff should be used, otherwise there would be bad suction when the tiler beds the tiles.

Screeding

When tiling or paving is required as a finish on floors, the floating for same is called screeding, whether it be for wood blocks, terrazzo or any other kind of floor composition. Screeds are formed and the bays filled in. If a floor has to be screeded for tiles the level is taken from the finished floor, and the thickness of tile and bed is usually $\frac{5}{8}''$ to $\frac{1}{2}''$ for tile, $\frac{1}{8}''$ for bed. This is taken below the floor level, and screeds levelled and put in to a suitable length for ruling off in between. When screeding floors, a very good method is to obtain $\frac{1}{2}''$ or $\frac{5}{8}''$ battens to form screeds.

These are bedded down to level dots. They are much better to work on with the filling-in rule instead of the soft screeds. After the bays have been filled in, the screeds can be taken up and the space filled in. One very important point when screeding floors is to clean the surface well, care being taken that no particles of plaster are allowed to remain on the surface. In addition, the floor should be well damped and brushed over with liquid cement and sand. The screeding material, if gauged 6 parts of sand and 1 part of cement, will be found quite strong enough. If screeding is required for tiled hearths, a short rule notched at each end, that is cut to a shoulder of $\frac{5}{8}''$ for tile and bed, can be worked off the finished floor.

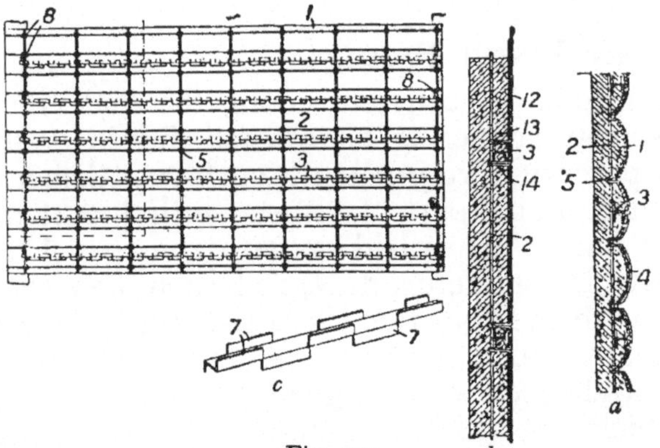

Fig. 114.

Patent Plastered Walls and Ceilings

A novel method of plastering walls and ceilings has been worked out by the Mitchell Holding Corporation of New York (British Patent, No. 244,329), Fig. 114. The plaster is applied to walls and like surfaces against a lathing consisting of a flexible fibrous board which may be corrugated, as in *a*, or flat, as in *b*, and reinforcing and supporting cross bars 3, provided on the front of the board, are held in position by members 2, 4 arranged in connected alignment on both sides of the board. The cross bars 3 have alternate flanges as *c*, extending in opposite directions on the inner and outer edges. The wires 4, in the case of corrugated boards, have loops 5 which pass through the crests of the ridges and are hooked or pass over the corresponding inner wires 2, the cross bars being held in the furrows of the board. When the board is flat, the bars 3 are retained by wires 12, of which loops 13, 14 extend through the backing and are bent around the inner wires 2. The backing is fastened in position by staples 8. The fibrous board may be coated, impregnated or otherwise treated with asphaltum, etc.

Shallow Partitions

When space is limited, partitions can be erected with the use of iron rods, $\frac{3}{8}''$ diameter, and reinforced expanded metal. These are very useful in lavatories for the construction of a series of w.c.s. The rods are fixed in the floor and ceiling about 1′ apart, and the wire fixed horizontally to the rods by threading wire through and round the rods, twisting the ends together. Fig. 114A shows the rods fixed and the reinforced metal lathing. The wire is covered with whatever material is desired, but a small quantity of lime and hair should be added in the first coat. When covering the wire on the first side, the opposite side requires strutting, as it makes it much easier to cover. When the first side is set the struts can be taken down. Partitions can be made less than 2″ when finished by this method.

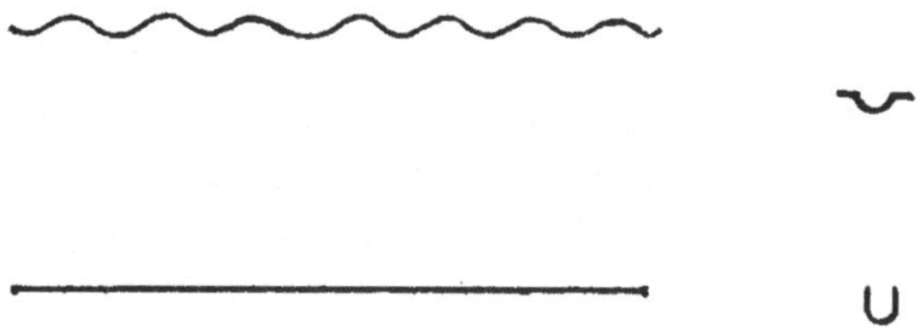

One-coat Lime-cement Plastering

Mr. John Dyre has published a system of one-coat lime-cement plastering which has been used successfully both for interior and exterior work. In the course of this book reference has been made to the use of hydrated lime in cement and other mortars for the purpose of retarding setting and making them more plastic. Mr. Dyre uses equal quantities, by measure, of hydrated lime and good Portland cement. If the cement is poor, rather less hydrated lime must be used. To complete, the three parts of sand are added. This will do for ordinary interior and exterior work. If a slower-drying mortar is required more sand is added.

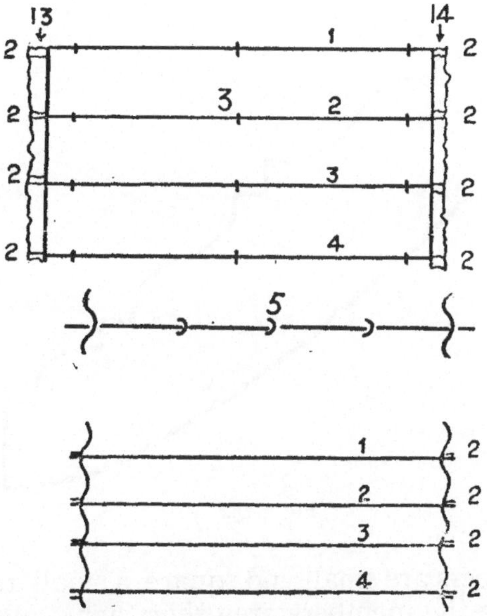

Fig. 114A.

If desired, colouring matter can be added, the usual precautions of shovelling together dry and passing through a sieve to ensure uniformity being taken. Water is added in moderation, as the hydrated lime ensures perfect hydration of the cement.

The mix is applied in the same way as blocked stucco. The wall is wetted if necessary, dots put in and screeds run. The whole wall is then rendered in and ruled off between the screeds. The coat is laid on ¼″ thick. Very soon after laying out, the coat can be floated up or otherwise finished. Owing to the slower setting and greater plastic qualities of the plaster, considerable variety may be introduced in finish, much as described in the chapter on "Cement Stucco."

CHAPTER XIV

RUNNING CORNICES AND MOULDINGS

Building-up the Mould—Running the Moulding—Forming the Screeds—The Running—Mouldings on Arches—Circular Mouldings Run with a Radius on a Bench—Ovolo Mouldings—Running Elliptical-shape Mouldings *in Situ*—Running Mouldings in Cement—Method of Running the Moulding—Materials for Running off Cornices—The Use of Driers.

In dealing with cornices and mouldings and the method of running these, it is necessary to note the difference in method of running interior plaster cornices and external cement cornices.

Building-up the Mould

As regards plaster cornices, when a sketch of the desired detail or section has been decided upon, as Fig. 115, it is traced off on to tracing paper so as to avoid damaging the original drawings. The tracing is pasted on to a piece of zinc if only a small piece of cornice has to be run, but if a large amount is required, a piece of sheet iron would be better, for, being stouter than zinc, the prominent members would not be so likely to get bent or damaged in use. When the paper is dry the profile is cut out with a small pair of snips, or cutters as they are sometimes termed, care being taken not to cut out the line of the tracing. Where the members are small and square, a small iron chisel will be found very useful; where the members are other than square, round chisels, such as carvers use, will be found handier. After the profile has been cut out, the zinc is flattened with a flat-headed hammer, and the back part of the zinc (that is the reverse side to that on which the paper with the section has been pasted) is filed down to take off the rough edge caused by the cutting of the profile. This done, the zinc profile is put in the vice between two pieces of wood, which stiffens the zinc and makes it much easier for the next process, which is the filing up of the mould. Square or half-round

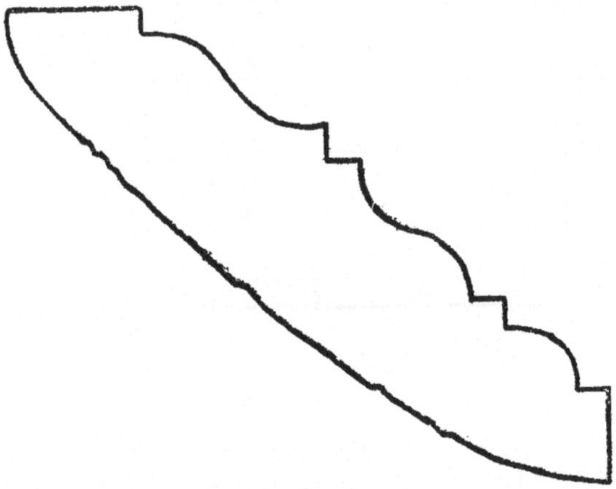

Fig. 115.

files are used for this purpose. Sufficient filing is given up to the line of the section of the moulding as traced off from the original drawing. It is for this reason care must be taken when cutting out the profile with the snips or chisel, that the line is not cut out, otherwise when filing up, the operator is likely to distort the correct detail. Fig. 117 shows the zinc with the section or profile of the cornice desired to be run, the dotted line showing the portion of zinc which is sufficient for the running mould, as it is termed after the filing has been completed. A suitable piece of wood is required, about $\frac{3}{4}''$ thick, which should be planed up, as the mould can be cleaned so much better than if it is left rough. This piece of wood is sometimes called the stock. When you have obtained this, lay the zinc on the section of the moulding from the ceiling line member to the wall member. Then remove the zinc and draw another line about $\frac{1}{8}''$ back from the original line. Fig. 116 shows the wood and the two lines as described.

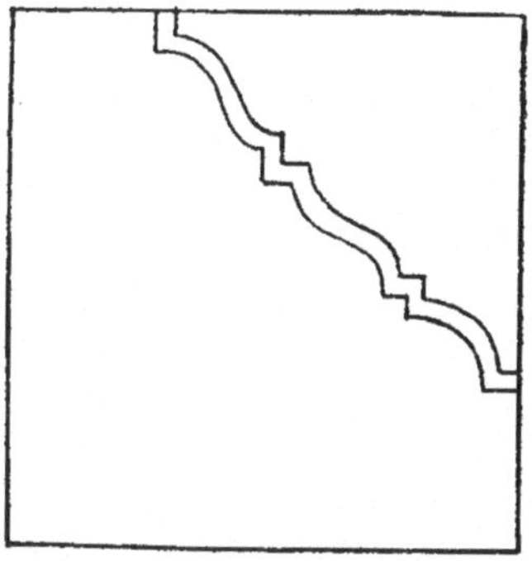

Fig. 116.

When placed in the vice, with a key-hole saw carefully cut away the back line on the splay downwards from the front edge about $\frac{3}{8}''$. The reason for so doing is to allow for the wood swelling when wet, also to permit the mould to run free and at the same time to render cleaning easy while in use. If this is not done, a considerable amount of trouble will be met with in running the required moulding, as the grit and setting of the material used in running the moulding would project in front of the section of zinc. After cutting to the two lines as described, holes are punched through the zinc, as shown in Fig. 117. Here it is well to emphasise the point that in cutting the wood for the section of cornice, the ceiling and wall line must be cut square, so as to allow the mould to run fair on ceiling and wall.

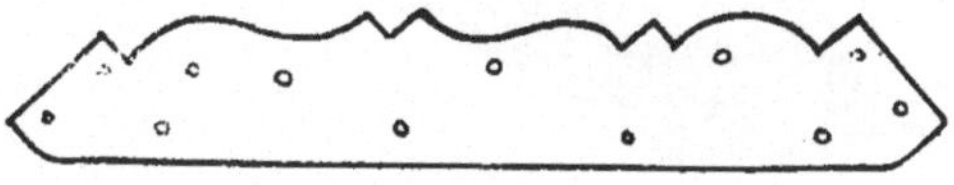

Fig. 117.

To return to our operations. The zinc or iron plate is next tacked on to the stock, with $\frac{3}{8}''$ tin tacks. We then require a suitable piece of wood, about $\frac{3}{4}''$ thick for what is termed the horse, the length of which should be about one and half greater than the depth of the cornice, though in large running moulds the thickness and length would be increased to allow for stays, or braces, to stiffen the moulding, for unless the mould is well stiffened, great trouble would arise in running the

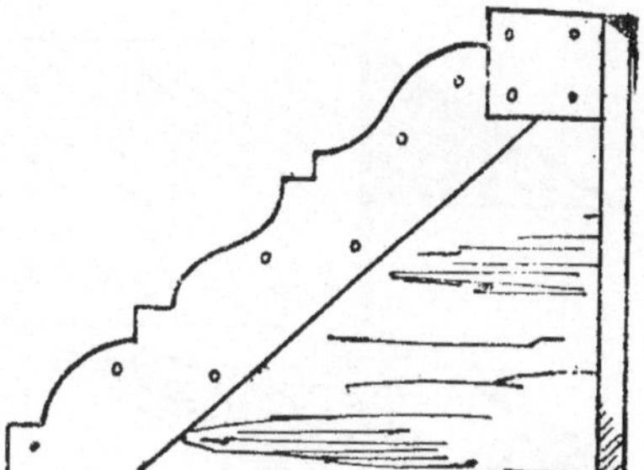

Fig. 118.

moulding when the plaster starts to swell. Fig. 118 shows the stock after it has been cut out and the zinc tacked on. Fig. 119 shows the horse; the groove in centre being cut out about ½″ to form a sinking to

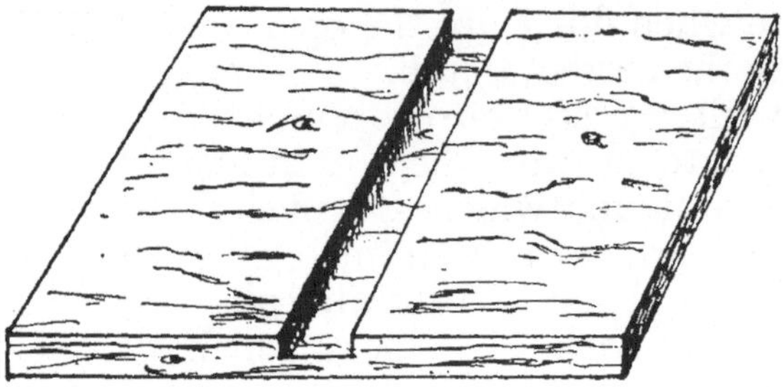

Fig. 119.

receive the wood with the zinc tacked on. Next, two strips of zinc, 1″ wide and about 3″ long, according to thickness of horse, are tacked on 1″ from each end of wood; tacking one side first and dressing the remainder round the front edge, and when fitted to the wood, tacking the other end. Fig. 120 is a section of the strips of zinc, which are termed slippers (in some districts shoes), showing the position they would occupy after being nailed on to the horse.

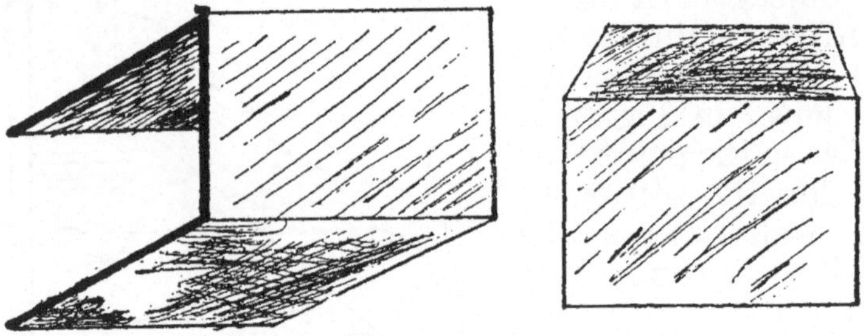

Fig. 120

Another piece of zinc is required to tack on the mould; this is called the nib; Figs. 118 and 119 show the zinc slippers on horse and nib.

The next process required is to place the mould in the sinking as shown in Fig. 119. It is temporarily nailed through the horse, care being taken that it is kept upright by trying the square to it. Then it can be made a fixture, but being particular to see that the horse is flush with the wall line of the section of moulding, otherwise it will cause the mould to jump when running, if it is full, or result in the mould gathering on the wall line, if it is back from the face.

To stiffen the mould, it is necessary to fix a brace on each side from the horse to the nib, care being taken that they are clear of the section of the mould, as these braces are used as handles. As already stated, the larger the mould the more stays or braces will be required to stiffen the mould batten. 1″ × 1″ stuff will be found a very useful thickness for this purpose, as it can be easily trimmed with the knife to fit the hand, which makes it better for working when in the process of running. Fig. 121 shows the two braces in prospective from the back of the mould. Fig. 122 displays the mould complete and ready for running. It is not necessary for the stock to bc of the same width as the horse, as the mould requires to be made as light as possible. The lighter it is, the easier it is to be run. It is advisable to make the horse as short as possible, so as to be able to push the mould close, thus reducing the amount of hand labour in working up the mitres.

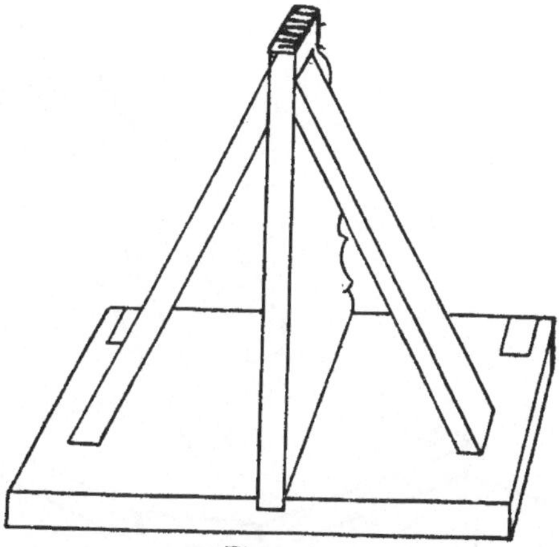

Fig. 121.

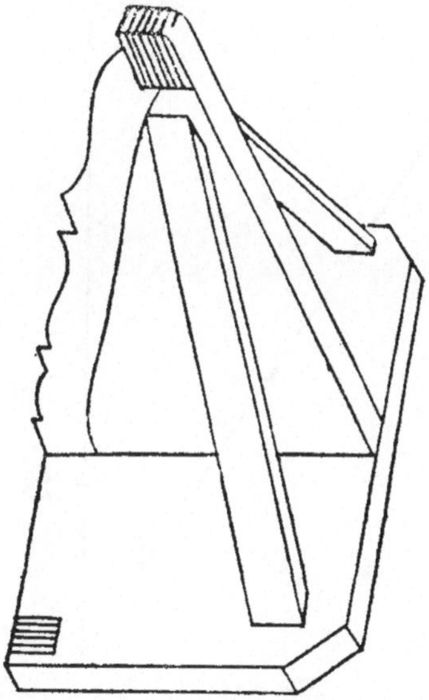

Fig. 122.

It will be noticed in Fig. 122 that the corners of the horse, and at the nib, have been taken off, which contributes to the lightness of the mould. The wood must be cut off square and evenly, for when fixing the rules for running this edge is required to run to the line snapped on ceiling, to straighten the running rules. In some districts the mould is run with the left hand and fed with the right, though in London and the South of England it is customary to run the mould with the right hand.

Running the Moulding

When the mould has been horsed up and the zinc slippers have been filed over to take the rough edges off, the following process will be proceeded with.

Place the mould up to the ceiling and against the wall, Fig. 123, and mark off the nib at A on ceiling, and the underside of horse at B on the wall. Another way to make sure the mould is upright is to mark the projection of the moulding on a square, also its depth, Fig. 124, then place the square up to the angle on the wall and ceiling and mark off the two points A and B. When this has been completed a line is chalked, usually with a portion of dry plaster, and then snapped to the two points as marked on ceiling and wall.

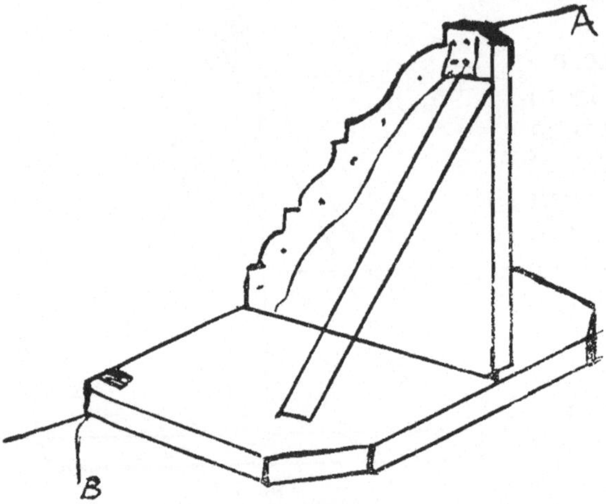

Fig. 123.

Forming the Screeds

Seeing that mouldings are usually run on the floating face, it will be found that plaster is most suitable to whiten the line with, as in most cases inside mouldings are run on thin screeds, which have to be formed. These are called running screeds. The process of forming these is as follows: A portion of lime putty and plaster is gauged up by placing a sufficient quantity of putty on the gauging board. No more than what the plasterer thinks he can use before it sets should be taken, and opened out into a ring, enough water being poured in with a gauge pot (a tin used for dipping water out of a pail for gauging purposes), to make a soft putty. It is advisable to gauge the materials as soft as possible, as it then takes longer to set and so gives more time to use. When once set, it is killed. If killed stuff (plasters or cements after setting) is knocked or mixed up again, the screeds would be soft and the zinc slippers of the mould would be "digging" in before the stuff is laid on. This is usually done with the laying trowel. A short rule about 6′ long is required, called a traversing rule, the edge of which it is necessary to damp with the water brush before use, to prevent the plaster from sticking to it; if it is not damped there would be a possibility of the dry rule pulling off the stuff as laid on.

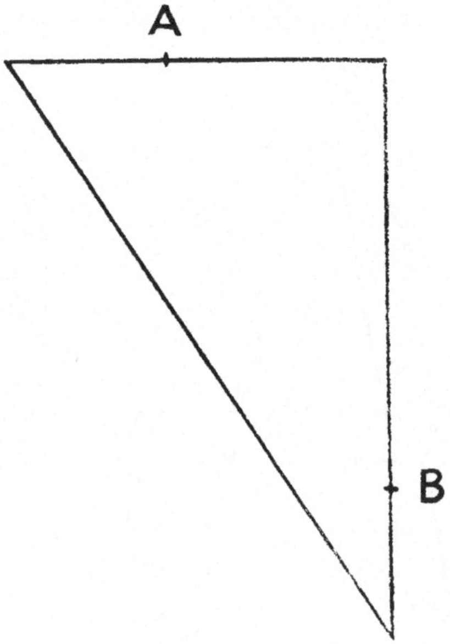

Fig. 124.

The Running

When all is ready, one man lays the stuff on the wall or ceiling, as the case may be, to the line, as previously described, and his mate (for as a rule two men work as mates) takes the traversing rule and straightens off the stuff to the thickness of little more than $\frac{1}{8}''$, as the screed is only required for a straightened and smooth surface to run on. The thinner it is kept the better, as a thick screed would entail more plaster being laid on in the setting coat of the ceiling or wall.

As soon as the screeds have been traversed with the rule, they are rubbed over with a cross-grained float, and, if necessary, a little water is thrown on the screeds with a brush. This is needed, as it takes off any rough places there might be, or any small places that have not been filled up by the rule. This done, the mould is again placed up to the angle of the wall and ceiling, or the square can be used as before described, to mark off the ceiling and wall line. A piece of black is then required, which can be prepared by putting a piece of wood in the fire and burning it partly through. On taking it out of the fire it should not be dipped in water, as that would spoil the effect of the black, but it should be allowed to die out itself, or if it is required at once, it can be put in the plaster box or sack. Another method of marking the line is to gauge a little weak plaster and a small quantity of Venetian red. This makes a very good substance for marking the line with. Whatever is at hand, a line covered with it is snapped on the wall and ceiling. Black or red are most suitable, as they show up so much better on the white screeds.

The next thing required is a running rule, which is usually 2″ wide and $\frac{1}{2}''$ thick, planed smooth. These run in various lengths and are pieced out to suit the lengths of moulding required to be run, care being taken that the intersections of these pieces are flush with each other, otherwise it would cause what is termed a cripple or a kink. If the rules meet in the angles, attention must be given that these intersect with each other; should this not be so, a great deal of trouble would be given in putting in the mitres, as the two mouldings would not intersect correctly. When the required lengths of rules have been cut, they are nailed up on the wall with small nails at the lower part of the rules (so that they do not catch the man's hand when running the mould), to the line as marked off by the mould or square (Fig. 124). The mould is then placed on the running rule, starting from the right-hand corner, care again being taken that the nib of the mould (Fig. 123) carries along to the line as marked on the ceiling. The rule on the wall is tacked up so that the ceiling line runs straight, for then, if the wall should be a little hollow, it is not so noticeable when the moulding is run.

Fig. 125 shows the position of mould on running rule when ready for running. When running plaster mouldings, in fact any description of mouldings, with the exception of very small ones, it is always advisable to core them out with plaster mouldings coarse stuff, which is composed of lime and sand with a small quantity of cow hair. This is gauged about 1 part of lime and sand to 2 of plaster. Before proceeding to gauge the material, it is necessary to muffle the mould, which can be done in various ways. One method, which I consider is the best if one has a lot to run, is to cut out the profile in zinc, tinplate or sheet iron, about $\frac{1}{4}''$ full of the original detail with the exception of wall and ceiling line, which would require to be flush to allow of the mould to run. This would not require filing up. When this has been cut out, it is tacked over the original profile to be run.

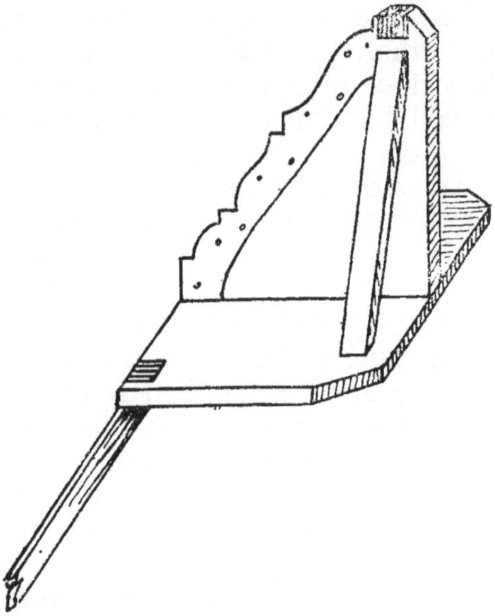

Fig. 125.

Another way, useful if only a small quantity has to be run, is to make a plaster muffle, gauging a small quantity of neat plaster and laying it on the front part of the mould, covering the wood and section of the zinc, and when it is set, trimming it down to about $\frac{1}{4}''$ in thickness in front of the profile of the zinc. Fig. 126 shows the plaster muffle after it has been trimmed. The back edge, that is the reversed side to the zinc, is chamfered off, downward, about $\frac{1}{4}''$, to allow of the mould running free. Any stuff which would adhere to it in the running could then easily be cleaned off with the trowel or small tool.

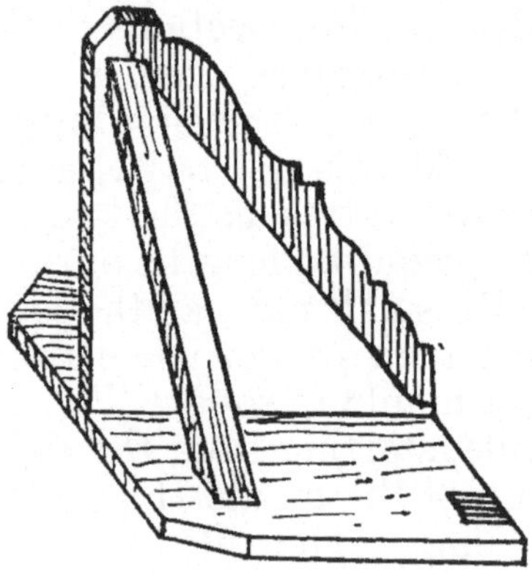

Fig. 126.

In cornices of 6″ projections or over, it is beneficial to adopt the method of coring out on what is termed Scotch bracketing. This method is the means of saving a lot of material, so reducing the cost of plaster decoration. Apart from that, much is gained by cutting down the weight added to the ceiling. With large mouldings or cornices it will be seen that it is a very useful method. For example, a moulding is required, as shown in Fig. 127; the desired projection has been noted on the wall and ceiling, and the length of the lath (A being the lath) is taken about 1″ back from ceiling and up from wall line, and the laths are cut to that length. Then with the materials gauged, as I have previously described for coring out, and with the gauging trowel, which is more handy to use as only a small amount of gauged stuff is required, sufficient stuff is laid on wall and ceiling to bed the laths in, these being kept about $\frac{3}{4}''$ apart. It must not be forgotten that the angles or mitres require coring

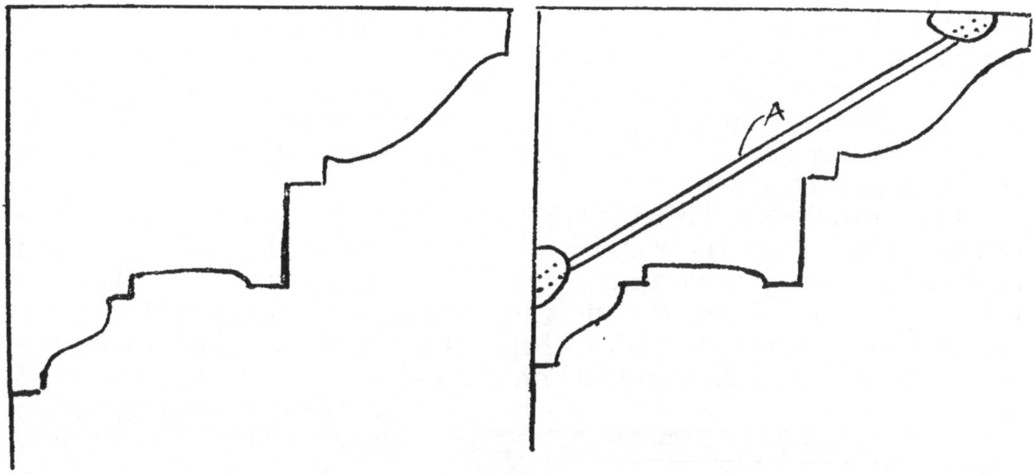

Fig. 127.

out with a short rule, as owing to the horse of the mould a certain amount from each angle has to be made out by hand. This can be worked off the core run by the mould, so that when the finishing of the mitres has to be done, it does not require a lot of finishing stuff, again a saving of fine material. Care must also be taken when coring out that no salvage or gathering of superfluous material is allowed to collect on the running screeds on wall or ceiling, or on the running rules, for if this occurs great difficulty in running the finishing of the moulding will be experienced. Fig. 128: A horse, B mould, C running rule.

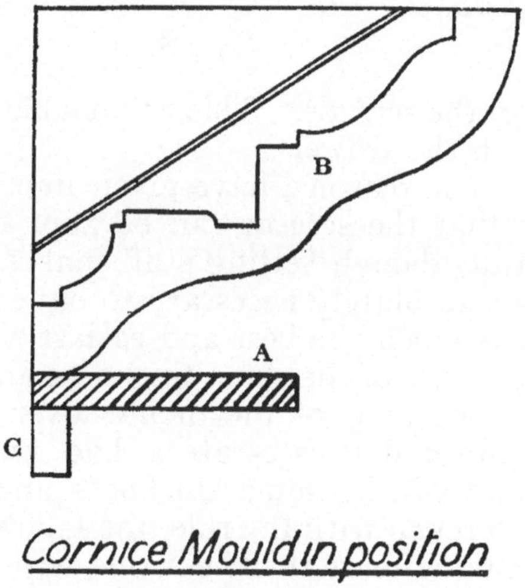

Fig. 128.

The coring out being finished, the muffle is taken off, if a zinc one. If it is of plaster, it is advisable to cut it off with a wood chisel, so as not to damage the profile of the moulding. The mixing board, or spot board as it is commonly called, is then cleaned off, ready for the finishing of the moulding. If it is required to be finished off in putty and plaster, enough putty is put on the board as is judged to be about the amount needed to run off, as it is easier working to lay on just enough to complete as near as can be done, with as few gaugings as possible. A ring is formed and sufficient water is added; that is to say, enough to absorb about two parts of plaster to the quantity of putty. Unless sufficient plaster is used, when the mitres have to be worked by hand with the use of the joint rule, it would be found that in the working of the joint rule on the finished moulding, the arrises and the members would get damaged. So it is of great importance that the amount of plaster should not be stinted. After the moulding has been run off, the rules are taken down and the nails removed. These rules should be tied up together so that they may be used again.

It is generally found after running mouldings that a certain amount of salvage occurs. By the running of the mould on the screeds, it will be found necessary to clean off, unless the running screeds are kept as thin as possible. Extra thickness of finishing coat is required in some cases; then a short nail, with no head, is put in the mould between the bottom members of the moulding on the wall line and the upper part of the horse, projecting about an $\frac{1}{8}''$. This cuts through the screed, so when the running is completed the screed is cut off to the line made by the nail; though if it is desired to take the screeds off, a small portion of lime putty is spread along the ceiling and wall line before the lines are snapped for the screeds. This will enable the screeds to be taken off quite easily with the trowel.

Horse, another view. Rabbet tacked on.

Fig. 129.

The reason I have given lime putty for forming the running screeds, is that the screeds can be kept much thinner than if done with setting stuff, though setting stuff makes much harder screeds to run on. It is not absolutely necessary to have screeds on the wall or running line, but it is much the best and easiest way for running mouldings, as unless the rules are of the same thickness and the floating very smooth and straight, the running of the mould over the joints would cause the mould to jump and thus create a bad line in the mouldings. So when one has rules of the same thickness and the floating is straight and not left too rough with the rule, the following method is adopted, which is termed running on a thickness rule with a rabbet. In this the mould is put in position on the rule, and about $\frac{1}{8}''$ from the wall, to allow for the thickness of the finishing coat to wall. Then the underside of the horse of the mould is marked where the edge of the running rule meets it. To this line a short piece of rule, about $1'' \times \frac{1}{2}''$ and the length of the horse, called the rabbet, is tacked on the horse (Fig. 129). The running of the moulding is then carried on as I have previously described. This method of running with a rabbet will be found very useful when running across chimney breasts or returns to recesses, etc. When running returns, or brakes, as they are sometimes called, a piece of ¾″ board is required, such as a floor board, but care must be taken that it is of sufficient length beyond the wall line to allow of enough projection to be run (Fig. 130). If a board is not available, the running rule on the external angle is carried beyond the wall line a little longer than the length of the horse, but before fixing it in its position a short piece of rule is cut and tacked on the back edge of the rule. At the same time two pieces of lath are put in between at each end, keeping the back piece of rule up above the front rule about 1″, for the mould to run against. It will then be found necessary to form a screed on the back rule, the same face as the wall line. That is the reason the two pieces of lath are put in between to allow for the thickness of screed.

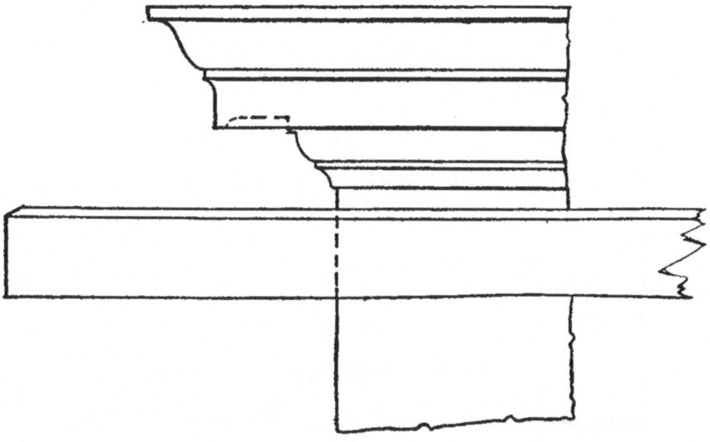
Running across a pier with floor board and rabbet

Fig. 130.

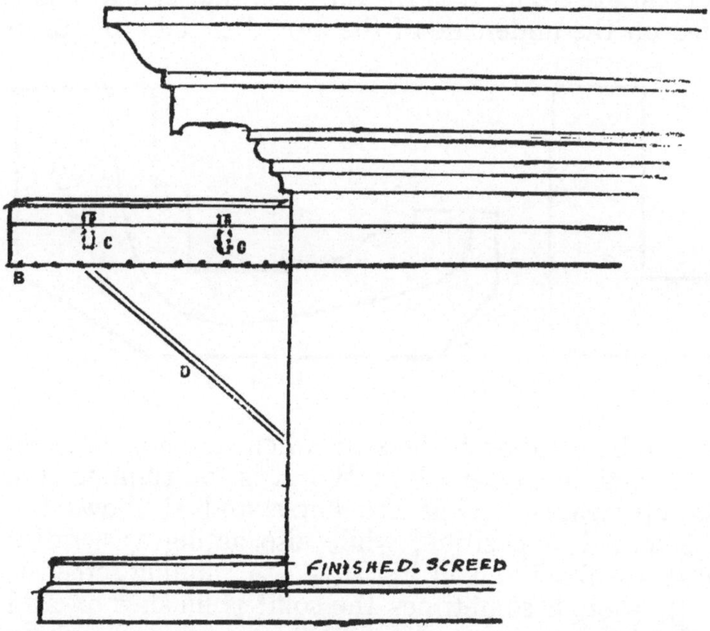

Fig. 131.

Fig. 131 shows a cornice running across a pier or chimney breast. B, the dotted line, is the position of the running rule across the face, and C shows the two pieces of lath in between the two rules, to allow of the screed to be carried through from the face line of the breast or pier D, a stay to support rule.

Hanging Moulds

A hanging mould will be found very useful for running bands to soffits of beams or mouldings, outside stringing to soffits of staircases, etc. For example, a soffit of a beam is required to have a moulded sunk panel, with plain band on the outside, as Fig. 132. The mould would be horsed up in the following manner. Fig. 133 shows the section required, which would be carried out by the process already described when cutting the profile and preparing the stock. In this case the wood has to be of a different shape to the stock of the cornice mould, as Fig. 134 will show. The dotted line denotes the zinc tacked on to the stock. A is the concrete beam, B is the nib, C is the screed, D is a thin piece of wood nailed on the underside of the mould to help to stiffen

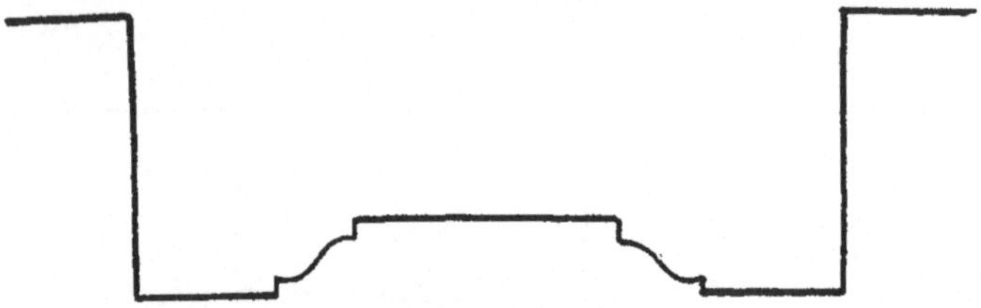

Fig. 132.

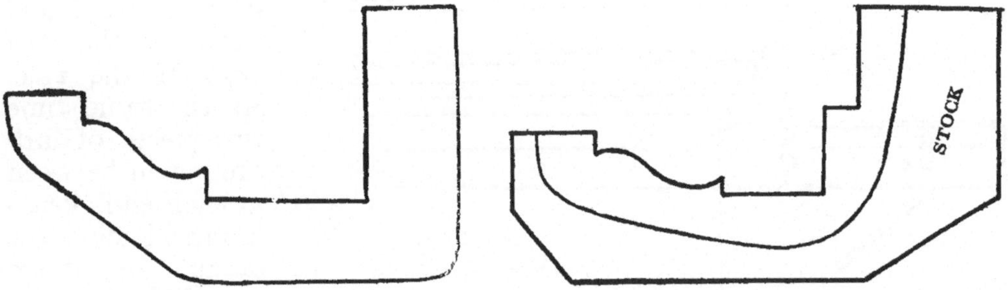

Fig. 133.

it, and at the same time to help to catch the stuff when running, and also to enable a brace to be nailed on to the nib at B. R is the running rule which the mould runs up against. G is the horse, and H shows the two braces to keep the mould in position, while also acting as handles and to stiffen it. These are fixed at I in Fig. 135. A running screed is required at C (Fig. 134), though sometimes the soffit is finished off and the mould run on on the finished face. In some cases, however, where there is good fixing on the soffit, a parallel nib rule is fixed to save a screed, and the same rule without refixing acts for running both sides.

The following method will be found very useful where ceilings are split up into bays, by girders or dummy beams; follow out the same line, that is if the cornice has to be run. The same running rule which is used to run the cornice can be utilised to run the moulded angle to soffit of beam, also on the same screed, though it would be advisable to try the rule with the straight-edge to adjust it to suit the centre of the soffit, because when one runs the cornice, as I have previously stated, it is run to the ceiling line. If this method of the hanging mould is adopted, the mould will run on the top of the rule and not against the underside, as in Fig. 134. In looking at Fig. 133, it will be seen that the stock of the mould is cut out to allow of it passing over the running rules, which should be nailed up with short nails and driven in flush, otherwise hands may be damaged in the operation of running. A hanging mould of this description is easy to run. As the mould runs on the top of the rule, its weight rests on the rule, instead of being held up by the hand, as in the previous method describing the mould being run on the underside of the running rule.

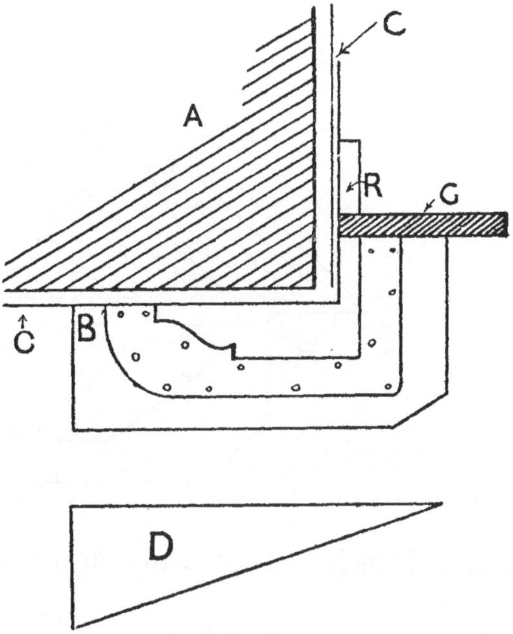

Fig. 134.

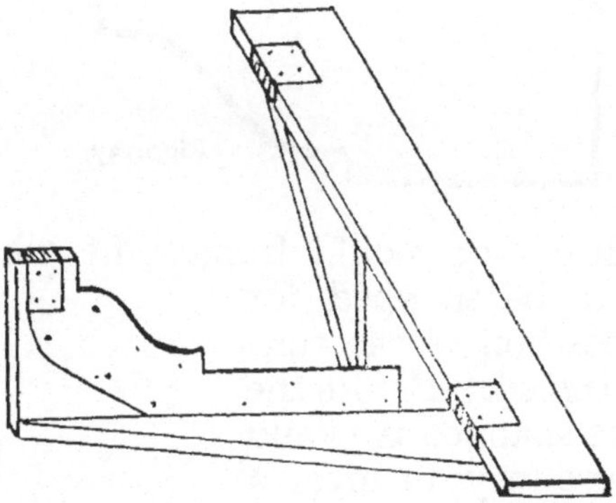
Fig. 135.

Fig. 136 shows how the mould would appear in its position for running. A is the section of the cornice, B the Scotch bracketing; C, the dotted line shows the profile of wood brackets, which are fixed when the mouldings are very large and then lathed. D is the concrete casing of girder, S the running screed; H the horse of mould, R the running rule. The dotted line round mould is notched out to allow of mould passing rule, S and R, screed or nib rule, whichever method is adopted.

Another way used in running panel soffits to beam or concrete girder

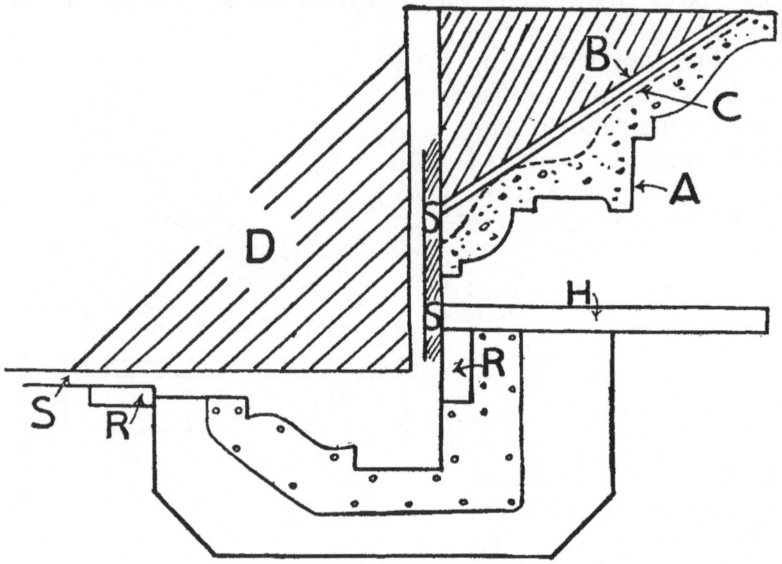

Fig. 136.

casings (which method I myself do not recommend) is a mould made to run the full soffit and moulding and arris. A screed is formed on one side of the girder casing, sufficiently high from the soffit to allow for the depth of the mould, the distance being arrived at by the following method: Square a line across the mould section, as shown in Fig. 137, by the dotted line and from A to A, this being the depth of the mould from soffit, which would be the distance up for the line to be snapped for the position of the running screed. Before the line is snapped it would be necessary to level a narrow screed across the soffit at each end, as shown at Fig. 138. Then with a short piece of rule held on this screed, a piece of lath cut to the length A to A permits of the correct distance being marked on sides

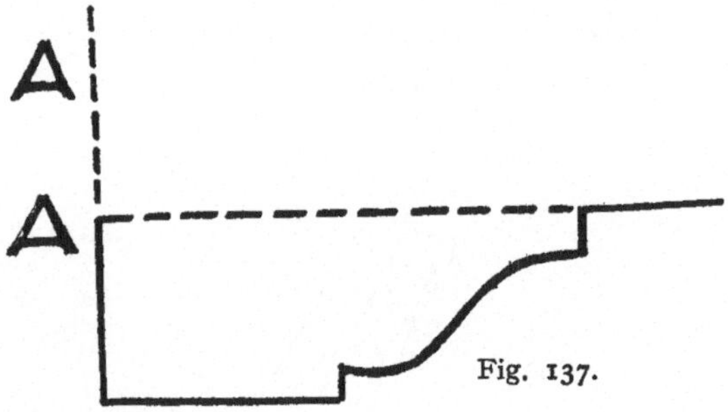

Fig. 137.

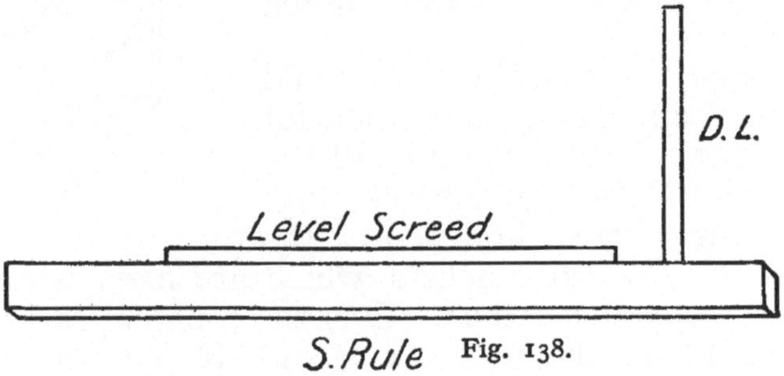

Fig. 138.

of beam. In Fig. 138 L S is the level screed; S R is the short rule; D L is the distance lath as from A to A in Fig. 136, which is a perspective view of the mould. In Fig. 139 N N is a sharp-pointed nail to cut off

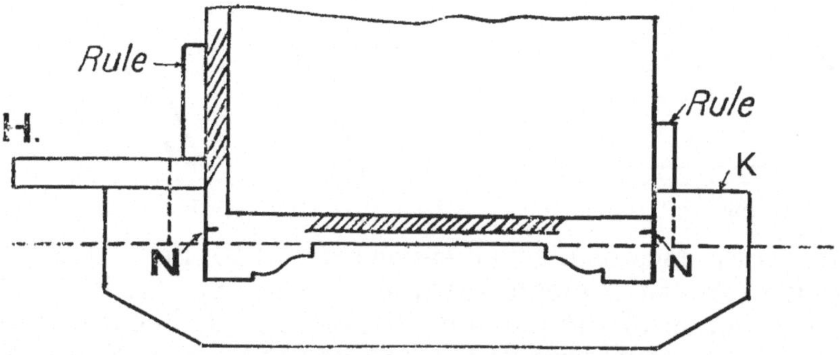

Fig. 139.

salvage to a sharp edge. H is the horse against the side of beam where the running screed would be formed; K is the nib which runs on the nib rule, its position being obtained by the same method as from A to A on the horse side. When the screed has been completed, the same method is adopted with the short rule and the distance lath to mark the position of the rule line. A line is then snapped, and the running rules are tacked up, and the running of the soffit is proceeded with. Fig. 139 is a perspective view of the mould in position. This method of running soffits and panel moulding to beams is found useful where the soffits are narrow, but if the soffits are large it is an awkward job.

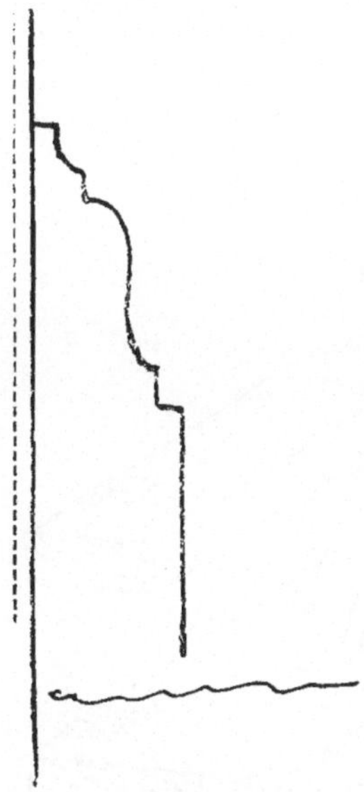

Fig. 140.

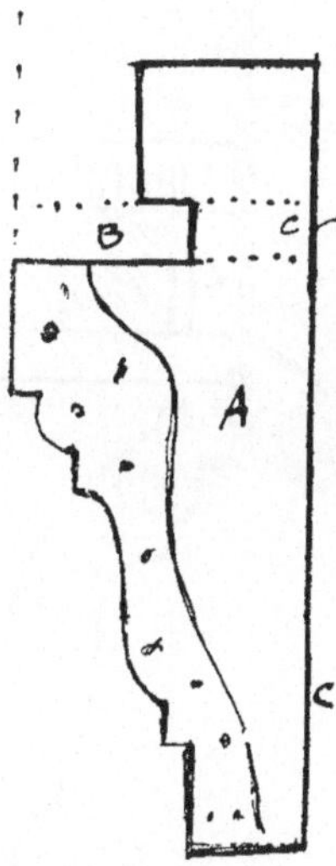

Fig. 141.

Another kind of hanging mould is the double horse mould; it is sometimes termed the double slipper mould. However, the use of the word slipper is rather misleading, as when working in the shop very often the horse of the mould is termed a slipper, because the mould slips along the bench on the horse. In the shop the term is "horse up the mould," and when working on the building the same term would be used, though it would be understood that zinc or iron slippers would be put on. As regards the term shoes, I always look upon a shoe as the iron or brass slot for gig-stick in running arches. This I will describe later when I am dealing with arches. The advantage of using double horse moulds will be seen by reference to Fig. 140, which shows a section of a moulded skirting, with a projection beyond the wall face. The dotted

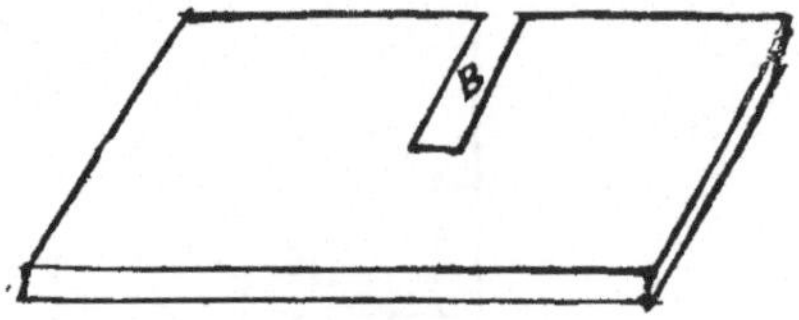

Fig. 142.

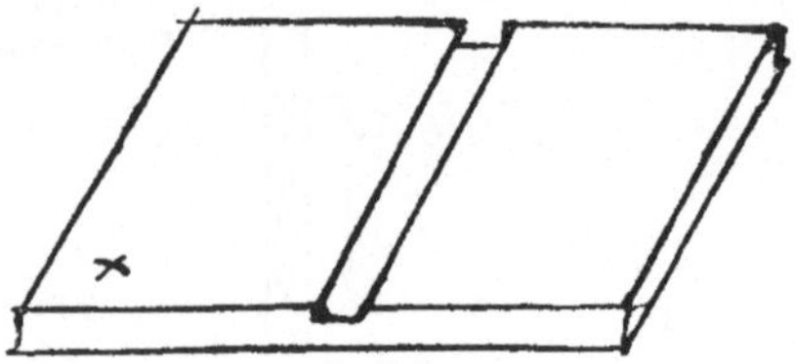

Fig. 143.

line indicates the face of the wall carried from floor line. Fig. 141 shows the various parts of the mould ready for horsing up. A is the stock cut out and the section of the moulding tacked on; B is cut out to dotted line to allow of narrow horse, in Fig. 142, to fit in C, which is the back line of horse in Fig. 141.

Fig. 143 shows the top horse, which is wider than Fig. 142. The reason why Fig. 142 is narrower than Fig. 143 is to allow of the two braces (which

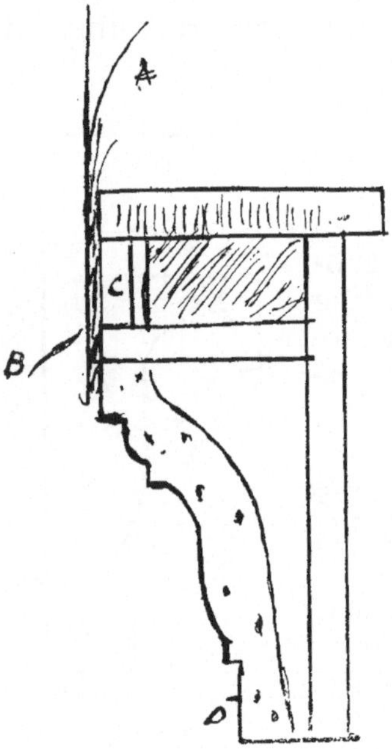

Fig. 144.

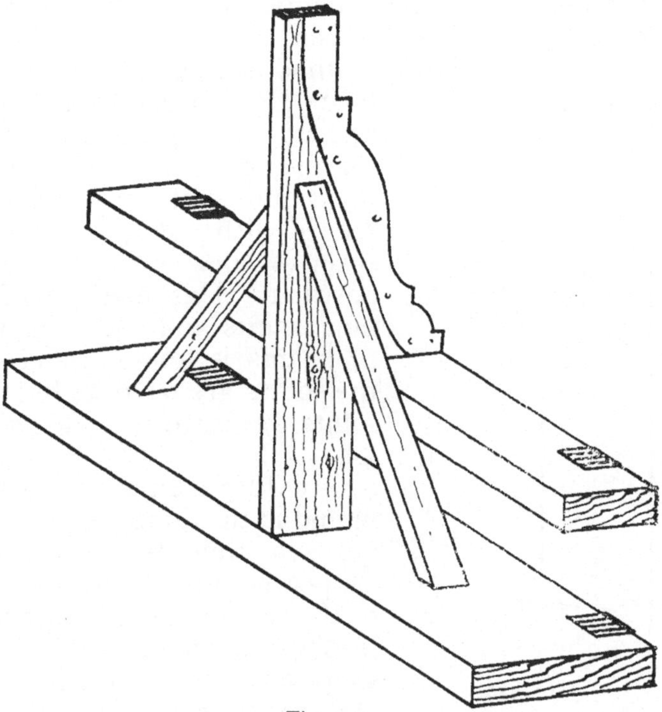

Fig. 145.

can be used as handles) being nailed on. Fig. 144 shows the position of the braces when nailed on the mould to back of C, in Fig. 141, and underside at *x* in Fig. 143. In Fig. 144 A is the face of the wall; B is the screed; C the running rule; D a sharp-pointed nail put in the mould to cut off the plinth to a sharp edge, so as to make an easier and cleaner job when finishing the plinth. Fig. 144 the mould is in position for running; Fig. 145 showing a perspective view of the mould upside down.

Fig. 146.

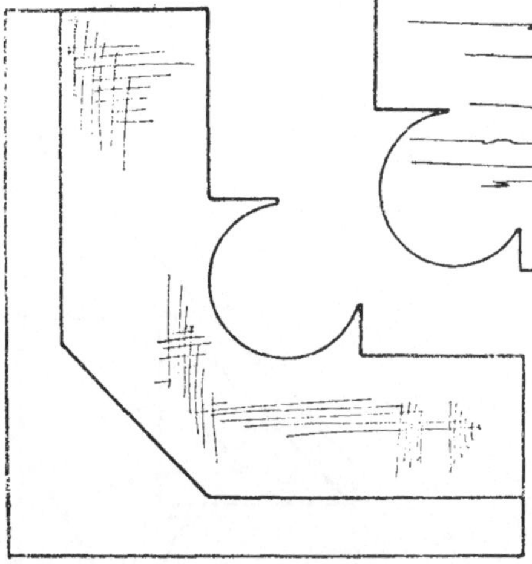

Fig. 147.

This method of the double horse, or double bearing mould, is very useful where there is no nib bearing, as it can be used either way, as before described, for skirtings with a projection or upright mouldings, such as running mouldings up to gables of a roof to finished tiles or slates, or running beads to tiled dados, or on staircases which have a marble stringing to stairs. In fact, it can be utilised in a good many cases to save the fixing of a nib rule. As one of the chief things one has to look to is the saving of unnecessary labour, and at the same time getting over the difficulty of running the various mouldings, it deserves special attention.

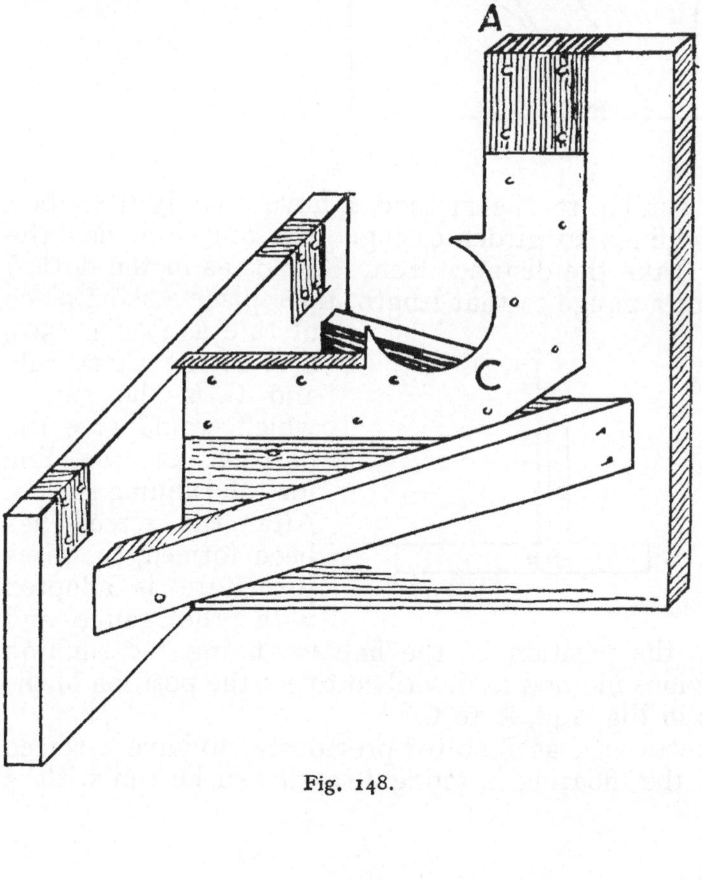

Fig. 148.

Angle Mouldings

In running a moulding of a staff bead, or ovolo moulding on angles either to reveals to windows or other openings, or on bullnose or chamfered angles, the same method would be adopted, so I will deal with a bead with two quirks.

Fig. 146 is the finished bead; Fig. 147 is the mould cut out and fixed on the wood ready for the horsing up. Fig. 148 shows the mould ready for use. The method of procedure for the running of

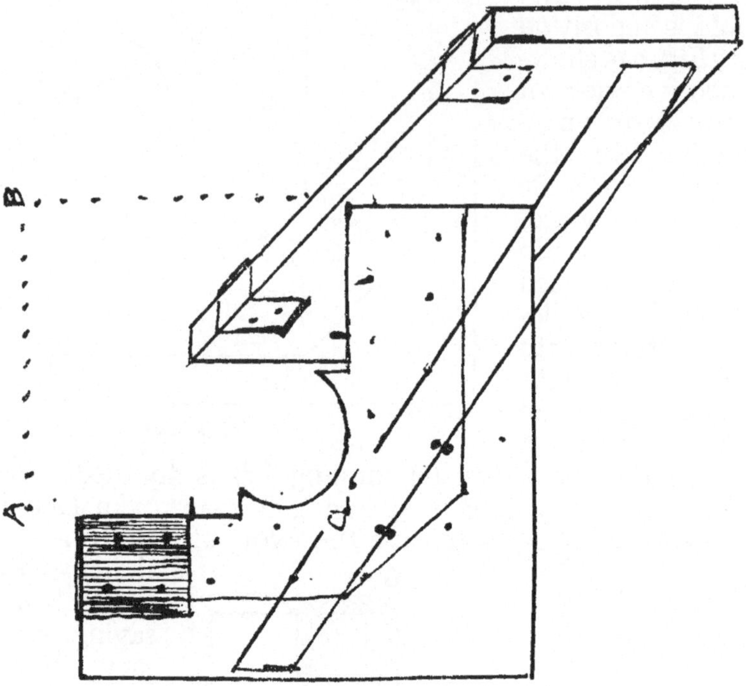

Fig. 149.

the angle moulding is similar to the method I have already described in the running of mouldings to girder casing. For angle moulds the method is as follows: Take the distance from A to B, as in the dotted line in Fig. 149, and cut a gauge to that length, then place a short piece of rule as in Fig. 150; S R is the short rule and G is the gauge, which would give the position of the line for the running screed. After the screed has been formed, the same procedure is adopted with the gauge and short rule, which gives the position of the line for fixing the running rule, repeating the previous method as described to get the position of the rule line of the nib, as in Fig. 148, A to C.

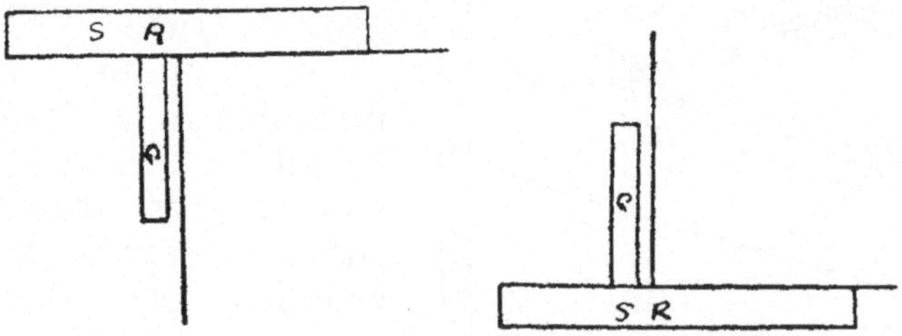

Fig. 150.

It is not always necessary, as I stated previously, to have a screed for running, provided the floating is straight, as it can be run with a rabbet tacked on the back of the horse of the mould. This method I dealt with when describing the running of moulding, etc., across piers.

One point that must be observed when taking the distance from A to B, Fig. 149, is to allow the thickness of finishing coat on the face at C, Fig. 148, as the nib of the mould running on the nib rule instead of on a screed, the extra ⅛″ would have to be added on the gauge from A to B, Fig. 149. Another small matter which must not be overlooked is the placing of two sharp-pointed nails in the mould to cut off the surplus stuff in running, and to make a clean edge to which to finish. In running a plain chamfer to an angle, as in Fig. 151, the same method would be adopted as in the running of the angle bead. When running angle mouldings on chamfers, the usual finish adopted to skirtings or to picture rails or friezes is what is known as a stop, shown in Fig. 152, though if an ovolo is run to window reveals or other openings, which do not interfere with picture rails, etc., they usually mitre up with each other. In some cases stops are put at about 2″ to 3″ from angle of reveal or skirting.

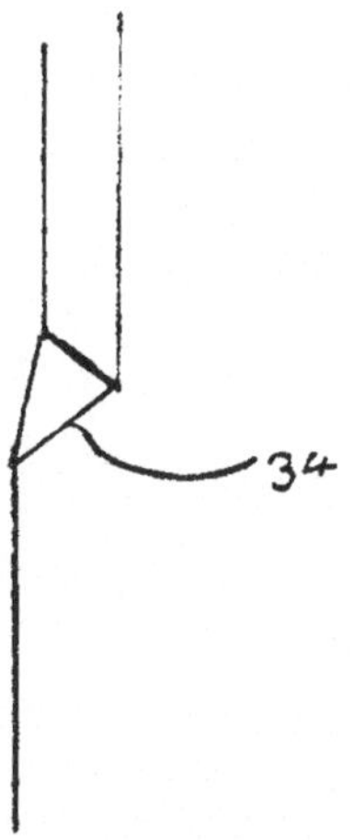

Fig. 151.

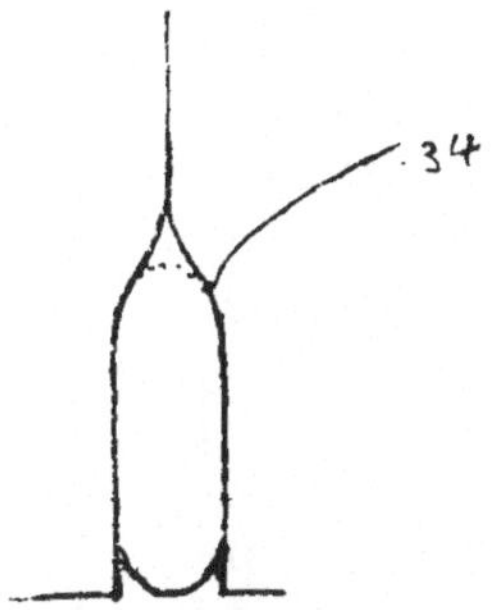

Fig. 152.

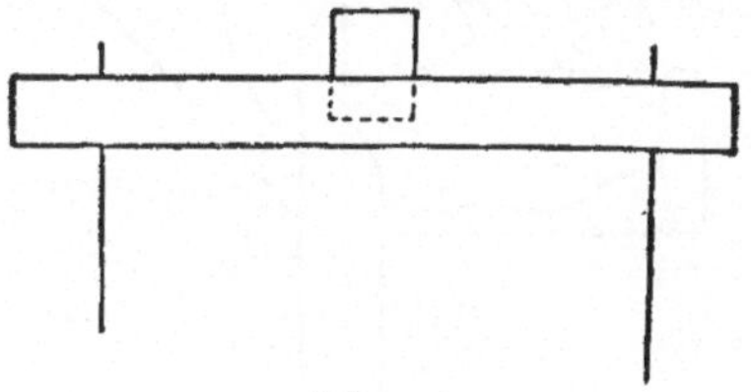
Fig. 153.

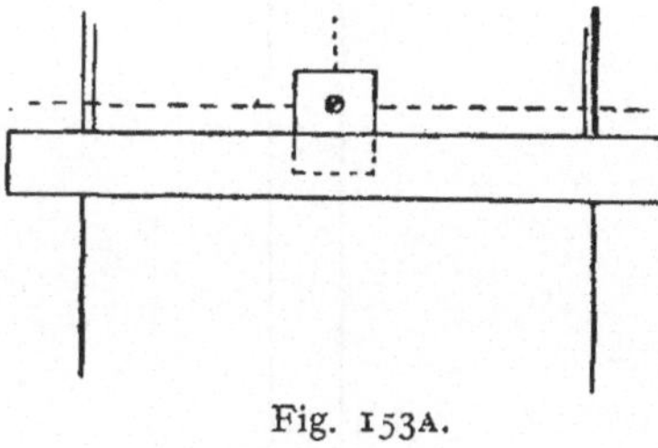
Fig. 153A.

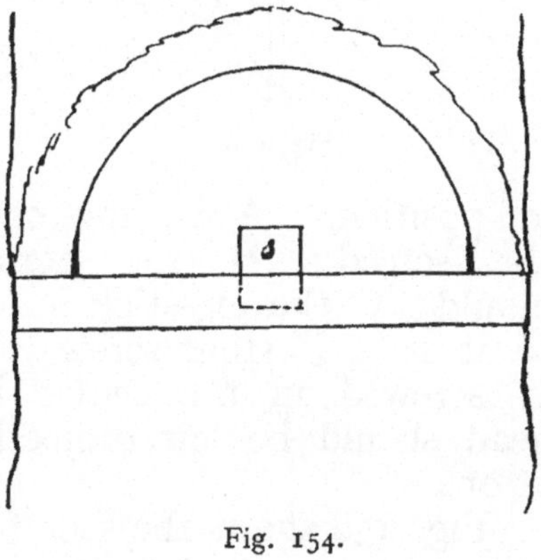
Fig. 154.

Mouldings on Arches

For arch mouldings and horsing up, and for running arches, whether they be of a moulded character or the plain arris, the same method is adopted. When an arch is required to be finished with a moulding or plain arris on the angle, it is necessary to obtain the desired curve. A correct one is obtained by taking a piece of ¾″ board about 7″ wide, such as a floor board. The length of board should be about 1″ longer than the width of the opening, which allows for the nailing of this board each side of the opening. The finished width of the opening is marked off on this board, but before fixing it mark off the centre. Tack on the board at this centre line a short piece of ¾″ board, not less than 2″ wide and about 9″ long, keeping it 4″ or 5″ above the level of the board at centre of opening, Fig. 153. When this has been completed, the board can be fixed about 4″ or 5″ below the springing of the finished arch. This allows for the horse of the mould to run below the springing. When running the arch, care must be taken that the board, which is termed the centre, is fixed level across the opening, and the correct centre is marked on the piece nailed on the centre. The correct springing line is marked on the T-piece, though the thickness of the screw which is used for the shoe of the gig-stick to turn on has to be allowed for, Fig. 153A. One thing must be borne in mind is the screeds for running the arch and also the upright moulding to opening. After the centre has been fixed, an upright screed is carried on each side of the opening, far enough back at the springing of the arch to allow for the upright screed below the arch to meet it, Fig. 154. When this has been completed, a screed is formed wide enough to receive the horse of the mould, traversing it off with a short rule bearing on the two upright screeds, as in Fig. 154.

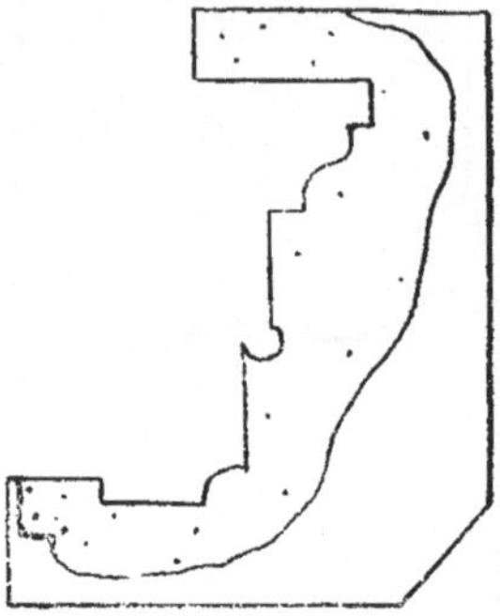

Fig. 155.

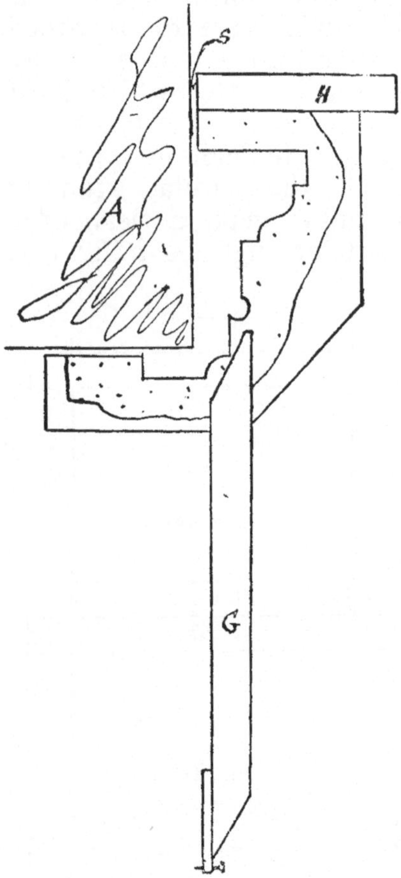

Fig. 156.

Having got everything in readiness for running, the next step is to prepare the horsing up of the mould. Fig. 155 shows the section of the moulding tacked on the wood; Fig. 156 shows the mould in position. A is the concrete; S the screed; H the horse of the mould; G the gig-stick. At the end of it is a ¾″ stout screw. When this is screwed in the centre board, the head should be left projecting sufficiently for the shoe (Fig. 157) to fit on.

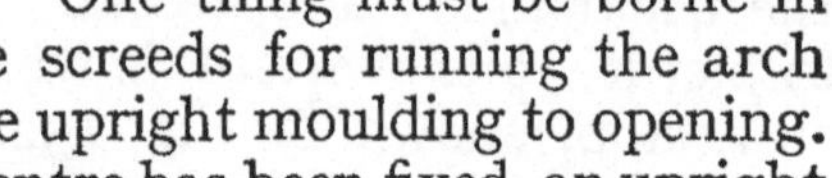

Fig. 157.

Fig. 158 shows the shoe fastened on the gig-stick. This is made of some suitable iron or brass, about 4″ or 5″ long, $\frac{1}{8}$″ thick, $1\frac{1}{2}$″ wide, with a slot about $\frac{1}{4}$″ wide, cut to within 1″ of each end. That is to allow of the plate being screwed on the gig-stick at each end. It should be hollowed out to allow of fitting over the screw, to permit of the turning of the mould. One end of the plate is made larger than the other, to allow of a stouter screw being used. If a larger arch has to be run, when fixing the shoe it is always advisable to put a screw in the centre of the groove of the shoe temporarily, so as to get the exact radius of the opening. When the correct radius is proved to fit the opening both sides and the finished line of the crown of the arch, then two other screws can be put in the plate, as shown in Fig. 158. Care must be taken that the gig-stick is made of a stout piece of wood, as it has to be notched out, as in Fig. 159, to fit each side of the mould, and sometimes great pressure is put on the mould, so it requires to be strongly horsed. Another point to notice is that the slot in the shoe is in a line with the members of the mould. Having followed out all the foregoing instructions, it is possible to proceed to start running the mould in the usual manner. If the same moulding is required to follow on with the upright angles, the gig-stick is taken off, as in Fig. 160, and the mould without any further alteration is ready to proceed with.

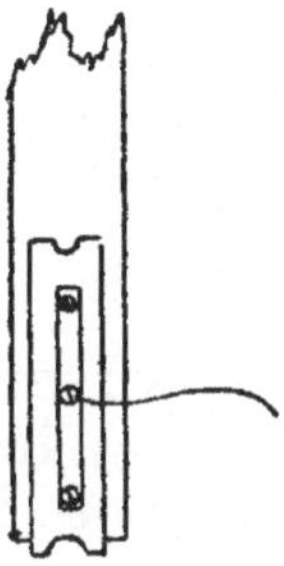

Fig. 158.

Fig. 159.

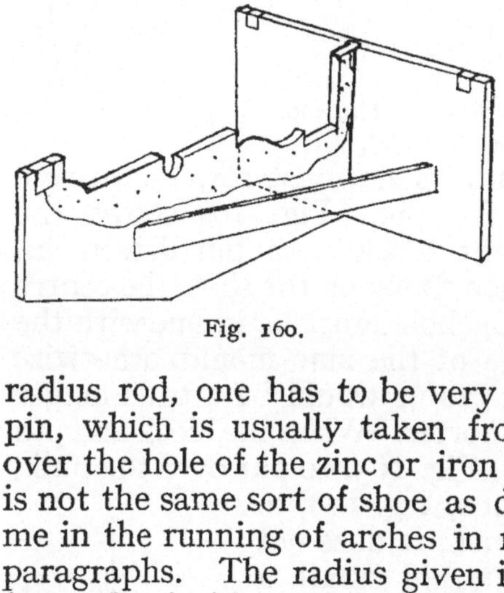

Fig. 160.

Circular Mouldings Run with a Radius on a Bench

When running mouldings by the radius rod, one has to be very careful in getting the exact radius to the pin, which is usually taken from half-way over the hole of the zinc or iron shoe. This is not the same sort of shoe as described by me in the running of arches in my previous paragraphs. The radius given is sometimes from the inside member of moulding and sometimes from the outside, and sometimes from the centre of the moulding. Fig. 161 shows the section of moulding and the radius as taken from the centre at 12″. After having cut the centre line marked off, then take a piece of wood about 2″ × 1″ in thickness and about 18″ long to nail on the mould to use as the radius rod. This can be cut

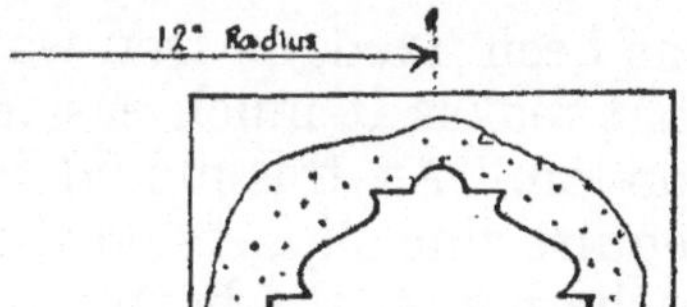

Fig. 161.

Fig. 162.

back, as in Fig. 162, to fit against the stock of the mould. The shoulder fits against the zinc on the front part of the mould. When this has been fixed, mark off the centre line of moulding on it and then measure the length of radius, cutting the radius rod off a ¼″ shorter than the actual length. This allows for the working of the shoe clear of the pin in block.

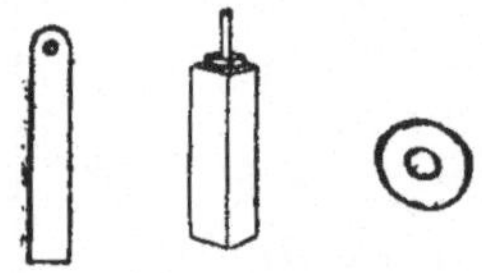

Fig. 163. Fig. 164. Fig. 165.

The next thing required is a piece of zinc or sheet iron about 3″ long and 1½″ wide, Fig. 163, in which a hole is punched in one end to the size of the nail (commonly used in running circle of this description), also a few other holes to allow of tacking the zinc to the rod, Fig. 163. A height block is next required. This can be a piece of wood about 1″ square. The pin is an ordinary wire nail with the head filed off. The height block must be securely fastened to the bench, and be of the height from bench to about ⅛″ less than the underside of the radius rod as tacked on to the mould. A few nails are driven round it in the bench and some canvas and plaster, called a wad. This must set, then drive in the pin. Fig. 164 shows the height block with pin driven in. When fixing on the shoe, the centre of the hole must be in line with the edge of the zinc mould, otherwise the radius would be a trifle longer or shorter. When the block and pin has been fixed, an iron washer is required, Fig. 165, to put over the nail; that makes it much easier for the zinc shoe, Fig. 168, to run on. As this mould is run on the bench, it does not require zinc slippers on the horse, for this would create a gathering on the bottom members, and with bench work one always has the bench well shellaced and greased. Fig. 166, the mould ready for running.

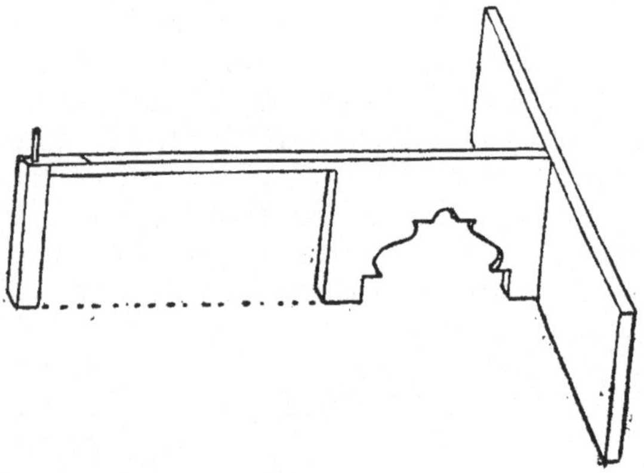

Fig. 166.

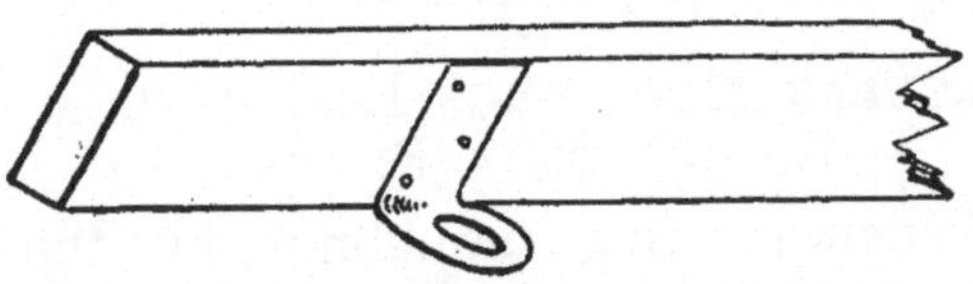

Fig. 167.

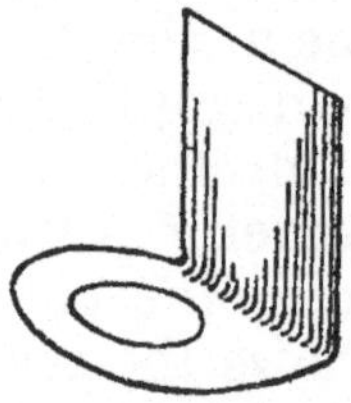

Fig. 168.

This method of the shoe is used when only a small circle is required, but the usual and best way on the bench is as follows. When the radius is fixed on the mould and the distance measured off, a piece of zinc or sheet is cut about 1″ wide and 3″ long and the hole punched through for the pin in height block. This is tacked on the side of radius rod and bent round the underside of the hole, just clearing the side and in line with the profile of the mould plate, even if it was necessary to tack a small block on the side of radius rod to get the required line. Fig. 167 shows the radius rod with shoe attached on side; Fig. 168 the shoe bent to shape before being tacked on.

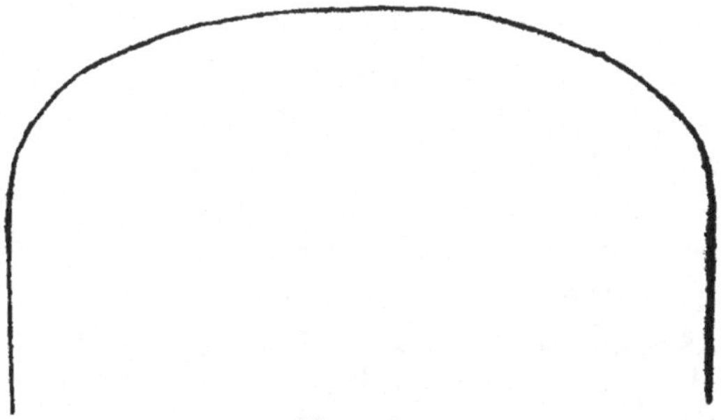

Fig. 169.

Elliptical-shaped Mouldings

The running of an elliptical moulding on plan or elevation requires quite another procedure to all other mouldings, as these are either run with a trammel or two pegs, or plates on the mould, over a rib or template. For example, it is desired to finish an opening with a moulding forming an elliptical head, as in Fig. 169. The methods of striking out the desired lines for the rib I will give later. A rib is made usually by the carpenter (though in the plaster shop the plasterer generally makes his own). It is of wood, about 1″ thick, so when the running of the mould is being done the wood does not twist.

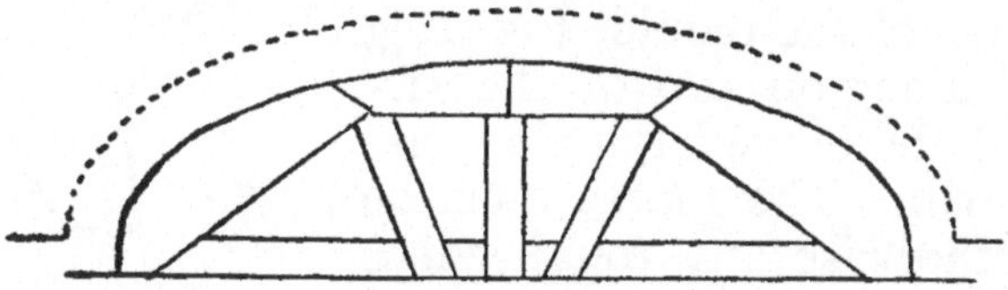

Fig. 170.

Fig. 170 shows the rib as it would appear when made and fixed in the opening. This is made smaller than the finished opening, to allow for the thickness from the bottom member of the moulding at A, Fig. 171, to underside of pegs or plate. Assuming that the opening has to have a moulding each side with a panel soffit, it will be seen that when the rib has been made (as in Fig. 170), it extends below the springing, to allow of the pegs or plates on the mould passing low enough to get the correct haunches of the arch. So when marking the springing on the rib one must bear in mind, when fixing the rib, that whatever the depth from the bottom member to the underside of pegs or plates may be, that distance must be allowed from the correct springing line as would be marked on the wall of opening.

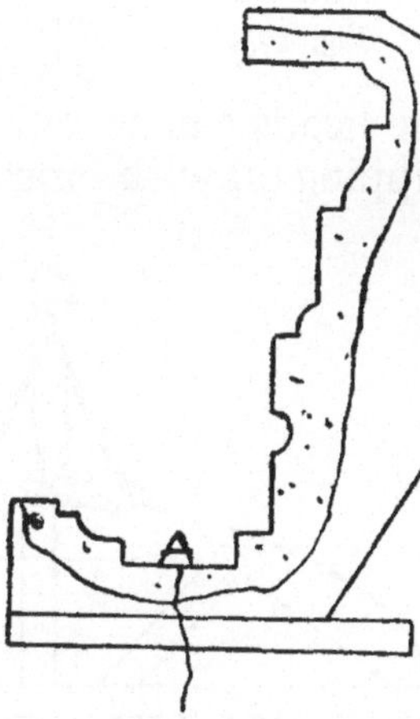

Fig. 171.

When fixing the rib, take the underside of the rib and level across to the back side of the opening at A, Fig. 172. Then to this line nail a block on the wall, sufficiently thick to allow the rib to rest on. Repeat this process the other side of opening, at B,

Fig. 172. This allows, when one side has been run, for the rib to be moved to run the other side, as it would not be necessary to make a double rib. A temporary piece of wood can be fixed across the opening at the back of the rib, to enable the rib to be made stiff with a few stays. When horsing up the mould (Fig. 173), the horse should be as short as possible, and the plates or pegs should be fixed close up to the centre of mould, so long as sufficient bearing is allowed for the mould to pass round the quick part of the curve without jumping. Two stout brass or iron plates about 3″ long and ½″ wide, fastened with three screws on top of the horse about ¾″ in front of the horse of the mould, Fig. 173, will allow of the mould passing along on top of the rib, the front part of the horse having a bearing against the rib. The plates must be filed smooth, otherwise any rough edges would cut into the wood rib and cause bad running of the moulding.

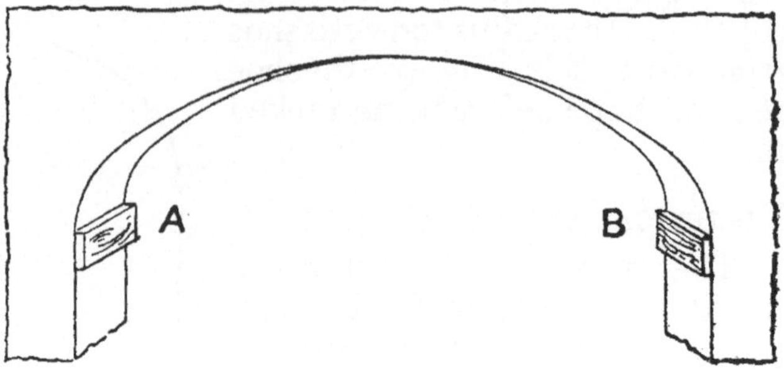

Fig. 172.

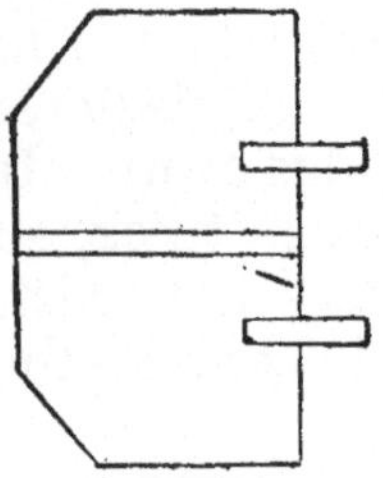

Fig. 173.

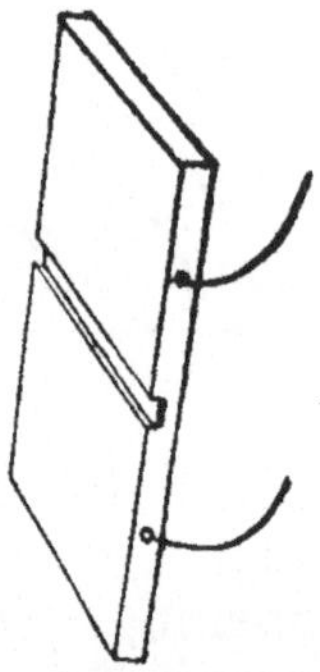

Fig. 174.

Another method is to have two circular pegs of hard wood and insert these into the horse of the mould. If good wood is not available, stout screws can be used, for after they have been screwed in the horse the heads can be cut off with a file, filing part way through each side and nipping off with pincers. Fig. 174 shows the two pegs or screws half the thickness of horse. If pegs are used, the extra depth must be allowed for when having the rib made, and an additional piece of wood would greatly assist in the running of the mould, if nailed on the underside of the horse, as it helps to make more bearing against the rib when running. A screed is needed for the nib of the mould to run against. The method for this I have already described in my previous section on the running of arch mouldings.

Fig. 175.

Fig. 175 shows the mould in position, looking at it from the back side of the running line. One must be very careful in setting up the rib in

SHOWING THE FINISHED ARCH WITH THE MOULD BEING TAKEN OFF THE T-PIECE.

position. If carried out as the instructions given, the young plasterer should have no difficulty whatever in being able to take on an arch of this description. Though I mentioned that ovals are run with a rib and trammel, what I have just dealt with is not an oval or part of one, as an oval is an egg-shaped figure, one end having a greater curve than the other, but the term is generally used by plasterers, and the term elliptical when the ellipse finishes at the springing, such as the opening with the elliptical finished soffit I have already described. As a rule, when oval panels are referred to one naturally looks for a complete oval, as it is termed, on wall or ceiling, which is not quite correct. But where the whole figure is complete, which would be elliptical on plan, it is generally termed an oval. I will again touch on the running of the oval with the rib when I am describing the running of plaster moulds later on in the book.

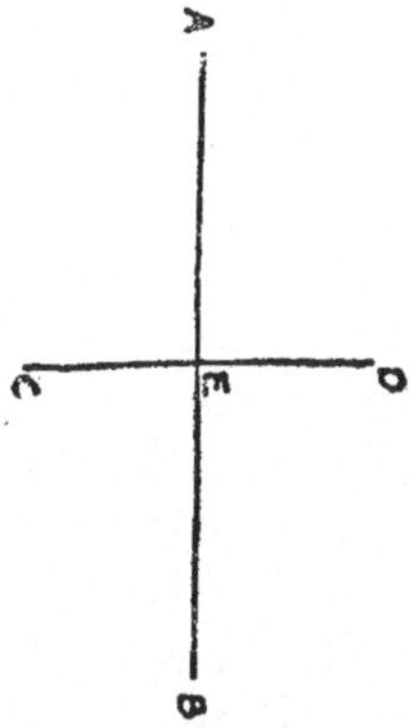

Fig. 176.

Running Oval Mouldings "in Situ"

Running of an oval panel moulding, either on the wall or ceiling, is best accomplished by the use of the trammel. If on a ceiling and the panel is a very large one, it will be found easier to run off part section by the use of the trammel and run the remaining parts from off the finished section. But one does not see many large oval panels *in situ*, on ceilings, so I will deal with the ordinary panel and the procedure for the running of same, if one has good fixing on the ceiling and the floating is straight, otherwise the rules forming the trammel would have to be straightened with thin pieces of laths between the rule and ceiling, and tightened up with a little putty and plaster. But it will make a much better job, and save time, to run a rough screed the line of the trammel to fix the rules to.

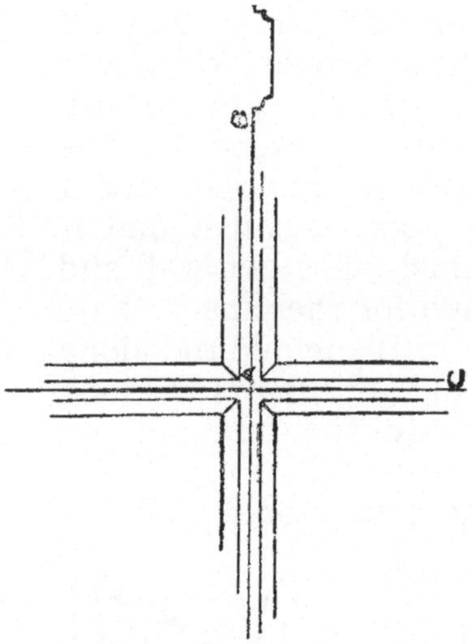

Fig. 177.

Before proceeding with the construction of the trammel, mark off the centre of the desired panel as follows: If on ceiling care must be taken that the centre is taken from each end of the ceiling and the line snapped from end to end. Repeat the process the other way of the ceiling, though one could set off the second line from the centres at right angles to the first one and carry the line across the ceiling, as it would be very hard to discern if the oval did not cut quite to the centre one way or the other. When this has been done, set off the desired length of the oval, Fig. 176,

A to B, then the width C to D. A to B is the greater diameter, C to D the lesser, which is usually a third less than the greater diameter. The next thing required is some $2'' \times \frac{1}{2}''$ running rules, which should be of seasoned wood, and planed, as the stuff does not cling so much to the wood when smooth. These are cut up in the following manner and nailed on the ceiling, making sure the heads of the nails are flush. Fig. 177 shows the centre line each way and the rules complete in position, which forms the trammel. Fig. 178 shows the long and short section of the rules when fixed one side of the centre line.

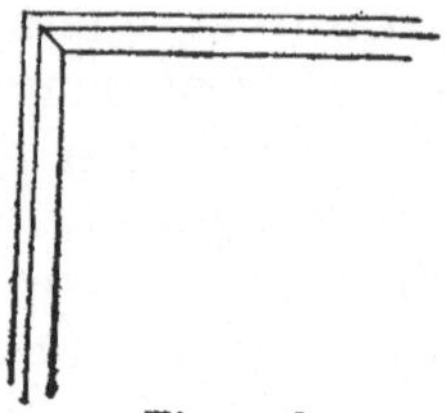

Fig. 178.

In Fig. 177 A to B is the inside of panel moulding of the major diameter. Now cut eight pieces of rule square at one end and to the angle of 45 degrees at the other, the length of these being a little less than C to D and A to B. This will give the desired lengths of the rules, which allows for the mould to run past the rules inside the moulding. Care must be taken that these are fixed central to the centre lines and that the space between the two rules forms a groove. The width of the groove depends on what sized pins one has in the pin rod.

Fig. 179.

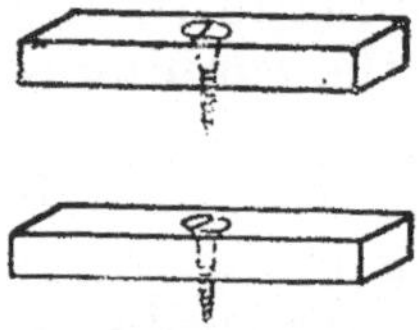

Fig. 180.

Two stout screws, or a round piece of iron $\frac{3}{8}''$ diameter, let in the rod will do. With screws the heads would need filing off, or cutting off with the hack saw. Fig. 179 shows the two pins or screws, whichever happens to be easiest to get. Many prefer a short piece of wood for the pin, screwed into the pin rod, but I prefer the iron pegs, as when one is using the wood pin on the bench, and the trammel rules get wet and swell, the groove being only made just wide enough for the pins to pass along, much trouble is encountered in the passing of the mould rod along the groove of the trammel. Fig. 180 shows the long wood pin which is screwed into the trammel rod with the head flush.

Fig. 181.

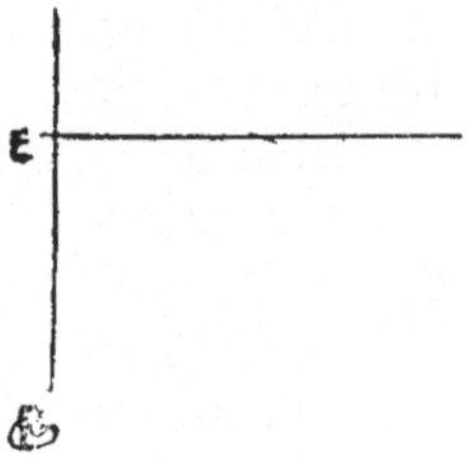

Fig. 182.

In running this mould one man should push the mould and his mate hold up the rod to allow it to pass along the groove of the trammel. Fig. 181 shows the pin rod with the two pins. That on the left being the major or longest axis. This pin is fixed at the distance E in Fig. 176 to B, that at the shortest axis is fixed at the distance E to C, but these pins are not usually put in the rod until the mould has been fixed on, so as to get the correct distance from the inside of the moulding to E in Fig. 176. Fig. 183 shows

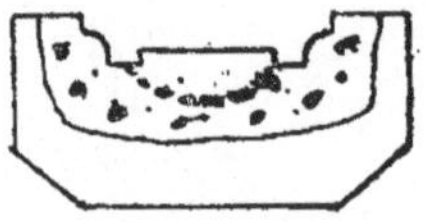

Fig. 183.

the section of the panel moulding with a sinking to allow for a bed of an enrichment, the dotted line showing the profile of same. Fig. 184 gives the mould attached to the pin rod, up in position for running on the ceiling. A is the ceiling line, B is the thickness rules forming the trammel, C the two pins, D the trammel rod and mould attached. When the ceiling has been finished, before the oval is run, the mould will run on the finished ceiling, but if not finished it will be necessary to form running screeds.

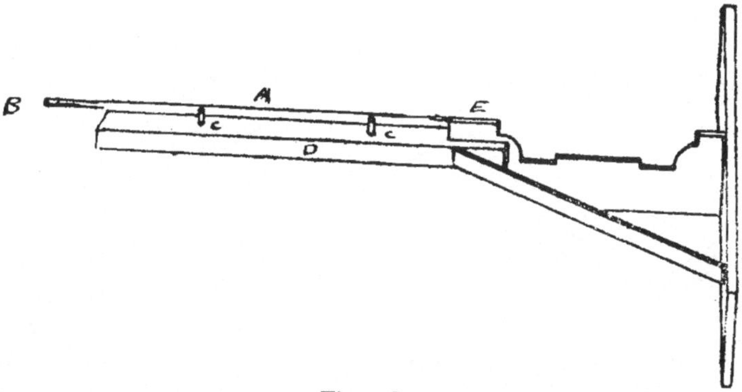

Fig. 184.

When fixing the mould on the rod, the mould will have to be above the rod, as E in Fig. 184. That is to allow for the thickness of the trammel rules on ceiling.

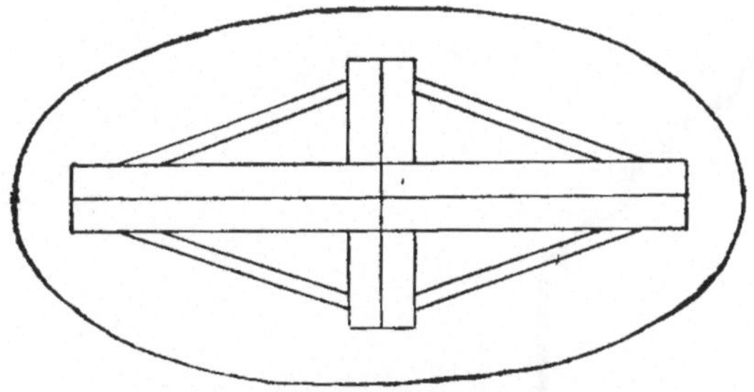
Fig. 185.

Another way to construct a trammel, which perhaps would be a little easier, if the fixing was not very good in the ceiling, would be as follows. Get a length of $\frac{3}{4}''$ floor board and cut off one piece a little shorter than the desired length of the oval inside of moulding. Next cut two pieces to fit in between the long piece and a little less than the width, Fig. 185 shows the three pieces, which can be laid on the bench or floor and fixed together with cross stays made from floor board ripped down about 2″ wide. At the intersection of the boards four iron angle brackets should be screwed on at the four junctions of the three pieces of wood, as in Fig. 186.

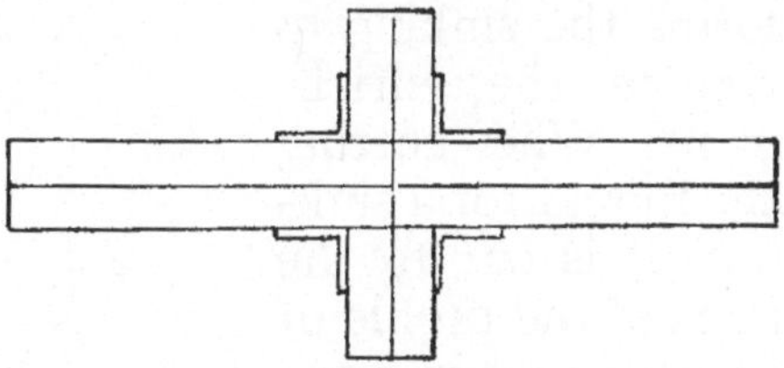
Fig. 186.

When the template has been securely fastened together, set off a centre line on each of the cross boards at right angles to each other. From these lines set off the desired width of the groove at the intersection and at the end. Cut off the length of running rule as before described and nail or screw them on to the template to the width of the desired groove. Thicker rules can be used, such as $\frac{3}{4}''$ batten, which would require planing. Some plasterers prefer the rules for trammels to lap (Fig. 187), instead of being cut to a mitre. I have had them both ways and seen little difference, though there is one advantage in having thicker rules for the trammel, the pins can then be longer, so giving more bearing in the grooves. This template can be fixed up to the ceiling complete. If the fixing is bad, it would be advisable to plug the ceiling so that the template could be fixed up in position with stout screws, and it could easily be taken down and used again when required.

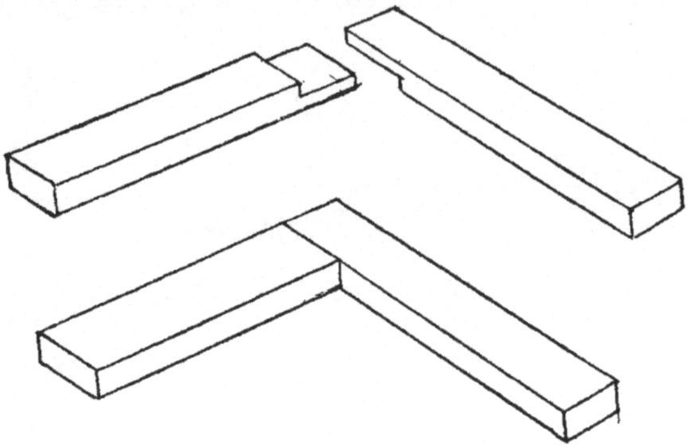

Fig. 187.

Running Mouldings to receive Enrichments or Bed Moulds

Sometimes one has a cornice or moulding to run with an enrichment to insert, as in Fig. 188, showing the profile of the egg and dart moulding. The other section of the mould plate gives the mould with the part which forms the sinking to receive the enrichment. When cutting the mould for a sinking, it is cut to the back of the profile of the ornament. Sufficient room is allowed for a thin bed of putty and plaster with which the ornament is usually fixed. Hence the term bed mould. Fig. 189 shows a cornice mould cut with a sinking to receive a dentil block, also in Fig. 191 a section of mould plate

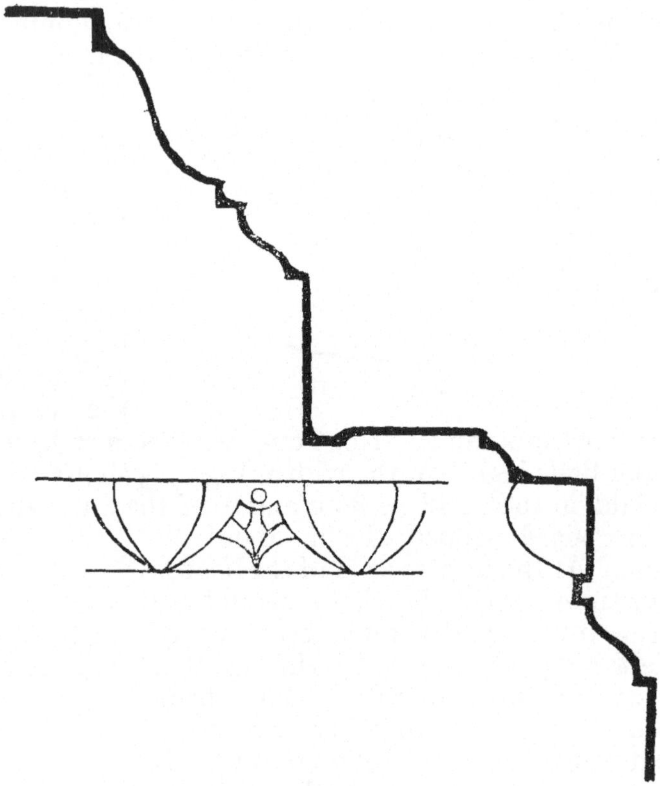

Fig. 188.

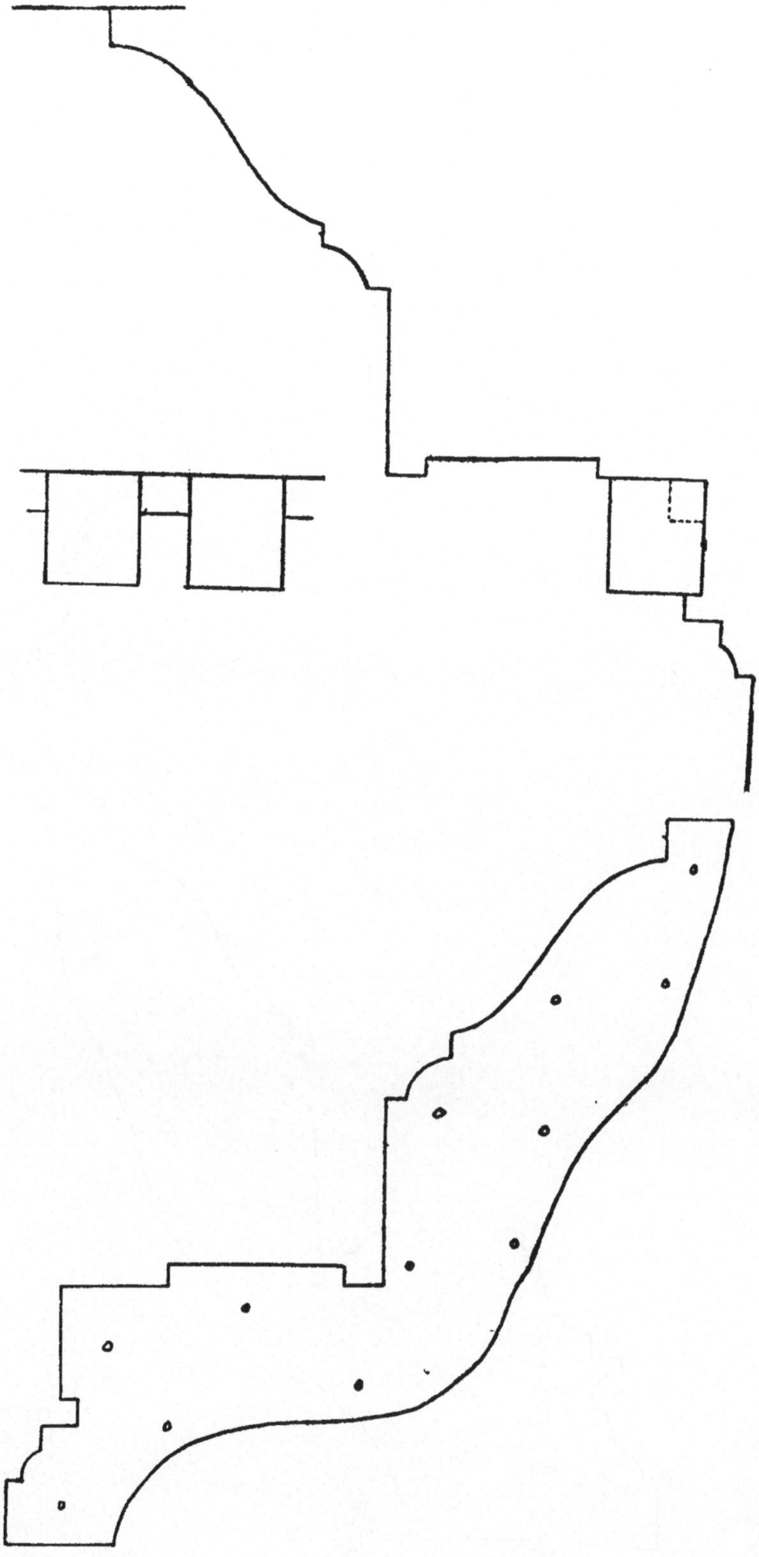

Fig. 189.

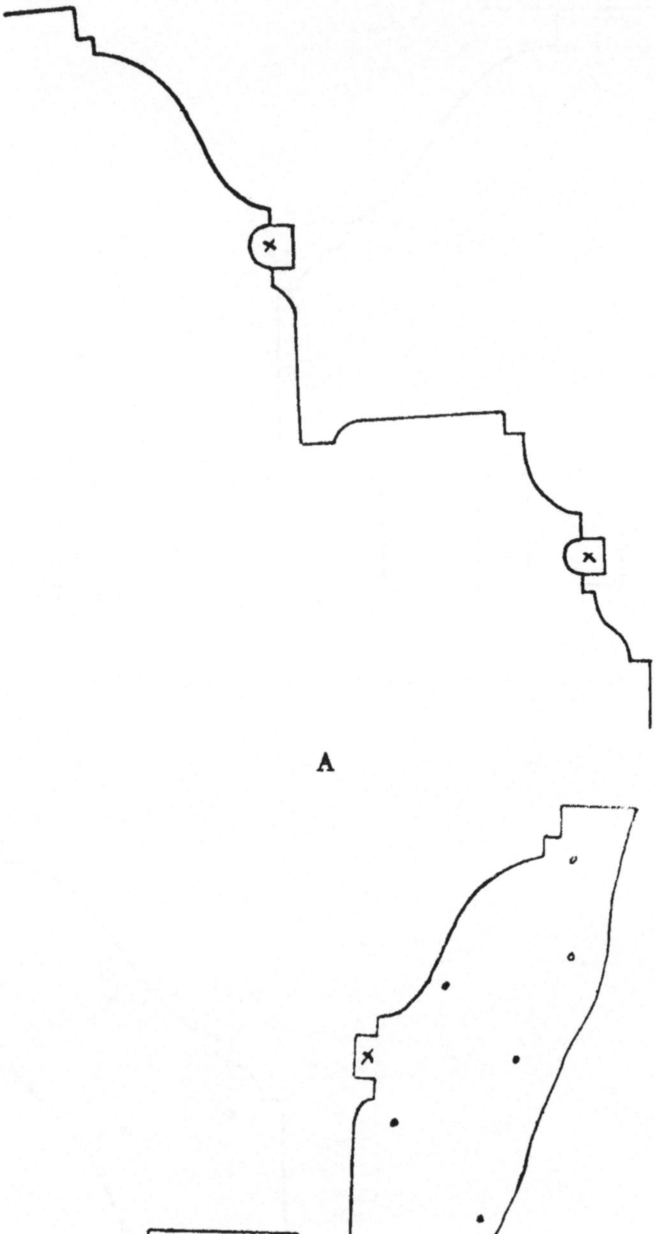

Fig. 190.

showing the profile of the dentil block. Fig. 190, A, shows a cornice mould with profile of bead and also bead and reel, and likewise the mould plate (Fig. 190, B) showing the sections which form the sinkings for the bed mould. Fig. 192 is a section of ceiling panel mould, with sinking to receive either a fruit enrichment or oak leaf, laurel and berry, as the case may be. Fig. 193 shows a cornice with a sinking in the soffit to receive a guilloche or scroll enrichment, also a frieze decoration, such as flutes or Greek key, which make a neat decoration. Fig. 194 shows a block cornice and the mould plate cut to receive the bed for block, also leaf enrichment.

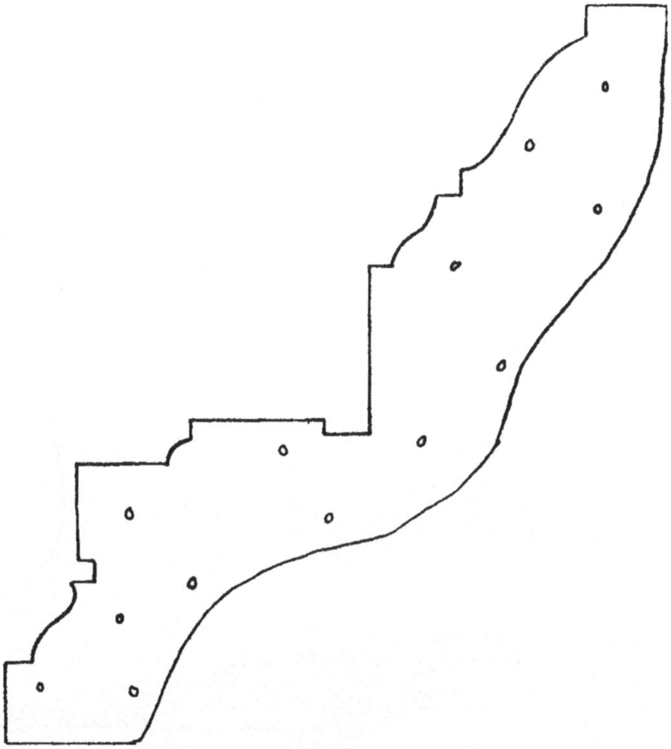

Fig. 191.

Running Mouldings in Cement

Quite a different method of running a cornice or moulding for external work is adopted to that for the running of an inside cornice. Take, for example, a large cornice required to be executed in Portland cement, as Fig. 195. A is the concrete laid to receive cement finish of the cornice, though sometimes the projection is built out in brickwork, sailing over, as bricklayers term it. Or we may have slabs of stone bedded on at the drip as Fig. 196. When the mould has been cut and nailed on the stock, as in Fig. 197, a line is squared up to clear

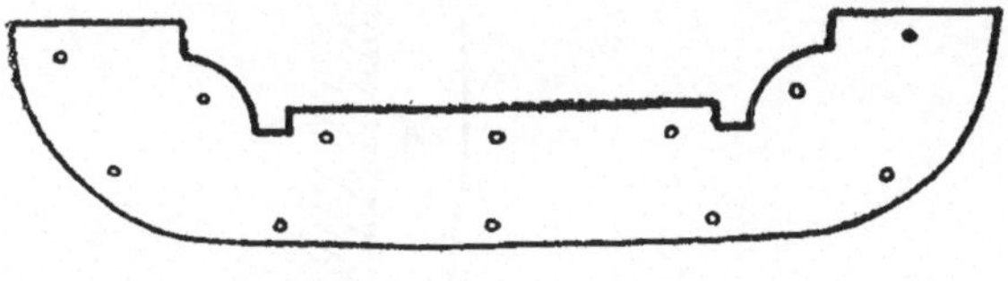

Fig. 192.

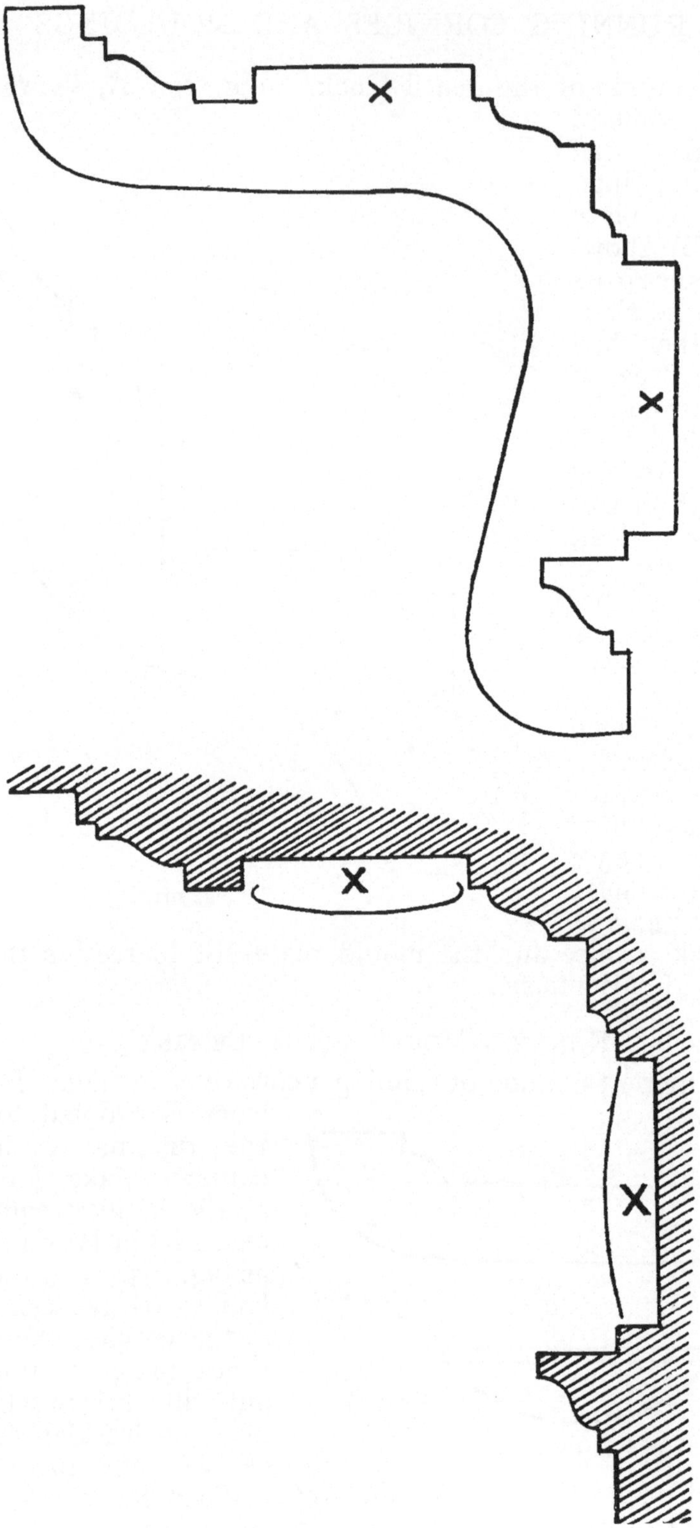

Fig. 193.

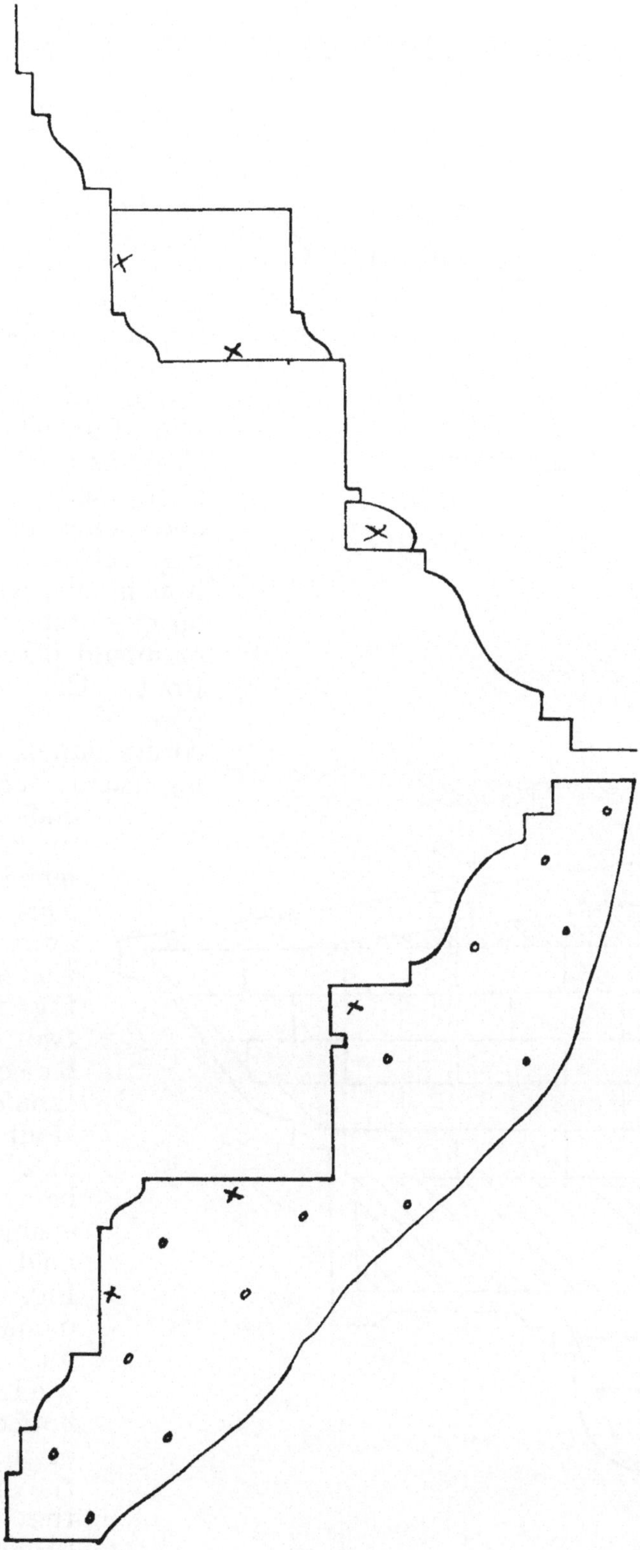

Fig. 194.

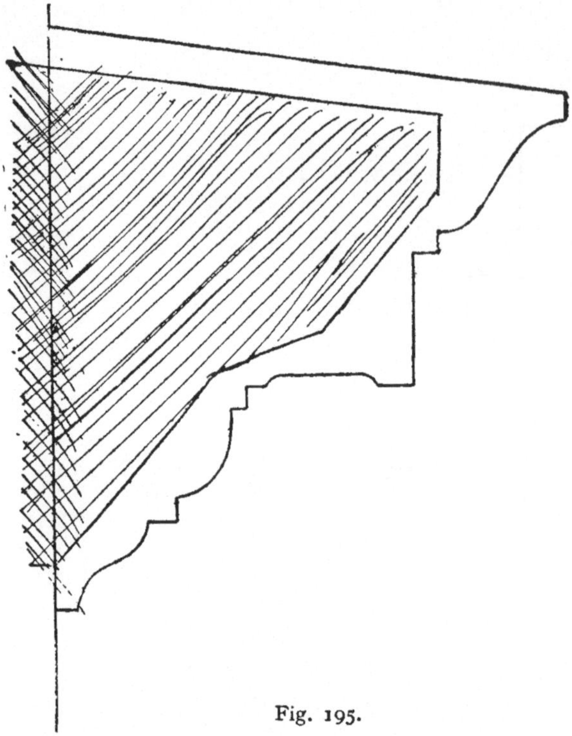

Fig. 195.

the horse, as will be seen in Fig. 200. This enables the plasterer when setting the mould to get the correct position. It will be noticed that the corners of the wood are taken off. The mould requires to be as light as possible, because in the process of running the surplus stuff, which the mould clears as it is being pushed along, falls on to the horse, and adds to the weight. Fig. 198 shows the zinc or iron muffle, which is nailed on over the original profile or mould plate, about ¼″ in front. This is for the purpose of coring out. The coring out of a big moulding usually takes some time and requires the mould to be run over it a great deal. This would wear away plaster, so a zinc muffle is used. Fig. 199 shows the two handles, or braces, which should be made of stout wood, such as 2″ × 2″, but can be rounded off to enable the hand to hold comfortably. Fig. 200 shows the mould in position for running, in which will be noticed the square or upright line on the stock, and also the plumb bob hanging down. The

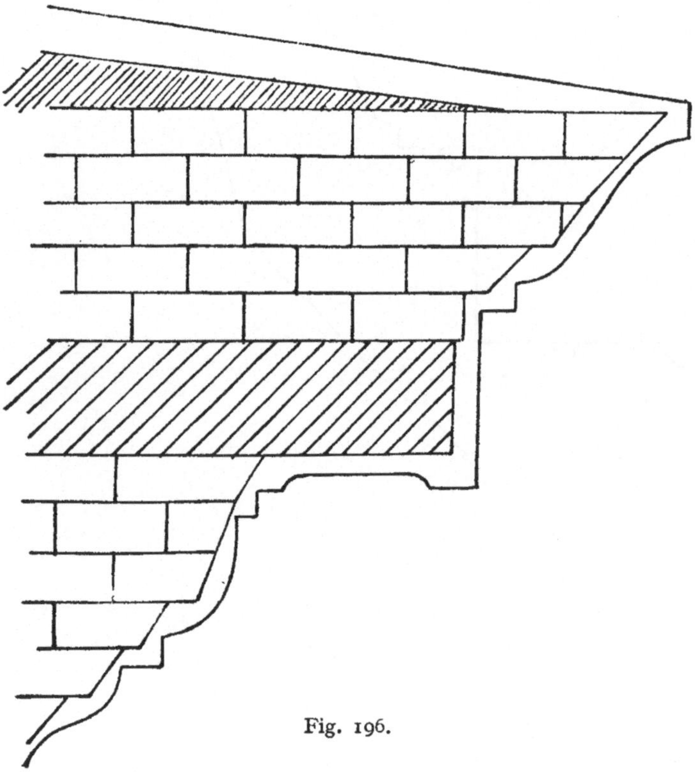

Fig. 196.

latter, when swinging vertically with the square line, gives the correct position of the mould. Figs. 201 to 204 shows a thin piece of wood (or zinc will do) which acts as a save-board, and is tacked on the mould

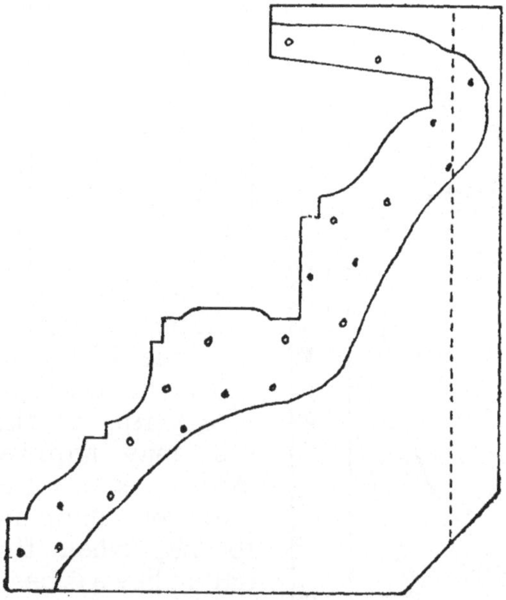

Fig. 197.

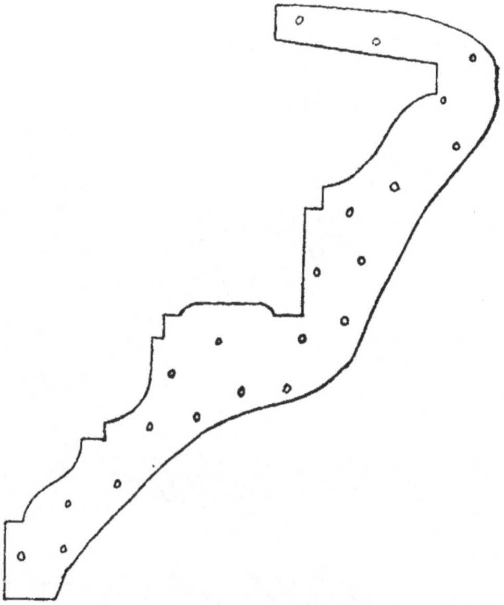

Fig. 198.

in front of the handle and to the front side of the mould. This catches the surplus stuff from off the top part of the moulding, and slides down to the horse of the mould instead of allowing it to fall on the scaffolding. As the mould is passed along, the material can be taken off the horse with the trowel. This enables the stuff to be used again, and also helps to keep the scaffolding clean. With this mould it is necessary that good stout iron shoes or slippers should be put on at the nib, and also the horse, as with the running of a cement moulding the sand soon wears the iron. If one runs the mould on a rabbet it is necessary to put slippers on the rabbet as well, as it assists in the running of the mould; otherwise the horse of the mould would wear, owing to the continuous passing along on the running rule, and also the sand material falling on the rule.

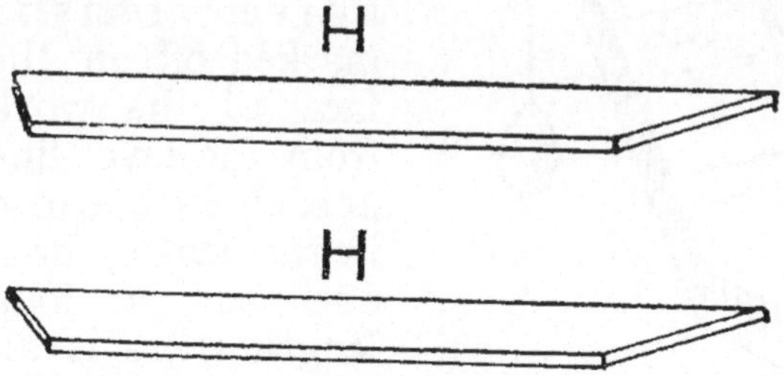

Fig. 199.

Method of Running the Moulding

When the plasterer has carried out all the foregoing instructions in the horsing up of the mould, the following procedure will then be carried out. Get the correct level of the bottom member of cornice; mark it on the face of work and carry a level line along the face to the distance which is required to run. Care must be taken if there is a narrow margin underneath, such as window head, etc., that the height of these is checked. If there is any variation in the level, a little adjustment can be made when fixing the running rules, by fixing above the line one end and below the other, or vice versa, as the case may require. Whichever way the variation happens to be, when the datum line has been carried through the distance is taken from the bottom member of the mould to the underside of the slipper on the horse. This is marked off on the face of the work from the level line at each end and also in the centre, for if one has a long length there is always a possibility of the line sagging. When this has been completed, a line is chalked and held on to the lower line, which would be the rule line, and at the same time it is necessary for one man to place his finger or thumb on the mark in the centre before the line is snapped.

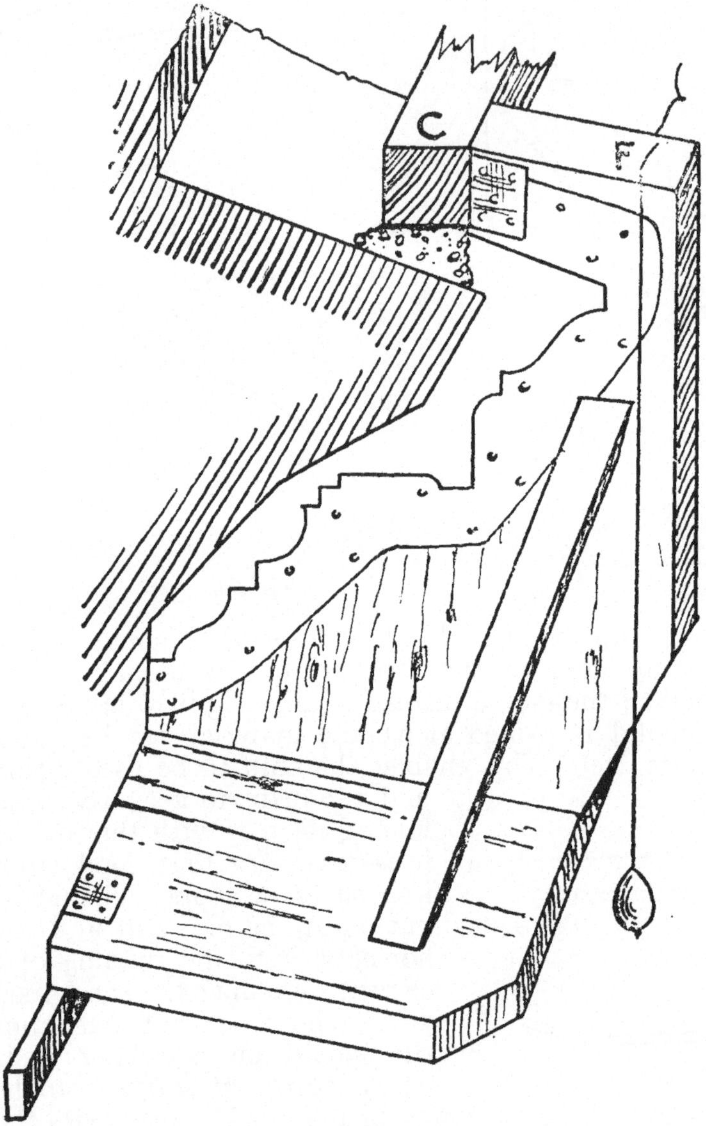

Fig. 200.

If it is intended to run the mould on a screed, the screed would have to be formed before the line is snapped, and also before the level line is carried along the face. If a screed is desired, the mould is held up in position and the underside marked in the same manner as I have just described, and the line snapped and the screed formed. In the case of outside work, it is formed with Portland cement, gauged with a small quantity of washed sand, though some prefer to use neat cement, but if the floating is green it does not adhere so well as if it contains a little sand. This would be about $\frac{3}{16}''$ thick, to allow for finishing, as cement finishing is usually thicker than inside work, as the finishing coat is ruled off, which is not usually the case with interior work. In some cases where one does not want to run the mould on a rabbet, and it is not convenient to form a cement screed owing to the amount of time it would take to set before the mould could run on it, a plaster screed is used in the same manner as it is adopted for inside running. It is well to arrange to form the screed in time to allow of it to set before it is required. When the line is snapped the running rules are nailed to the line, taking care if there are any joints that the intersecting rules are flush with each other. Then the mould is put up in position, as in Fig. 200, seeing that the plumb bob hangs clear on the mould, which can be done by driving in a lath nail over the top of the mould at F. This enables the position of the nib at C to be adjusted with the nib rules. If floor boards are available, these make good rules for the nib, as they are much better to bed in position, and also to weigh down with bricks, as unless the nib rules are bedded firmly the mould will "chatter" or spring. If floor boards are not

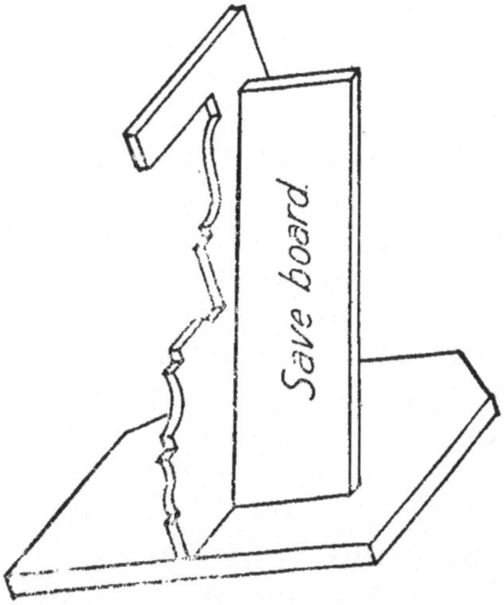

Fig. 201.

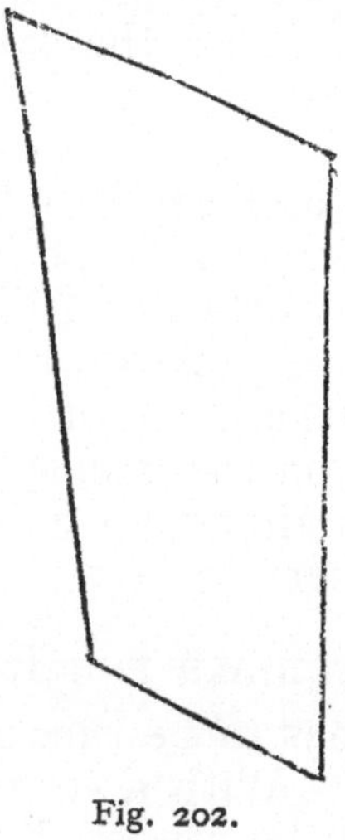
Fig. 202.

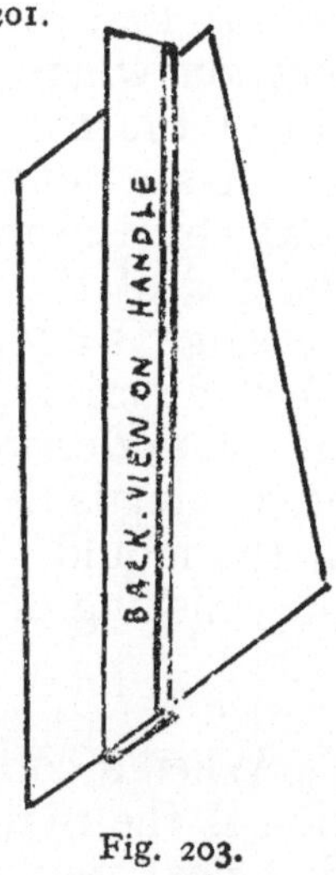

Fig. 203.

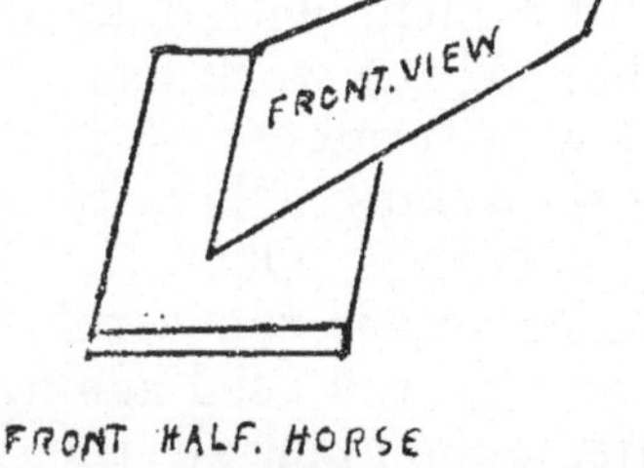

Fig. 204.

available 2″ × $\frac{3}{4}$″ battens can be used. I suggest these because the nib rule is the chief thing in running an outside cornice, as the nib rule has to be bedded up to suit the nib of the mould, and, as a rule, the top of the concrete or brickwork of the weathering of the cornice is rough and uneven, thus entailing much work in fixing the rules, if they are thin.

When bedding the nib rules, gauge the cement and sand rather weak, say 8 parts of sand to 1 of cement. The reason for so doing is that when the cornice is run it is necessary to chip off the bedding of the rules as the underside of nib rule comes up flush to the weathering on the cornice when running.

Reverting to putting the mould in position, in Fig. 200, place the nib rule on the top of weathering or projection of cornice, against the nib of the mould as shown at C. At the same time put a small quantity of cement and sand, as previously described, about 18″ apart along the nib rule. The nib rule is then laid on these dots, and the mould is put in position each end of the length of nib rule, tapping it backwards or forwards to suit the hanging of the plumb bob. Bricks can be laid or bedded on the nib rules, which will keep the rules in position when set, and at the same time make it firmer to push the nib of the mould against. A line can be stretched end to end; or if one has a long straight-edge it can be tried on the rule and checked again by the mould and the hanging bob. If the cornice projects over the end of the building the same method of procedure for running over the edge is adopted, as shown in Fig 130 in the early part of this chapter, when describing running across breasts or piers with the rabbet tacked on the underside of the horse of moulds. When the rules are fixed, the next thing is to tack on the muffle, Fig. 198, about $\frac{1}{4}$″ in front of the plate of the mould. This allows for a fair coat of finishing when running off the finishing of the cornice.

Materials for Running the Cornice

When a cornice has to be run, one of the chief matters for consideration is the material. With a cornice such as I have just described, the first coat requires to be rendered all over the concrete or brickwork, as the case may be, and should be gauged at the rate of 3 parts of sand and 1 of cement with 1 lb. of hair well beaten to each gauging. After the whole of the work has been covered, it requires to be well scratched with a wire scratcher, and, if available, a rough wire brush passed over before scratching. This takes off the smooth places left by the trowel when rendering. One point I must mention is to lay clean sacks on the scaffolding under the cornice. This enables any material that falls to be picked up and used again.

After rendering all the background, if possible allow it to stand awhile to pick up, but not too long, otherwise one would lose the suction. It can then be carried on with the roughing out at the same rate of 3 parts to 1 as before; but it is not necessary to add any hair. If there should be any large uncovered place, which often happens in the roughing out, owing to the weight of stuff, or perhaps to thickness on the drip, usually a little extra cement is added by the plasterer on the scaffolding. One great point in roughing out a cement cornice on outside work is to well scratch each coat, especially the smooth places left by the trowel.

When the roughing out has been completed, the muffle is taken off the mould plate and the mould passed over several times to see that there are no obstructions, such as any soft places that may have sagged after the mould had been last passed over. When it is found to be all clear and ready for the finishing coat, the finishing stuff is gauged up. This is composed of 5 parts of washed sand to 2 parts of cement. It is very important that the sand should be measured in the pail or box, and also the cement. In addition to this, the plasterer should have a small box on the scaffolding, so that when the bulk of the gauging has been completed, any small amounts required can be gauged by the plasterer as he requires it. When running off the finishing, the material that falls on the sacks is picked up and used up in the roughing out of the weathering, which is made to pitch well out from the building.

While the running off of the cornice is proceeding, much trouble arises from pieces falling out from the most prominent members. When this occurs, if a portion of cement and sand is mixed up dry and a quantity pressed into the holes caused by the pieces falling, this will greatly assist the plasterer in filling out the moulding. If this method is adopted, the dry parts of the cement and sand must be scratched off with the back of the trowel, otherwise the wet material will not adhere.

In running a cement moulding, it must not be expected to get it run off in the one day, as it requires time, and, I might add, patience. When the finishing coat has been well laid in all over the cornice, and the members well filled out, it is not advisable to put the stuff on the members with the trowel, but to throw it on, at the same time spreading the stuff as much as possible, so as to well cover all the parts of the cornice. By this means it is possible to get it all filled out at the same time, which helps to get the body of the material all about the same texture.

After all the members have been filled out and the arrises formed up as much as possible (which will be assisted by the use of small wooden floats and forming by the hand), and having proceeded with the filling out of the cornice, it will be necessary to prepare the driers. These are composed of 1 part of washed sand and 1 part of cement mixed together dry. If the weather is damp and cold, it is necessary to have the sand dried, as the cold and damp affect the setting of the cement in the running of the cornice; but in warmer weather it will be found that the dry cement and washed sand will act as desired.

The Use of Driers

When running off the moulding, throw the soft stuff all over it with the trowel, spreading it as much as possible. Repeat this two or three times or more if required, so as to get a soft and even wet surface all over. Then with some mixed driers, as described above, either in a small box or in the operative's apron, which is usual with the plasterer, though some use the hawk, throwing the stuff on with the trowel. I myself like to throw it on with the laying trowel with the use of the hawk. When the mould has been passed over several times, and the required wet surface has been obtained, throw on the driers, either with the hand or trowel, as it requires very light treatment, though one need not be quite so gentle when throwing on the driers in the first operation, as it greatly depends upon how firm the material of the cornice is. One very important thing in the use of the driers for finishing is that on no account must any more wet material be used on the moulding after commencing with the driers. Again, when pushing the mould, it is necessary to go as lightly as possible and keep the nib of the mould from jumping. This greatly adds to the desired finish.

The above procedure is used when a stone-like finish is required. If a tooled stone finish is required (that is to say, to imitate stone which has been faced up with the chisels, etc.), a different treatment is needed, which I shall describe when I am dealing with "External Plastering."

PAINTED PLASTER CEILINGS OF THE LATE EIGHTEENTH CENTURY.

Mouldings and ornament commonly in medium relief.

THE MODERN PLASTERER

A PRACTICAL WORK ON PLASTERING, DECORATIVE PLASTERWORK, ROUGH-CASTING, CEMENT-WASHES, PLASTIC FLOORS, ARTIFICIAL MARBLES, ETC.

BY

W. VERRALL

TEACHER OF PLASTERING, NORTHERN POLYTECHNIC, LONDON; FIRST-CLASS HONOURS AND MEDALLIST IN PLASTERING, CITY AND GUILDS OF LONDON INSTITUTE, ETC.

WITH SPECIAL CONTRIBUTIONS BY

HERBERT JEANS

AUTHOR OF "THE PERIODS IN DECORATION"

AND

J. TURNER

MEMBER OF THE NATIONAL ASSOCIATION OF PLASTERERS, GRANOLITHIC AND CEMENT WORKERS

EIGHT COLOURED PLATES BY

H. GUY HOLT, F.R.I.B.A.

WITH EIGHTEEN HALF-TONE PLATES AND 343 DIAGRAMS IN THE TEXT

VOLUME II

CAXTON PUBLISHING COMPANY, LIMITED

CLUN HOUSE, SURREY STREET, LONDON, W.C.2

CONTENTS

VOLUME II

CHAPTER I

SUITABLE SANDS FOR LIME AND CEMENT PLASTERS

CHAPTER II

PANEL MOULDINGS

CHAPTER III

MOULDING AND CASTING

CHAPTER IV

JELLY MOULDS

CONTENTS

CHAPTER V

PILASTERS AND COLUMNS

CHAPTER VI

CEMENT HEARTHS AND FLOORS

CHAPTER VII

EXTERIOR PLASTERING AND WATERPROOFING

CHAPTER VIII

FIBROUS PLASTER

CHAPTER IX

REPAIR WORK

CONTENTS

CHAPTER X

THE RESTORATION OF PLASTER CEILINGS

CHAPTER XI

PLASTERING BY SPRAYING

CHAPTER XII

PORTLAND CEMENT STUCCO

CHAPTER XIII

MODELLING IN PLASTER AND PLASTIC CONCRETE

CHAPTER XIV

MISCELLANEOUS DECORATIVE PROCESSES

CHAPTER XV

GESSO

CHAPTER XVI

SURFACE TREATMENT OF PLASTER AND CEMENT

CONTENTS

CHAPTER XVII

ARTIFICIAL MARBLES AND GRANITE

CHAPTER XVIII

CONCRETE AND PLASTIC FLOORS

CHAPTER XIX

SPECIFICATIONS

LIST OF PLATES

VOL. II

THE MODERN PLASTERER

VOL. II

CHAPTER I

SUITABLE SANDS FOR LIME AND CEMENT PLASTERS

Nature of Sand—Simple Tests—Colloids—Ferric Hydroxide—Tumbler Test—Other Impurities — Grading — Sieve Analyses — Washing — Qualities — Mechanical Action — Chemical Action.

CONSIDERING the endless variety of manufactures carried on in this country, the question might arise: " What commodities are in greatest demand in the production of finished articles? " Nine people out of ten would immediately answer, " Coal and Iron."

The demand for these raw materials is so obvious as to require no thought on the matter, but there are other materials which, whilst not quite so patent, nevertheless enter into the production of so many finished articles that the casual observer would be astounded. Of these one of the most interesting is sand. From glass-blowing to stock brick making; from the manufacture of tiles to the important duties it fulfils in an iron foundry, sand is put to a multitude of uses.

Sand varies from fine dust-like material to coarse granular grit. Pure white sands are not uncommon, but when shades of yellow, brown and red predominate, the colour generally indicates the presence of iron.

Sand is found largely on the sea-shore, whence it may be blown inland to form sand-hills and dunes. Mixed with shells, etc., it covers large portions of the sea bottom. It is found along the courses of river estuaries and on the shores of lakes. On the surface of desert countries sand is the prevailing material.

It is, of course, common knowledge that certain classes of sand are better than others for any given purpose, but it is astonishing how seldom any inquiry is made or any trouble taken to ensure the right sort being used.

Nevertheless, this matter is of paramount importance to all using cement and plaster. How often it is that no inquiry is made into the source of sand supply—price being the only factor considered! When trouble occurs complaint is made—not that the sand is impure, but that the cement or lime is at fault, and it is only after considerable cost and trouble taken by the cement and plaster manufacturer, who knows that his material is beyond reproach, but feels it is due from him to prove it to the customer, that the sand is tested and found wanting—or rather that it contains substances that are not wanted.

Having this in view, Messrs. Broad & Co., Ltd., of Paddington Green, conferred with Messrs. Cafferata & Co., Ltd., of Newark-on-Trent, manufacturers of plasters, who have pointed out the chief dangers to be avoided in sands, and suggested a few simple tests to discover the suitability of sands in preparing plasters.

It is with the kind permission of Messrs. Broad & Co., Ltd., that we give here this very informative essay.

Simple Tests for Sands

The fitness of a sand for use with a gypsum cement, plaster, Portland cement or with lime plaster, depends mainly upon its freedom from colloidal and finely divided matter.

Colloids

A colloid may be roughly defined as any substance which is non-crystalline, and which is made up of particles, the diameter of any one of which does not exceed ·0001 mm. In other words, the state of division is almost unimaginably fine. While many substances (e.g. clay, loam, marl) may take a colloidal form, the one most frequently met with in sands is hydroxide of iron, or, as it is usually called, ferric hydroxide. This substance has the chemical formula $Fe_2(OH)_6$.

The presence of colloidal matter in general, and of ferric hydroxide in particular, in any quantity in a sand renders it unfit for use with a gypsum cement or plaster, with Portland cement and with lime. The effects caused are as follows:

The setting properties of a gypsum cement or plaster, and of Portland cement, are markedly impaired. If such a sand has been used, the work does not harden properly, and after drying will be found to be friable and easily broken. With lime, again, the effect is much the same, and shrinkage cracks may occur in addition.

Ferric Hydroxide

The colour of ferric hydroxide varies from yellow to brownish-orange and orange-red. Its colouring power is considerable. Sands, therefore, containing it in any quantity are always strongly tinged with yellow, orange or orange-red.

The following tests, which are quite simple, will give a good indication of the presence of ferric hydroxide and colloidal matter in general:

Take a glass vessel, such as an ordinary tumbler. Pass some of the sand for test through a sieve having 8–10 meshes to the inch. Introduce some of this sieved sand into the tumbler to a depth of 1″. Pour in water until the total depth of the mixture of sand and water is 3″. Shake well and leave to stand for fifteen minutes. If the sand is good the mixture will rapidly separate, and the appearance in fifteen minutes will be as shown in Fig. 1, with a clear line of demarcation between the water and the sand. A little cloudiness in the water may be disregarded. Such a sand, provided it is not an unwashed sea sand containing much salt, may be trusted absolutely.

Fig. 2 gives the appearance of a sand that must be regarded with suspicion. In such cases the water will hold fine stuff in suspension and slowly deposit it in the form of a definite layer or band on the top of the sand. The thickness of this " band " may vary from 0 up to $\frac{1}{4}''$ or more. The " band " may be made up of finely divided sand, colloidal matter of a clay or loam nature, colloidal ferric hydroxide or a mixture of the lot. None of the above is good, while the last named is positively dangerous.

The colour of the sand will give a good indication of the presence

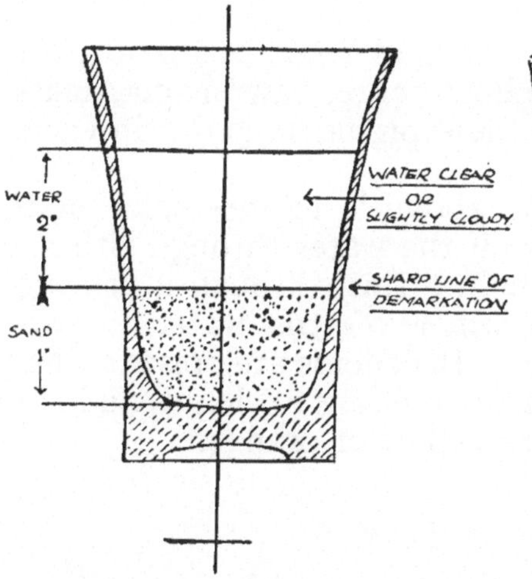

Fig. 1.

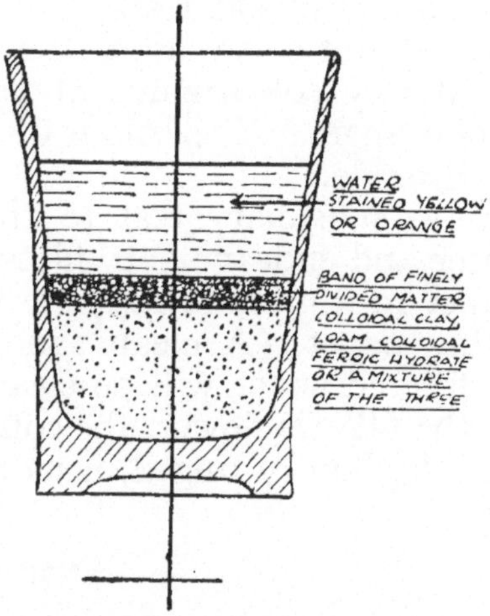

Fig. 2.

or otherwise of ferric hydroxide. If suspected, allow the tumbler to stand until the water above the sand is fairly clear. This may take an hour or more. Then carefully pour off all the water, disturbing the " band " as little as may be.

Now pour off the whole of the matter composing the band into a shallow vessel, such as the lid of a tin. This is quite easily effected, the bulk of the sand remaining firmly in place. Put the lid in a hot place to evaporate the water, or boil off the water over a gas-jet. When the water has almost gone and the material is becoming pasty, cease the heating and allow to cool. Take a little of the pasty matter on the tip of the finger, and rub between the finger and thumb. If the feel is one of sliminess and stickiness, then colloidal matter is certainly present, and the sand is quite unfit for use in its present condition. If the feel is decidedly gritty, and the thickness of the band not above

$\frac{1}{8}''$ thick after fifteen minutes' standing, the sand may be taken as passable. It is well, even then, if plastering work is in hand, to put up a trial patch of the plastering material and sand in the proportion it is decided to use. Note the time taken to harden and the strength attained. This is the most important test of a doubtful sand.

To recapitulate this test, place in a tumbler 1″ of sand and 2″ of water, giving 3″ in all. Shake and leave standing fifteen minutes. A good sand will give no band. An indifferent sand may have a band $\frac{1}{16}''$ thick, and the material of the band be greasy, or $\frac{1}{8}''$ thick and the material gritty. These should be regarded as the limit allowable. Put up a trial patch, if in any doubt.

A bad sand will give a band from $\frac{1}{8}''$ up to $\frac{1}{4}''$ or over, the colour will probably be yellow or orange, and the material composing the band will be slimy and sticky to the touch. Reject or thoroughly wash such a sand and repeat tests.

If any doubt exists as to the colour of ferric hydroxide, ask to see samples at any colourman's of yellow ochre, raw sienna, mars yellow and mars orange. The colour of these pigments is due in each case to ferric hydroxide.

A useful subsidiary test to the above is to mix some of the sand with water and endeavour to filter off the water through either a proper filter paper in a glass funnel, or, failing this, through a cone composed of blotting paper. A clean sand will allow the water to pass easily, and the filtered water will be clear. If colloids are present, the passage through the filter is slower, the colloids tending to clog it, and the filtered water will be faintly tinged, usually yellow or orange.

Tumbler Test

A further useful subsidiary test to the "tumbler test" to ascertain the presence of ferric hydroxide may be quickly made. Put a pinch of the dry sand on a piece of sheet iron (the lid of a tin will do), and heat to redness over a gas-jet or a fire.

During the process of heating most sands will darken in colour, those containing ferric hydroxide more than others. Remove the tin and allow it to cool. Place the sand thus calcined on a sheet of white paper. Put alongside it a pinch of the dry sand which has not been heated and compare the colours of the two. If much ferric hydroxide be present, the unheated sample will be yellow or orange-brown, while the one which was heated will be very decidedly red. In bad cases it will be a deep colour comparable to Indian red. Nearly all sands redden somewhat, and some skill is required to determine the safety of the sand by this method. It must only be regarded as a check to the "tumbler test."

A sand having a decided red colour will contain iron, but it may not be colloidal. Usually such a sand is suitable if the red is not due to red clay or marl. The " tumbler test " will demonstrate this.

Other Impurities

There are a few other impurities occurring in sands which are dangerous. Happily they are rarely met with:

Magnesium salts. Usually in the form of magnesium sulphate (Epsom Salts): Cause salting or efflorescence on the surface of plaster or cement.

Common salt, found usually in imperfectly washed sea or tidal water sands (sodium chloride): Quickens the set of gypsum cements and plasters, and causes salting and efflorescence with all cements and plastering materials.

Sodium Sulphate (Glauber Salts): Quickens the set of gypsum cements and plasters, and causes salting and efflorescence with all cements and plastering materials.

Note.—Magnesium and sodium sulphates occur frequently in bricks and show as salty deposits on the surface. If present, they *always* cause trouble with *any* plastering material whatsoever. Therefore, cut out salty bricks before plastering and replace with sound ones.

Grading

There is another property which a sand should have when used for plastering purposes. It should be well graded—that is, it should be composed of comparatively large, medium and small particles. The smallest particles should, however, be retained on a sieve of forty meshes per lineal inch.

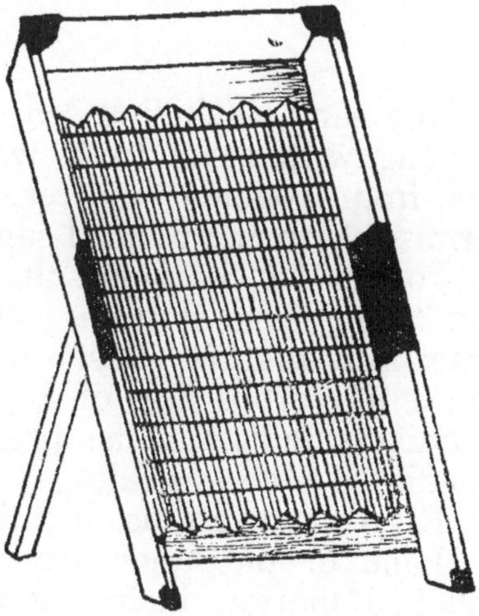

Fig. 3.

The suitability or otherwise of a sand on this point is shown by a rough " sieve analysis." Before dry mixing prior to plastering, the whole of the sand should be passed through a sieve of about ten meshes to the inch.

Practice in this respect varies somewhat, but the above may be taken as average. There is no objection to using a coarser sieve, provided the plasterer can work the sand with larger particles in it. Assuming, then, that the sand has been sieved through a ten-mesh sieve, of the remainder, the bulk should be retained on twenty and forty-mesh sieves. None should pass the sixty. The reason for this is that we have found that *anything fine enough to pass a forty-mesh sieve begins to act much as a colloid does.* Fig. 3.

Sieve Analyses

Sieve analyses of three sands may be as follows:

No.	On 20 Sieve. p.c.	On 40 Sieve. p.c.	On 60 Sieve. p.c.	Through 60 Sieve. p.c.	Remarks.
1	40	60	Nil	Nil	Ideal.
2	20	60	20	Nil	Good.
3	5	20	30	45	Too much fine. Result probably weak.
4	80	20	Nil	Nil	Too coarse. Will work lean, but will get strong.

Beware, therefore, of sands containing too much fine matter. Although they may work nice and fat, the result never justifies the ease of working. They usually contain colloidal matter.

It must be clearly understood, however, that a sieve analysis is useless for determining the presence or otherwise of colloidals. By their very finely divided condition colloids are essentially sticky or gelatinous. When the sand is dried they agglomerate, clinging together and to the larger particles, and only separate when wet. Their presence may, consequently, remain undetected by the sieves. A dry sand which is *full of friable lumps may, therefore, at once be suspected of containing much colloidal matter.*

Washing

Any sand, including sea sand, may be fitted for use by a thorough washing with clean fresh water. Indeed, sea sand is usually quite free from colloidal matter, but only the faintest trace of salt is permissible in the washed sample.

To test for common salt (sodium chloride), proceed as in " tumbler test," and leave standing until the water is clear. Pour off the clear water into another glass. Add to this water a few drops of nitric acid, and a few drops also of a solution of silver nitrite. A distinct white cloudiness indicates the presence of common salt. Only the faintest cloudiness is permissible.

River sands are usually the most suitable.

Land or pit sands are the worst offenders under the heading of colloidal matter, including ferric hydroxide. If they show up badly in the " tumbler test," nothing but a drastic washing will render them suitable. The washing water must carry over all the colloids and most of the remaining fine matter. Before being used they should unmistakably pass the " tumbler test," if they are to be incorporated *in full proportion* as plastering ingredients or as the sand constituent of a Portland cement concrete.

If a sand has been supplied as a washed sand, the "tumbler test" will at once show how well (or badly) the washing has been carried out.

Qualities of Sand

In connection with plastering, sand may be looked upon as a "filler," a substance which extends the more expensive lime or cement, and therefore an undue use of sand in proportioning, or the mixing of plaster and mortars, must be avoided, hence the stress usually laid on this point in specifications.

Mechanical Action

But sand fulfils other functions than that of a filler or extender; its mechanical influence is to induce and hasten crystallisation, and so hardens and strengthens both plasters and mortars.

It therefore follows that size of grain has some importance in making up the quality, hardening properties and durability, of plaster. Speaking generally, it is advisable that the sizes of the grains should not be uniform. If they vary within reasonable limits, they form better fillers and hasten crystallisation. It may be said that fine sand is preferable for the quick-setting hydraulic limes and cements; coarse sand for fast limes. Coarse sand is also suitable for coarse stuff and for Portland cement mortars for floating. Fine sand will be needed for top dressings and fine work generally, especially if a fine polished surface is required, and for moulding where delicate ornament is used.

Chemical Action

Apart from mechanical influence, it must be remembered that when sand is intimately amalgamated with moistened lime or cement, a chemical reaction is set up between the two substances, slow but sure. It is not merely that the cementitious matter is induced to crystallise and set; there is a true amalgamation, a give and take between the two substances, forming a new one. So the actual nature of the sand is of importance.

There are broadly two classes of sands: the siliceous, with its sub-division of quartzite, and the calcareous. Now, though the amalgamation of the silicate and quartz with the lime or cement is very slow, it does take place, and tends to strengthen the plasters or mortars. On the other hand, the action on calcareous sand is more rapid and thorough; it is disintegrative, and only adding another form of lime does not tend to increase hardness or durability. In time, it rather increases porousness. For good-class work, therefore, calcareous sand should not be used, but the siliceous kinds from river or pit preferred.

Clean artificial sand, derived from the dressing of stone in masons' yards and quarries, is excellent, especially if rich in quartz grains. This class of material, however, should always be washed, to get rid of the impalpable dust, clay and other impurities.

When washing sand, it should be well agitated in plenty of water, and allowed to soak for some time before it is passed through the sieves. It must, of course, be dry before it is mixed with lime, and more particularly with hydraulic limes or cements, otherwise proper mixing will be impossible, and premature setting will be induced.

Sea sand, and that obtained from tidal rivers, must also be very thoroughly washed to remove the alkaline salts, otherwise efflorescence, the appearance of white patches, is sure to be seen sooner or later. And the presence of salt also tends to induce dampness by absorption of moisture from the atmosphere.

Dr. Brysson Cunningham, M.Inst.C.E., Lecturer at University College, London, says that "ground ashes are an admirable substitute for sand, and mortar made in this way gives better results than sand mortar in the testing machine."

Colour, however, must also be taken into consideration in connection with plastering, especially when white cements are used as a final and dressing coat.

A rough-and-ready, but effective, test for clean sand is to rub it between the damp fingers hard; it should leave no stain.

CHAPTER II

PANEL MCULDINGS

Raised Panels—Thickness Rules—Floating the Panels—Sunk Panels or Raised Margins—Panel Ceilings—Panel Mouldings for Walls—Fixing Fibrous-plaster Panel Mouldings—Panel Mouldings with Circular Angles—Fixing Circular Pieces—To Set Out an Elliptical Panel without a Trammel—Semi-arches or Segment Arches—Raking Mouldings.

THE use of mouldings for the forming of panels to decorate plain work, either for interior or exterior jobs, is very interesting for the young plasterer. This work includes sunk and raised panels. The raised panel gives the stiles or margins the appearance of being sunk, though the margins are only just the ordinary thickness of plaster.

Panel mouldings are formed to soffit of beams, on ceilings and walls; some being plain and some enriched. The process of running is the same as for other mouldings. Screeds are formed and running rules fixed, but a little variation is necessary in the floating of walls for the panels. If the panels are sunk, the walls are screeded and floated out in the ordinary way, and the stiles or margins filled out to the thickness required afterwards. If the panels are of the same face as the stiles or margins, the walls are floated out in the ordinary manner, as the mouldings and panels and margins are to be of the one face.

RAISED PANELS

When the panels are raised, the walls are plumbed each end with the bob and gauges, the dots put in and the intermediate dots ranged through horizontally, top and bottom. A wide screed is formed up each angle of the wall, and also along the top and bottom, and on the top screed a level line is carried. The top screed should be below the bottom member of the cornice (if any), as this line can be used in the fixing or running of the cornice. Fig. 4 shows the process in commencing the floating for the panelling; A and B are the two angles or ends of panelling; C and D the ceiling and floor line; E, F, G, H are wide screeds, the thick line running through F being the level line. With a pinch rod (this is a rod of any fixed length) take the distance down each end marked with a cross, and snap a chalk line through H at the bottom, the dotted line showing it snapped. Having done this, the margins and panels are set out to the required size on the top level line, shown at A in Fig. 5. Sometimes the sketch or detailed drawing for the plasterer indicates that margins are to be 5″, and that two panels are required. If such is the case, the length of the wall is taken and the margins are divided for

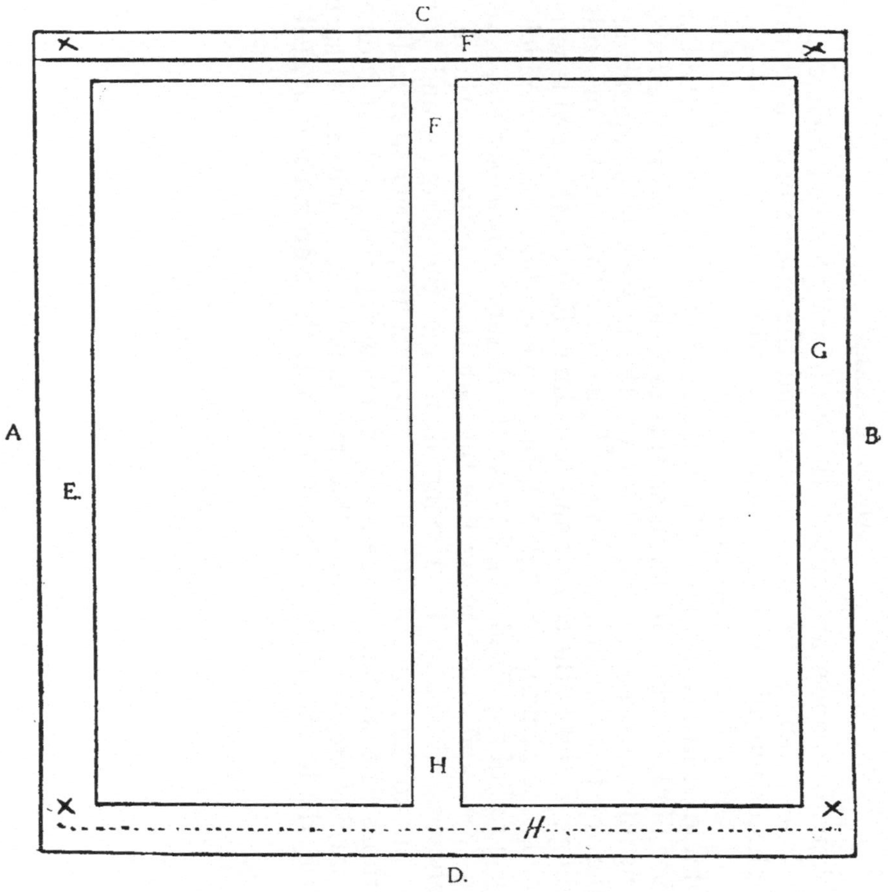

Fig. 4.

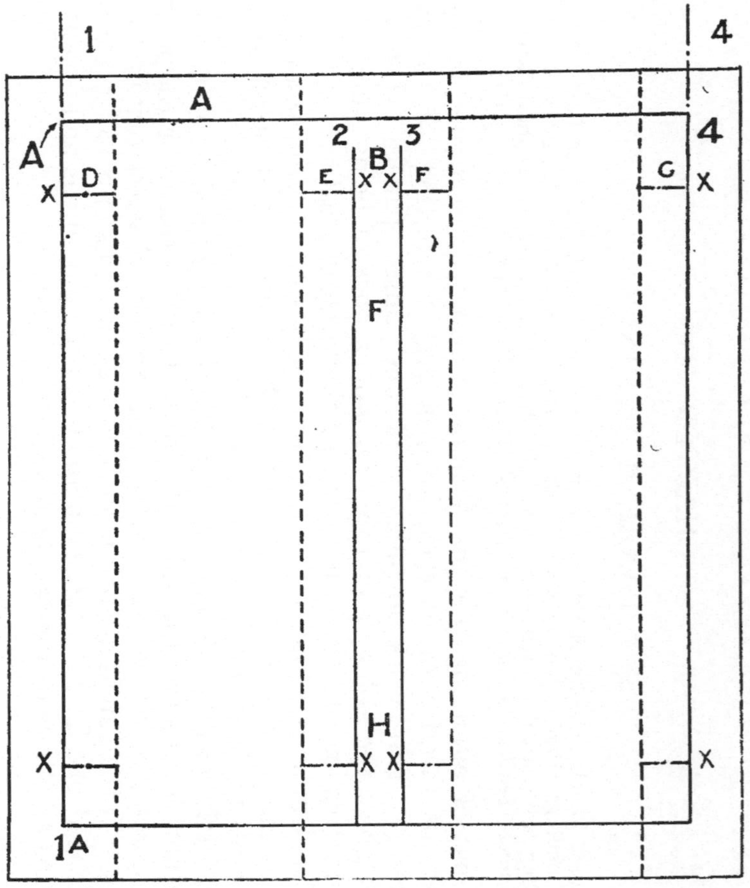

Fig. 5.

the two panels. The centre margin, B, having been fixed, a wide screed is traversed off from the top and bottom, as F and H in Fig. 4. Having set out the margin (which in this case governs the size of the panels) on the level line, as 1, 2, 3, 4 shown in Fig. 5, a plumb mark is carried to the bottom line, as H in Fig. 4, with the use of a small square. Fig. 6 shows a perspective view of the plumbing. A is 1 in Fig. 5, which the

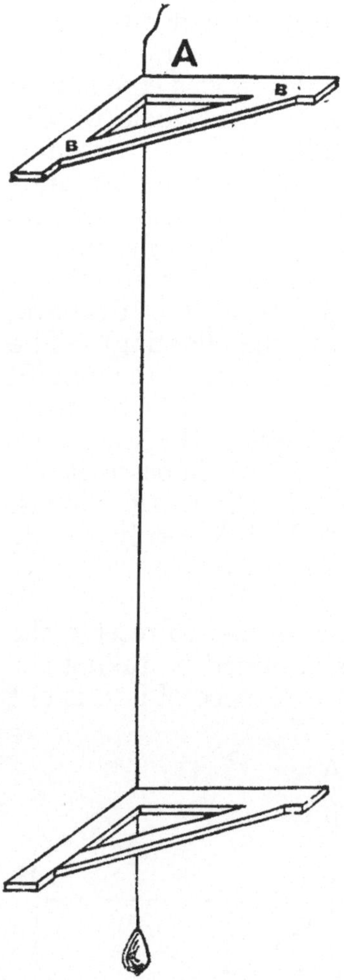

Fig. 6.

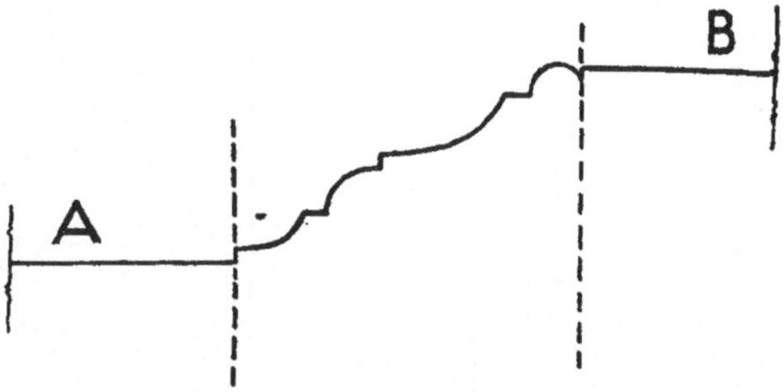

Fig. 7.

square as B is put to and the plumb bob hung vertically against the square. Another square is put on the bottom screed up to the hanging line (allowing it to hang free and clear of the square); this gives the plumb line at 1A in Fig. 5. This method can be carried out to secure the remaining marks at 2, 3, 4, or a pinch rod can be taken from 1, and the distance 2, 3, 4 marked off on it from 1 on line A. Having marked off the lower margins, a chalk line is snapped from top to bottom, extending it above the level line to ceiling, also to floor, so that the position of margins can always be seen. Having reached this stage, it is necessary to refer to the detail drawing for the width of moulding; Fig. 7 showing the section of moulding, A being the broken line of the margin, B of the panel. The dotted line shows the actual width of moulding which is to be run. To make it clearer, the width of moulding has to be taken so as to carry on with the floating of the panels. The width being taken, it is marked off on the level lines from 1, 2, 3, 4 and 1A, 2, 3, 4 in Fig. 5, a chalk line is snapped from top to bottom from these numbers. The dotted lines in Fig. 5 show the snapped lines.

Thickness Rules

Thickness rules a little less than the finished panels (to allow for the finishing) are tacked to these lines. Fig. 8, showing the thickness rule (thickness rule is any rule of the required thickness) which should be fixed inside of the panel mouldings as Fig. 12.

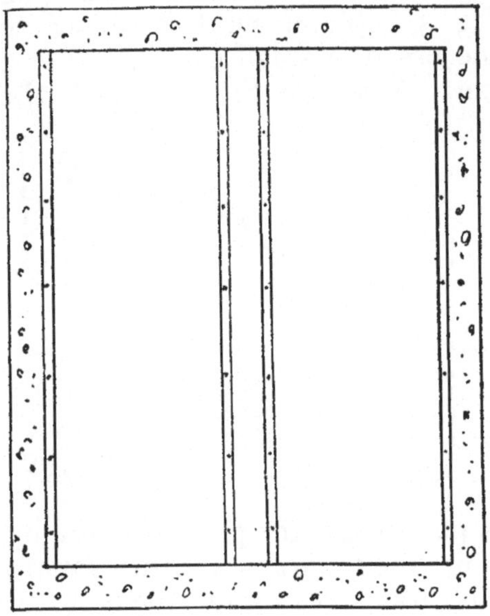

Fig. 8.

Floating the Panels

The plaster is laid on between the thickness rules and ruled off with a suitable length rule or straight-edge, the thickness rules in this case acting as the screeds. The whole of the panels being filled in, the tops and bottom are cut off the same distance from the finished margin as the vertical distance. When the plaster is set, the thickness rules are taken down (not forgetting the devilling or combing of the floating). The next step is the horsing of the mould, which is done in the ordinary way, but it must be noticed that the stock is slightly different in shape, as it has to be cut to suit the raised panels. Fig. 9 shows the stock cut and the mould plate tacked on ready for the horsing-up. Fig. 10 shows a perspective view of the finished mould ready for running.

Thc mould having been completed, screeds are formed to receive the nib and horse. The position of the nib screeds is obtained by taking the distance of the dotted line in Fig. 9, from A to B. A piece of lath is cut to this length and from the vertical lines 1, 2, 3, 4 this distance is marked off at D, E, F, G in Fig. 5. This is repeated a little above the line A, and then a chalk line is snapped vertically from each of these points. If screeds are required for the horse, the distance would be taken from the dotted line, A to C, in Figs. 9, 10 and 11 and taken back from the vertical lines marked with crosses. If screeds are used for the horse to run on, care must be taken that the position marks of the plumb line top and bottom are not erased. Screeds are not always necessary for the horse to run on, providing sufficient rules are available for the job and of the same thickness, as with good floating the mould can be run with the use of the rabbet, which is described in the "Running of Mouldings."

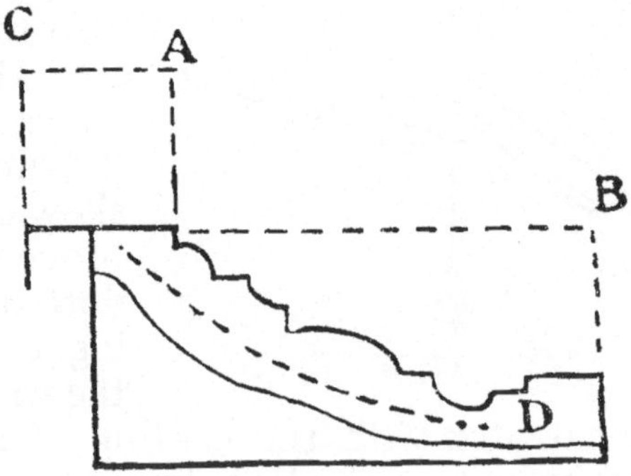

Fig. 9.

If a series of centre panel mouldings, as B in Fig. 5, are to be run, it is a great advantage to have parallel thickness running rules, these being fixed to the centre of the margin, and the mould being horsed-up to suit the remaining distance, which is to make the required width of margin. This enables both sides of the moulding to be run with the one rule. Care must be taken that this rule is fixed centrally to the centre of margin. A left and right mould are necessary in running panel mouldings. What is meant by a left and right mould is shown in Fig. 11; this is the opposite hand to Fig. 10, which runs up the right side. With panel or other mouldings having upright mouldings intersecting, the tops are run off first; for if the upright were finished first, when running the tops the finished upright would get splashed. When cutting the mould cut an extra square at D in Fig. 9, which does not appear in the section of the moulding panel side in Fig. 7. The reason for this is, when the moulding is run this forms an arris to work on when filling in the finish face to panel.

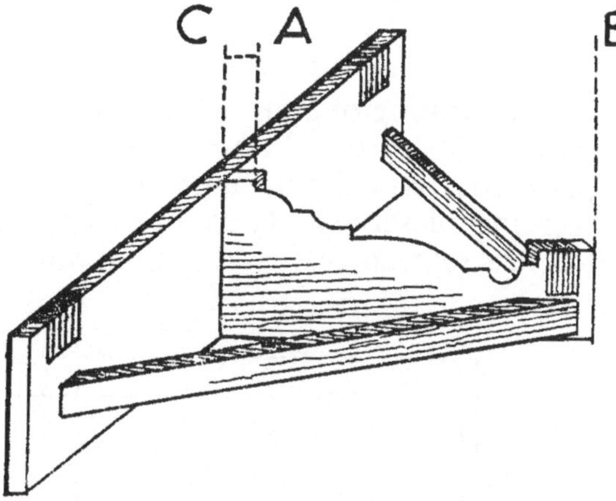

Fig. 10.

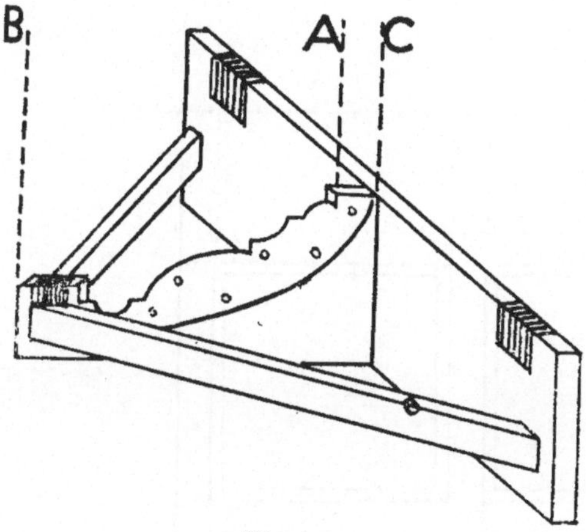

Fig. 11.

Sunk Panels or Raised Margins

Having dealt with the running of raised panels we are in a position to proceed, for the forming and running of the sunk panels is carried out in exactly the same manner. The only variation in these panels is in the floating, as the whole of the face is floated out; then the margins are set out, and the thickness rules tacked on and the margins filled out. Fig. 8 shows the thickness rules fixed, which are filled out with the use of a suitable rule. When fixing these rules care must be taken that they are fixed far enough back to clear the

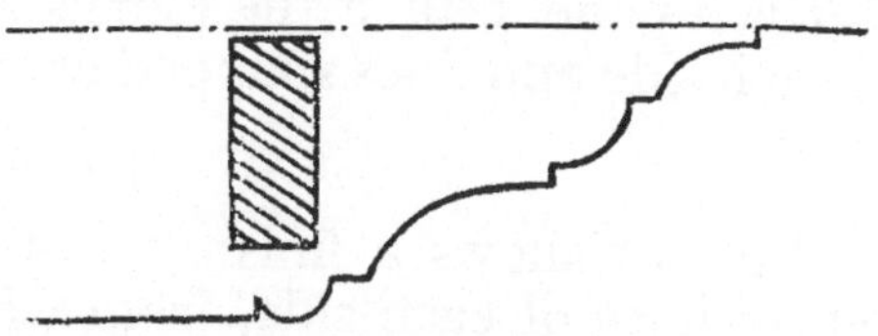

Fig. 12.

members of the moulding, otherwise the floating would have to be cut off to clear the mould when running.

Panels, Margins and Mouldings on Plaster Face

In the forming of panels with mouldings, either on the ceilings or walls, where the face of panel and margins are level, a very different method is adopted, that is if the ceilings or walls are finished in Keene's. There is no forming of screeds, therefore there is a saving of time and material; so it is less expensive. The ceiling and wall are floated and finished, and

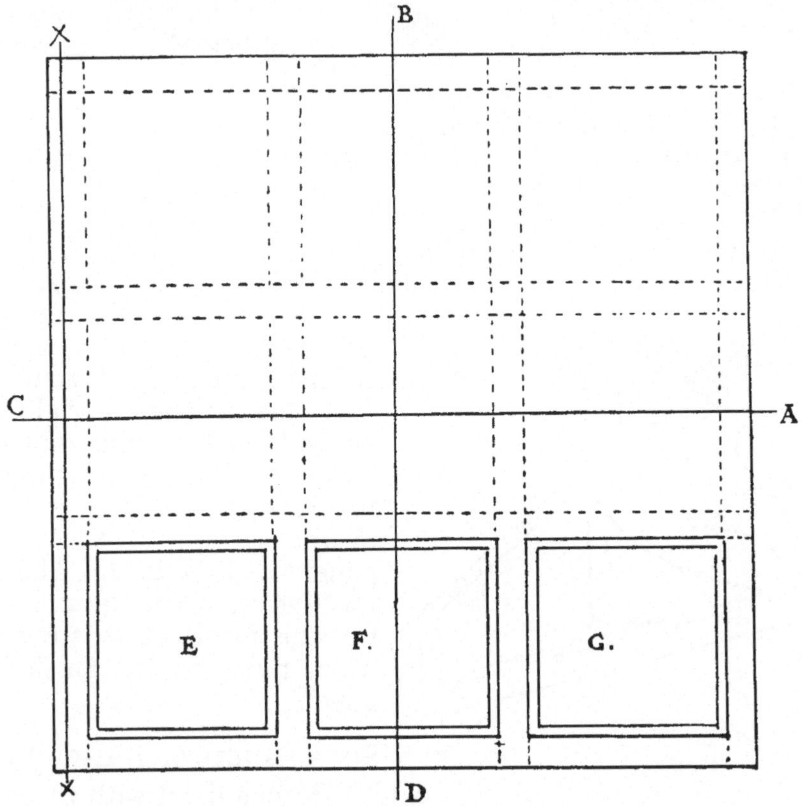

Fig. 13.

if it is a panel ceiling the panels and margins are set out from the centre of each side and lines snapped across.

Panel Ceilings

Fig. 13 shows a finished ceiling to be panelled. A, B, C, D are the centre lines of each side, from which the margins and panels are set out to whatever size is required. The dotted lines show the snapped lines as the line of panels and margins. E, F, G show three panels when finished. Having set out the ceiling, the next matter requiring attention is the mould. This is slightly different to the raised panel mould. This panel being of the one face, the mould is cut and horsed up with the nib and horse of the same line. Fig. 14 shows the mould cut and tacked on the stock. Fig. 15 shows a perspective view of the finished mould. This mould having to run on the finished plaster face, the zinc slippers must be let in nearly flush with the wood of the mould and horse, otherwise the mould would gather when being run and cause a salvage to collect which would have to be cleaned off. The next thing necessary is to take the distance from A to B in Fig. 15 (B in the figure should actually be further to the right ⅛″), on a piece of lath and set this distance back from the snapped lines, as shown by the two crosses in Fig. 13. This gives the position where to fix the running rules. This is repeated for each of the other panels. Before commencing the running, it is necessary to mark the width of the moulding on the ceiling, and with a wood chisel cut through the finished face so as to form a key for the mouldings.

Fig. 14.

All the mouldings should be run, and the mitres put in afterwards. If the moulding is of a large size it should be muffled, and if it is being done in Keene's it can be cored out with 2 parts of Keene's to 1 of sand or 3 of sand to 1 part of Portland cement; ceiling panel mouldings are sometimes formed with putty and plaster. When run with this material, it should be cored out with lime and hair coarse stuff, and finished off with strong gauged putty and plaster.

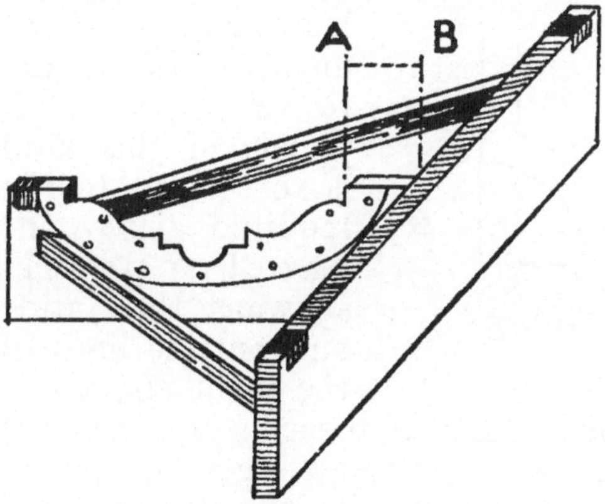

Fig. 15.

Panel Mouldings on Walls

When the panel mouldings are to be run on the finished face, exactly the same process is adopted for the walls as is carried out with the ceiling, with the exception that the panels and margins must be set out on the level line, top and bottom, as described in the previous paragraph on panel mouldings and in Fig. 5. If there are any intermediate cross panels and margins, these would have to be set out on the vertical lines at 1 and 4 in Fig. 5. When fixing the running rules, the same method would have to be adopted with the lath gauge from A to B in Fig. 15 to arrive at the position of the rule line. After the rules are fixed, it must not be forgotten that the face of plaster has to be cut through to form a key. Mouldings formed on finished plaster are usually finished with Keene's, after being cored out with Keene's and sand or Portland and sand.

Fixing Fibrous-plaster Panel Mouldings

When fixing fibrous-plaster mouldings on floated or finished plaster face, the same method is adopted as in the preparation for running mouldings on the finished face. If on the ceiling the margins and panels are set out and lines snapped as in Fig. 10, and the mouldings cut to the length required before fixing. If the panelling is on the walls, the same method is adopted as in Fig. 13, the panels and margins being set out top and bottom from the level lines formed at A and C and the lines snapped as before described and the mouldings fixed to these lines.

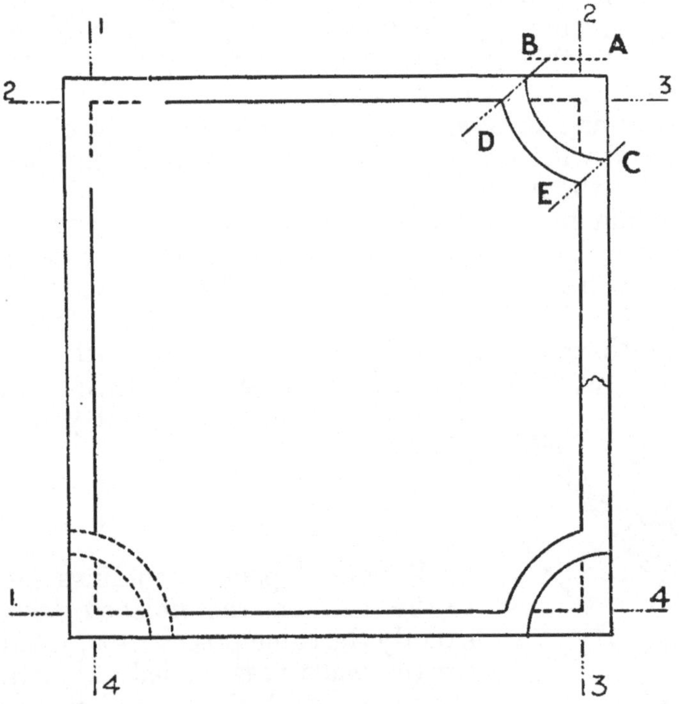

Fig. 16.

Panel Mouldings with Circular Angles

Panel mouldings are sometimes required having quadrant or part circular angles instead of being square. Fig. 16 shows a panel having quadrant angles. The method to be adopted with panels of this kind is to set out to the required size the panel or panels showing the inside and outside lines of the members, which simply means the width of the moulding is taken from each square angle of the panel and the lines snapped at points 1, 2, 3, 4, repeating 1, 2, 3, 4 gives the other side position for snapping. Having got so far, the distance is set off as to the radius required for the circular part, which is taken from A to B; then with the compasses the arc is described as shown. This intersects the outside line of panel at B and C. The compasses are extended the width of the moulding, and the arc described from D, carrying to intersect the inside line at E. Then a line is drawn from B to D and C to E, carrying the line over the width of the moulding both sides, which gives the mitre cut of the upright and circular moulding. This is repeated at each angle, which gives the distance to run the straight mouldings. After the straight mouldings have been run, it is necessary to fix on the radius rod as described in running circular mouldings on the bench. A complete circle is run as shown in Fig. 17, and to get the correct part of the circle that is required, a centre line is set out with another line at right angles to it. A second line is set off the width of the moulding, as shown by the dotted line, which, it will be noticed, corresponds with lettered angle of the panel in Fig. 16.

When these lines have been completed, the run circle is placed to the centre lines and cut through at A, B, C, D. Each piece is then placed to the dotted lines and cut to the mitre which will intersect at B, D and C, E, Fig. 16.

Fixing Circular Pieces

Bedding fixing (or planting, as it is called) is done by roughing the back of each piece with the knife or wood chisel, and damping the pieces with the water brush as well as the bed. Having gauged a small quantity of material, it is placed on the back of the piece, which is pressed well into its position, care being taken that the mitres intersect correctly.

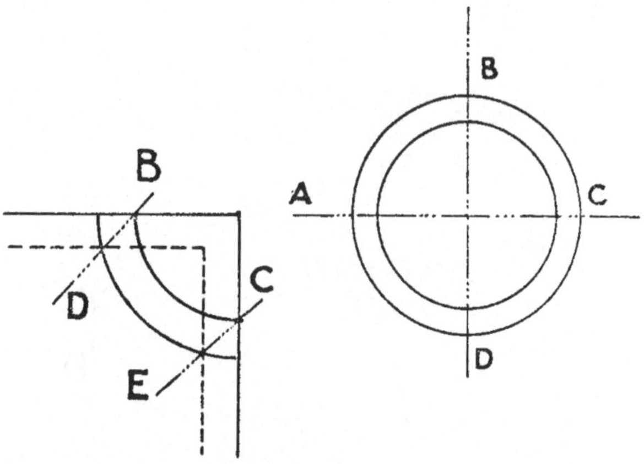

Fig. 17.

To Set Out an Elliptical Panel without a Trammel

An elliptical panel (which is an oval having its major axis one-third larger than its minor axis) can be set out without a trammel (the elliptical compasses usually employed for drawing ovals). To do this you must proceed as follows:

Draw a line *a* to *b*, the greater diameter to the desired length of the panel; then bisect it and set off the centre line *c*, *d*. This, being the lesser diameter, is made a third less than the line *a*, *b*. Again bisect each half of the line. This will divide *a* to *b* in four equal parts and give the centres *e*, *e*, which are the centres for describing the ends, as from F to F1 and F1 to F2. Next, from the centres C and D describe the flat curve from F to F1 and F to F2, which completes the ellipse. The moulding is run to these two radiuses, cut off to the desired length and fixed to the lines as set out.

It is, however, far better to set out the lines with a trammel, as the intersecting parts of the described line would appear more or less to have a kink, or be crippled as the plasterers term it.

Another method of setting out a flat elliptical panel, or an arch or rib, for casting purposes is carried out as follows, shown in Fig. 19. Taking

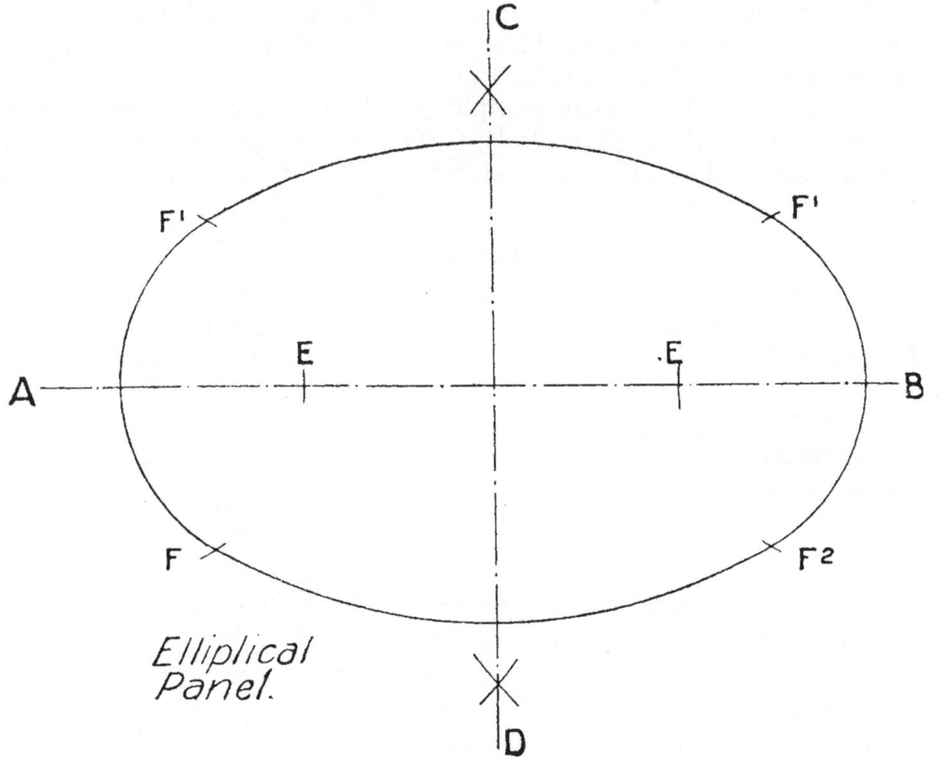
C
F'
F'
E
E
A
B
F
F2
Elliplical Panel.
D

Fig. 18.

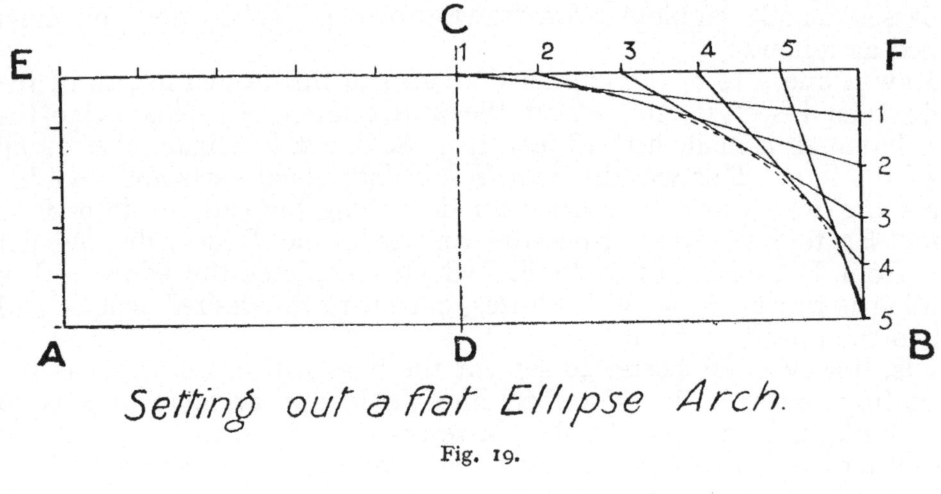
C
1
2
3
4
5
E
F
1
2
3
4
5
A
D
B
Setting out a flat Ellipse Arch.

Fig. 19.

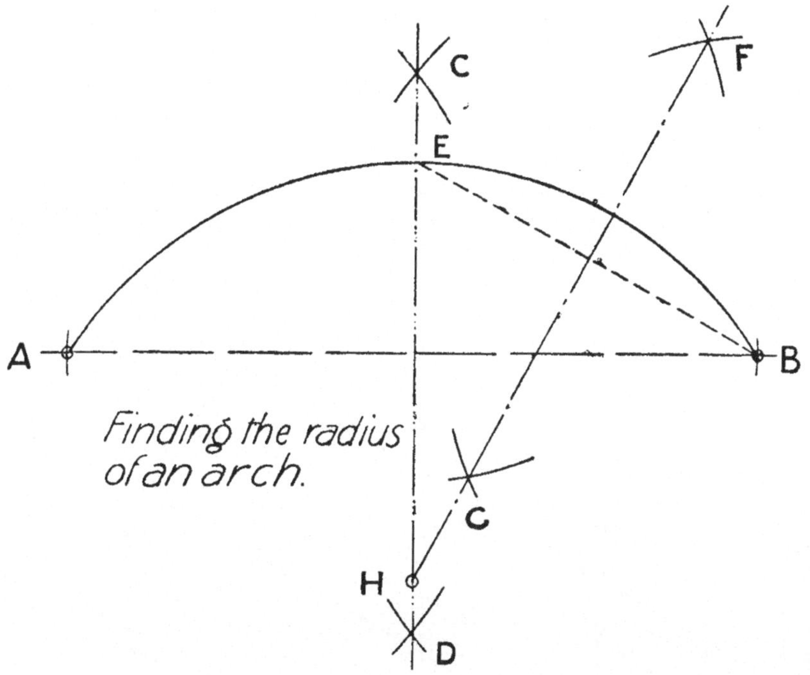

Fig. 20.

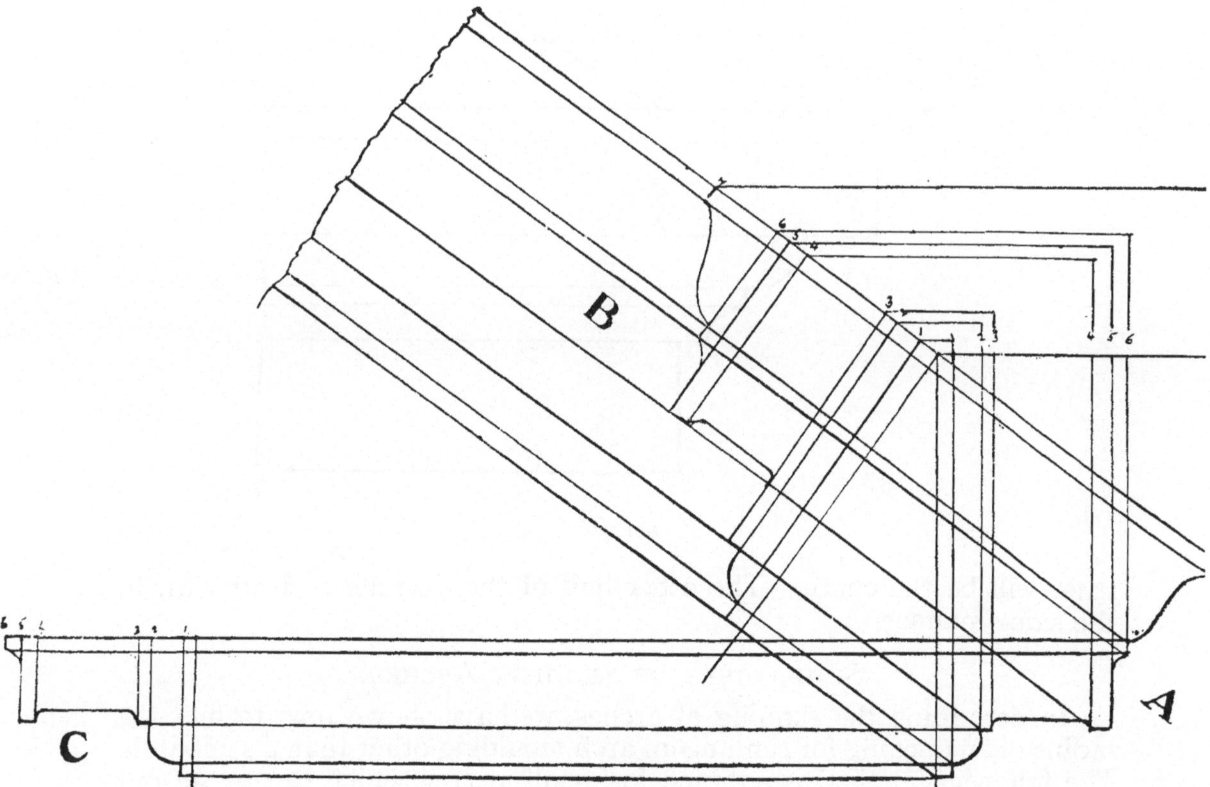

Fig. 21.

A to B, the distance required C, D is the half width, or rise, required, which is set off parallel as E to F. Divide C to F in any number of parts, as shown by 1, 2, 3, 4, 5. Then divide F to B in the same number of parts and draw the intersecting lines 1 to 2, 2 to 2 and so on. This forms the curve. It will be found that the more parts divided the

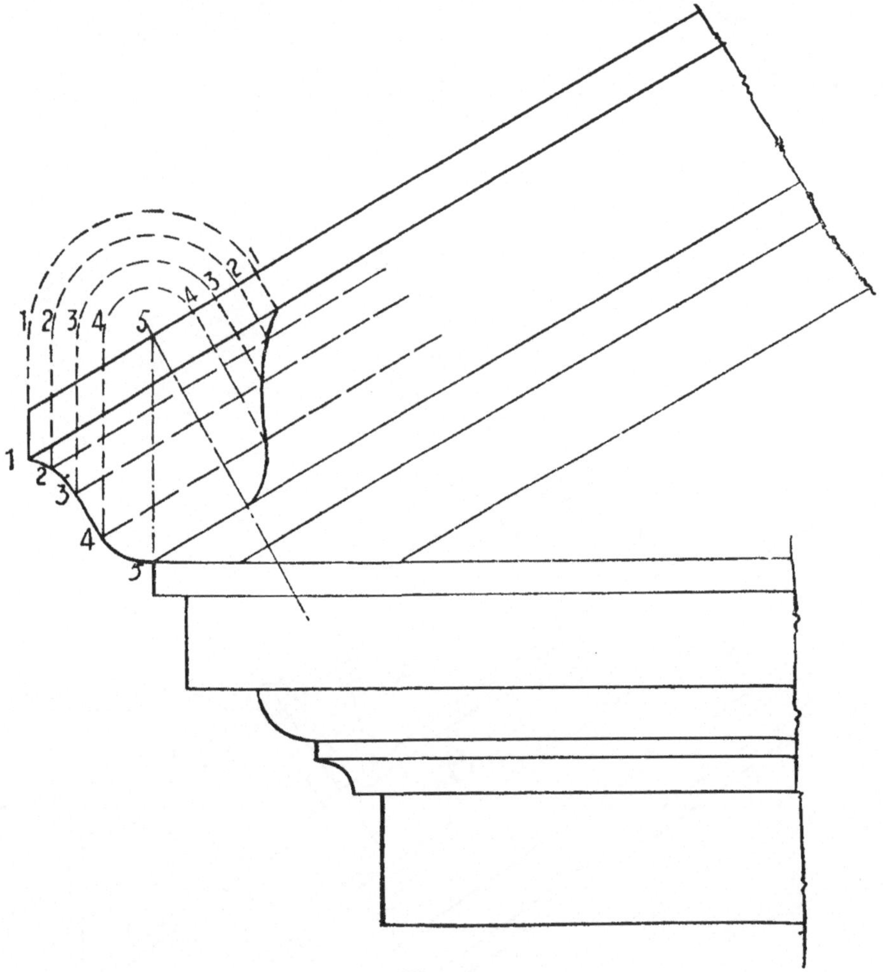

Fig. 22.

truer will be the curve. The other half of the opening is dealt with in the same manner.

Semi-arches or Segment Arches

In describing the running of arches, we have shown how to find the radius of an opening for running an arch moulding other than a semi-arch. The following method is adopted for small arches, which can be set out direct on the wall or bench, or even on the floor, as shown in Fig. 20.

Draw the line A to B, the width of the opening, and bisect it with the vertical line C, D. Then set off the rise of the opening at E, and draw the raking line to B. Bisect E to B, extending this to where it meets

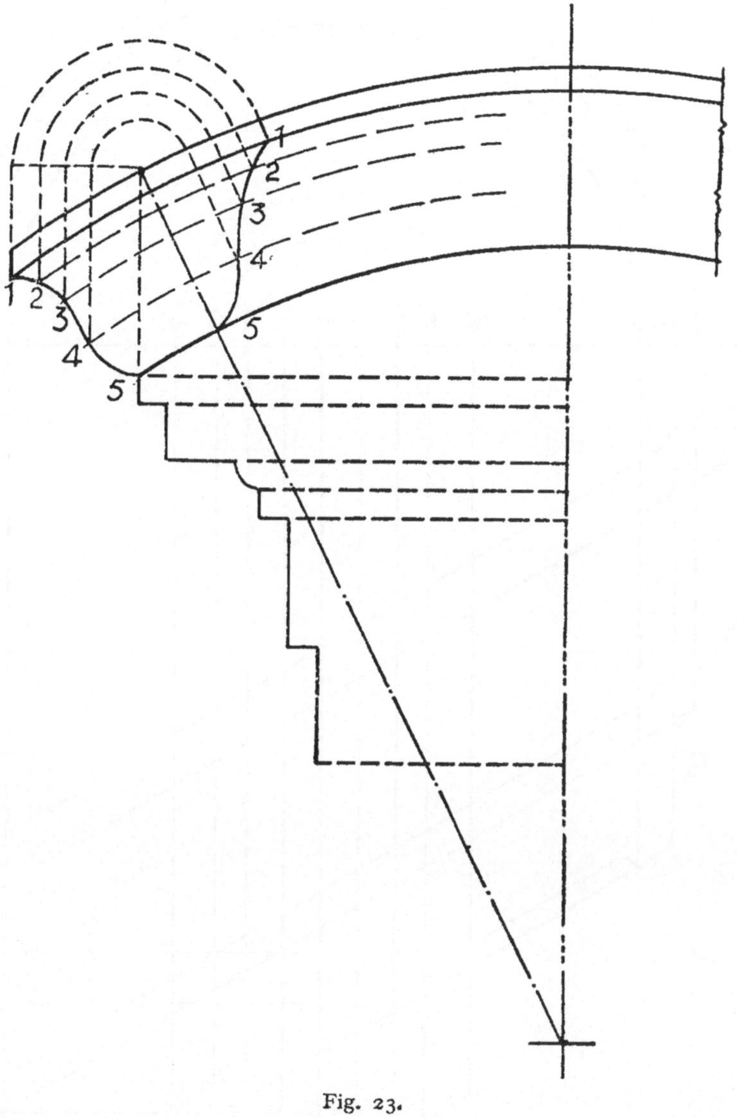

Fig. 23.

the vertical at H. With the distance from H to E describe the arc, which cuts across A, E, B.

A very useful method to find out the length of the radius rod, where it is not convenient, or where the opening is too large for the setting out required, is by the use of certain figures quite simple to apply. For example, a panel or an arch has an opening of 5′ 6″ and the rise of

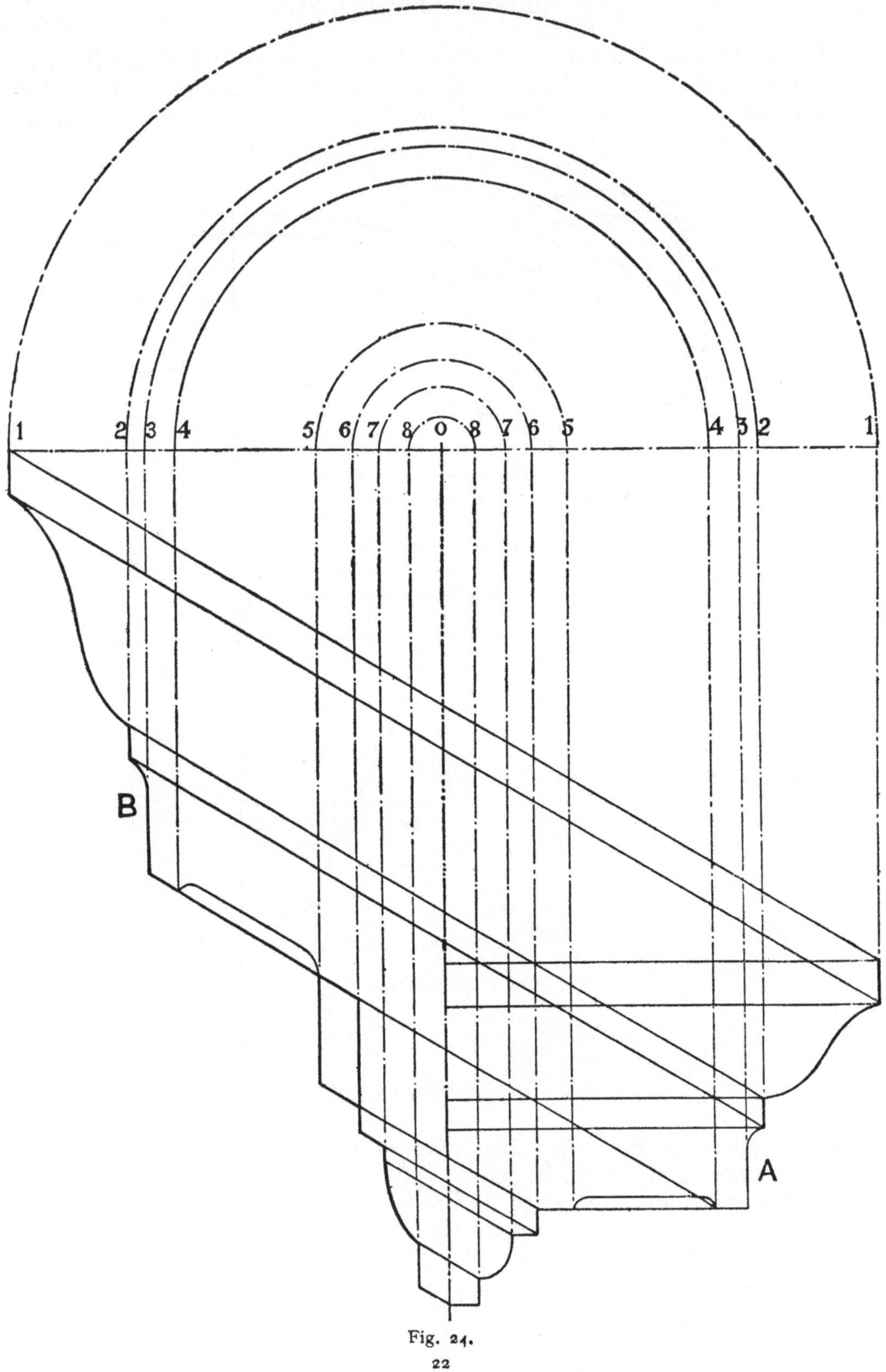

Fig. 24.

half width of panel is 1′ 6″. Take the square of the half opening in inches, and divide by twice the *rise*, or half width, and add the half rise. Thus:

Half opening 33″
Twice the rise 36″ 33
33
——
99
99
——
36)1089(30¼
108
——
...9
Half rise 9
——
12)39¼
——
3′ 3¼″ radius.

Raking Mouldings

Mouldings when intersecting from the horizontal to raking require developing by lines radiating from the horizontal profile. This will be seen in the various diagrams. Mouldings of this description are usually met with in interior plaster work on overdoors or on staircases, where the mouldings carry down the soffit of the staircase and intersect with mouldings running across the level again. We also see examples in external work on pediment heads, either complete or broken (as it is termed) when the raking moulding returns in itself. In Fig. 21 is the horizontal moulding from which the raking moulding is developed. B is the section of the raking moulding, C is part section, which is sometimes continued horizontally and returned in itself back from the returned end of the raking moulding. This is shown in Fig. 24, where A is the horizontal, B the returned end of the raking moulding.

In some cases only part of the moulding is developed, as shown in Fig. 22; the raking part finishing down on the top of the horizontal. But the top section of the raking part of the return moulding requires developing, that being the only part to intersect with the horizontal section of the return pediment head. The full development of the raking section is shown in Fig. 24 at B. Fig. 23 shows a circular head with the top section developed as in Fig. 22, and the remaining part of the section would finish down on the horizontal. Circular mouldings of this description are run with the radius rod or gig-stick, as described in the running of arches, though in some cases a wood rib is cut out and fastened in position, and the mould passed up and run with the use of wood pegs, which is described elsewhere.

CHAPTER III

MOULDING AND CASTING

Types of Moulds—Solid and Skin Jelly Moulds—Plaster Moulds—Cutting and Horsing Up Loose-piece Moulds—Cutting the Profile and Plate—Filing the Profile—Horsing Up the Mould—Muffling the Mould—Coring Out the Mould—Preparing Core for Running Off—Running Off the Moulding—Preparing Mould for Casting—Stops to Joints of Casts and Mitres—Making the Stops—Wood Stops—Filling In or Casting Moulds—Position of Wood in Casting—Reverse Mouldings—Forming Plain Shaft of Pilaster or Girder Casing—Forming the Cheeks—Fixing the Cheeks—Straight Fluted Pilasters—Running the Mould—Running the Return—Mouldings Run on a Core—Mouldings in Coloured Cement—Method of Preparing the Mould—Fixing the Reverse Mould—Filling In—Returns to Mouldings or Window Sills, etc.—Special Construction for Exterior Cornices.

Types of Moulds

For casting purposes various types of moulds are required, these being adapted to the size and character of the piece to be cast. Some moulds are quite simple, but others are more complicated, for what is needed in one case can be done without in another.

In plaster moulds with a cove as part section, or with an undercut member, loose pieces are required, while in some cases, where the section of moulding has an enrichment, part plaster moulds are used and part jelly insertion. What is meant by a jelly insertion is that the ornament is made up separately and moulded in a run case, a bed being allowed in the plaster mould to receive the jelly mould taken off the ornament.

Then, again, there is the jelly mould, either solid or in the case, taken from a model made up, and a case made and the jelly poured in the case by means of funnels.

Solid jelly moulds are generally used if only a small number of casts are required. In solid moulds the model is prepared and a fence put round, just above the top part of model, so that the jelly covers to about $\frac{3}{4}''$ thick.

Skin moulds are used for large models where only one or two casts are needed. The skin mould requires a light case to keep the jelly in position. With a skin mould the model is prepared and a clay fence put round the edge to keep the jelly up together. The jelly is poured over the model (not too hot) and drawn up over the most prominent parts with the hands, sufficient jelly being covered over to clear the model to about $\frac{3}{4}''$ thick. As the jelly grows cool it remains in position which allows the plasterer to get a fairly good coat all over. When the jelly is quite cold (any undercut places are cut off to free the plaster case), it is given a thin coat of shellac and dusted over with French chalk ;

when dry the whole of the jelly and surrounding parts are well greased over and a light frame of wood and canvas and gauged plaster is made over the jelly so that when the mould comes off it lays flat in the case for casting from.

Another type of mould, such as a beam or girder casing, or a pilaster mould, has a reverse face run and loose cheeks or sides, which can be taken away after the mould is filled in to allow the cast to free. Reverse moulds are not always necessary, as the mould can be cut to run the section on what is called a core. In this case, the mould is horsed up and muffled to about 1″ thick, and run off, and then the muffle knocked off, the mould being run over the core, using canvas and wood to strengthen the moulding to allow of its being taken up. This method is used when not more than one cast is required, such as for a circular panel to a bull's-eye, or a short length of cornice. Reverse plaster moulds, where they can be utilised, are much the better for casting, as they do not get out of shape like the jelly mould. With a little repairing occasionally, plaster moulds enable the plasterer to obtain a good many hundreds of feet of casts.

Plaster Moulds

Plaster moulds are formed by the running of the reverse section or profile of the moulding. With some mouldings the reverse can be run and cast without any trouble. Other mouldings require loose pieces

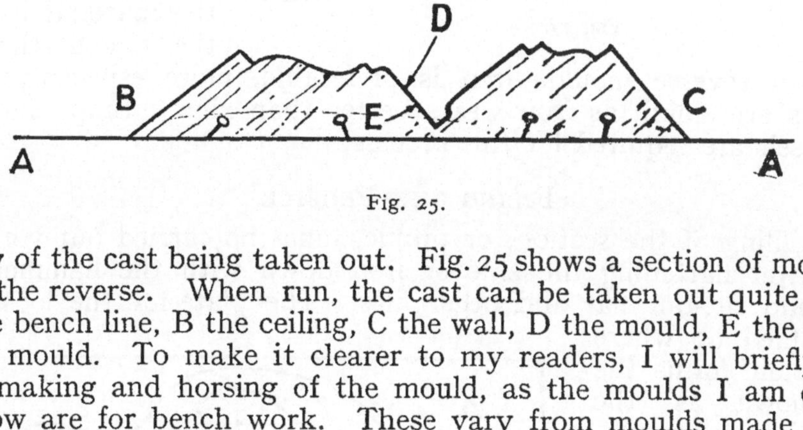

Fig. 25.

to allow of the cast being taken out. Fig. 25 shows a section of moulding cut to the reverse. When run, the cast can be taken out quite easily. A is the bench line, B the ceiling, C the wall, D the mould, E the plaster core of mould. To make it clearer to my readers, I will briefly refer to the making and horsing of the mould, as the moulds I am dealing with now are for bench work. These vary from moulds made to run *in situ*, though the actual cutting and filing up are exactly the same.

Cutting and Horsing Up Loose-piece Moulds

The section of the moulding to be run is traced off the detail given. Fig. 26 is the required moulding, in which provision will have to be made to form a loose piece (that portion of the mould which is movable and is considered necessary by the plasterer to allow of the cast freeing itself out of the mould). The loose piece is marked out on the tracing, which will be the loose piece in the mould when ready for casting. A is the loose piece.

The method of proceeding when a plaster mould is required is as

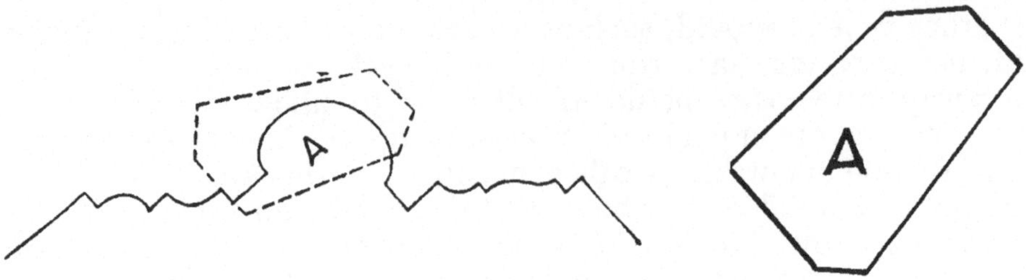

Fig. 26.

follows: the full section is traced off and the tracing stuck on a piece of zinc or iron plate, marking the position of the loose piece for the fixing of the plate. Fig. 27 shows the full tracing—A being the plate which is fixed on to form the bed for the loose piece; B the wall and ceiling line to form striking-off line when casting.

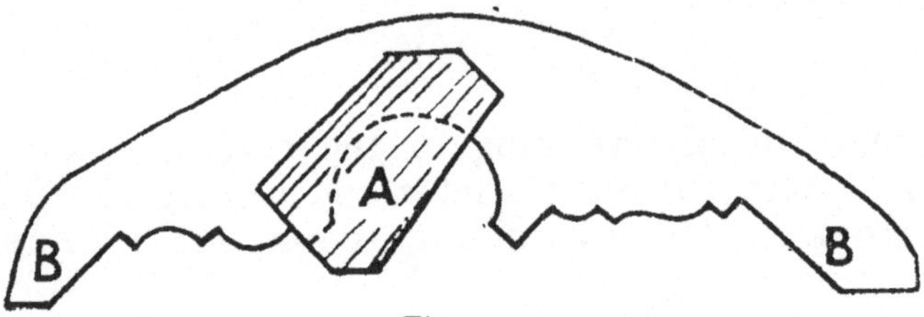

Fig. 27.

Cutting the Profile and Plate

The method of cutting depends on the material used and the size of the members. In reverse moulds iron is preferable, more especially if the members are small, as iron wears better than zinc. Snips and small iron chisels are required for this process.

Filing the Profile

The filing of the section, or profile, must be carried out very carefully (after flattening the zinc or iron down with the hammer) with half-round rat-tail, flat and other files. The plate for the loose piece is only filed up where the portion forms the bed, shown as the dotted line in Fig. 30.

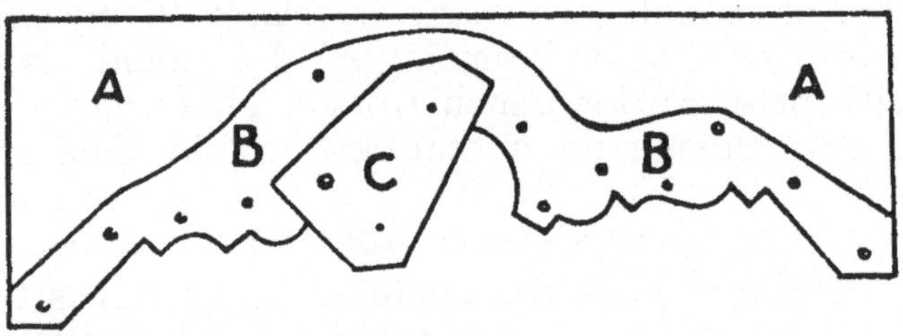

Fig. 28.

Horsing Up the Mould

Fig. 27 shows the iron and loose plate ready to fix on the stock: A being the loose plate, B the striking-off lines of wall and ceiling as shown on the dotted line. Fig. 28 is the mould plate tacked on the stock, not forgetting the stock has previously been cut back to allow a chamfered edge about ¼″ clear of iron for running purposes. A is the stock, B the mould plate, C the plate for bed of loose piece. Fig. 29 is a perspective view of the mould horsed up ready for running, the dotted line shows the bed plate fixed on over the full section of moulding. As a reminder, the slipper or horse (this being a shop job) is determined by the size of the stock. Usually the length is about one and a half times that of the stock, care being taken that the stock is nailed on square and upright on the bench; A the bench line.

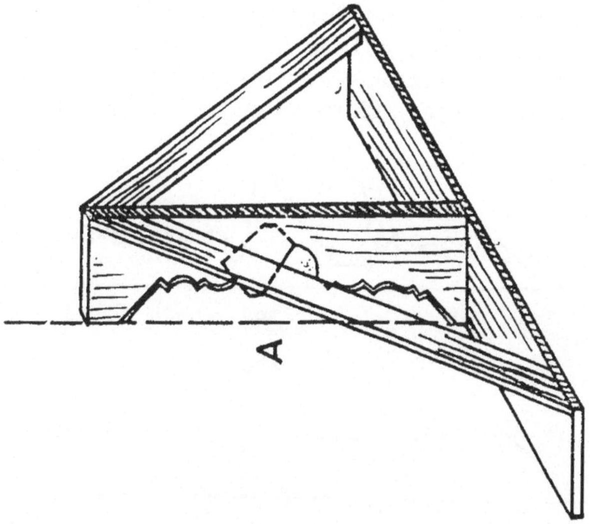

Fig. 29.

Muffling the Mould

After having horsed up the mould no zinc slippers are necessary, for the mould runs flat on the bench. That is the reason why the horse is sometimes termed the slipper.

The next thing necessary is the muffle. This is gauged plaster placed round the edge of the mould plate and pressed back on the back edge of the stock. But before doing this it is necessary to drive in, round the edge of the stock, ½″ tin-tacks. This prevents the muffle from swelling off. Sufficient thickness of plaster must be put on to allow of being trimmed back to about $\frac{1}{16}$″ in front of the iron profile. The shaded part in Fig. 30 shows the muffled part, which is kept clear of the plate forming the bed. It is to be noted that the loose plate is not muffled. This allows a clean bed to be run when coring out.

Coring Out the Mould

Coring out is the term applied to the portion of the running which is done with the muffle on the running mould. Disused plaster moulds are broken up and used for this purpose. This is a saving of plaster, and also makes it much easier for the plasterer to run off the core, as the more core used in forming the mould the less plaster is required; therefore the less swelling is attached in running.

Running large mouldings without first muffling the mould would often prove disastrous, owing to the swelling of gauged plaster.

The core having been run, it is advisable to allow it to stand for any swelling that may arise, as swelling always follows where plaster is used in a body. One very important point must not be overlooked in coring out, this is to drive in large nails in the bench in a slanting position. All old plaster being used for the core is tightened up by gauging very thin, weak plaster in the pail or bowl, and sprinkling over the old stuff.

Preparing Core for Running Off

The muffle having been knocked off and the loose plate removed (care being taken that the profile is not damaged in this process), the core is hacked with a stout knife or chisel. This makes a key for the finishing thickness; without this the finishing coat would swell away from the core. Care must be taken not to damage its bed for the loose piece. The run bed is varnished with shellac several times, until a good polish has been obtained. French chalk is then passed over the bed, and it is well greased with tallow to permit of the loose piece leaving when run off. Fig. 31 shows the bed and the core run off.

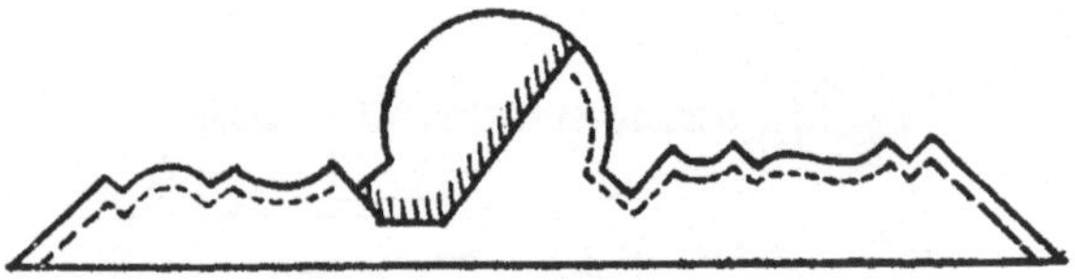

Fig. 30.

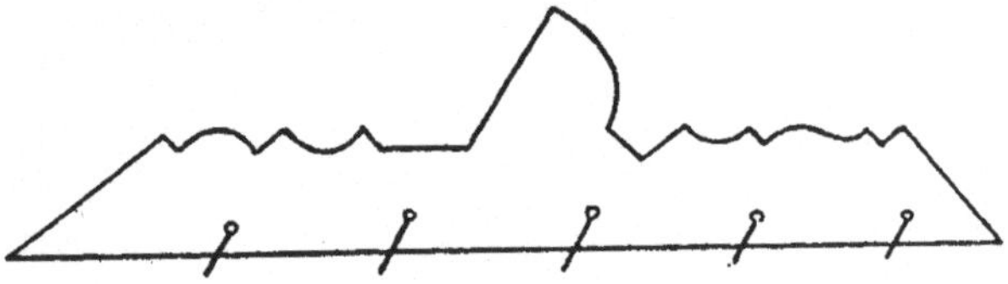

Fig. 31.

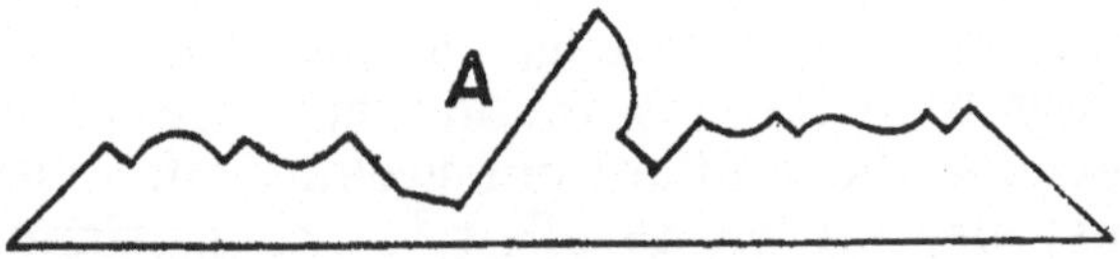

Fig. 32.

Running Off the Moulding

Before running off the moulding it is necessary to cut a piece of canvas the length of the mould, and of a sufficient width to nearly fill out the loose piece. This should be well soaked in plaster, twisted round rope-fashion and laid in the bed, at the same time passing the running mould over to see that it clears. The plaster in this case should be gauged round (a thick cream), as it makes a stronger loose piece. This is an advantage, because it requires lifting out of the mould sometimes to clean, etc., and unless strong there is a possibility of its breaking. It is for this purpose that the roped canvas is put in. Fig. 30 shows the mould run over and the position of the loose piece, the shaded line being the back part fitting against the bed at A in Fig. 32. The dotted line in Fig. 30 is the core as shown in Fig. 31.

Preparing Mould for Casting

When a reverse moulding has been run to use as a mould to cast from, it is necessary to prepare it before casting. To do this the plasterer examines the mould, and if there are any burrs (rough parts) on the plain members, the small joint rule is passed over. The next process is to shellac the face of the mould, the varnish being applied with a soft brush. The shellac should be on the thin side for the first coat, as it adheres to the plaster better, penetrating the deeper. It is very quick drying, so when starting varnishing it should be continued until the mould is finished. If the shellac is put on too thick it is liable to peel off, for the plaster being damp the thick gum would only form a film on the surface When the shellac gets thick, for it thickens when not in use, it can be thinned down with methylated spirit.

When the first coat has dried another coat is applied, which can be a little thicker than the first. Three coats in all are needed, but each application must be allowed to dry before another is added. It is necessary to have a good polish on the mould, as this prevents the absorption of the grease when applied. Having allowed the shellac to get dry, a small quantity of french chalk is passed over the mould with a soft brush, brushing it well into the members. This helps to fill in any small places the shellac may have passed over, and in addition it dries up any small wet part of the varnish.

The mould, having been well brushed over, is ready for greasing. This is very important, as unless the mould is well greased before being filled in, the cast would adhere to the mould. The greasing prevents close contact. The grease must not be too thick; on the other hand, it must not be too thin. It is composed of Russian tallow and about 50 per cent. of linseed or sweet oil made up in a can and stood over the gas or fire till it is dissolved. It is allowed to stand to cool before use, and then is applied with a brush and well rubbed in, using a piece of canvas in addition to smear the small members. Sufficient grease should be put on to make it discernible on the surface.

Stops to Joints of Casts and Mitres

Before commencing filling in the mould, which is the casting process, stops are made to fit in the mould to cast up to, at each end of the mould, or to whatever length of cast is required. Stops are not always necessary, as the required number of casts are measured up and the mould cut off square each end to suit the number of casts required. A piece of board is greased and fixed up each end, thus forming a trough. If stops are to be used, they are made before the casting is started, and moved along in the mould to whatever length of cast is required.

Making the Stops

This is how to proceed in making stops: The mould having been greased, lay a sheet of paper over one end of the mould so as not to splash it when making the cast for the stops. Gauge sufficient plaster and lay over the members of the mould about 6″ or 7″ wide and about

½" thick. Soak a strip of canvas, double the width, in plaster, and lay it on the plaster, pressing it into a fair thickness with the gauging trowel. More plaster is added until it is the required thickness, which is usually

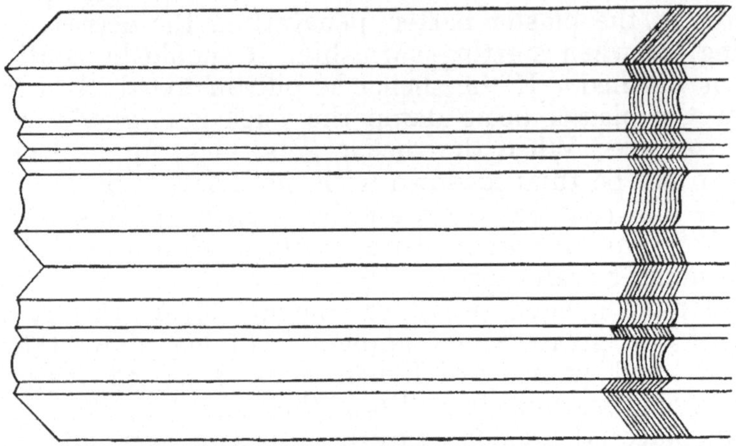

Fig. 33.

2" to 3". When the plaster is set, it is taken up and a piece cut off and cut through the centre (with the saw) from ceiling to wall line. This gives a square end for left and right hand of the mould. The stops are varnished with shellac, greased and put back in the mould, a small nail being driven in to keep them to the required length. Fig. 33 shows a reverse mould ready for casting. Fig. 34 is the mould showing a right-handed stop and mitre stop for left-hand internal mitre.

When casting the piece for the stops, sufficient should be cast for

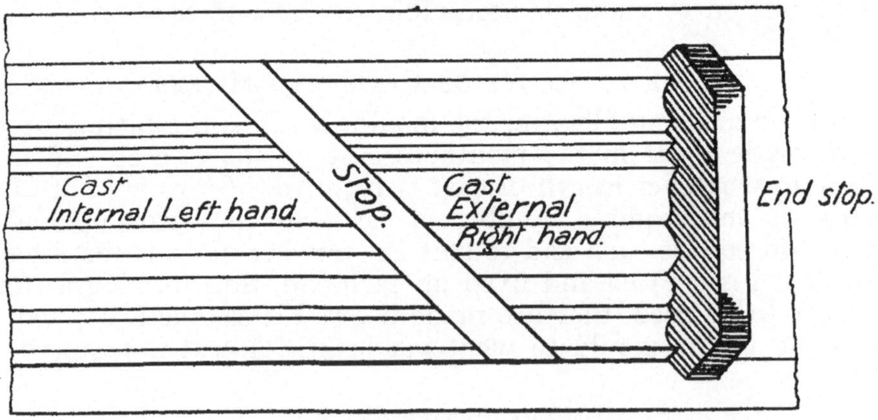

Fig. 34.

the two end stops, internal and external. Having cut off the piece for making the square stops, the remaining piece is set up on the bench, the wall line being perpendicular. To ensure the piece being set up in its correct position, the best method to adopt is to square a piece of board up off the bench the depth of the moulding, securing this by a piece of batten nailed to the bench at each end. Then take the distance of the projection from the board and mark it on the bench at each end, and nail a lath or piece of wood of 1″ by ½″ to these marks. Next put the piece of moulding between the lath and the board, if it should be deeper than the moulding. This puts the piece in its correct position. Having done this, carry a square line across from the wall line to ceiling line, then carry the distance of the projection from the square line of the ceiling to whatever angle is required.

To make it clearer, start from the right-hand end of piece and mark off as described; then saw through from wall to ceiling line on the projection as marked, keeping the saw upright when cutting. This gives a left-handed internal mitre. Mark off about 2″ and cut through again.

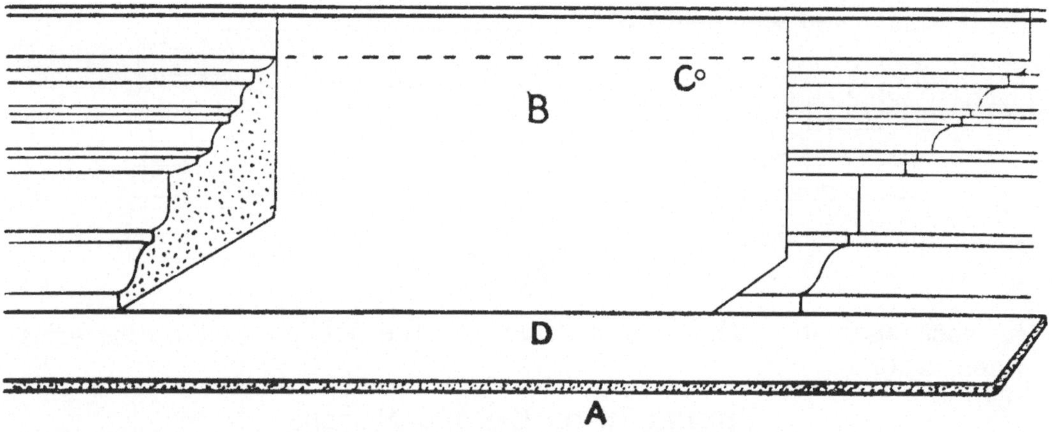

Fig. 35.

This gives the piece for the mitre stop. Repeat this at the other end, which gives a right-handed internal mitre stop.

When casting the mitre, stops are placed in position according to the mitres required. Casting one side of the stop gives the cut to the cast of an external mitre, and the other side gives the internal cut. This also applies to both stops, left or right hand.

Fig. 35 shows the piece set up on the bench; looking through the piece A is the bench; B the board, the dotted line; C the depth of the cornice; D the projection lath nailed on the bench. Figs. 36 and 37 are perspective views of the piece with the left and right projections marked.

Another way of getting the mitre stop is to set up a square line off the mould with 1″ by ¼″ wood, and a cross piece from projection, and strike off with the long joint rule a thickness of clay laid on the mould, and then cast the stop from it. This is repeated for whatever stop is required at angles. Fig. 38 is the stop for a right-handed external mitre, making the stop against the clay, which is the finished side of the stop; therefore when casting, the end of the cast would be up against stops, C its ceiling line and W its wall line. A the stop drawn away from the clay; B stop for right-handed internal.

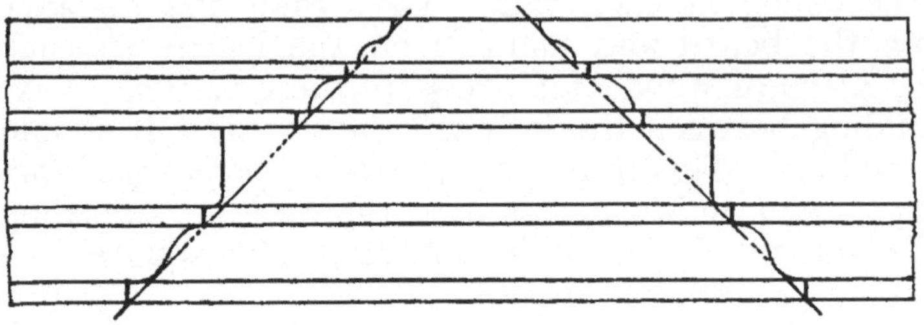

Fig. 36.

Wood Stops

When only a few casts are required, it is not necessary to make stops, and all that is needed is a piece of board at each end of the mould, kept in position by a short strut nailed in the bench. But if large numbers of mitre casts are required, it is a saving of time and material to make mitre stops

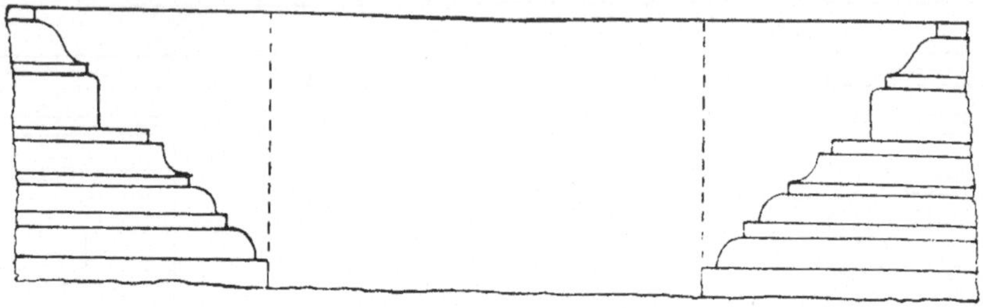

Fig. 37.

to cast against. This saves time on the job as well as material when casting.

Filling In or Casting Moulds

Filling in is the term commonly applied to casting. Before gauging the plaster for filling in the canvas, wood laths and brackets, or whatever wood is required, should first be cut. The plaster is prepared in pails (unless the casting is very large, when baths are used) in two strengths, namely, firstings and seconds. The firsts or firsting (first coating) is so treated that the stage of setting commences by the time the seconds or the second coating is applied.

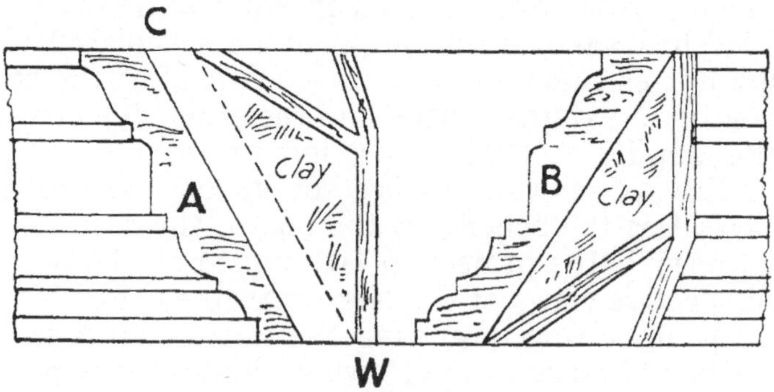

Fig. 38.

PLASTER CEILING AND COLUMNS IN BANK PARLOUR.

PLASTER CEILING AND WALL DECORATION, ENTRANCE HALL, DERBY HIPPODROME.
(Marshall & Tweedy, Architects.)

Size, in a liquid form, is always used in casting (unless it is a very small mould). This is to retard the setting of the plaster. It is mixed in the gauging water, first testing to ascertain how long it requires to set. The retarding is to allow sufficient time for the plasterer to fill in the moulding after laying in the canvas and wood. It must be clearly understood that sufficient size should be put in the firstings to allow of time for it to be used. More must be added in the seconds to permit of the canvas and wood being put in. The aim is to fill in the whole of the mould before the seconds have set. This gives time for the cleaning off of the edges of the mould, and whilst the cast is setting the plasterer prepares the wood and canvas for the next cast.

Sufficient plaster should be gauged for the filling in of the mould. It should be gauged as thick cream, as this has the making of strong casts. Plaster when gauged thin for casting makes very bad casts; they are unreliable and apt to buckle up when being lifted out of the mould.

Filling In

When preparing for filling in, the firsting and seconds are gauged up in the order named. The reason for so doing is that as soon as the first coat is applied the seconds is ready to follow on, without waiting, as sometimes the first coat sets quicker than anticipated, and the seconds must be applied before the firsting is quite set, otherwise it would not adhere and the cast would be a failure.

The first coat is applied with a fibre brush, called a splash brush. specially made for shop work. It is well rubbed in all over the members. This rubbing also takes a lot of the grease off the mould. When the mould is well covered, not less than $\frac{1}{8}''$ thick, the remainder of the material gauged is splashed all over it. Then the plasterer washes out the brush and bowl in the water tub, rinsing the brush in clean water. If this is not done; the brush will get hard and prove a nuisance when using the next materials. By the time this operation is completed, the firstings would have picked up enough (that is, set sufficiently) to carry on with the seconds. The seconds are well splashed over the first coat and well rubbed in with the brush. It is important to remember to keep the striking-off edges, shown in B, Fig. 27, which is the ceiling and wall line, clean by passing a short piece of lath along periodically.

Having well brushed in the second coat, the first layer of canvas is stretched over from end to end. This should be several inches longer at each end than the mould, and this also applies to the width. It is well pressed in with the fingers, but not so hard as to go through the first coat, otherwise it would show on the face of cast. The canvas must be longer and wider than the mould, because the ends require to be strong for standing up the cast, so that the ends of the canvas are turned in and they make strong ends.

As regards the width, laths are put in the mould at the wall and ceiling line, and then the spare canvas is turned in and under the laths. This makes a strong edge for wall and ceiling line when fixing. The ceiling and wall line wood having been put in, it is well brushed with more plaster. The remaining wood (which is generally 1″ × ¼″ laths in long lengths) is laid in the mould, being well pasted with plaster. This causes the canvas to adhere better. Sufficient laths are set out to make what is considered a strong cast. After well pasting all the wood and any dry places of canvas, also turning in the ends (as previously mentioned), another length of canvas is required. This should be the actual length and width of the mould. It is well brushed over with plaster, at the same time pressing it well in with the fingers to ensure contact with the first coat of canvas. The brackets, which are pieces of 1¼″ lath (cut to fit across mould), are laid on from ceiling to wall line. These should be dipped in the plaster before being laid on, and are spaced out about 1″ × 3″ apart, or according to the size of cast. They are covered with canvas, cut wide enough to lap round and under the canvas in the cast. The brackets bind all the long laths and canvas together, giving strength and rigidity, and when the plaster is set the cast can be lifted out without any fear of breaking.

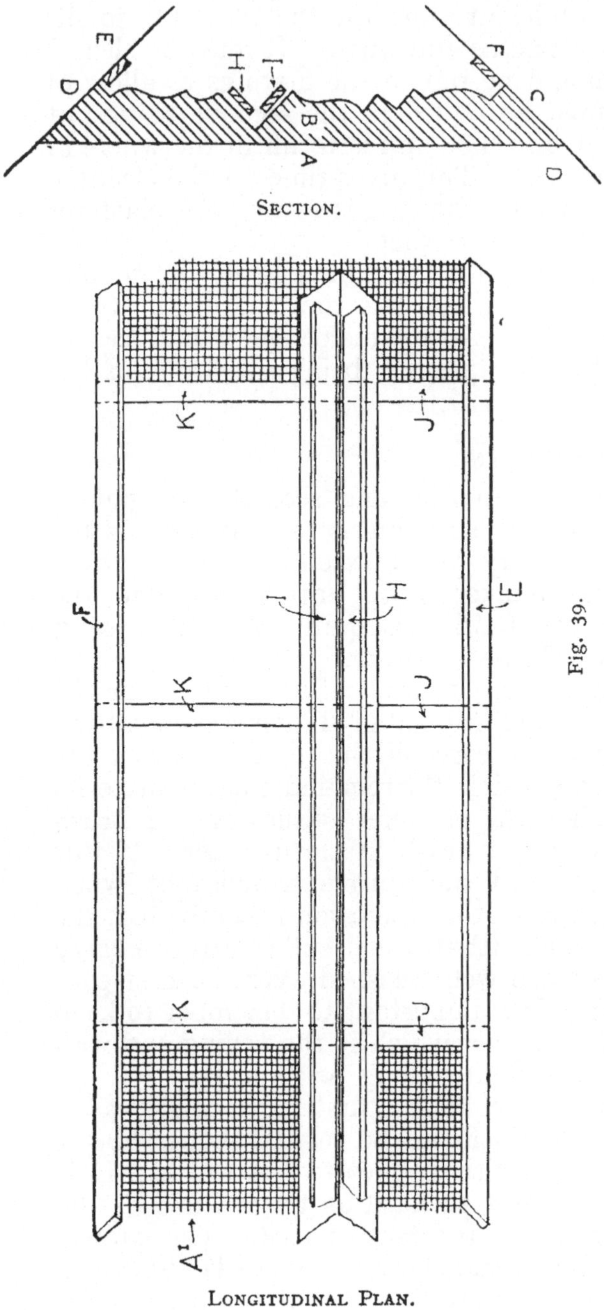

Section.

Longitudinal Plan.

Fig. 39.

Position of Wood in Casting

The position of the wood when filling in the mould is one of the chief points in casting, as the casts require to be light yet strong, and only sufficient wood should be put in as is considered strictly necessary. Wood must always be put as near the bottom edge of the ceiling and wall line members as possible, otherwise when handling the casts the top and bottom members get very much damaged, which entails much work in making good after the casts are fixed. Moreover, the fixings for the mouldings are the two edges, whether wall mouldings or cornice. The intermediate laths are put in prominent parts, such as the drip, and are put in on edge, being much stronger this way.

Fig. 39 shows a plaster mould with the canvas, laths and the cross brackets to prevent cast opening out: A being the bench line; B the mould; C and D the striking-off lines, which must be kept flush to the rake of line shown, otherwise the cast would not set tight against the wall and ceiling; E and F are the ceiling and wall laths, which must be kept to the square line as shown by the section at E and F. H and I are the two main laths, which give strength to the body of the cast. This is seen at section J; K are the cross brackets which, when backed with canvas when filling in, connects all the long laths together. When properly set, the cast is lifted out of the mould and stood up on its end. When casting cornices, the cast should be stood upright against a good background and tied together, which helps to keep them straight.

Reverse Moulds

Reverse moulds are so named because the method of using them is the reverse to that usually adopted by the plasterer in the forming of plaster running moulds. The running mould for the casting mould is cut from the inner profile of the detail of moulding, which is the reverse to the ceiling of the ordinary running mould. By this method the mould is made by running direct with the reverse section of the moulding. Reverse moulds are expeditious and effect a saving in the general casting of plaster work, and also in the making of artificial stone, as the reverse moulds are the principal features in the casting of imitation stone. Pilasters plain fluted straight or with an entasis are cast from the reverse profile of the moulding. However, in pilasters with an entasis the making of these is better executed by making up the model and moulding therefrom.

Forming Plain Pilasters or Girder Casings

When a plain pilaster without capital or base is required, or a plain casing for girders, the same method is adopted, whatever the size of face or returns. The bench must be well cleaned, and any holes filled in and varnished over (a little wider than the job to be prepared), and when dry dusted over with french chalk. If a shaft or casing is required

10′ × 1′ 4″ × 4½″, a straight line is set off on the bench and another 1′ 4″ from it. These are squared off at each end to the 10′. These are marked in with indelible pencil. As it is necessary to make two 4½″ plaster cheeks, which are to form the returns, a piece of 1″ wood is nailed to one of these lines; 4½″ is measured off. Then a thick shop straight-edge, usually about 2″ thick, is fastened down to the bench. These two pieces and the bench are well greased.

Forming the Cheeks

The next step is to cut a piece of canvas a little over 10″ long and about 1′ wide, ready for the filling in. Having gauged about a pailful of round plaster, the space between the two rules is filled in. The piece of canvas is soaked in the plaster and laid in between the rules. In addition to this, any rough canvas can be soaked in plaster and placed on the thick side of the rule. Then the first piece of canvas is turned in, more plaster is added if required to fill out from the thick rule to the thin one. The thick edge is used, because when it is placed on edge it stands quite firm on the bench. When this piece is set, one rule is removed and the gauging trowel pressed in under the piece. This loosens it and enables the plasterer to slide it away from the other rule. The rule is replaced and greased as before, and filled in. This gives the two cheeks which form the sides for the 4½″ returns.

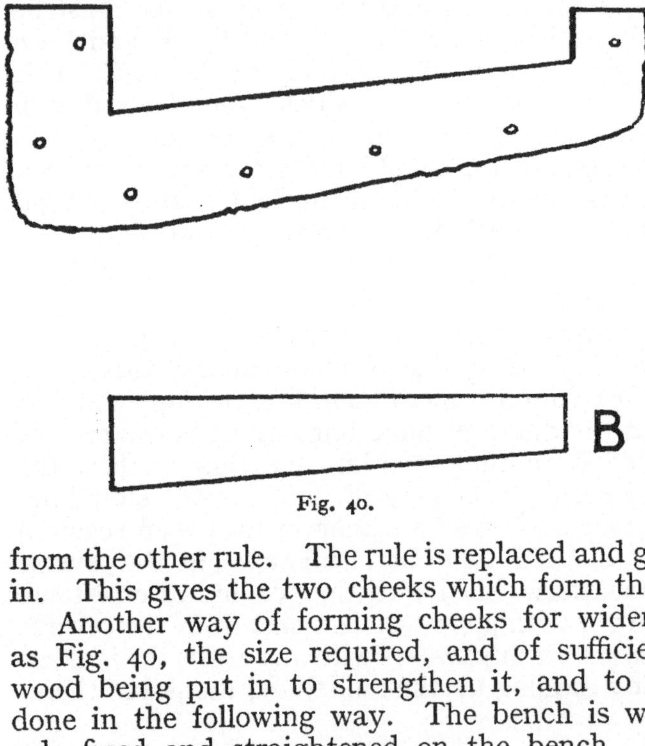

Fig. 40.

Another way of forming cheeks for wider sides is to cut a mould, as Fig. 40, the size required, and of sufficient thickness to allow for wood being put in to strengthen it, and to run two lengths. This is done in the following way. The bench is well greased, and a running rule fixed and straightened on the bench. The length is marked off and a small nail driven in where the moulding is run at the start. This keeps the plaster in position when running off. Sufficient plaster is gauged to complete the running of the length, or as near as possible. A piece of 1″ × ½″ lath is put in after a layer of canvas has been soaked and put on the bench, passing the mould along the running rule. The canvas is pushed in to clear the mould, lapping the canvas over the wood. When the plaster is set the piece is gently prised up by pressing the gauging trowel under it. It is pushed clear of the rule and another length run. B the run piece.

This method is usually adopted for side faces to girder casing, etc., or for plain pilaster with wide return. Turning back to the first method of forming the cheeks of the 10′ pilaster, the two $4\frac{1}{2}''$ cheeks are cleaned up by passing the thin piece of steel over them to take off any rough places. A piece of canvas is rubbed over and then the two pieces are given three coats of thin shellac. Having cut the two sides off to the 10′ and squared off both ends, they are well greased on the face and top and bottom edge, and then set up to the two lines as marked off at 1′ 4″, testing the position with a lath cut off to the required width. The two sides are squared up off the bench and a lath tacked across at each end and the centre. A temporary piece of wood is also tacked on the bench to keep the pieces in their correct position for the next process.

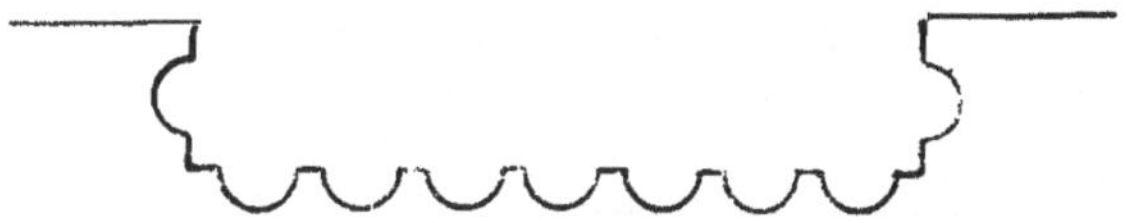

Fig. 41.

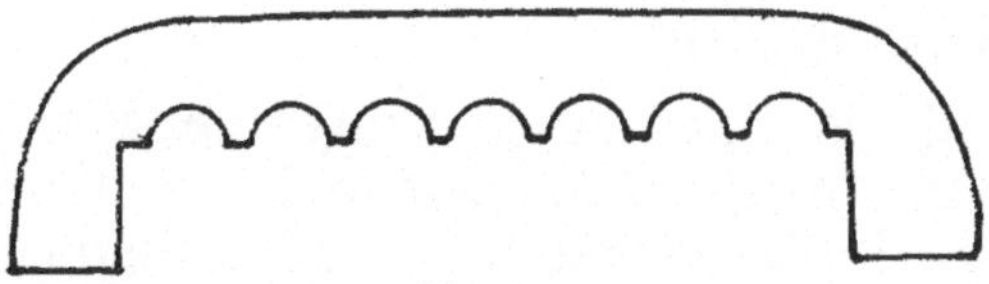

Fig. 42.

Fixing the Cheeks

Joggle holes are cut in the bench at each end and about 2′ apart. These are merely holes made in the bench with a chisel, and are not unlike mushroom heads. The holes are dusted and smoothed out and varnished with shellac and then greased. The cheeks being in position, plaster is gauged up in the bowl, and pieces of old clean canvas are soaked and pressed into these holes and against the back of the cheeks, so that when set they lift out with the sides, and can always be replaced in their correct position. When both sides have been joggled, they are lifted out and the joggle holes varnished with shellac. Before being put back they should be greased, in case that when filling in the mould any plaster should work its way underneath. The ends of the mould having been squared off, a piece of thin wood is greased and fixed up at each end. This forms a stop when casting.

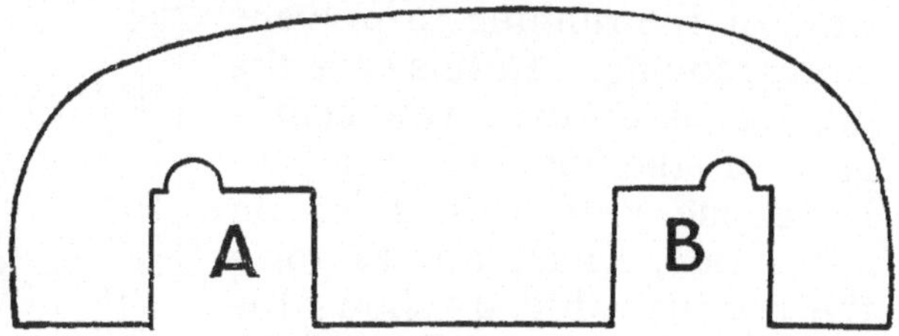

Fig. 43.

Joggle holes are not always necessary. They entail a certain amount of time, apart from damaging the bench; so in the place of joggle holes nails are driven in skew-wise at the bottom of the cheeks, leaving the heads out to enable this being drawn with pincers. With wide cheeks, struts are necessary from the bench to nearly the top. These are nailed, and a straight-edge held inside the cheeks to make sure the inside edge is straight, care being taken that the cheeks are squared upright from the bench.

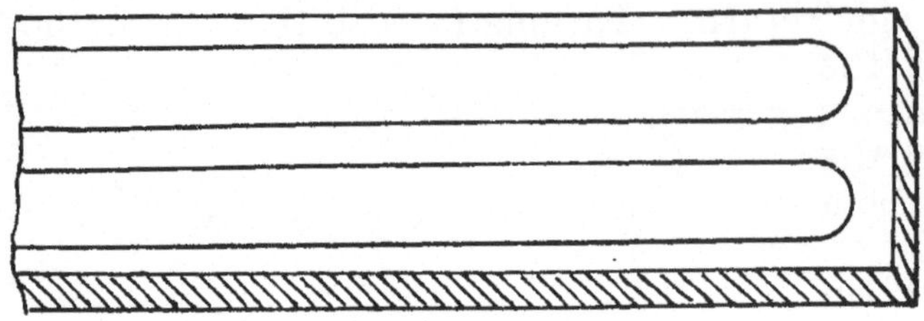

Fig. 44.

Straight Fluted Pilasters

In the running of straight fluted pilasters the reverse or inner profile is traced off and horsed up in the manner described for bench running. Fig. 41 shows the section of a pilaster with small returns, and Fig. 42 the reverse mould section cut for running. It will be noticed that the angle line is carried down lower, which gives the thickness for mould, as well as the bearing for the nib and slipper.

Running the Mould

When running the mould the rule is fixed and straightened and nails driven in the bench crosswise, the mould cored out and run off as described previously in the running of plaster moulds. This disposes of the front face, which at the moment is the reverse, but before running the face it is an advantage to run the returns so as to occupy the one part of the bench and also utilise the one running rule.

Running the Return

Before running the returns, a loose plate is fixed over the centre of the mould, or that part which is not required. This enables both returns to be run at the one operation. The bench should be well greased, and a small nail driven in at the start of the running to prevent pieces moving. In this case the plate is fixed over the centre part of the mould, so as to leave one flute with fillet on either side, which are to form the returns. Fig. 43 shows the plate fixed at A and B, and the two returns. When running the returns, it is necessary to have narrow widths of canvas soaked in plaster laid in the running parts. The plaster for this should be gauged as round as possible, and if necessary a little size water added.

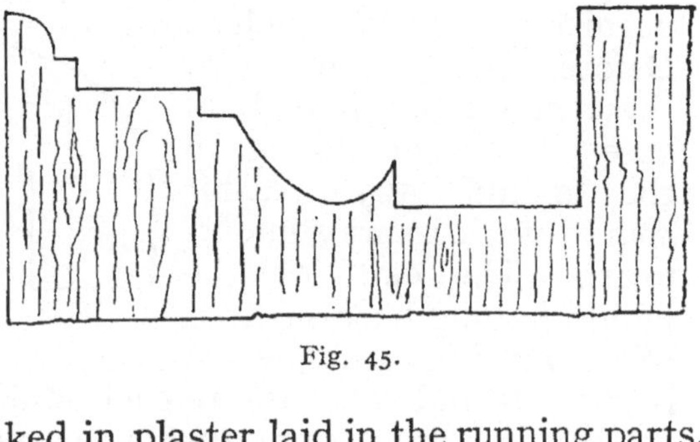

Fig. 45.

The returns having been run, the ends are cut off and the ends of the flute chiselled round at top and bottom to the face of the fillet line, which makes some circular end to the flute. The front face, which can now be proceeded with, is run off, and the ends squared and cut off. The ends of the flutes are trimmed round to the fillet in the same manner as the returns. Fig. 44 shows the end of flute chiselled off. The reason for having return ends to the flute top and bottom is that the fillet forms a band or margin to the top member of capping or base mouldings.

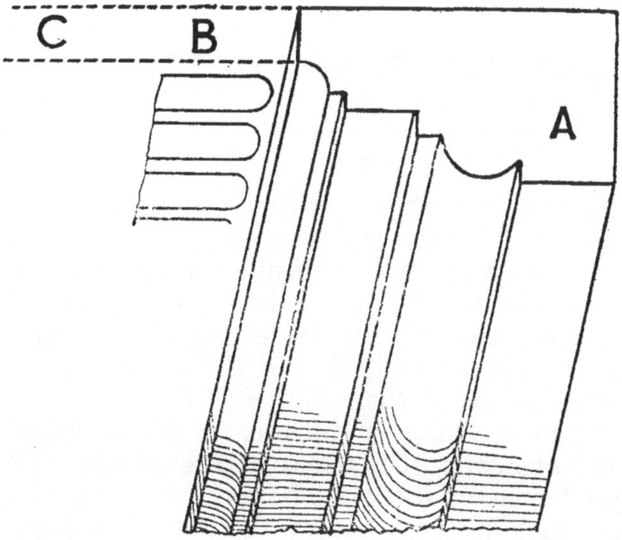

Fig. 46.

With the base mouldings the section is cut in the same manner as the shaft; that is to say, the reverse section. It is horsed up, and sufficient is run to form the face and the two returns. Before running this piece the bench must be greased so that when it is run it can be taken up. Two pieces of this are cut off for the two returns. The returns are placed upright against the piece which is to form the face of base and marked to the section. The return pieces are scribed to fit in the section of the base Fig. 45, and carefully chiselled out to fit. The fine pad-saw or drag can be used to assist in this operation. The base mould is coated with shellac and greased, then the returns are put up in position squared up from base. Before doing this the underside must be scratched with the knife, as the scribing is not done to a close joint. This done, and the returns fixed in the correct position, plaster is gauged and pressed through, under and where the return has been scribed.

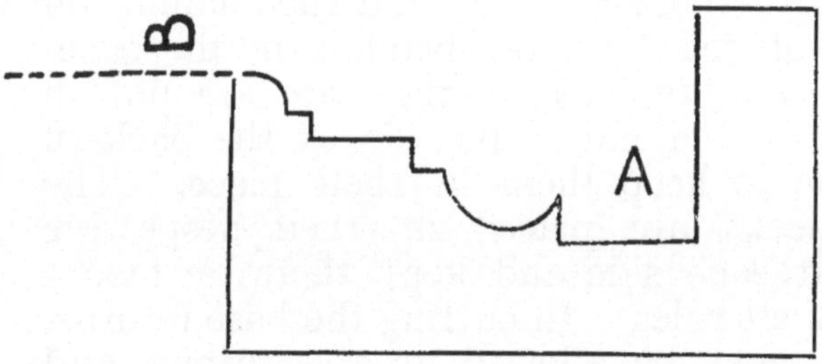

Fig. 47.

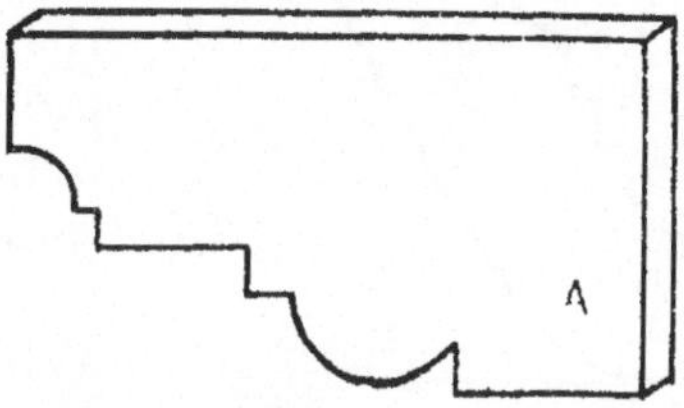

Fig. 48.

Fig. 47 is the run section of the base, the dotted line being the fillet face of pilaster. Fig. 48 shows the scribed return in position, fitting over A in Fig. 47; B the dotted line the same as Fig. 46, C the top part of return cheek to pilaster. Both returns are treated in the same manner and sufficient plaster should be gauged and put on the back of the returns as will form a shoulder, for these returns are made to slide along the base mould to fix in their correct position. Fig. 49 shows the end section at capping with the two returns of pilaster in position, kept firmly fixed against the mould by nailing blocks, or a thickness rule, to the bench against them. A is the face mould, B and C the cheeks or returns, D and C the thickness rule.

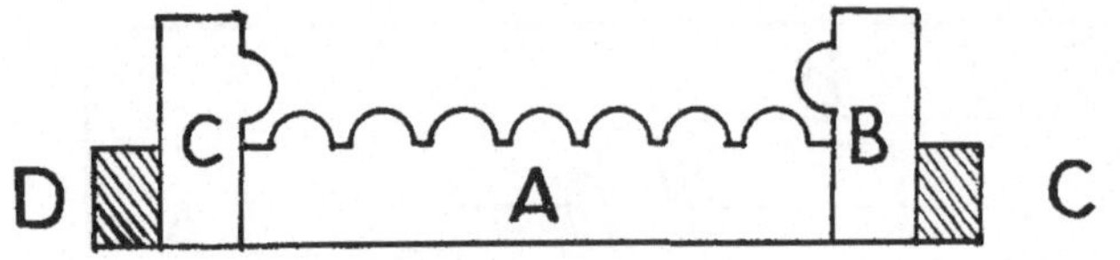

Fig. 49.

When running a pilaster with a base mould attached, it is usual to run at the end of the bench, as when the base mould is placed in position the face line of fillet is flush with the top part of hollow to base moulding. This necessitates a lower bench level for the base to rest on, owing to the projection of the base section. This is easily got over by nailing a fillet on the side of bench and an upright piece of the floor as a leg, and then nailing on a piece of board for the bedding of the basement. The returns to the base are put in position and a nail driven in at the back of each one to keep them in their place. The long cheeks are placed in their respective places at each side and kept there by blocks or thickness rules. In cutting the base returns, enough must be allowed for the scribing and the extra length for the returns to pilaster.

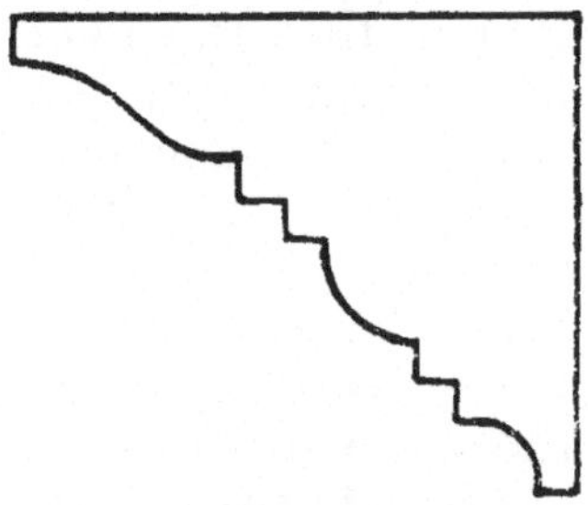

Fig. 50.

Cappings to pilasters are usually cast separately, and the base method is adopted when it is only a small capping; only the face mould of pilaster is brought up higher off the bench to allow for the thickness of capping mould. Some benches are made thick to allow the cutting in for jobs of this description. When a capping moulding is required, such as Fig. 50, it is traced as in Fig. 51. It is horsed up, and a length run off sufficient to make the face and returns. The inside width is marked off, and the projection, which is the width of the moulding, is added, and then cut through.

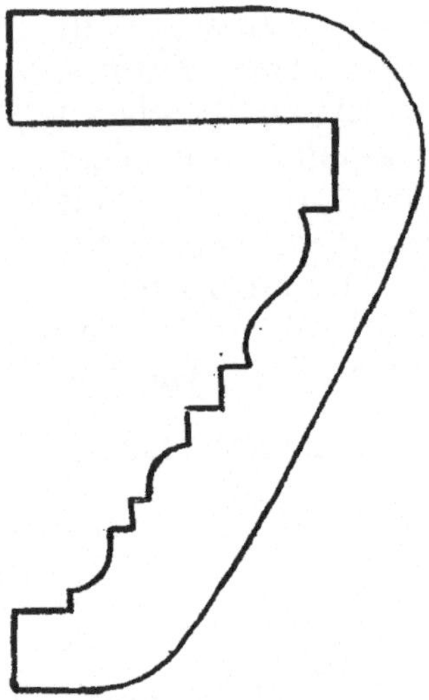

Fig. 51.

The ends are cut off square, left and right; then the length of return to capping is marked off and the projection which is simply marked off to the angle of 45 degrees and cut through. This is repeated on the other side, and then we have the two return pieces for the cap, which

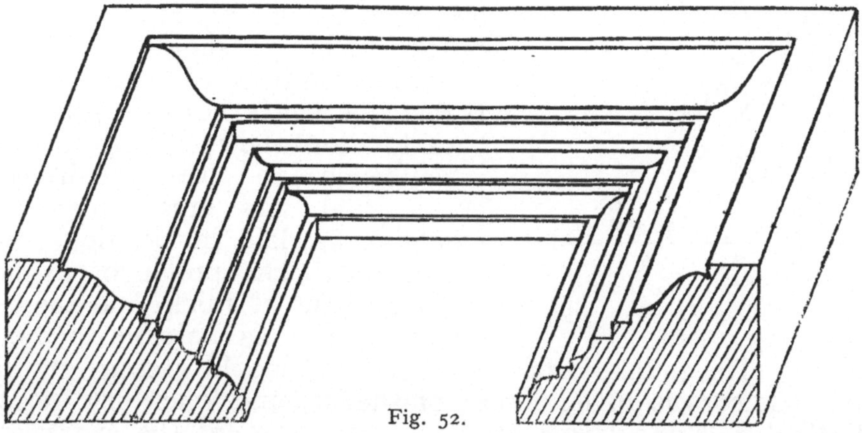

Fig. 52.

is mitred up together as in Fig. 52. It is then given a coat of shellac and prepared for casting.

This method is for running a reverse capping mould separately, and not combined with the shaft and base. If the capping mould was required to be cast in conjunction with the shaft and base, it would be traced off as Fig. 53, and the returns scribed and set up as described in the base and return of fluted pilaster.

Mouldings Run on a Core

Another method of forming cornices or mouldings on the bench besides the casting in plaster and gelatine moulds, is to run mouldings

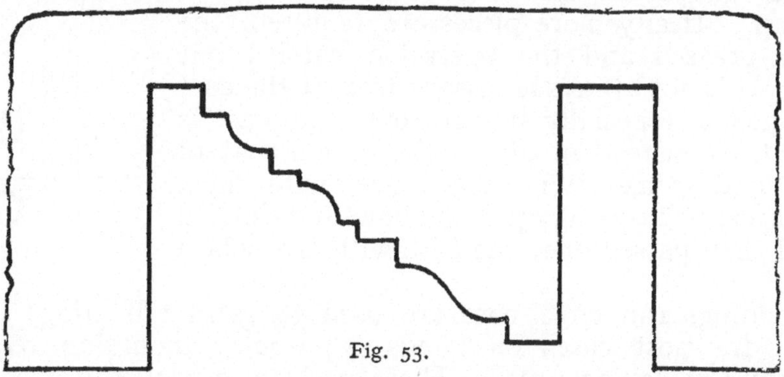

Fig. 53.

on a core. This practice is applicable only when one length or perhaps two are required or a circular ring, or any piece of moulding that is required without being run solid or cast from a mould. This is called running on a core. But this operation is not the same as running a core for a running off of moulding for casting purposes.

This kind of work is carried out in the following manner. The mould is cut, and in this case the outer profile is required. This is the same section which is cut when tracing off for the running of a model on the bench. Fig. 54 shows a section of a moulding run on a core; A is the profile, B the core which should be splayed at the back to allow the moulding to lift off, C is the bench line. Nails are driven in the bench, and any small pieces of old moulding are laid in position ready for coring out. The profile of moulding is muffled about 1″ thick, and the core run off as smooth as possible to the length required. The muffle is taken off and the mould greased at the same time. The core is given a coat of shellac, and when dry well greased with thick tallow. Two pieces of canvas are cut the exact length of the core, and with width enough to pass all over 1″ × ¼″ laths. By adding a little size water the plaster is gauged round. This plaster is brushed over the core thinly. Then a piece of canvas is soaked in the plaster and stretched and spread over all, but keeping it well inside the running line, so as not to impede the running. The mould is passed over to make sure all is clear. Laths are put in where necessary, such as for wall line, etc., and then a second piece of canvas is soaked and spread over to clear, as before described. Fig. 30 shows the three pieces of 1″ × ¼″ laths in position. Any remaining stuff can be used in building up the moulding, occasionally passing the mould over and continually gauging till the moulding runs off clean. When the plaster is set, the two ends are cut off square, a small priser is put under the ends and the run cast lifted up. If any more pieces are required, the core is greased and the operation carried out again. It is not always necessary to coat the core with shellac, especially if only one run cast is required, as instead of giving the core a coat of shellac and grease, it is coated over with thick clay water. This is clay put in a bowl and slurried up and then passed over the core with the splash brush.

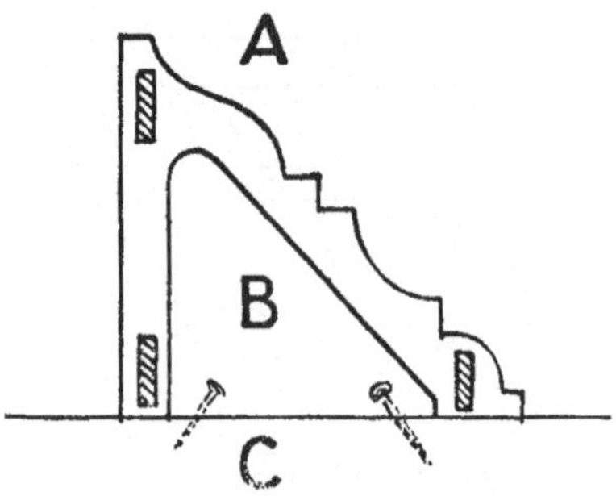

Fig. 54.

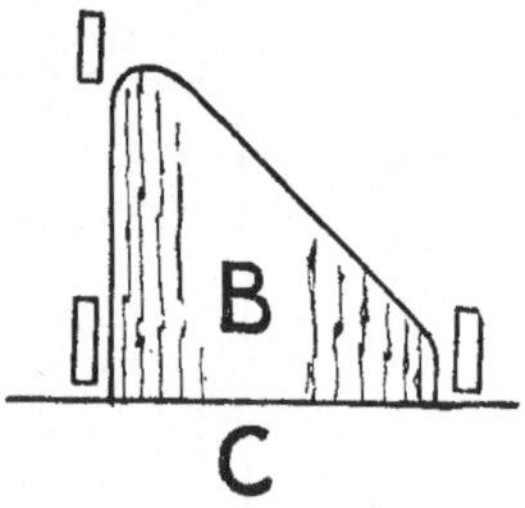

Fig. 55.

Mouldings run on a core are used to great advantage in circular work. In most cases one finds a circular moulding required for ventilators in ceilings, etc. These perhaps occur only once or twice on the job, and the moulding may be with an undercut section. For such work a reverse mould with a loose piece would take up much time.

Mouldings in Coloured Cement

Only plain moulds should be cast *in situ*. Blocks with dentils or more elaborate ornaments should be produced in the shop and fixed separately. Special features are also carried out on the job, but I will deal with this in "Moulding in Cement."

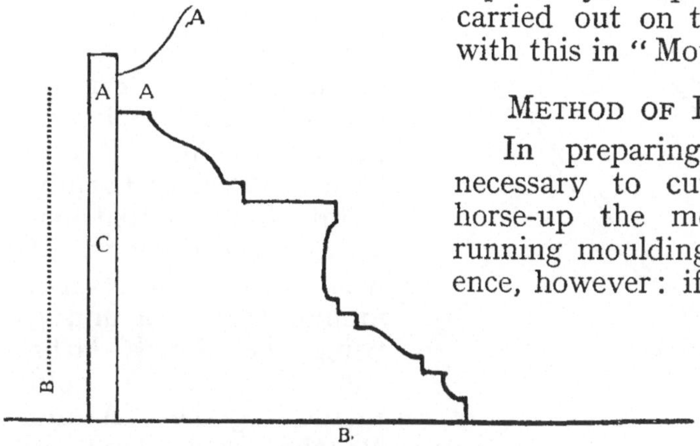

Fig. 56.

Method of Preparing Moulds

In preparing large moulds it is necessary to cut out the detail and horse-up the mould, as described in running mouldings. There is this difference, however: if the moulding is run on a bench, as in this case, the reverse mould is required to fix up in position. The following illustration will make the point clear: Fig. 56, B bench, A being the nib line as in Fig. 58, B bench level, C a board squared up from bench a little higher than the distance from B dotted line to A in Fig. 56. This board is straightened and stiffened at the back by pieces of batten nailed against it and to the bench. If the board is rough at A, Fig. 57, a thin plaster screed is wiped on it. The reason for doing this is that the nib of the mould runs against it when being run. If the depth is too great to find a suitable piece of board, a narrow board can be used by nailing pieces of batten as legs to lift it up to the required height, as in Fig. 57, where A is a narrow board, B legs and C struts at the back to stiffen it. Fig. 58 shows the mould horsed in running position. This is a perspective view, as it is seen when running. A is the stock, B the horse, C the running rule on bench, D and E the braces as handles, which in this case must be nailed on the top of the horse at F, and as near to the nib as possible. This nearness to the nib prevents the mould chattering when running (chattering is when a mould instead of running smoothly keeps trembling, which forms ridges on the members). The stiffer the nib the better. The mouldings are run in definite

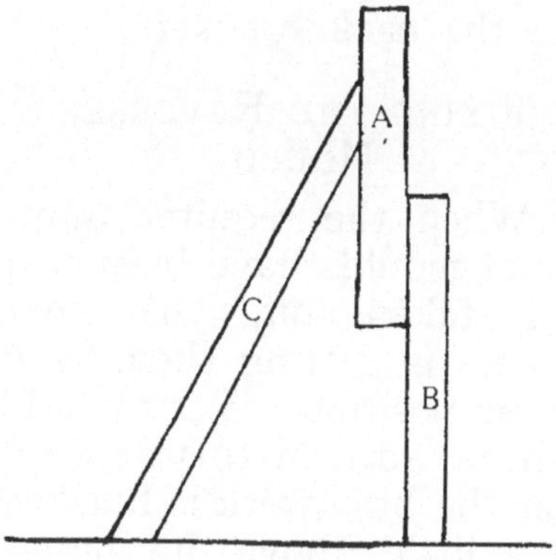

Fig. 57.

lengths (about 8′ would be a convenient length to mould and to fix in position). After the mould has been prepared as described in moulding and casting, the moulding proceeds; the firsting, the seconds and canvas are applied, and the wood laths and brackets added. Fig. 59 shows the end section of the cornice after it has been moulded and ready to lift off, the dotted line showing the thickness of what is now the reverse mould, 1, 2, 3, 4 and 5 being the 1″ × ¼″ laths run through to hold mould together; A and B the back brackets which hold the whole material intact, and C and B a canvas wad laid on the back of mould and round the brackets to strengthen the intermediate parts of the mould. Fig. 60 shows a perspective view of the back of the mould, with the laths in position and the back brackets.

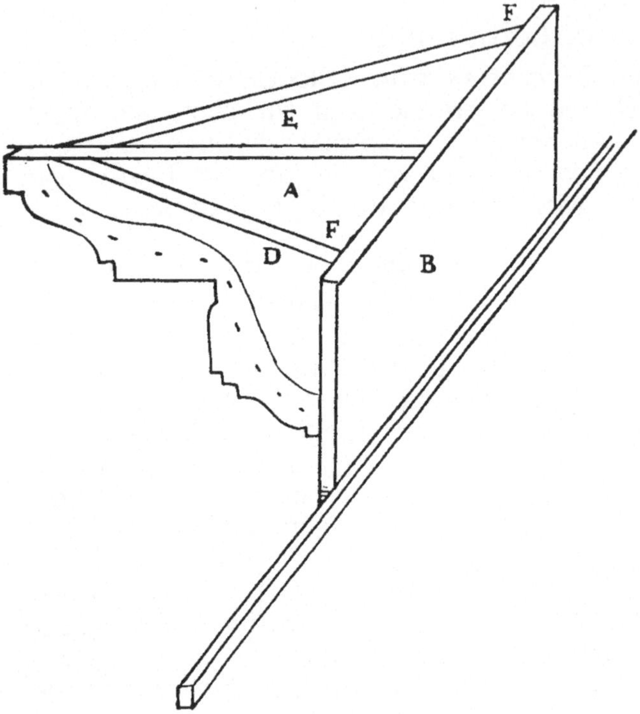

Fig. 58.

Fixing the Reverse Mould

When the required number of moulds have been cast and taken off, the next process is getting them fixed up in position. I may add that in work of this description the brickwork is finished off a little above the underside of the bottom member of the cornice. The bottom member on wall is marked off at each end, the thickness of the plaster moulding being allowed for, and a line is snapped from end to end. Two-inch wood blocks or fillets are nailed on the wall to this line (this is for the moulds to rest on). A depth and projection rod is fixed at each

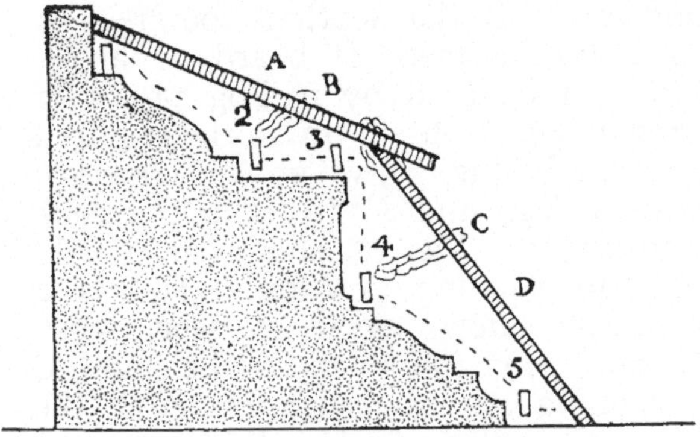

Fig. 59.

end, as C to A and C to D in Fig. 61. A line is fastened and stretched from one end and made fast at the other. Then the required number of moulds are put up in position, resting on the wall blocks or fillets, and the nib members at A, Fig. 56, fixed to the line. The whole of the moulds having been checked to ensure their being in the correct position, they are made a fixture. Great care must be taken that the whole are securely fixed, as in the filling-in process there would be a heavy weight against them. Having securely fastened them, and where necessary added struts to further strengthen the job, it is ready to fill in. The moulds must be well varnished with shellac and greased before use. Having examined the moulds and found the intersections are flush with each other and all parts of the moulds have been greased, they are wiped as free as possible from grease and then well dusted out with french chalk. Leaving the dusty material on the surface, the colouring material is got in hand. When a large amount has to be covered, it is necessary to mix sufficient for the job. The colouring can be mixed dry with the cement and put in bags ready with the sand when required. It should be gauged 2 parts of cement and 1 part of washed sand, and should be laid on at least ¼″ thick.

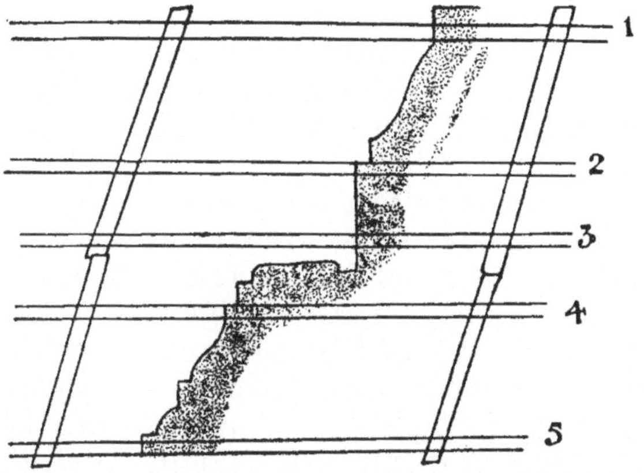

Fig. 60.

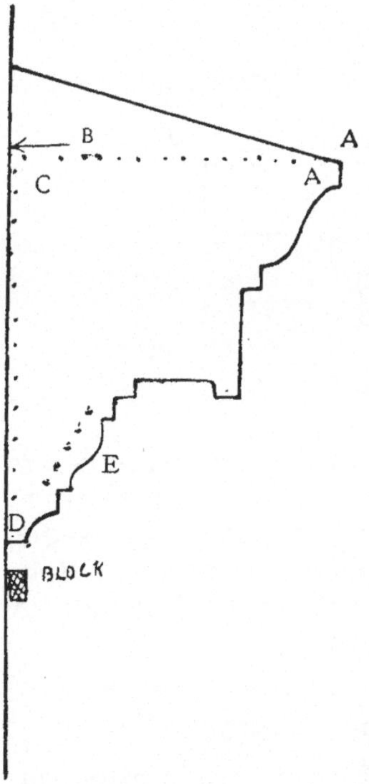

Fig. 61.

Filling In

Having gauged the coloured material, which must be fairly pliable and not too soft, it is spread evenly with the gauging trowel on the mould, care being taken that a uniform thickness is spread all over. When this coat has been completed and has got a bit steady (that is, firm enough to spread the next coat over without going through on to the face), ordinary cement and sand, gauged 3 parts of sand to 1 part of cement, is spread lightly over the back of the coloured material. It must not be pressed too hard, or it will penetrate through the coloured material. The whole of the mould having been well covered, the next process is the filling in of the remaining part of the mould with concrete. The necessary steel bars are put in for reinforcement, and the whole filled out up to and including the weathering, which should be given a good sharp fall to front edge of moulding. Care should be taken when filling in that the concrete is not thrown in roughly to disturb the cement surface. The filling out should be done in a regular manner. When the concreting has been completed, it should be allowed to stand several days to set before the reverse moulding is taken down. When sufficient time has elapsed, the moulding is taken down and parts where necessary are touched up, as soon as possible before it has been exposed long to the air. The whole of this description of coloured plaster concerns the reinforced concrete method with cornices. If joints to imitate stone are required, it is quite a simple matter, as it is only necessary for a saw cut to be put in the reverse mould at the points required.

Returns to Mouldings or Window Sills, Etc.

For returns to cornices, sills, etc., on reverse plaster moulds the projection is taken from the wall line to nib, as in Fig. 62, where A is the wall

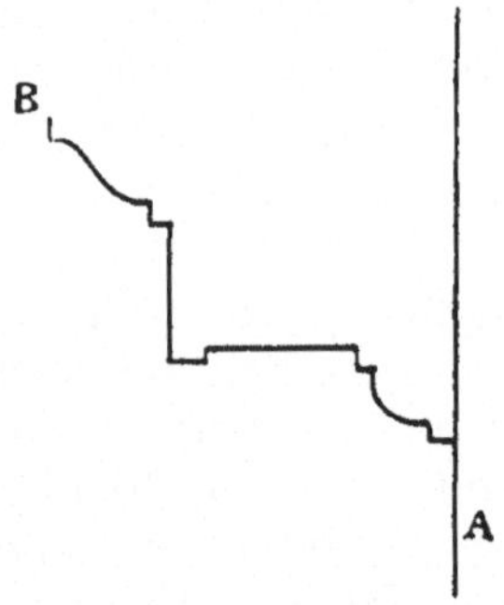

Fig 62.

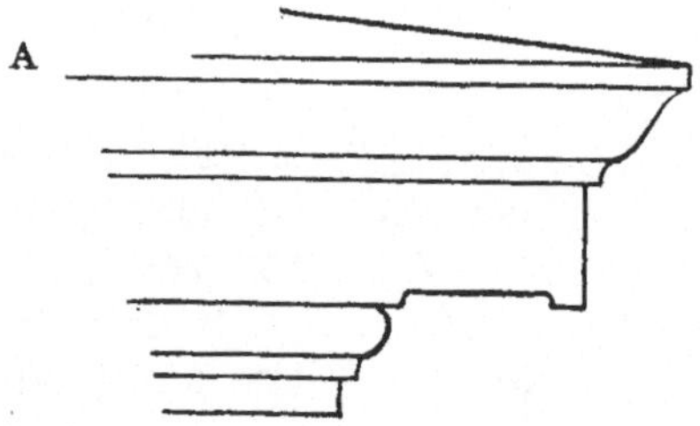

Fig. 63.

line, B the projections at nib. The moulding is cut through, care being taken that it is fixed up temporarily in its correct position when cutting, otherwise it would not cut a true and correct mitre. The same procedure is adopted for the return. Whatever length is required it is allowed for, and the projection added from that as at A and B in Fig. 62 and 63. This shows the cut for a left-handed return.

Special Construction for Exterior Cornices

When a main cornice or moulding is required to a façade after the building has been constructed and no provision has been made, the following method can safely be adopted without a great deal of trouble.

While it saves labour and material, it also lessens the weight on the front of the building. It makes a good, sound, practical job.

Malleable iron rods about $\frac{3}{8}''$ diameter, HY rib wire, expanded metal, etc., will be required. The iron rods are cut off in lengths and bent as near as possible to the required section of the moulding, as in Fig. 64; A being the desired cornice, B iron rods bent in a vice to the required shape and bent about $1\frac{1}{2}''$ down at C to let in brickwork, D the HY rib, which is similar to expanded metal and has ribs with narrow flanges to lap over at joints, Fig. 66. Its average length is about 8′, and is 13″ wide. E is reinforced expanded metal. This is like ordinary expanded metal lathing, but has V-shaped ribs passing through it about 6″ apart. It is sold in sheets 8′ or 9′ long and 2′ wide, Fig. 65.

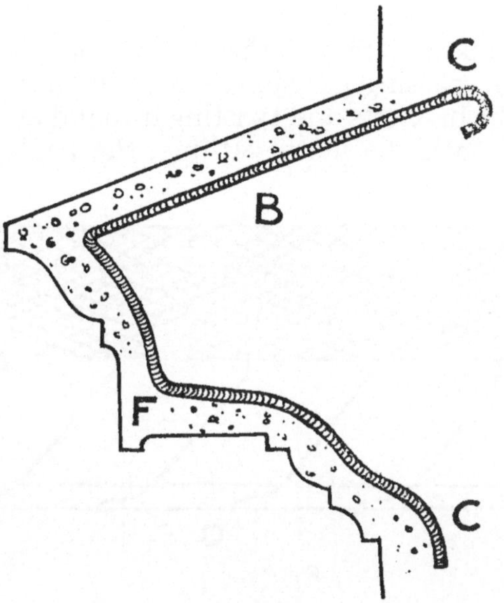

Fig. 64.

The position of the desired cornice is taken, and a level or datum line is also taken about 6″ below the bottom member of the cornice (this is also necessary for the fixing of the rules for running). The position of the bottom member is marked on the wall, also the top of weathering, not forgetting to allow for a sharp fall. Then about $1\frac{1}{2}''$ up from bottom member, and $1\frac{1}{2}''$ below the top of weathering, snap a chalk line at the two places from end to end. The line is then spaced out to about $1' \times 1'\,2''$. This gives the position where to fix the iron brackets. Holes are cut in the brickwork about $1\frac{1}{2}''$ in and $1\frac{1}{2}''$ downwards, which allows for the brackets, as marked in Fig. 67. A bracket is fixed at each end, with strong gauged cement and sand (dusting out and well wetting the hole to receive the bracket.) When the two end brackets have been fixed, a line is fastened and stretched from end to end, at F in Fig. 67. Fig. 67 shows a perspective view of the brackets and section of wire turned up from the weathering and in the joint of the brickwork at B; A showing the position of the fixing line. After the brackets have been

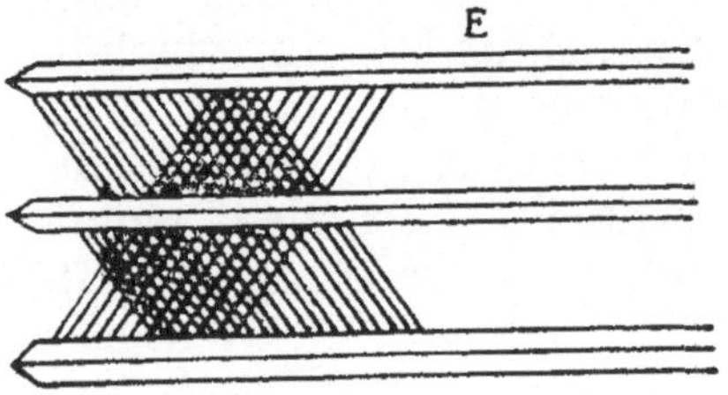

Fig. 65.

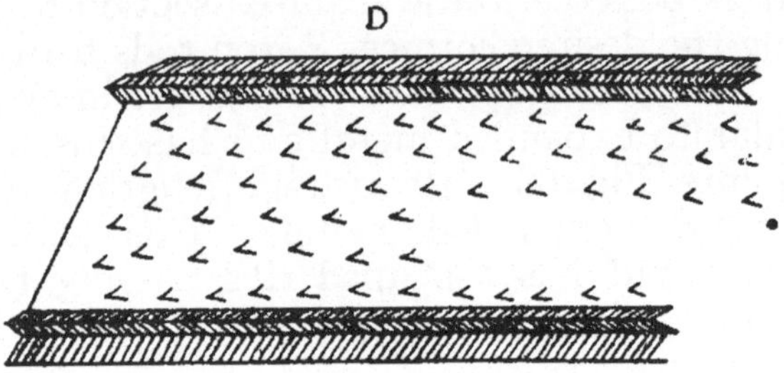

Fig. 66.

fixed and the cement allowed to set, the HY rib, or other metal lathing, is laid on and fixed by passing galvanised wire in short lengths through the metal and round the brackets and twisting it round at the ends. When covering the brackets with the metal lathing, the best and soundest job

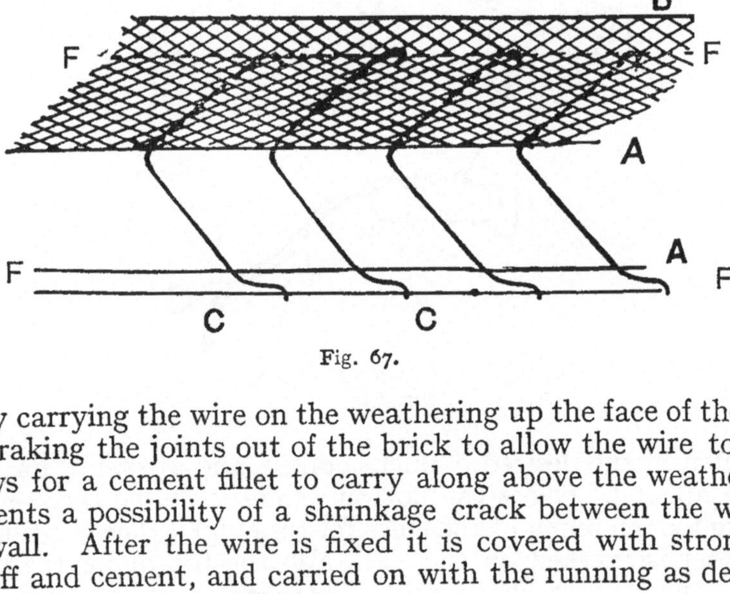

Fig. 67.

is made by carrying the wire on the weathering up the face of the building about 3″, raking the joints out of the brick to allow the wire to turn in. This allows for a cement fillet to carry along above the weathering, and thus prevents a possibility of a shrinkage crack between the weathering and the wall. After the wire is fixed it is covered with strong gauged coarse stuff and cement, and carried on with the running as described in the running of cement mouldings.

SECTION OF PLASTER CEILING OF THE EARLY GEORGIAN PERIOD, REMINISCENT OF WILLIAM KENT'S STYLE.

SECTION OF EARLY GEORGIAN PLASTER CEILINGS.

Borders and ornament in medium relief ; gilded lining, or lines in rich yellow, panels by artists or copies of classic subjects.

CHAPTER IV

JELLY MOULDS

Peculiarities of the Jelly Mould—Preparing the Mould for the Case—Preparing the Case and Model for the Jelly—Dissolving the Gelatine—Pouring Gelatine to Form a Case Mould—Stopping the Holes—Taking off the Case and Preparing the Gelatine Mould—Preparing the Jelly Mould for Casting—Preparing with Alum Water—Preparing the Mould for Varnishing—Painting the Mould with White Lead—Casting from Jelly Moulds—Filling In—Varnished Moulds—Painted Moulds—Skin Moulds in Gelatine—Forming the Skin Moulds—Cases for Skin Moulds—Solid Gelatine Moulds.

JELLY moulds are formed by quite a different method than that adopted for the plaster mould, though in moulding the cutting and horsing up of the mould is practically the same. With this exception, the profile is cut from the outer section of the moulding, whereas in the plaster it is cut from the inner section. Then, again, the section of the moulding for the jelly mould has to be cut with a sinking for the bed mould, which is described in the running of mouldings in Chapter XIV, Vol. I. Another little variation is the moulding line, which serves the same purpose as the striking-off lines in reverse plaster moulds. Fig. 194, Vol. I, shows a section of a cornice with two enriched members and the sinking or bed for the two enrichments at *x* and *x*. This mould, when horsed up ready for running, is called "The running mould for the model."

The moulding is run to whatever length is required, in exactly the same way as running the plaster reverse moulding, though it may be added that when preparing the bench for the running, care must be taken in fixing the running rule that it is straightened with the long straight-edge and made quite secure. If necessary, canvas wads should be stretched against the rule and lapped round and under the bench, as in running mouldings on the bench the running rules are apt to get loose. Loose rubbish and old plaster is used in the coring out.

As previously described, the mould is muffled and run off. After it is run off and examined for any rough or defective places, a length of the required ornament is bedded in position and prepared for moulding, including varnishing with shellac, dusting out with french chalk and oiling. Having done this, a shield is put up against the ornament, either with a piece of wood or clay. The sides are treated in the same way, keeping the shields or fence away from the enrichment far enough to allow of a 3″ thickness of jelly. To prevent the fence moving when the jelly is poured in, several pieces of clay should be pressed up against it. Having poured on the jelly, it is allowed to stand to cool, and when it is quite cold the fence is taken away. If wood is used for the fence, it must be well greased to prevent the jelly from adhering to it. Both ornaments are treated in this manner, and when the fence is taken away the jelly should be well cleaned out with turps, using as little as possible; then drying out the mould with soft paper by pressing in the mould and stabbing it in with a dry tool brush.

Having dried the mould as far as possible, it is dusted out with french chalk, and then washed out with strong alum water, using about ½ lb. to 1 pint of water. This is heated and the alum dissolved. The alum water is well brushed in the various parts of the mould, and should be warm when applied. Sufficient lengths of ornament are cast as may be needed for the model. The casts are fitted to the bed where required, and trimmed at the joints to ensure a proper fit. The whole of the ornament is stopped in with plaster, so that no particle of jelly can run in underneath (when being poured), and the model is well coated with shellac. Before bedding the enrichment, the back part of the cast and bed should have a coat of shellac. This prevents absorption by the plaster when being bedded. A moulding line or ground, which forms a thickness and striking-off line, is required at each end of the model. This is formed by chiselling down from the profile of the moulding about 1″.

Preparing the Model for the Case

Having applied the shellac and dusted the mould with chalk, it is brushed over with a small portion of linseed oil, ready for the clay thickness. The clay is well beaten together and then cut off in about ¾″ slices, with a piece of wire having at the ends two pieces of wood about 6″ long for handles. A clay board is made about 9″ wide, and a piece of ¾″ wood tacked on each side to form a ruling off. This board should be dusted with french chalk to prevent the clay sticking to it. Clean newspaper is laid over the model, and at each side of the moulding line a piece of 1¼″ lath is tacked down temporarily. The rolls of clay are carefully laid over the model, keeping it clear to about a little under half-way back from the front edge of the 1¼″ lath. The clay is carefully pressed on top of the paper to an even thickness, as the clay when taken away will be replaced by the jelly. When laying clay on models for plaster cases, it is necessary to note any parts of the model that are under-cut, as thin strips of clay are placed on the paper at those parts before laying the clay on. This is to allow the case when made to leave freely; if the precaution was not taken there would be a possibility when lifting off the case of damaging the model. The clay should be made as smooth as possible, using the fingers to work in a little water. Where the funnels are to be placed for pouring the jelly, an extra piece of clay about ¾″ thick, the size of the funnel, is added on top of the layer, and the edges smoothed down. This forms a shoulder to the case, which is cut round after the case is taken off.

A clay rim about $\frac{3}{4}''$ thick and of V shape is raised at each end of the moulding line.

Making the Case

The clay having been smoothed over and the edges cleaned back from the wood, it is now ready for the making of the case. The case, which has to resist the strain and pressure of hot gelatine, must be made very strong. Wood battens 2″ × 1″ are used, one piece at each side of the model running over about 3″ to 4″ each end. This is placed upright on its edge and forms the framework of the case. Intermediate battens are spaced out in between, one about 3″ from each end, and the others about 1′ 3″ apart. Canvas is cut the exact length of model, and usually a full width of 30″ is used for ordinary size cases. Firstings and seconds are gauged on the same lines as described in casting, only more size is necessary, as cases take longer to make, therefore more time must be allowed for the setting of the plaster. The first coat of canvas is well pressed round the stout battens and another piece of canvas is cut about the width of the model and length to ensure a good, strong case. A full width cut up in four is generally used as a long wad, after being soaked in the plaster to put all round the edge to make a level bed for the timbers on edge. Large canvas laps are cut and soaked in plaster, and laid over the cross timbers and lapping well on to the case. The back of the case should be ruled off straight and flat, so that when taken off it will lay flat on the bench. The sides of the case are well cleaned off to the model, and the case is marked with indelible pencil. The reason for marking is to ensure that the case goes back in the same position as before taking off.

Preparing the Case and Model for Jelly

The plaster forming the case having set hard, is now ready to lift off. The overlapping pieces of the outside timber are used for lifting purposes; a short piece of 2″ × 1″ being used as a lever by placing under the ends alternately and lifting. When loose, it is held at each end and lifted off; care being taken that it is lifted clear of the model, so that no damage is done. The clay is taken out and cleaned from the case, and the case placed where it can be prepared for replacing. The model is cleaned off the paper and clay removed, and then dusted over with french chalk. It is then oiled with a soft brush; this time with boiled oil, which is considered more suitable for gelatine. The case is well cleaned, and any rough parts dragged off with the scratch tool, a piece of rough canvas being used for rubbing off any clay marks. The position of the funnels is according to the number required, usually one at each end, and about 1′ 3″ from end of case, the others about 2′ apart, according to size of model. These holes are cut through, besides air holes, which are very necessary. It is essential that the plasterer should not be sparing with these, otherwise air locks would form, which prevent the jelly from running, resulting in an incomplete mould. A rimer is used for the air holes, which are made at the highest part of the case. The funnels, on the other hand, should be placed in the lowest parts. Slots are made in the outer edges of the case with a knife, roughly about 1′ 4″ apart. The greater the provision made for the escape of air in gelatine moulding the better.

The air and funnel holes having been cut, the case is varnished several times with shellac; when dry, dusted over with french chalk and well *greased* with thick tallow and linseed oil. The model having been treated as previously described, the case is ready to put on. Referring back to the two 1″ × 1¼″ laths tacked on the two outer edges, these are taken off when the clay and case are taken off, and when the mould and case have been prepared, two clean laths are put on in the place of the two taken off. The reason for so doing is that the jelly adheres much better to the clean wood. They are tacked down very lightly with small brads. Some plasterers prefer to rough the same two laths, replacing them in the case.

Laying on the Case

The case should be laid on the model very carefully, the position being noted by the two marks made with the indelible pencil. The case should be well wadded down to the bench, driving in a few long nails crossways, to help to make the wads secure. If overhead room is available wood struts can be used, as the case must be well secured. The funnels are placed in position, and held by putting a piece of clay round. The wads should be about 1′ 4″ apart, and should be kept clear of the V-shaped slots or edges, both sides, and also the air holes. Plaster is put round the funnels to help to secure them, and also round the edges of case to the joint of model. This prevents any leakage of the jelly. Having carried out all the instructions as previously described, the plaster and wads are allowed time to become thoroughly hard before pouring in the jelly, which is called "pouring the jelly."

Dissolving the Gelatine

Gelatine, which is a superior kind of glue, is manufactured from bones, hoofs, hides, cartilage and fish skins. It is extracted by means of acid, by boiling and washing in lime water. It is then strained, cooled and dried, and is usually sold in thin oblong-shaped cakes. It is very brittle, and requires immersing in cold water before being put in the boiler. It does not dissolve, but expands to the consistency of jelly. Good quality gelatine will break hard and tough after having been well soaked, whereas inferior will break easily and leave a smooth edge. When the gelatine has been soaked it is put in a large galvanised or iron-jacketed boiler to dissolve; though if only a small quantity is required, a pail containing the jelly can be put in a larger pail half filled with water, and put over a gas-ring. In the large jacketed coppers or boilers the container with the jelly fits into the water vessel, so that the heat of the fire or gas heats the water, and this indirect heat dissolves the gelatine to a liquid without fear of burning it. A small quantity of hot water is put in with the gelatine to start it dissolving. High temperature is best, so as to dissolve the gelatine as quickly as possible. It must be stirred continually with a wooden stirrer, about 2″ thick and of convenient length to get hold of above the top of the boiler. When all the gelatine has been dissolved, and there are no lumps left, it is ready to take out of the container. Galvanised pails are used, or a hand-bowl. It must not be forgotten that all utensils used in handling hot gelatine (as well as the plasterer's hands) must be well greased with tallow.

Pouring Gelatine to Form a Case Mould

The pouring of gelatine to form a mould in a case requires a certain amount of experience, as the knack is only gained by practice. The jelly requires to be at a certain temperature to allow it to run or flow all over the model. If too hot it will draw the grease and shellac off the model and the case, with the result that great difficulty would be experienced in getting off the case. Apart from that, there would be the damage to the model when removing the jelly. In some cases the mould has to be poured several times when a large quantity of cast work is required. If the jelly is too cold it will not flow, consequently a bad mould results. It should be estimated what number of pails of gelatine will be required for each job. This number is taken out of the container and allowed to stand to cool (usually outside of the shop). This will generally take over half an hour, though the time depends on the weather and the season of the year. Jelly is improved by stirring it up several times while cooling, as when standing a skin forms on the top, and each time it is stirred the skin causes the jelly to get thicker.

Whilst the jelly is cooling small balls of clay, about the size of walnuts, are prepared. These are required for the stopping of the air holes made in the case to let out the air.

Having examined the funnels to make sure they are secure, that the holes leading to the model from funnel are clear, and that sufficient balls of clay are available, the next process is the pouring. When pouring a mould the plasterer must have assistance to stop the air holes, also to watch the case for any leakage of jelly, which sometimes happens.

It is also necessary to get well above the funnels, as the higher the pail is held when pouring the better the results, as it gives greater pressure and force when passing through the funnels into the case.

Pouring should be commenced from one end, and as soon as the jelly passes the next funnel and begins to cover the model, it is poured from the second funnel, and so on till the jelly is seen rising up the last funnel, which should be allowed to rise till the whole of the funnels are filled up to a level just above the mould case.

Stopping the Air Holes

The air holes made in the case to allow the air to escape naturally also permit the jelly to escape, so when pouring the jelly these holes are watched at the back part of the case, behind the pourer, as it will be seen that as the jelly escapes air bubbles out at the same time. As long as these air bubbles appear the holes should not be stopped.

This should be especially noticed at the ends or any high part of the case. If the holes were stopped too soon, the mould obtained would be full of " blubs or blisters." When all the air holes have been stopped, and sufficient jelly has been allowed to remain in the funnels, any jelly that is over is put back in the container, the boiler covered over with a sack, and if no more is required the gas or fire lowered. By being covered up the jelly keeps warm, in case any more is needed.

Taking Off the Case and Preparing the Gelatine Mould

After having allowed the mould to stand sufficient time to become cold, the funnels are taken off by chopping away the plaster and removing the clay. Bending the funnels over, a sharp knife is required to cut the jelly. If the jelly is hard, the case is quite ready to be taken off. All the clay from the air holes is taken off and put in a bowl or pail, and returned back to the clay pit or box and covered over with a wet sack for future use. The funnels having been greased before use, the jelly is easily pushed out of them with the end of the hammer handle. All loose pieces of gelatine are picked up and returned to boiler. The funnel holes are trimmed round free of the case to allow of the case being lifted. All the wads are cut off and any struts removed.

The next step is lifting the case. This should be started very carefully, from one end, by using a short piece of wood upright and another piece as a lever. While the plasterer is prising at the end another man is needed to press the jelly down at the nearest funnel hole to him by using the handle of his hammer. This process is repeated at the other end, and whenever any looseness is felt in the case it is continued until the whole of the case becomes loose. Having loosened the case, which should be put in position near by, a place is prepared either at the back of the model or in front on a board on two stools. The case is then lifted off and wiped out with canvas, and any places requiring shellac are done over again, and when dry either a small quantity of chalk is brushed over or a small portion of dry plaster. If any part of the case is left bare of shellac, owing to it having been melted by the hot jelly, and not covered with fresh shellac before dusting, there is danger of the mould becoming mildewed. The case being lifted off, it will be seen that little tips of jelly have been formed where the funnel and air holes were. These are trimmed off with a sharp knife, down to the level of the back of the jelly mould. After this has been done, some dry plaster or french chalk is rubbed over the back of the mould. This cleans off much of the grease which will be found on the back. This is generally done with a piece of canvas, as it helps to take off any portion of shellac that has come off the case. The next step is freeing the jelly from the model, by commencing at one end by pressing a thin chisel or knife under the 1″ × ¼″ lath, which has been tacked. The knife or chisel is passed under the lath both sides and lifted slightly upwards to allow the plasterer to get his fingers under, remembering the small brads which the laths are nailed with. Each side of the jelly is lifted as high as possible before anything else is attempted, so as not to distort the mould or damage the model. Having loosened as much as possible of the mould, care being taken that the lath on each side is not broken, it is necessary to have sufficient help in the lifting, so that it is commenced from one end and carried along.

If the enrichment is very deep, the plasterer would have to put his fingers well under the jelly and on the face of the model to help to free it ; otherwise there is a possibility of the mould tearing. The mould all being free, it is cautiously lifted off and put in the case. This should be carefully done, as very often damage is done in this process. Any enrichments which have become loose are re-bedded, nails are removed from the 1″ × ¼″ laths, and the mould replaced, adjusting here and there.

Preparing the Jelly Mould for the Casting

The mould having been properly set in the case, the next thing is to prepare the mould for casting.

The mould is prepared in various ways. In the case of new gelatine, the mould is given a coat of pure varnish or is pickled with alum water. With old gelatine moulds, the usual practice is to paint them with white lead and turps, using a small quantity of terebene for drying purposes.

The mould should be well cleaned out with a small quantity of turps, and any shellac adhering should be scraped off. The mould being well cleaned out, it should be dried up as much as possible, using newspaper and the stabbing process, as I have previously described. It is then well dusted out with french chalk, care being taken that any under-cut parts in the ornament are well cleaned out.

Preparing with Alum Water

The mould being cleaned, it is now ready for pickling, which is done by washing it out with alum water. The alum, about ½ lb. to 1 pint of water, is dissolved in a vessel over a fire, and when of a temperature which can be borne by the hand, the liquid is splashed over the mould.

With a large tool brush more is brushed over to cover every part of the mould. Having given it a fair douching, it is well brushed out, and then stippled with the tool brush, so as to allow a certain amount of alum to remain on the face of the mould.

Preparing the Mould for Varnishing

New gelatine, when used for moulds, is sometimes coated with varnish, instead of pickling with alum water. When this is done the mould is first cleaned out with turps and dried up with paper, which helps to absorb the moisture and dry the mould. When this has been done, it is ready to receive the covering of varnish. Good painters' varnish is laid on with a clean, soft brush. One coat is sufficient if only a small quantity of casts are required, but if a large number are required, two coats should be given, and after each day's working the mould should be touched up again with the varnish.

Painting the Mould with White Lead

Gelatine when used a great deal loses its strength and elasticity. To obviate this it is painted with white lead, turps and terebene. The mould is treated in exactly the same way as when prepared for varnishing. No french chalk is necessary, as when cleaned out with turps the paint follows on and dries. Three coats are necessary, allowing each coat to get dry before another is added, and the last coat should be allowed to stand at least twenty-four hours before the mould is used.

It will be remembered that in the first part of this chapter I mentioned that 1″ × ¼″ laths should be tacked on the model. This is one of the chief aids in keeping the jelly mould in its position, as the gelatine adhering to this prevents the edges of the mould from turning up. These laths require nailing to the case when the final preparations of the mould are completed. The nails are drawn when the mould is being prepared for casting, and after each day's working of the mould the laths are nailed down again. In addition to nailing down the laths, clean dust sheets, or newspapers, should be covered over the mould when not in use. This keeps any dust or dirt from getting on the face of the mould. If jelly moulds have any deep sides or parts to them, struts of 1″ × ½″ laths are put against these parts and slightly wedged from one side to the other.

Casting from Jelly Moulds

The method adopted for the filling in of a jelly mould is much the same as for plaster moulds, with the difference that moulds of this kind have to be treated with particular care. For example, when filling in a plaster mould the " firsting " is well rubbed on the face of the mould, whereas in a gelatine mould it is brushed on very lightly. Indeed, in the parts containing the enrichments the brush must not be used more than can be possibly helped, as it would distort the edges of the jelly forming the ornament. Therefore, on these parts the "firsting" is splashed in with the brush as much as possible. The plaster is gauged as described previously. Firstings and seconds are used with the laths and brackets and canvas, the same as for plaster mould. The one great thing which has to be studied in the filling in of gelatine moulds is that the size water is tested, and only sufficient used to ensure enough time to fill in. It should be so gauged that by the time the whole of the mould has been filled in the plaster of the seconds should be well on the way for setting, and when set it should not be left longer than necessary before being lifted out of the mould, as the plaster when setting creates a great heat; therefore it has a great effect on the jelly.

Filling In

The mould having been pickled with alum water, or otherwise treated, and allowed to stand to become dry, it is prepared for casting. It is well dusted out with a dry tool brush to remove dust and any alum crystals. Linseed, machine, sweet or rape-seed oil is next used to oil the mould, which is done with a soft brush, only sufficient oil being put on to smear it. If a large quantity was put on, it would collect on the face of the cast and cause it to look yellow. A fresh mould after pickling is dusted with french chalk before oiling, as this helps to clean the mould, but it is only dusted out this once with the chalk. The mould being oiled, the next process is the gauging of the plaster, which is the same as described before, only alum water, or dry alum, is added to the firsting, together with size water, to counteract the quick setting which the alum would cause. Alum is used to harden the face of the jelly; it also causes the plaster to set quicker on the face, which helps to preserve the gelatine. Only a small quantity of alum should be used.

Having filled in the mould, a small quantity of the seconds should be left on the bench. This gives the plasterer an idea as to how the plaster is setting in the mould, though the plaster brushed in would set quicker than the portion on the bench.

When the cast is set it begins to swell, and then it should be taken out of the mould by slightly prising one end and then the other. One man lifts up one end and another places his fingers under the cast at the outside parts where the 1″ × $\frac{1}{4}$″ laths are, and as one lifts the other passes along with his fingers both sides, and then the cast is lifted out, care being taken that after it is lifted it is kept on its edge. If kept flat there is a possibility of its bending in a circular shape. The cast is examined and any part touched up if necessary. When the cast is lifted out, the mould must go back in its proper position in the case. While the mould is allowed to stand to get cool, the wood and the canvas are prepared for the next filling in. When sufficient time has elapsed, the mould is oiled up as before, and the filling carried on again.

Another method is sometimes adopted in the gauging. Instead of alum, or alum water, being put in the firsting, alum water is splashed over the mould or sometimes dry alum is sprinkled over it through a piece of fine canvas.

The one method is as good as the other, though I prefer to sprinkle the powdered alum over immediately each cast is taken out.

Varnished Moulds

Varnished moulds, when prepared for the filling in for the first time are dusted out with french chalk, and then oiled up as described in the preceding section. There is no difference in filling in.

Painted Moulds

Moulds painted with white lead paint when prepared for casting are first dusted out with french chalk, and then oiled—with any of the previously mentioned oils. They are filled in as already described. These moulds require painting again after several casts have been taken, as being old jelly it is not so durable as the new: therefore it requires more attention when working. Not much alum should be used on moulds of this description, as the alum would affect the paint and cause it to peel, so it is advisable to gauge the firsting as round as possible and put in plenty of size water to retard the setting.

Skin Moulds in Gelatine

Skin moulds are very useful when only a few casts are required, such as cartouches, shields, caryatides, etc. The clay model is varnished with shellac, several coats being given so that it forms a fairly hard surface. It is then dusted over with chalk, and a clay fence put all round it, several inches in height (I might add that the clay model is laid on the bench for this process). This is to keep the jelly up together, and a small quantity of plaster is put against it. It is then oiled over with boiled oil, and the jelly taken out of the copper and allowed to get cool.

Forming the Skin Mould

The gelatine being fit to use, the plasterer well greases his hands, as in this process the jelly is poured over the model by the use of the hand bowl, and is drawn over the prominent parts by hand. The whole of the model is covered, and as it cools the jelly forms a skin. The high parts are continually kept covered, until about $\frac{1}{2}''$ in thickness is obtained all over, though in some parts it will naturally get thicker. The jelly having become cold, the clay fence is taken away and the gelatine given a coat of shellac, including all round the outside of the bench; french chalk is dusted over when this is dry and then oiled.

Cases for Skin Moulds

Skin moulds being very thin require light cases made to keep them in position. The plaster is gauged, firstings and seconds being used.

After the firsting is laid on, the whole of the jelly is covered with canvas several inches longer than the mould. A quarter width of canvas is then soaked in plaster and used as a wad all round the outside of jelly, and the surplus canvas turned over this. By this means a good, strong edge is provided for the outside of the case. Pieces of batten are used to form a frame, and canvas is cut the required width to connect these to the first or second coating of canvas. When set, the case is taken off and prepared in the same manner as in the case of a poured mould, the skin mould put back in the case and dealt with as described either for pickling or other treatment.

Solid Gelatine Mould

Heavy moulds are sometimes formed with gelatine, and are called solid moulds, being chiefly used for flat ornament, or perforated ventilating panels, or small girth panel mouldings. These kinds of moulds are generally used when only a few casts are required.

The preparing of the model is carried out in exactly the same way as in preparing other models, as regards shellac chalk and the oiling. The covering of the model with jelly is much the same as in the skin moulds. A clay fence is put round it, about 1″ higher than the model, keeping the clay in position by using the gauged plaster. If the model is a long panel moulding, a thin lath is tacked on the bench about $\frac{3}{4}$″ each side clear. Then a plain board, which should be well greased, is fixed up each side of the model and round the end, $\frac{3}{4}$″ clear of sides. If the wood is required to remain on the jelly, which is advisable in some cases, the two thin laths are done away with and the plain board not greased. By leaving the wood on the side, the mould is kept straight, as it can be laid on the bench, and wood blocks here and there nailed on the bench and the mould put against them. Whatever model is being moulded, the clay fence or wood board must be put round and well secured with plaster. Having fixed the fence round the model and taken the jelly out of the boiler (remembering that it must be allowed to stand to cool), it is poured on the model, starting from one side and continually pouring till it passes over, enough being used to give not less than $\frac{3}{4}$″ thickness.

When the jelly is found to be nearly cold, the clay shield used should be taken away, but not till the jelly is firm enough to stand alone. The thickness of the layer makes easy removal necessary. With solid moulds there is much heat caused by the quantity of jelly used, and if left to get thoroughly cold it would be found extremely difficult to get the mould off the model, therefore it is very important when forming solid moulds not to let the gelatine become too cool.

These moulds are prepared in the same manner as other jelly moulds when required for casting.

CHAPTER V

PILASTERS AND COLUMNS

Columns—The Classical Orders : Greek Doric, Ionic, Corinthian, Tuscan, Composite—Drawing Profile of Columns—Entasis—Plastering Plain Diminishing Columns—Collars—Plumbing the Dots—Fixing Collars—Pressed Collars—Floating the Column—Finishing—Rim Method—Fixing Mouldings with Rims—Forming Diminished Fluted Column—Making the Moulds—Setting Out the Flutes—Casting the Flutes—Column Moulds without Flutes—Forming Diminished Fluted Columns—Plaster Collar Flutes and Plaster Reverse Flutes.

In dealing with pilasters (flat and engaged in the wall) and columns (round or oval and detached), we are touching on one of the most interesting tasks of the plasterer. They are often fluted, the fluting consisting of a series of channels running up the face of a column or pilaster. Sometimes these channels are filled by a staff which runs up to one-third of the height, and when this is the case such columns or pilasters are said to be cabled or reeded. The setting out and the forming of the pilasters or columns are difficult matters if the plasterer has not had much experience of this kind of work, as great care must be exercised not to over-do the entasis (which is a slight swelling in the middle of a column or pilaster). Apart from their flatness, pilasters are similar to columns, their parts having the same names and standard of measurements, and being diminished and fluted on the same principle. If placed close behind columns, they should not project more than one-fourth part of their diameter ; this gives regularity to the returned parts of the capital. A pilaster with a semicircular face is known as an engaged column.

Columns

Columns are round or cylindrical pillars, resting on a base and crowned by a spreading mass called the capital. The shafts are diminished in diameter as they rise ; some examples have a slight swelling in the centre, and the diminution at the top is seldom less than one-eighth, and not more than one-sixth, of the lower diameter. The unit of measurement in drawing is the module, which is one-half of the diameter of the column at its greatest width, just above the base, or in the case of the Greek Doric (which has no base), at its lowest point.

The Classical Orders

The Classical Orders became an integral part of building early in the sixteenth century. By an order, it meant the column and its superstructure. It includes : 1, a plinth or podium, consisting generally of a series of steps, but sometimes of a high pedestal with a base and capping; 2, the base, shaft and capital; 3, the entablature, forming a lintel immediately above the capital, and 4, the frieze, or broad band, frequently decorated, and the cornice or boldly projecting top members.

The orders are five in number. Three of them—the Doric, Ionic and Corinthian—are called the Grecian Orders, being used and perfected by the Greeks. The Romans modified these for their orders, and produced two others of less importance, namely the Tuscan and Composite.

Greek Doric

The Greek Doric is a massive column divided into twenty flutes. It is without a base, and has a very simple capital, with a square abacus. The architrave is plain, and the frieze is divided into triglyphs and metopes. The triglyphs consist of three projecting vertical fillets. The metopes are square spaces between the triglyphs and may be filled with sculpture. The Greek Doric has no enrichments, and it is eight times its diameter in height, inclusive of base and capital.

The Ionic Order

The Ionic Order is slender in its proportions. The column has a base, and the capital is ornamented with volutes or spirals, which are the most characteristic features of the order. The shaft is fluted with twenty-four flutes, which are separated by flat fillets. The entablature is light compared with the Doric entablature, and the frieze is plain. Dentils and egg and dart enrichments are often used. Its height is nine times its diameter.

Corinthian Order

The Corinthian Order resembles the Ionic in its general proportions. Like the Ionic, its distinguishing feature is its capital, which is ornamented with rows of acanthus leaves, with small volutes at the angles and flowing towards the centre of the abacus. The mouldings are enriched with various leaf ornaments, modillions and dentils. This rich order was the favourite of the Romans, and they treated the acanthus in a more luxuriant manner than the Greeks. The height, including capital and base, is ten times its diameter.

Corinthian Order on Pedestals

If this order has to be set out with a pedestal, the total height is divided into thirty-two parts; then one part will give the module, and of these thirty-two parts the columns will take twenty, the entablature five and the pedestal seven.

Pedestal and Base

The pedestal is seven modules high, which is rather more than one-third the height of the column. This is an exception to the general rule as to the height. The die of the pedestal is two squares high, and the plinth of its base may be increased from four parts to eight parts. The shaft of the column is fluted with twenty-four semicircular flutes, separated by fillets generally terminating top and bottom.

Entablature and Capital

If the height of the entablature is divided into ten parts, three parts are given to the architrave, three to the frieze and four to the cornice. The height of the capital is two modules, six parts, the vase being two modules and the abacus six parts. The shaft of the column has twenty-four flutes and the face of pilaster has seven.

Tuscan Order

The Romans occasionally used a variation of the Doric, which is known as the Tuscan. This order is the simplest and most solid of all the orders. It is devoid of all ornament, is massive in construction and is similar to the Doric divested of its triglyph and metopes. It is the first order according to the Romans, and it is most easily executed. The column, which is plain, diminishes one-tenth to one-quarter of its diameter, and its height is seven times its diameter or fourteen modules, inclusive of base and capital, and the module is divided into twelve parts. If the total height of the order is divided into five parts, four are given to the column and one to the entablature. The base and capital are each one module high, leaving twelve modules for the shaft, which is cylindrical one-third of its height from the base and diminishes to one module seven parts immediately below the astragal of the capital. The shafts of Tuscan columns are never fluted.

The Tuscan Order on Pedestals

The pedestal is rarely used with this order. To draw the column with a pedestal, divide the total height into nineteen parts; the three upper parts are given to the entablature, the four lowest to the pedestal, and the twelve intermediate parts to the column. The height of the column is divided as before.

Entablature and Capital

The height of the entablature is one-quarter that of the column, or three and a half modules; of this one module is given to the architave, one module and two parts to the frieze and the remainder to the cornice. The height of the capital is one module, which does not include the astragal. The diminution of the shaft reduces it to one module and seven parts, and the abacus, which is square on plan, has a projection at the top of five parts on each axis.

The Composite

The Composite Order is so called because it is composed of the Ionic and Corinthian. The capital is enriched with acanthus leaves, like the

Corinthian; only it has two rows and is distinguished from the Corinthian by having the large volutes and enriched ovolo of the Ionic capital. The column, including the base and capital, diminishes one-eighth part of its diameter and is ten diameters in height. The distinguishing features of this order is the capital and the rich detail of the entablature.

Pedestal and Base

These differ from the Corinthian Order only in the mouldings of the pedestal; sometimes plain but very often enriched. The shaft of the column is fluted, as in the Corinthian Order, with twenty-four semicircular flutes, the fillets between the flutes being two-sevenths the width of the flutes.

Entablature and Capital

The proportions of the entablature are similar to those of the Corinthian. The capital is set out similarly, but the volutes are quite different. The volutes are drawn as for the Ionic Order.

Drawing the Profile of Columns

There are several ways of profiling the shafts of columns. The following method is adopted when the column has its greatest diameter one-third of its height. The height and upper and lower diameters of the shaft being determined, draw a horizontal line at A about one-third of the height of the shaft. Describe a semicircle on this line where the shaft begins to diminish, and project the point B vertically on to the semicircle at B. Divide the remaining part of the shaft above A into six parts; divide the A to B into the same number and project vertically as shown. Fine pointed nails are driven in at the intersecting points from A, 1, 2, 3, 4, 5, 6, and a flexible piece of wood, or thin 1″ × ¼″ iron, is bent to these points, letting it take a natural bend. While this is in position, pins can be driven in at the back to keep it in its place. A pencil line is then drawn to the curved line

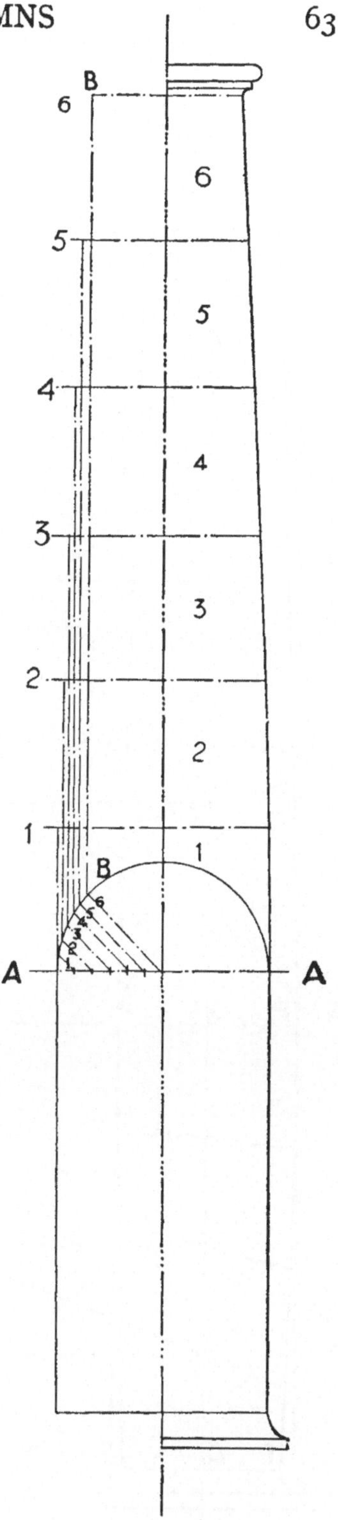

Fig. 68.

made by the bent rule or iron. The whole line is then transferred from top to bottom on to the board, which is to form the entasis rule for the plastering of the column.

For proportions of Columns, see Orders under Glossary.

The Drawing and the Setting Out of the Entasis of the Swelling and Diminishing Columns

Taking the data given below, at A draw the horizontal line P O; then through the point M draw the arc M R with radius equal to P O; join R where it cuts the axis to M, and extend this line to meet the horizontal through A at O, from which point draw as many lines as necessary for the required number of points, always making S T equal to P O.

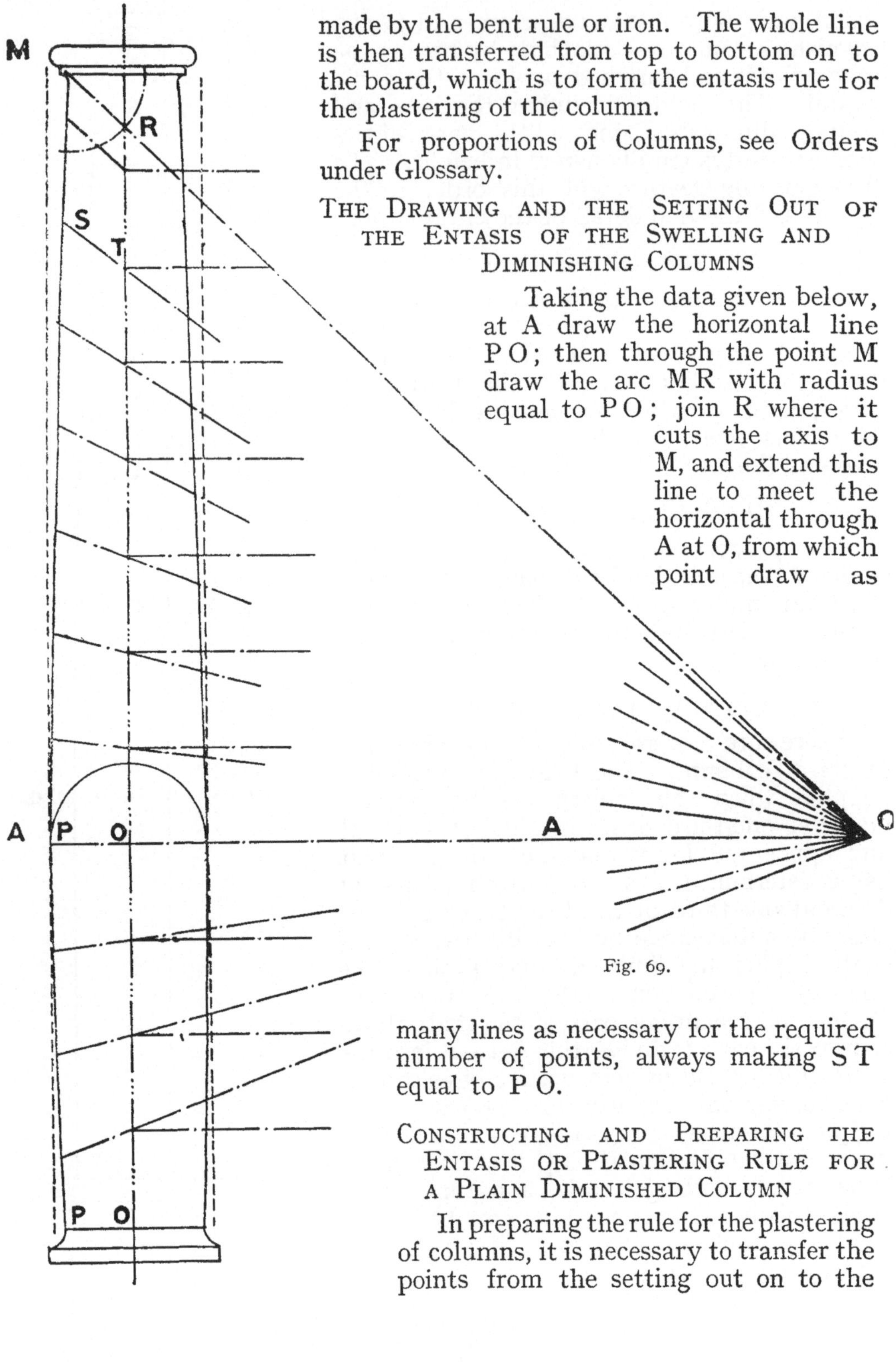

Fig. 69.

Constructing and Preparing the Entasis or Plastering Rule for a Plain Diminished Column

In preparing the rule for the plastering of columns, it is necessary to transfer the points from the setting out on to the rule. If the column is set out on paper on the bench, it is quite an easy matter to put vertical lines on the rule and lay the paper to the vertical lines, which should be carried 1″ above and below the entire length of the shaft of the column. The points are then pierced through, pins driven in and a flexible piece of wood bent round, and the remaining contour added. Or the whole of the original setting-out line can be pricked through and a pencil drawn to these points on the rule. Having transferred the line, a narrow-bladed saw is necessary for cutting out the contour on the rule, which must be very carefully done so that a regular and even surface is cut, otherwise a true swell and reduction would not appear on the finished face of the column. If more than one column has to be worked, it would be beneficial to have a piece of sheet iron cut to the line of the column and nailed on to the diminished rule. This enables the plasterer when floating to cut the stuff off cleaner, and also avoids the wearing of the wood, thus preventing losing the uniform diminish. When utilising the sheet iron as a mould plate, the back edge of the rule should be chamfered back as in the stock of a running mould.

Plastering Plain Diminished Columns by the Entasis Rule and Collar Method

To enable the plasterer to carry on with the plastering of the column before the caps or bases are fixed, the collar method is adopted, in which plaster collars are made and fixed at top and bottom of the shaft to work the entasis rule on.

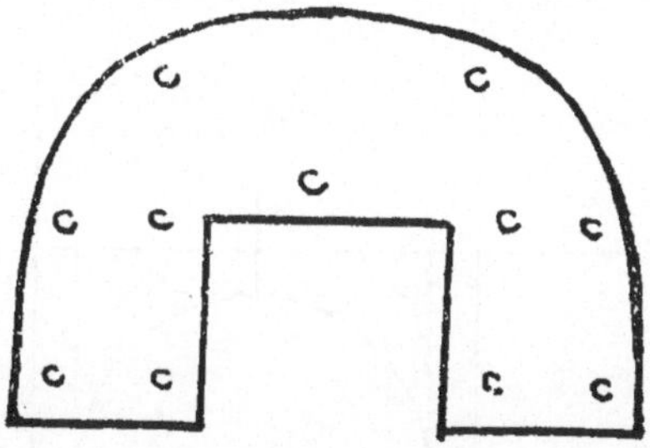

Fig. 70.

Making the Collars

A piece of zinc, about $1\frac{3}{4}''$ square, is required, with a piece cut out of it about $\frac{3}{4}'' \times 1\frac{1}{4}''$. This is horsed up in the same way as for running circular mouldings on the bench; fixing a radius rod on to it $\frac{3}{16}''$ less than the top half diameter of the finished column; taking the radius from outside of the mould at A in Fig. 71 to the pin-hole at B; Fig. 70 shows the zinc cut out. Fig. 71 shows a perspective view of the mould fixed up ready for running the collars on the bench. When running the collars, the plaster should be gauged round (which is well soaked and stiff), and a piece of canvas well soaked in plaster is put in the collar when running, keeping it clear of the mould. This strengthens the collar for handling purposes. The bench should be well varnished with shellac and greased so that the collar can be taken up when run. Two small brads should be driven into the bench, leaving a small part protruding above the bench. This keeps the collar in position should the plaster swell up before finishing the running off. After running this collar another one is required with the same mould, only the radius is extended at the pin-hole to $\frac{3}{16}''$ less than the half dimension or diameter at the base of the shaft. This collar is run in the same way as the previous one, using the same pin block. Cut the two collars in half, marking each respective half.

Another method of applying collars is to run them as previously

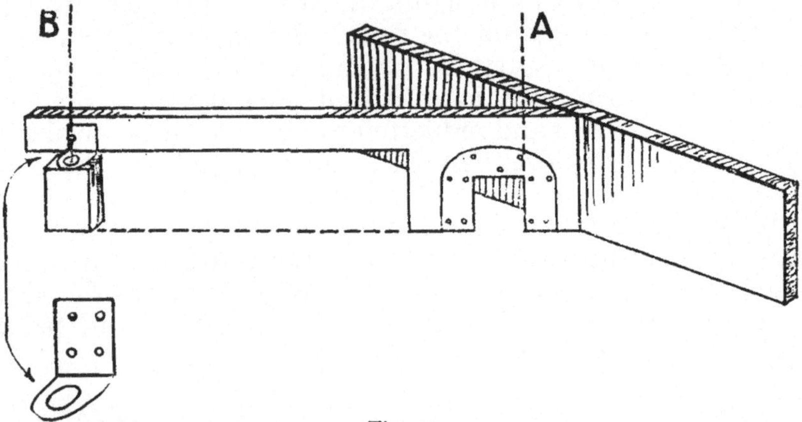

Fig. 71.

described with the radius of the full half diameter of top and bottom. When these collars are fixed and the entasis rule is used for floating, it is necessary to muffle it, which is easily done by tacking on a piece of 1″ × ¼″ (base) lath as used in making casting cornice. This method I consider is the best, as when finishing the columns all that is necessary is to take off the lath used as the muffle.

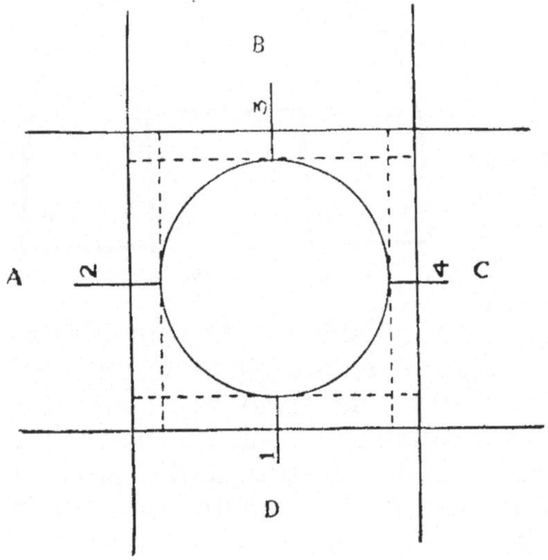

Fig. 72.

Plumbing the Dots—Fixing the Collars—Pressed or Squeezed Collars

In preparing for the fixing of the collars, it is necessary to get the centre in the correct position on the columns. Fig. 72 shows on plan the intersection or junction of beams; A, B, C, D being the soffits, 1, 2, 3, 4 the four centres. From these centres mark off on each side the half diameter or radius of the top collar. Carry the line across to each one, as shown by the dotted line. This gives the face line of the run collar. Extend 1, 2, 3, 4 to the dotted line, as shown, and from these points a plaster dot is plumbed ¾″ back (the ¾″ being the thickness of the collar) to just above the depth of the astragal moulding. Having plumbed the

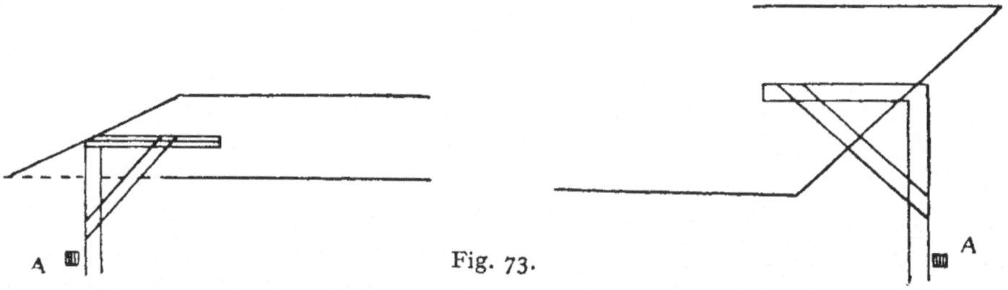

Fig. 73.

1 to 4 dots, the level position of the collars is arrived at by marking the depth (which would be the bottom or underside of the collar) on the square and placing the square on the soffit of each beam, testing the beam as to being level and transferring the mark on to the centre of each plaster dot. Fig. 71 gives a perspective view of the square on the soffit and against the plaster dot A.

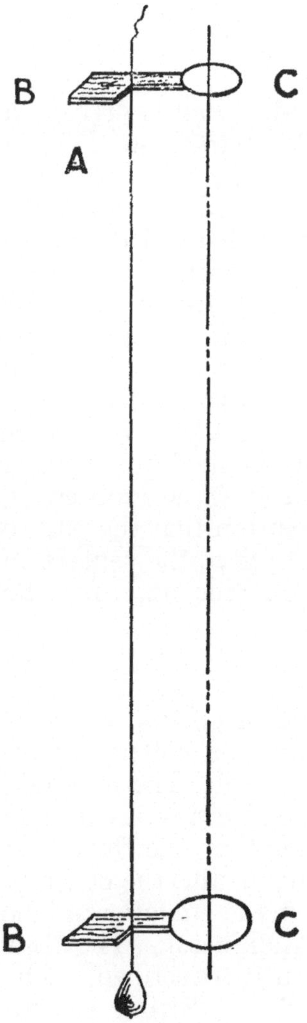

Fig. 74.

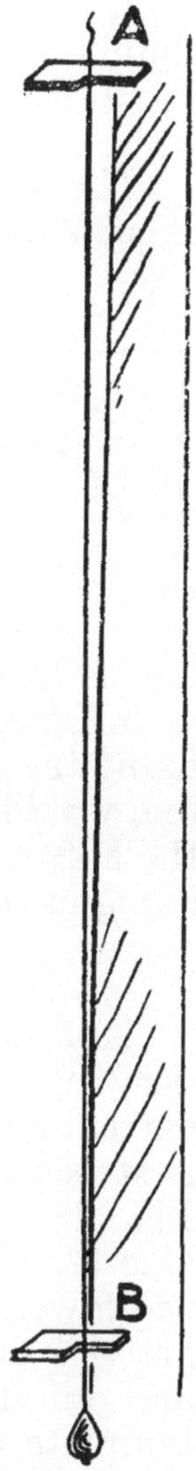

Fig. 75.

Plumbing the Lower Dots

Having marked off the centre on each dot, the next thing is to plumb the lower dots for the position of the collar at the lower or base diameter. In this the two gauges and plumb bob are necessary, but the two gauges are not of the same length (as described in the previous chapter on plumbing dots with bob and gauges), as the column being either diminished or provided with an entasis diminished the gauges would have to be cut to suit whatever half difference there was in the diameter at the necking or base, while allowing the bottom gauge to be long enough to permit the plumb bob to swing away from the brick, or concrete, construction of the column. Figs. 74 and 75 show the plumbing of the dots. In Fig. 74 we have the profile of a column, A being the line and bob, B gauges, C the two rings or collars fixed. Fig. 75, A and B the plaster dots which the inside of the collars fit against. If the columns were of a very great height, which would entail the plastering of them in stages, requiring intermediate collars, the collars would be run with the radius whatever the half diameter happened to be for the part of the column to be plastered. The gauge would be cut to suit (which would be taken from the setting out of the column), and the dots plumbed in for the fixing of the ring or collar as before described.

Another method can be adopted when putting in the four plaster dots as Fig. 73. This is done with the square instead of being plumbed down. The square is put on the soffit with the depth mark as previously described, and put to the centres at 1, 2, 3, 4, allowing the $\frac{3}{4}''$ back for the thickness of the collar. But in this method, should the soffit of the beams vary, not being level, the position of the dots would not be accurate; so to ensure the correct lines being obtained, the method of plumbing the dots is much the best.

Fixing the Collars

The dots being completed and the depth obtained, a nail is driven in at each of the four centres, and the half collars are held up to these. Whilst the collar is being held in position, the plasterer gauges a small quantity of putty and plaster and tucks in the back and underside of collar. When one half is fixed in the correct position, the other half is held up against it and fixed in the same manner. The same proceeding is adopted in fixing the collars at the lower diameter, with the exception that the plumb line would hang from the centres of the top collar to give the centres of the bottom dots, to denote the position where to put the nails for the fixing of the collar.

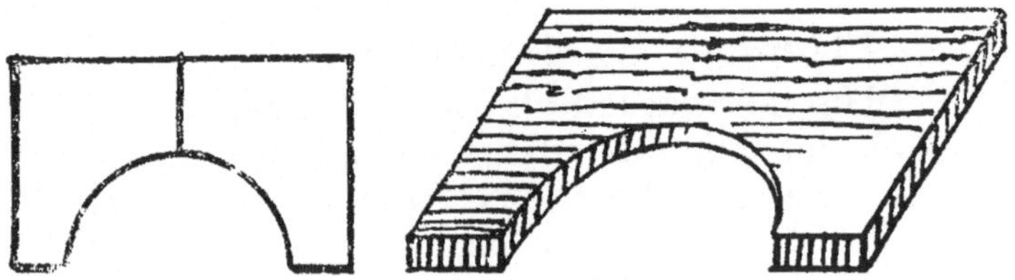

Fig. 76.

Pressed or Squeezed Collars

Another method can be adopted for the forming of collars, which is very useful where there is no convenience to make the collars. It is carried out with a wood template made from a $\frac{3}{4}''$ board. The template is struck out the size of the top half diameter of the floating, as seen in Fig. 76; one diagram showing the face view and the other the thickness. One template is made to the small or top half diameter and the other to the lower. The inside of the template should be made smooth, and be greased or smeared with linseed oil (to prevent stickiness). The dots are put in top and bottom to the centres previously described. The template is held in position to the centres on the dots. Strong gauged plaster, with a small amount of lime putty added (to prevent swelling), is pressed or squeezed in between the template and core of column. When this is set, the template is taken away and the operations are carried on to the remaining parts to complete the collars. Care must be taken that the template is the exact half of the diameter and is held in the correct position.

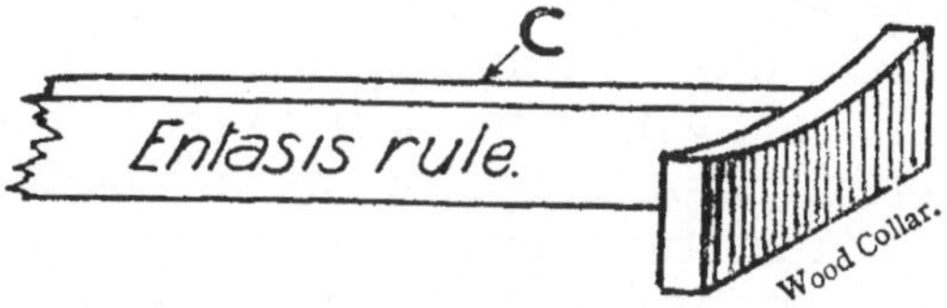

Fig. 77.

Any intermediate collars required would be treated in exactly the same manner as previously described, the template being made to suit the diameter required.

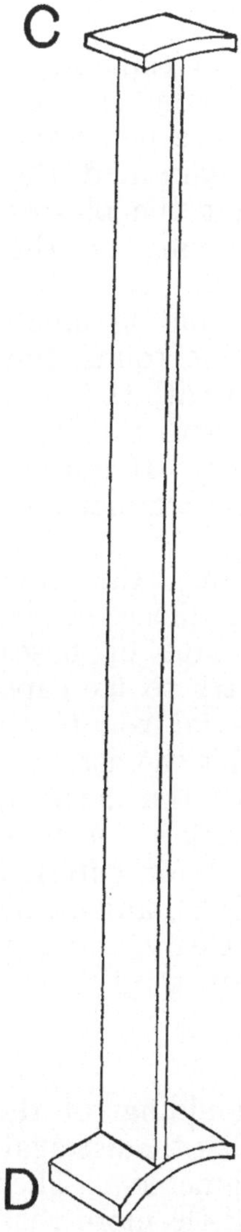

Fig. 78.

Floating the Column

The collars being completed, the colunm is now ready for the plasterer to commence the floating of the shaft with the entasis rule. But before doing so, if a small horse or collar is fixed to the rule, it will greatly assist in the working. This is a wood collar $\frac{3}{4}''$ thick and about 6″ in length, cut to the radius of the plaster collar and fixed flush on the rule as Fig. 77. Fig. 78 shows the two collars fixed on the rule, C and D being the collars. When working the rule, it is necessary to have one man at the top and one man at the bottom. Great care must be taken in working the rule, so that both men work in unison with each other and work the rule round vertically, otherwise the correct and uniform shape of the column

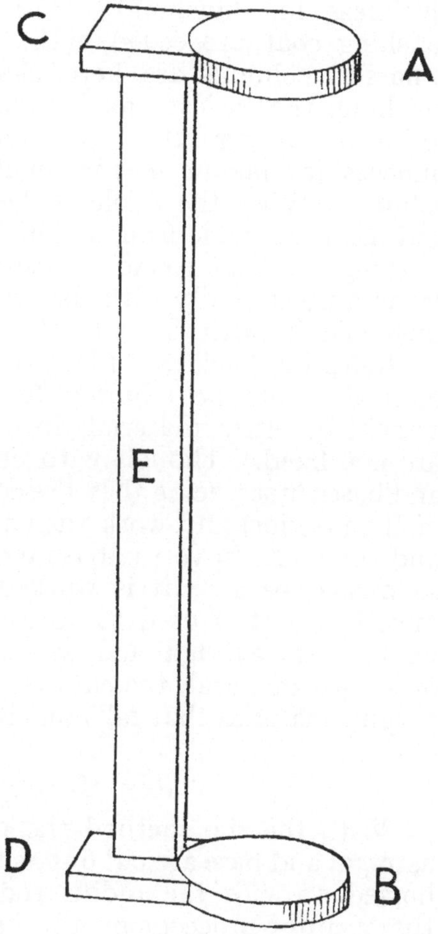

Fig. 79.

would not be obtained. Fig. 79 shows a perspective view of the entasis rule in position for working round the plaster collars when floating; A and B plaster collars, C and D wood collars on entasis rule, E the rule. The shaft is floated with cement and sand if the core is of brick or concrete. If the core is composed of wood and laths or metal lathing, a small quantity of coarse stuff would be necessary to add to the cement and sand. Three parts of sand to 1 part of cement make a good, strong floating coat for a job of this description. After the rendering and floating has been completed, the devil float is passed over it to give the necessary key to the finishing, which also assists in the forming of the floating.

Finishing the Column with Keene's

Where the entasis *rule is not muffled with the lath* the wood collars are taken off the floating rule and brought forward about $\frac{3}{16}''$ (this enables the rule to come forward from the floating face and give the necessary thickness for finishing). The collars are nailed on again and the finishing coat proceeded with. Care must be taken that the plaster collars in columns are kept clean when floating. For laying on the finishing, the easiest way is to lay the stuff on part way round from top to bottom, filling out to the full face of the rule, thus avoiding much unnecessary labour in continually having to work the rule round the column. When the whole of the column has been filled out with the rule, and has been made firm, a thin coat of soft stuff is laid on with the float, working the float round to the complete circle of the shaft. It is then scoured up carefully with the cross-grained float in the same manner and then laid in with the trowel.

Referring back to " Plastering Plain Diminished Columns," the same method would be adopted for columns required to be plastered in cement or other material, in exterior work, when the caps or bases are not fixed. Though with cement work it is customary to fix caps and bases first (seeing they are composed of a different material from those of the interior) and work the entasis rule on the fillet under the astragal and the fillet above the base moulding. This is called the rim method, so named because it is worked on the rim, either to necking or base moulding. It is used to special advantage when adopted for external work, as the astragal and base moulding can be cast and fixed, and, being of a hard material, the entasis rule is worked on it quite easily, and any surplus material that falls on either moulding is easily cleaned off.

Rim Method to Plain Columns

With the rim method the profile of the full-sized mouldings of the astragal and base are cut out as in Figs. 80 and 81. A,B,C is the astragal, horsed up with the radius rod to suit the two half diameters. Then the operator proceeds on the bench to run in plaster a little more than half the circumference of both diameters. When run, these are cut to the exact half circumference, and then the mouldings are piece-moulded in two sections, with a back saddle as Figs. 82 and 83 (which allows the two pieces to lay on their backs to fill in), finally casting the required member of necking and base mouldings. The material used for the casting must be of the same as what will be used for the finishing of the shaft. The square plinth to base can easily be formed after the column is finished.

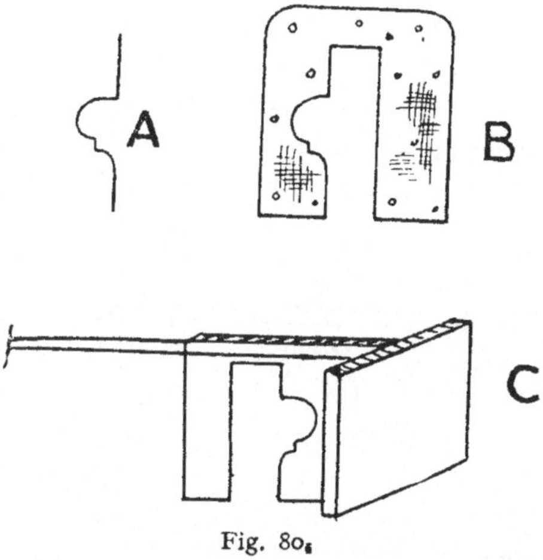

Fig. 80.

Fixing the Mouldings with Rims

Having marked the centres of the base and set out the position of the base and the required height, the base is bedded in position with cement and a small quantity of washed sand, care being taken that the two halves fit tight together and exactly level. A nail or dot is plumbed up to the required height and to the underside of rim or hollow under necking mould. This can be done with the two different-length gauges to suit the diminishing difference and plumb bob and line, or a plumb rule can be placed against the back of the entasis rule, which will give the upright, also the conformity at the centre of the base and necking moulding.

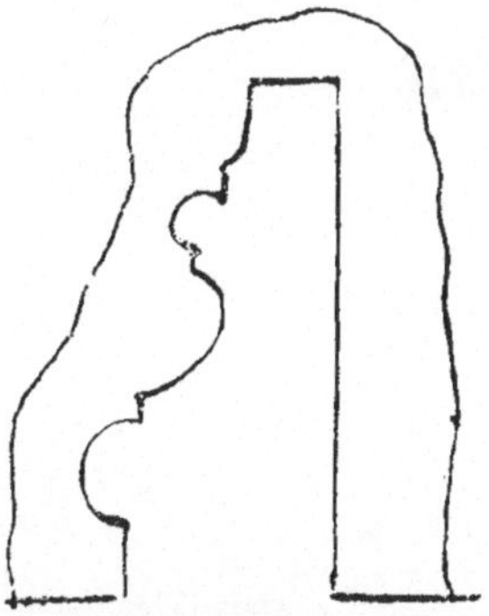
Fig. 81.

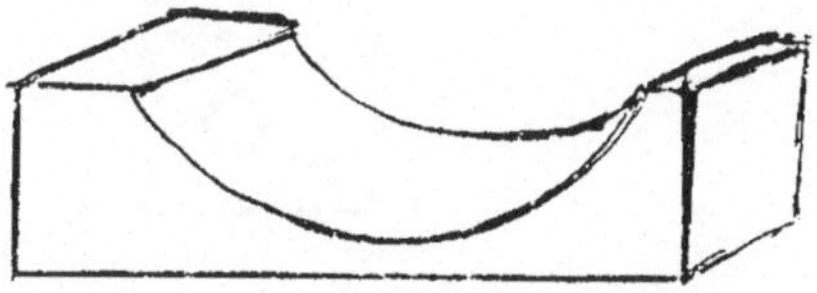
Fig. 82.

The operator repeats the dots to the three centres. These dots, or nails, give the face line to fix one part of the necking moulding. If there is any plain part attached above the astragal it must be fixed upright. Having fixed one half, making sure it is in its correct position, the other half only requires bedding up to a close joint. As a check for the parallel between the base and astragal moulding, a pinch rod is used.

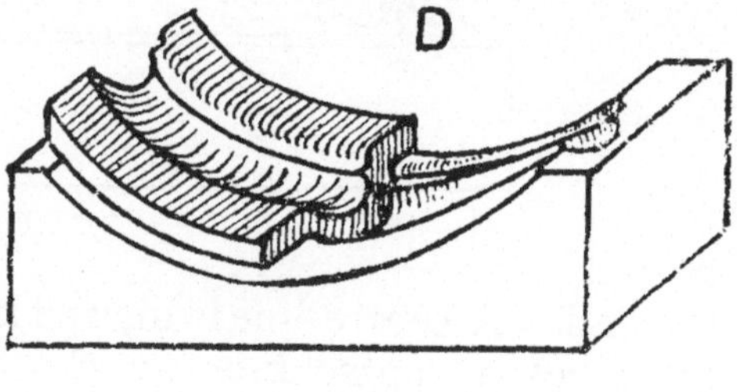

Back Saddle with piece.
Fig. 83.

Forming Diminished Fluted Columns

The forming of diminished flutes for columns is much discussed among plasterers, as there are various ways of carrying out the work. Whichever way is adopted, it entails a considerable amount of time. In plaster columns which are made in the shop, a plain mould is made with the reverse entasis rule fixed up on a frame and swung on two pivots, one at

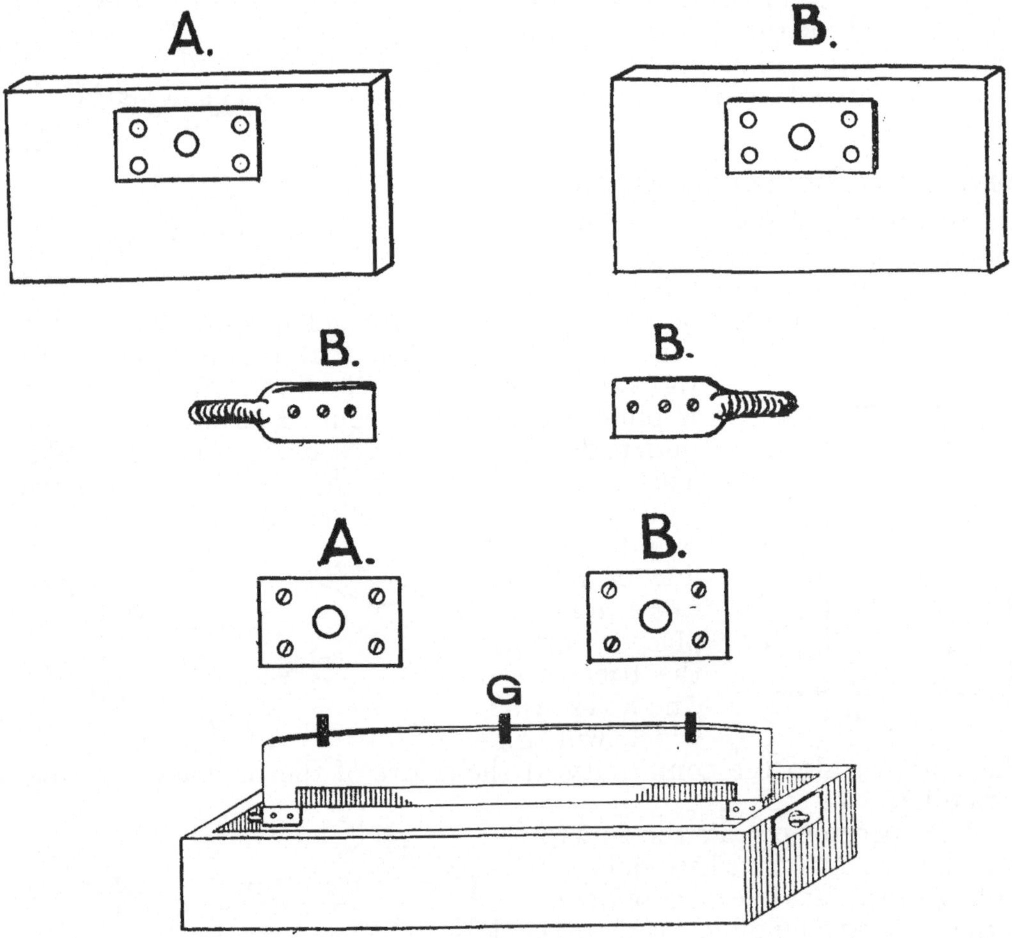

Fig. 84.

each end. A centre line is incised lengthways in the plaster of the mould. This gives a centre line on the thickness piece which is cast for the carving of the flute, Fig. 86.

Making the Mould

For the mould a wood frame is made about 3″ wider than twice the width of the entasis rule, and 2″ deeper. Brackets are made the shape required inside, to allow of 1″ × ¼″ wood being nailed for wide lathing. Two end pieces of ¾″ board are required, with a plate attached for the pivot on rule to swing. The centre line is marked on the entasis rule, and one pivot screwed to it. A stout piece of wood is screwed on the rule to stiffen it, and two cord handles are fastened to it, as this assists the plasterer in swinging the rule when forming the mould. When the rule is put in position, the second pivot plate is put in the slot and screwed

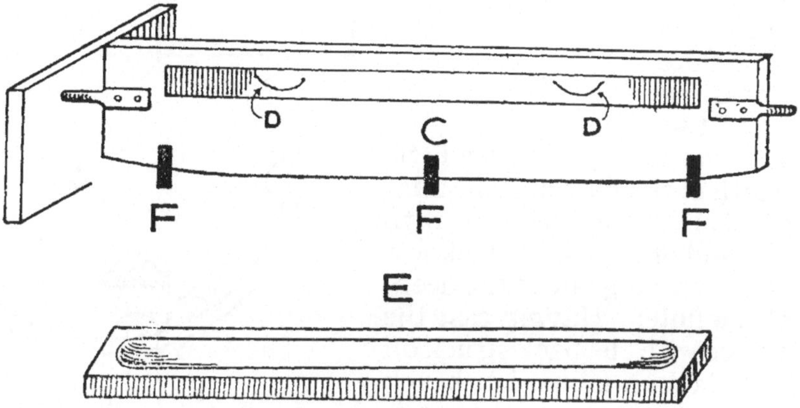

Fig. 85.

on. The reason for screwing on only one plate is that the exact width between the two end boards is between the two shoulders of the pivot, therefore the rule must be in position before the second one is screwed on.

When making moulds for plain columns, a mould plate is necessary to be nailed on the rule. When muffling the mould a piece of 1″ × ¼″ wood is nailed on to act as a muffle. After the inside is lathed (which is done before the mould rule is permanently fixed), it is covered with canvas soaked in plaster and stretched over and left rough for a key for the plaster in forming the mould. When forming the mould, the first thing necessary is to have a gauging of plaster to level off the top edge all round the frame. Care must be taken that it is

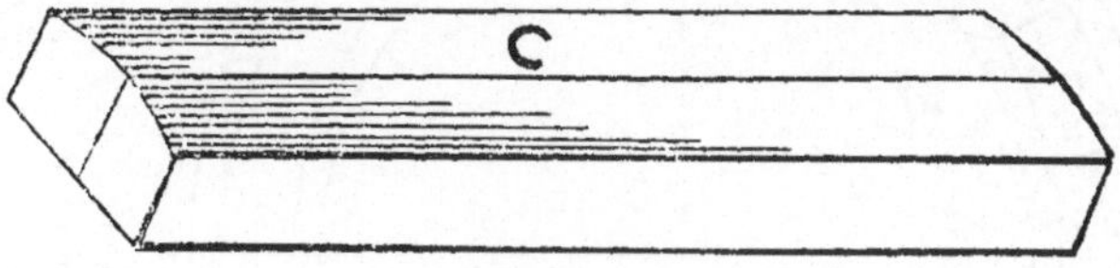

Fig. 86.

level all round to the exact half diameter of the shaft. Fig. 84 shows a perspective view of the frame with the entasis rule upwards; A and B the two end pieces of ¾″ board with the iron plate showing a hole ¾″ diameter or smaller (according to size of column rule) for the pivot, B the two pivots. Fig. 85 shows the rule downwards, C the entasis rule with pivots attached, D the cord handles fastened to the stout piece of wood, E one flute, as it would appear ready to lay in the mould, F the mould showing sinking for string to lay in to tie the flutes to save falling off when taking cast of column out of the mould, same are seen projecting from the rule in G. When using the entasis rule for the mould for fluted columns, it is necessary to allow on the diameter for the thickness of flutes laying in the mould. One flute is made by casting a length or two halves out of the column mould a little over the width of the flute and the fillet each side. This should be cast with neat plaster and of sufficient thickness to allow of the carving out of the deepest part of the flute. Having cast the thickness piece with the back struck off straight, it is laid on a board or bench and set out. Fig. 89 shows the piece.

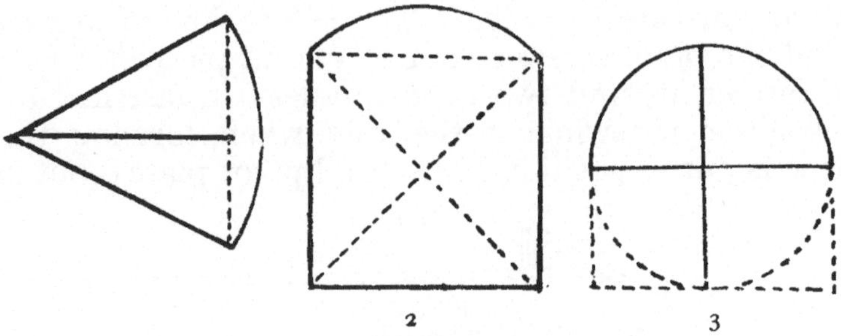

Fig. 87.

Setting Out the Flutes

Flutes are set out according to the various orders and divided into

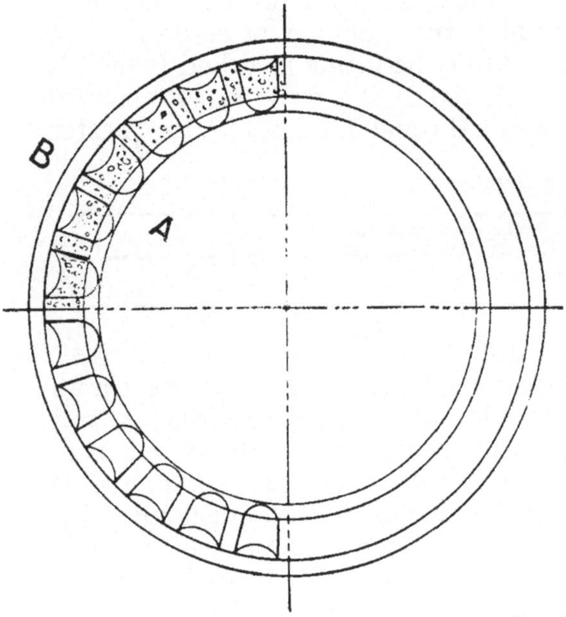

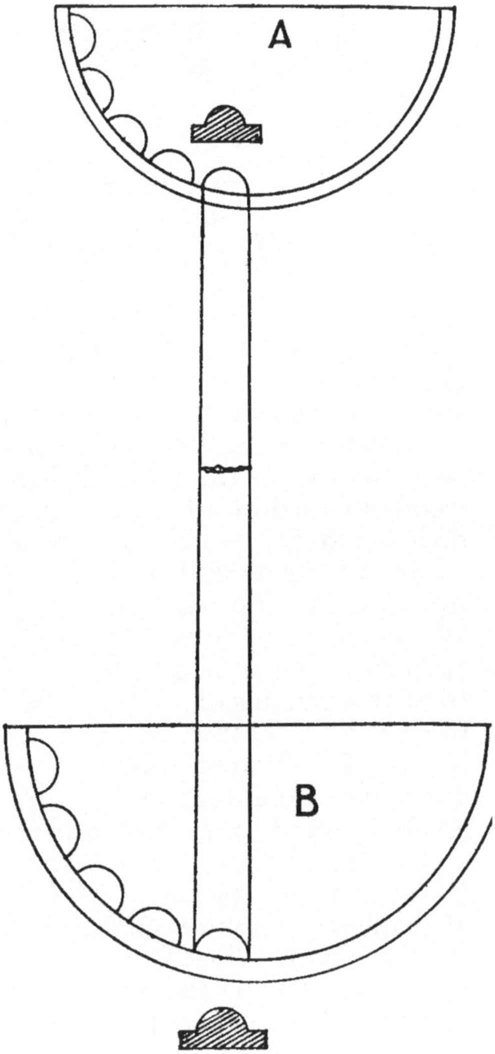

Fig. 88.

twenty, twenty-four and sometimes twenty-six. There are two kinds of flutes used: one works up to an arris and one to a fillet. The fillet one is described by an equilateral triangle, the sixth part of a circle, one the fourth, and the other the half circle, 1, 2, 3, Fig. 87. To set out the flutes the top diameter under the astragal moulding is set out to the half circumference and divided into the number of flutes with arrises or fillets as required, taking the centres and dividing to each centre of the arris or fillet forming the flute to the radius as required. This is repeated at the lower diameter above the base moulding. In Fig. 88 the flutes are set out with the fillets. A the inner flutes being the top diameter, B the lower; these are transferred on to the thickness pieces (which have been cast out of the mould), using the centre line, which would be on the piece formed by the incised mark in mould. Fig. 89 shows the thickness piece with centre line and the half fillets each side of the flute. The flute is carved out with gouges and chisels, and when finished it is moulded and the required number cast and laid in the mould. A sinking is formed in the mould so that a piece of string can be inserted to tie the flutes to the cast, so that when lifting the cast out of the mould the flute does not fall off. When carving the flute, the exact half of the fillet must be carved so that when the two are put together they make the exact size.

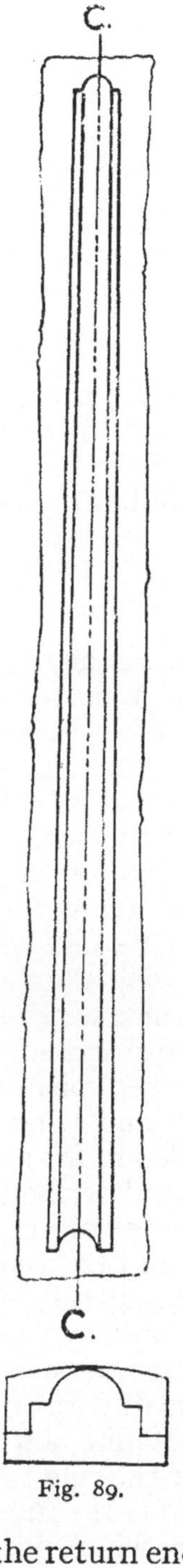

Fig. 89.

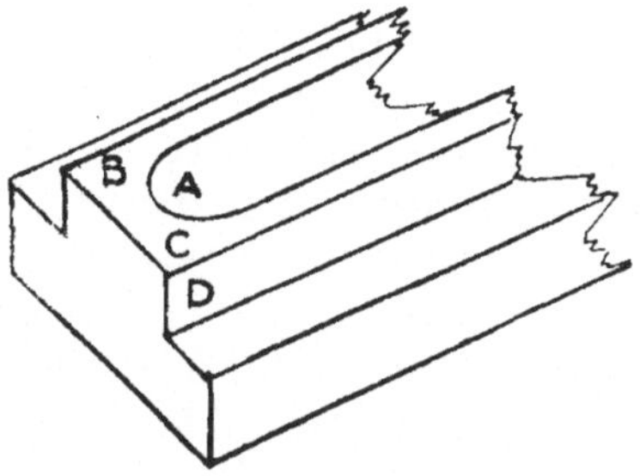

Fig. 90.

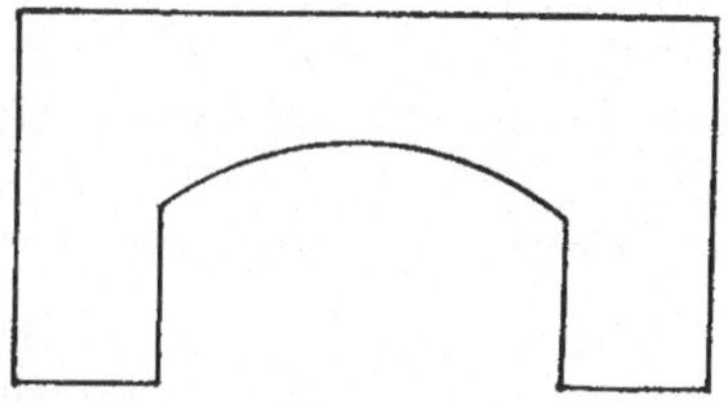

Fig. 91.

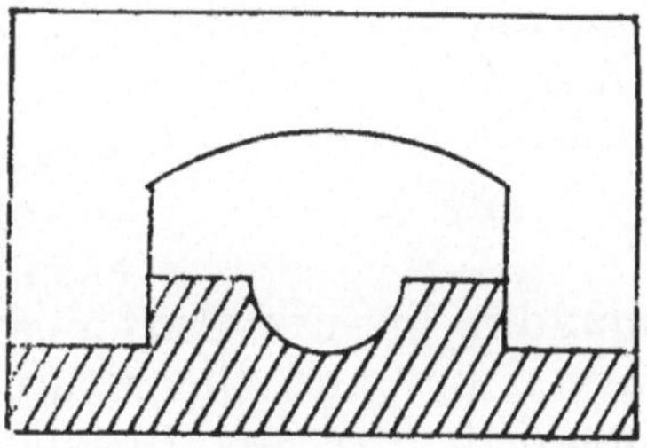

Fig. 92.

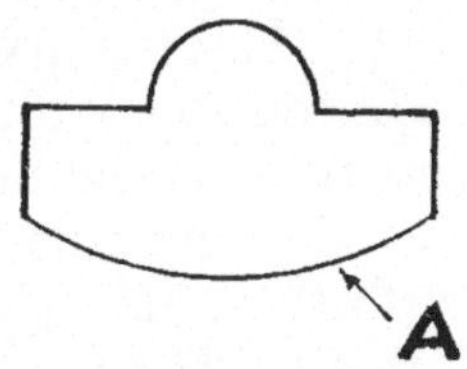

Fig. 93.

Casting the Flute to Form the Mould

Having carved the flute and made the return end, it is varnished with shellac and prepared for casting. Fig. 90

shows the section at the returned end; A being the flute, B and C half fillet, D sides for striking off the cast. When casting the flute, a wood template or gauge is necessary for striking off the sides and top to fit the hollow of the column mould. Fig. 91, the template is drawn along the bench and against the sides as D in Fig. 94. Fig. 92, a perspective view of the template over the centre section of the flute. When the flute is being filled in and the plaster stiffens up, the gauge or template is drawn over it. Fig. 93, the end section of the flute when cast which is laid in the mould of the column A, showing the circular bed to fit in mould. Fig. 94, section of flute mould.

Column Moulds without Flutes

The making of a mould for plain columns is carried out in exactly the same manner as previously described when working with frame. But the entasis rule with mould plate would be the exact half diameter of top and bottom, whereas for the fluted mould the entasis rule has to be a little larger to allow for the thickness of flute moulds added.

Forming Diminished Fluted Columns for Exterior Work

When forming flutes in columns for exterior work, it is necessary to fix the necking and base moulding with a portion of the flutes attached. A model of a little more than half the circumference of the base and necking mould would be made, with about 3″ of plain shafting to allow for the forming of the flutes. The thickness of the model at the portion for the flutes should be at least 1″ thick to allow for carving and depth. The model would be cut to the exact half of the circumference, allowing a little each end, and trimmed back for a thickness for the moulding line and also for casting. The flutes are set out as before described, carved out and the returned ends formed. The model is then piece-moulded in plaster (to allow of the cast leaving the mould), and when prepared either with shellac or seasoned with linseed oil, is filled in for as many casts as required. When the casts are hard, they are fixed as described in the fixing of collars. The position of one being fixed, the remaining one can be fixed either with the dot method or the plumb and entasis rule. Having fixed the necking and base moulding, the floating rule is required to form the shaft of the column, and is made to work round on the rim part of the necking moulding and the hollow and square at the base. The floating rule should be cut clear of the necking and the small portion of flute. Fig. 95, floating rule in position (cut clear of the moulding and flutes). The position at A over the necking forms a rebate for the rule to rest on. In addition to this two small wood collars can be fixed to the rule to assist in its working round the shaft. B shows the collars which are

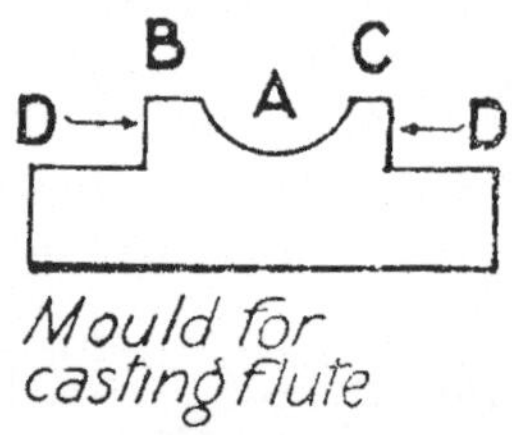

Fig. 94.

fixed at C and D. The main contour of the shaft being floated, the entasis rule is made to the full contour of the column, the dotted line A and C in Fig. 95. Fine material, such as washed or sifted sand, is required for the next operation. The gauged stuff is laid on the shaft and the rule worked round, and when filled out to the full face it is allowed to stand sufficient time to get firm. A chalk line is held from top to bottom fillet, and brought forward to clear the entasis, and carefully snapped. This gives a little guidance for the position of fillets and flutes, which can be checked with the compasses to one-quarter of the column. A mould plate is tacked on the entasis rule, the wood being chamfered back. It is then used for the working out of the flutes, working the rule on the top and bottom plaster flutes, the chalk line being a guidance for the centre part of the fillets. Small half round and part circular floats of various length are necessary for the working up of the fillets and the flutes. When the fillet and flutes have been completed, the short lengths of flutes used for the forming should be dragged over with a steel drag to allow of some of the same material to be covered over. If the column is very long, intermediate collars with plaster flutes would have to be made and the entasis rule cut accordingly.

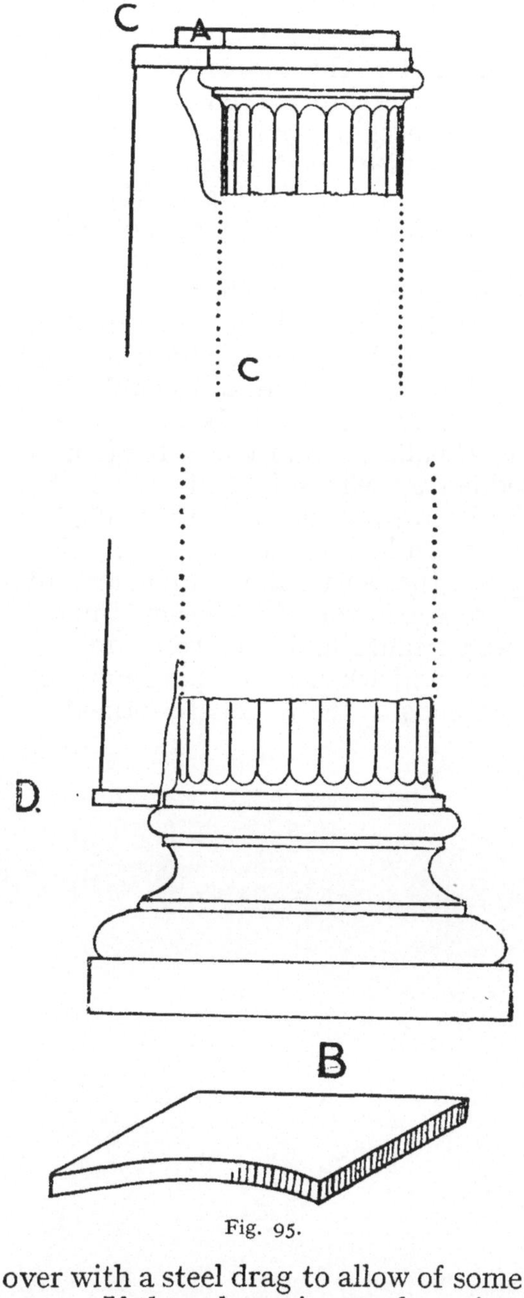

Fig. 95.

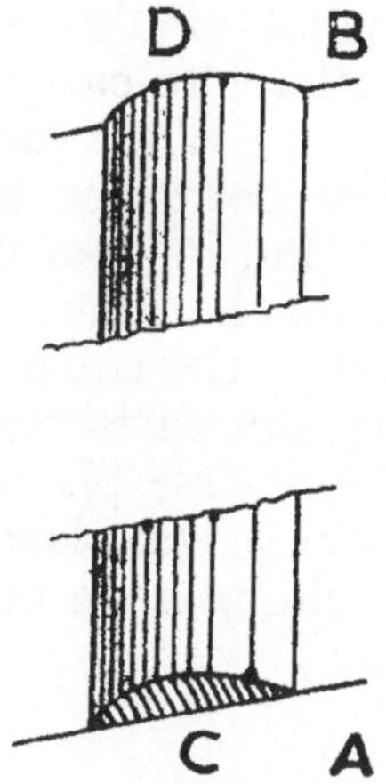

Fig. 96.

Plaster Collar Flutes and Plaster Reverse Flutes

Another method can be adopted with fluted columns for external work. The top and bottom collars are run in plaster about 3″ wide, and thickness according to the core of the shaft. The radius being of the half diameter of the finished shaft, the flutes are carved out and the collars fixed a little above and below the necking and base mouldings. A plaster flute is made by running a short piece of the top and bottom diameter and setting it up on the bench as a drum. In Fig. 96 the drum is made up, A and B is the bench line, C and D the short pieces of collar. The centre being filled in with plaster, the entasis rule is worked over. The flute is carved out and the required number cast with strong gauged plaster reinforced with canvas. If several columns have to be made it would be necessary to cast one flute and mould it several times so that several moulds would be available. This saves time and enables the plasterer to cast the flutes in Keene's, so that they could be used several times. When carving the flute in the 3″ collar, 1″ must be left plain, as these plaster or Keene's flutes are bedded up with soft stuff to fit in the top and bottom flutes. The spaces in between each flute being filled in forms the fillet, and the entasis rule is worked round on the 1″ plain part of the collar. When sufficiently set, the plaster flutes are taken out, and any parts requiring touching up are gone over with the small floats. The reverse plaster flutes should be well cleaned and oiled ready for use again. The plaster collars are taken out, and when the necking and base moulding have been fixed, the returned ends of the fillets are formed.

CHAPTER VI

CEMENT HEARTHS AND FLOORS

Grano—Finished Face to Concrete Hearths and Floors—Granite Chipping Finish—Cement Mortars—Coarse Stuff Setting and Running Lime for Putty—Setting Stuff—Fireproof Mortars—Quick-setting Mortars—Waterproof Mortars—Hints on Various Mortars—Coloured Plaster—Colours—Working Coloured and other Finish Faces—Stucco—Jointing—Method Jointing—Stone Finish—Stuke Finish.

GRANO is a term commonly applied to a hard flooring material composed of Portland cement and granite chippings. This and other Portland cement mortars are used for hearths and floor finishes. The early processes of screeding in these instances are similar to those already described. But the finishing requires full description.

FINISHED FACE TO CONCRETE HEARTHS AND FLOORS

When finishing a cement hearth, the opening is filled in with cement and sand and ruled off with a short rule across the opening on the finished floor. When sufficiently set, it is trowelled off and a V-shaped quirk cut in round the edge to the floor boards; this prevents the cracking of the edge caused by the shrinking of the boards by the wet cement. Cement hearths are always trowelled off smooth.

With cement floors, the levels are obtained as previously described in dealing with screeding, and rules for screeds are bedded. But the disadvantage of the screed rules in a finished floor is the taking up and filling in the spaces after the whole of the floor is laid. As the surface has to be gone over with the trowel after it has been laid, the difficulty is got over by laying a few bricks at convenient places and laying a board across, on which the plasterer kneels in order to fill in and do the necessary trowelling.

Cement and sand floors are usually laid in sculleries, areas, forecourts, etc., where there is not a great deal of hard wear. They should not be less than 1″ thick, and gauged not more than 3 parts of sand to 1 part of cement. Sharp sand should be used.

When laying floors, it is necessary to put down as much as possible in the morning, so as to allow of the finishing in the afternoon. The hand float is passed over to close in any rough places left with the rule when filling in. It is then trowelled off, but not smoothed, as cement floors wear smooth, and if trowelled too much become very slippery; so at the last part of the trowelling a very good finish is obtained if the trowel is passed over flat instead of on edge. When finishing treads and landings in grano and cement on a staircase, carborundum powder is sprinkled over the surface and trowelled in the finished face. This counteracts a certain amount of smoothness which arises in the wearing of the material. Iron or steel filings are also very useful for sprinkling over the surface of a cement floor when trowelling off.

Granite Chipping Finish

If a cement-finished floor is required which is to stand much traffic or hard wear, the best materials for a floor of this description is granite chippings, of which there are several grades, such as fine, medium and coarse.

Some years ago I had many bathroom floors to execute, which had to be brought up to the wood floor level. This meant a good thickness from existing concrete floor. As it was not necessary for further concrete to be added to the existing floor, old bricks were broken up (producing what is known as hard-core) to about 3″ thick and levelled down to dots. Then coarse granite concrete about 1″ mesh, gauged at 4 to 1 of cement, was placed over the brick rubble, the broken bricks being well damped first. This was about 1″ thick, and on top of this fine and medium granite mixed, and gauged 3 to 1 of cement, was laid about ½″ thick and then trowelled off. It was found that this method made a good solid floor, as the coarse granite filled up all the rough places in the brick rubble, which also formed a strong key for the finishing coat, while it lessened the weight on the floor.

In laying grano floors, the same method is adopted as for other materials, with this exception, it is not ruled off tight down to the screed, but after passing the rule over the screeds several times, it is necessary to pat the rule instead of ruling off, as this brings the cement up to the face. This makes it much easier to trowel off, and makes a much better job. Granite floors of ordinary requirements are, on an average, about 1″ thick and gauged not more than 5 parts of granite to 2 parts of cement. One important point in laying grano floors is that they must be trowelled off the same day. Under no circumstances must water be used on it the next day to trowel it up, as this only causes a certain amount of fat to come up on the surface, which, after all, is only killed cement. The cement sets during the night, and, naturally, if water is put on in the morning, the cement is killed. So it is far better to let it remain the next morning, even if it looks rough, as it will wear smooth in time.

When laying grano paving which has to stand much traffic, it is necessary to lay it about 1½″ to 2″ thick. In jobs of this description, coarse and medium chips can be mixed and used together, with not more than 3 to 1 of cement. This should be well mixed dry, then turned over and slightly damped. It is laid down in bays to allow of about ¾″ of finish. As the plasterer fills in with the coarse, he follows on with the fine, so that the whole forms itself into one homogeneous body. When gauging for the finishing layer, the medium and fine can be mixed and used 2 to 1 of cement. This needs to be gauged stronger than usual, as it is required to stand harder wear.

It should be remembered that the most important thing in order to secure satisfactory results with all kinds of cement floor laying, whether it is with cement and sand or grano, is to get as much as possible laid in the morning, but not more than can be finished in the afternoon.

Cement Mortars: Lime and Hair, Setting Stuff, Lime Putty

There are various kinds of mortars which have the setting powers of cement, and are known as mortars whether the mixture has any sand or not, or whether it be composed of lime or cement. Some mortars are ground in a pan with heavy rollers, known as a mortar mill. These are called milled stuff. Milled stuff is mostly made of old building stuff ground up together and lump lime added after it has been slightly damped to cause it to slack. Seeing that the materials used are composed of old mortar, plaster, bricks, etc., a great amount of lime is not required, but if the mortar is to be made of sand and lime it would require 1 part of lime to 3 parts of sand. Such mortar would be unnecessarily expensive, and quite as good stuff can be produced by adding ashes or clinkers from large furnaces. When using lump lime, it is beneficial to slake it before putting in the pan, as this causes the lime to powder more and helps to economise. All slaked limes have a greater affinity for water (owing to a chemical attraction) than limes which are ground. When mortars are only composed of sand and lime, the following method is adopted. The required sand is opened out into a ring (in the previous proportions as described) and the lime put in. Sufficient water is then thrown over it to slake the large portions. The whole is next covered up with the sand and allowed to stand for a day or two. After this, the whole is turned over. It is an advantage to pass it through a screen or riddle to take out any stones or any pieces of lime core which have not slaked and, if left in, would be liable in time to blow, making the finished face of the plaster look blotchy.

Making Up Coarse Stuff Setting and Running Lime for Putty

When mortar is made up like this, it can be moistened and used as required. It can also be made up in this manner and used as coarse stuff. After it has been through the screen or riddle, it is opened out in a ring and water poured in, and after the hair has been well beaten, it is put in the ring and distributed with the hair hook. Before putting in the hair, it is well to soak it in a tub or pail of water, for it is apt to get lumpy when put to dry.

If coarse stuff is required, it is wet-up and put in a heap and allowed to stand for several weeks. In fact, the longer it stands the fatter and better it gets, as every small portion of lime has a chance to slake during the period it lays before being used. The usual practice of making up coarse stuff is to have a tub or galvanised iron tank in which to slake the lime. The tub or tank is half filled with water, the lump lime put in and allowed to remain, after it has been stirred up with a bat-shaped board about 4′ long, such as a piece of 1″ floor board. When the lime starts to boil, it is again well stirred, which greatly assists the slaking. If it thickens up too much, more water must be added, but it must still be allowed to boil, otherwise, like milk, it curdles, and then if the lime was to be used for putty it would make very inferior work. This is what is called running lime.

After the lime has been well boiled and stirred, a frame is laid on the sand. The stand consists of two pieces of wood with a leg attached, triangle-wise across the sand to rest a sieve of a $\frac{1}{16}$″ mesh. The lime is then taken from the tank or tub and poured into the sieve, and the core which remains in the sieve is thrown away, care being taken that it is well away from any materials that are likely to be used. When sufficient lime has been poured in, the hair is added and dealt with as before described.

On jobs where the space is limited, a coarse-stuff bin is made by utilising scaffold boards nailed together to form a frame, and the lime is run and poured in the sieve on top of the frame, the hair and sand thrown in and larried up with the larry and hair hook. This is much the best method of making up the coarse stuff.

Setting Stuff

When setting stuff is required, the same method is adopted. The lime is run as before, but with a $\frac{1}{32}$″ mesh sieve as this line is required for running cornices, etc., and the washed sand is added in the proportion of 2 or 3 parts of sand to 1 part of lime to suit requirements; for if the lime is very fat, more sand can be added. When making up coarse stuff or setting, the bin method is the best, as it keeps the material together; while in the running of lime for putty, it can be poured through the sieve into the bin, and when this is full it can be covered up till required for use. In addition to this, when more setting stuff is required the putty can be taken from the putty bin and put in the setting-stuff bin and sand added. If superfine putty is required from the bin, always take it from the farthest corner away from the sieve.

Fireproof Mortars

A good fireproof, or rather fire-resisting, mortar can be made. The following is a very old method. Take 2 parts of slaked lime to 1 part of blacksmith's fliers or ashes (this is obtained from the back part of the forge and is very fine), mixed together with sufficient water to form a thick paste. This will form a mortar that will set nearly as hard as iron, and will be found very useful for bedding in firebricks or parging the inside of copper or furnace flues, etc.

Another kind of fireproof mortar can be made as follows: Take ordinary clay, such as used for making bricks, and mix it well with water in which horse or cow dung is soaked, or soot from the ordinary chimney can be added. To 4 parts of this mixture add 1 part of air-slaked lime (that is, lime slaked by exposing it to the air and absorbing the moisture of the atmosphere). This must be passed through a fine sieve before mixing.

Quick-setting Mortar

If a mortar is required which has quick-setting properties, the following mixture will be found good, but it must not be exposed to the air before being used. To 1 lb. of bone ash (that is, bones that have been calcined, or in other words burnt till they become ashes), add 1 lb. of air-slaked lump lime and 7 lb. of clean sand, which can be passed through a fine sieve if allowed to dry first. When the aforesaid proportions have been made up, they must be thoroughly mixed and kept well covered up in a dry place till needed for use.

Waterproof Mortars

Another mortar which belongs to the common run of materials, but if 1 part of Portland cement is added makes a very good waterproof mortar, is made as follows: To 1 part of lump lime slaked add 2 parts of ashes of any description (even the ordinary house ashes I have used with success), which must be thoroughly mixed dry and passed through a sieve. The sifted ashes are mixed dry and wetted up when required, the 1 part of cement being added at this stage.

Very good waterproof mortar can be made with ordinary road sweepings off macadam roads. It was formerly much used, especially in the country, but owing to modern methods of road surfacing it is going out of fashion. If lime stone is available, this can be crushed or ground. Whatever is used, whether it be lime stone or road sweepings, to three bushels of sifted material add 1 gallon of air-slaked lime. Mix well together and thoroughly dry. Drying can easily be done with a few pieces of sheet iron raised up off the ground on bricks and a fire lit underneath. When dry, an old iron boiler or tank is required to heat the materials. When heated, add 4 gallons of ordinary tar, and a small portion of burnt bones will improve the strength. This composition will be found very useful for covering damp walls.

Slaked lime with a solution of green vitriol instead of water. The necessary quantity of green vitriol is dissolved in warm water, the lime slaked in the usual manner with this mixture and then mixed with the sand.

Hints on the Various Mortars

The various mortars described should be put on in the one coat.

The following can be added in the use of ordinary lime mortars, which greatly assists in the hardening:

Crushed oyster or other shells.
Burnt bone ash.
Common beer; common clay.
Blood from the slaughter-house.

Coloured Plaster

For coloured plaster finishes the preparations for the backing or floating is carried out in the same manner as for ordinary plaster finishes, whether it be the ordinary floating of cement and sand or lime and hair coarse stuff. The coloured finish is dealt with in the same manner as in ordinary finished setting. But if the floating is of coarse stuff, the setting must be applied before the floating gets too dry, otherwise it will make a very bad finished face.

For ordinary work on middle-class buildings, the following will make a very good backing for a coloured finish. Take 6 parts of sand to 1 part of lime putty and 1 part of cement. If this is well mixed before gauging, it will be found to set very hard. For ordinary work these are quite suitable proportions.

In most cases in exterior work the finishing coat of coloured plasters is left with the float.

The following materials and mixing make a very good finish: Take 2 parts of dry washed sand and 1 part of air-slaked lime, which should be passed through a fine sieve; add the required colour and well mix the whole of the material. It is necessary to gauge a small portion to see if the required tint has been secured. If, as an average, 1 lb. of colour to every bushel of mixed material is used, more colour can always be added to deepen tints. When gauging the coloured material for finishing, it is necessary to gauge the materials in a banker. This is a gauging board about 4′ square, with a back and two sides about 6″ high (a piece of floor board is very suitable), nailed round the edge. In this way the material is kept together when being gauged.

The coloured finish is laid on as in ordinary setting, but instead of being laid down with the trowel, the float is used and the plaster rubbed or scoured up, and left with a float finish without the use of the water brush. That is the reason why the floating must not be allowed to get too dry.

If a trowelled face finish is required, it will be necessary to have fatter stuff, such as 1 part of sand to 1 part of lime. But trowelled face plasters are not so effective as the finished face left by the float.

Another method can be adopted in finishing the ordinary floating with a coloured face. Take 2½ parts of washed sand to 1 part of lime putty. This is best mixed by taking the putty when it is freshly run and in a creamy condition, pouring it into the sand in the banker, and then adding whatever colour is required.

Colours

Forge ashes produce black; forge ashes and white marble dust pearl greys; green enamel paint mixed up with white marble powder and lime water greens; forge ashes and marble dust ground in lime ashes gives brown; litharge or red oxide of lead reds; 2 parts white marble dust, 1 part of lime and ½ part oxide or carbonate of copper blues; crushed spar or glass is sometimes added to obtain a sparkling effect.

The following further hints will be found useful. Mixing black and red gives brown. Yellow ochre and venetian red gives buff. Yellow ochre straw colour. Black and blue pearl grey. Light green and black will furnish dark green. Lamp black and indigo blue gives silver grey. Yellow and venetian red orange. Blue and lake gives purple. Venetian red and black chocolate. Venetian red salmon pink. Lake rose colours.

Method of Working Coloured and other Finish Faces

With coloured plasters, the most important thing is the mixing and the gauging of the material. It should be measured out in a box or pail. When being used, also the floating must not be too dry, as the finishing must be done without the use of water. If it is necessary to add moisture, this should only be done by damping the float with the water brush.

A very useful finishing float can be made by fixing a layer of cork on to the float face. Another useful article is an old soft felt hat, which can be put over the face of the wood float and tacked round the edges to keep it on. This being so much softer than the wood, brings up a far better stone-like surface. A piece of carpet can be used in the same manner; but felt is much the best.

Stucco

Of late, coloured plasters seem to have brought stucco into use again. This dates back a great many years. Old stucco was composed of air-slaked limes, putty and sand. Since the introduction of new materials, such as Keene's, Parian white cement and silver sand, the name Stuke has been introduced. There is now on the market a product known as the New Stuke, which makes a very nice plaster finish. It comes from the manufacturers in bags ready for use. The background is floated in the usual way with cement and sand gauged 4 to 1 parts of cement, and the finishing coat put on in the same manner as the finishing coat of Keene's, which I have previously described.

When running mouldings with Stuke, the mouldings are cored out in the same way as for outside cement work and run off with the neat Stuke. It is used as if running a cement cornice, with the difference that the driers are left out. Plasterers quickly find out the knack of how to use it. If a polished face is required, it will be much the best way to cast the mouldings in the same manner as fibrous plaster mouldings. But if it has to be run, the treatment will have to be on similar lines to running tooled face cornices in cement, which will be dealt with in external work.

If casting with Stuke, and the cast is required for a polished face, when taking it from the mould it should be dusted over with putty powder, using a very soft brush and brushing on only a small quantity. When the cast is nearly dry, a cloth containing a small quantity of sweet oil is rubbed over the surface.

After the Stuke has been finished, whether it is cast work or done *in situ*, and it has been allowed to dry, a good polish can be obtained by melting down ordinary lard and adding a small quantity of paraffin, and when cool brushing the work over with a clean brush and then polishing up with a soft cloth. In addition to the lard and paraffin, white wax melted down and more paraffin added greatly assists in the polishing.

There are many kinds of Stuke in use at the present time, several plastering firms making up their own and selling it under special names. The floating for all the various kinds of Stuke is cement and sand, which should stand till it is dry before being finished, as the less suction there is in the finishing the better; for after the finished face is done, the jointing is required.

Jointing

Jointing is carried out to imitate stone masonry.

As the finished work sets very hard, it requires a sharp-edged tool to penetrate the finished face. The handle end of an ordinary file makes a good tool for jointing, if the joints are marked out to show, but if the joints have to be cut out and filled in, a square-ended joint is necessary. It must be pressed in the finished face when making the joints, otherwise the joints will have a very ragged edge.

In order to facilitate the putting on of the finishing coat for pointing, it is well to delay the hardening of the surface. It is possible to do this by thoroughly soaking the floating with water from the gauge pot. It is always advisable, if possible, to mark out the joints the same day as finishing. When setting them out, a good plan is to mark the depth of the stones on a rod, and at each end of the wall, ticking off the centre with the point of a pencil or nail, and then snap a chalk line to each one. With the vertical lines a plumb line is started at the centre and the remaining vertical lines gauged from it.

When marking out the joints, it is not necessary to have the joints of one size; for in the building of stone work various-sized stones are used, so long as they are in proportion, as in a block of stone of 12″ in thickness it would naturally be between 2′ and 3′ long. Another thing is that upright joints should not be vertical with each other, as in quarrying stone short lengths occur which have to be used up, and, moreover, it is bad masonry to have vertical lines in the bond. Therefore, it is not necessary for the vertical joints to follow each other.

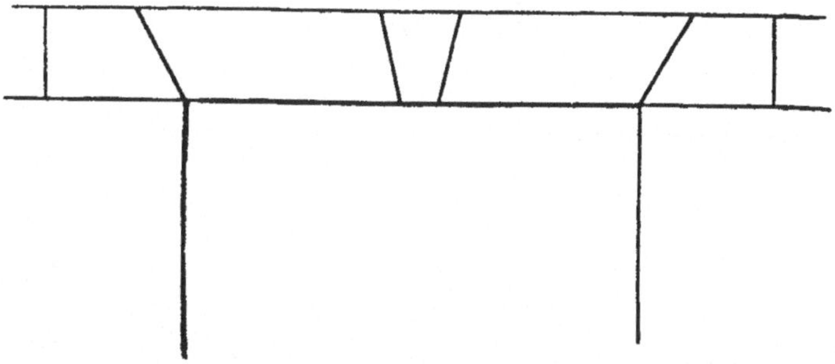

Fig. 97.

Method of Jointing

When jointing, one very important matter is to see that the horizontal line takes up with the level of the heads of windows or opening, so that the skewback joint (which is the radiating line of an arch from the centre) shows the full depth of stone, as in Fig. 97. It is not necessary when setting out the joints to windows or doorheads to have key stones, as stone carries its own weight across any ordinary size opening, as in Figs. 98, 99, 100 and 101. Such construction would be quite in order. When joints are required to be filled in, which can be done with white Keene's, or Parian or plaster, they are marked out with the jointer about ⅛″ wide and the same depth. Then, with a small quantity of whatever material is used, the joints are filled out with the use of a plasterer's small tool. After a fair amount has been filled in, a fine drag is passed over to take off any surplus or salvage, which always occurs when filling out the joints.

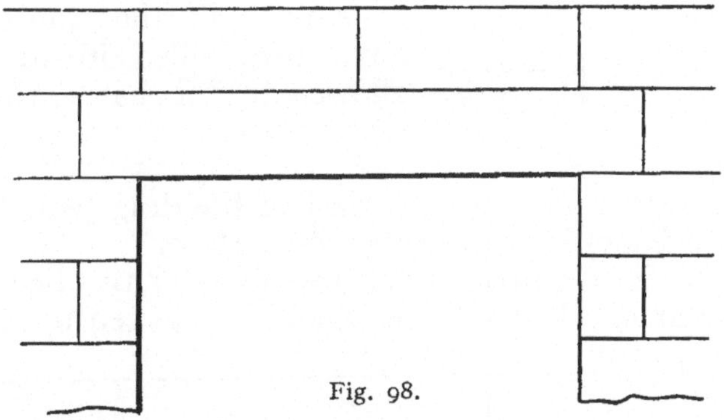

Fig. 98.

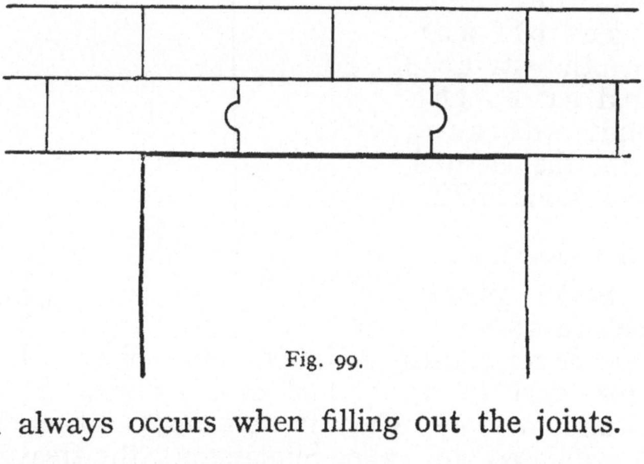

Fig. 99.

Stone Finishes

There are various ways of finishing Stukes besides polishing, and it entirely rests with what kind of stone finish is required. If a sawn-stone face, fine drags are used on the surface after it has been left by the float for a few hours. If a tooled face, the joint rule is drawn over the face after it has been rubbed up with the float. This closes in the sand, and, in addition, smooths the surface, which gives the face an appearance of a block of stone tooled up with a mason's chisels. A very useful tool can be made for this finish with a piece of three-ply wood, about 14″ or 15″ long and about 3″ wide, cut off to a taper at one end. This greatly assists in closing in the surface prior to using the joint rule.

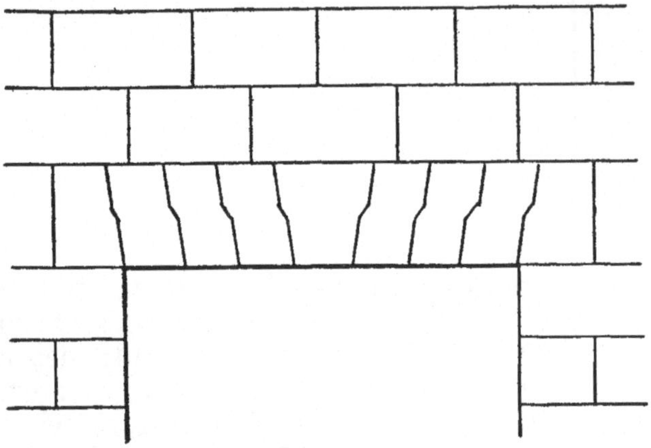

Fig. 100.

After all the joints have been filled in and allowed to set hard, the whole of the surface is gone over very lightly with a fine drag, and after this has been done the back part of the drag (which is usually a smooth surface or edge) is drawn over the surface.

When a sawn-stone finish is required, after the joints have been dealt with (either left as a joint or filled in), the whole is dragged over, and, in addition, when it is dry carborundum stone is rubbed over the surface. This takes off any smooth surface, and gives the sandy surface, with the desired sawn-stone finish.

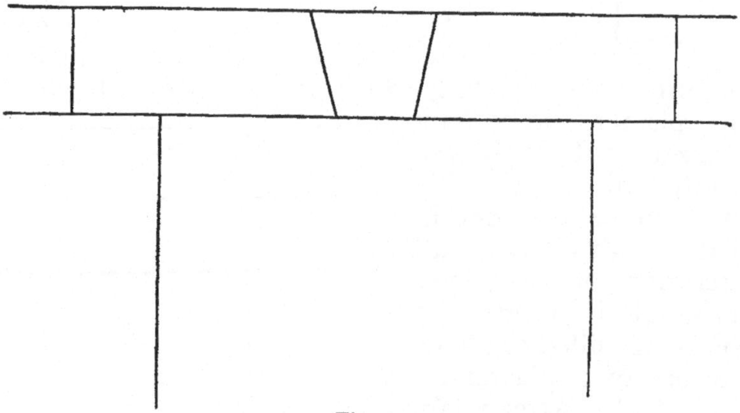

Fig. 101.

Stuke Finish

Stuke plaster finish is very suitable for staircases, banks and offices. It has a good hard finish, quite equal to stone, and it can always be rubbed down with carborundum stone when it gets dirty. There is another plaster finish to staircases, etc., called Simla stone, the treatment of which is similar to that for White Atlas Cement. This can be used for exterior as well as interior purposes. I can conscientiously recommend it for a good plaster finish, both for interior or exterior work.

When finishing Stuke, it is important to keep the steel tools clean, and if required to touch the surface of the finished face, they should always be damped with the water brush before being used. Another point is that before cutting out or marking the joint, the stones should be checked to see that they are correct, otherwise if a joint is wrongly cut and has to be filled in, the correction is always discernible on the finished face. When floating for Stuke, it is necessary to measure the proportion of cement and sand, so that only fresh stuff is used. If any is left overnight and used next morning, it is liable to crack and, in addition, will create a difference in suction, which would make it very bad in finishing.

CHAPTER VII

EXTERIOR PLASTERING AND WATERPROOFING

Plastering on Brickwork—Exterior Dots—Forming the Screeds—Floating—Fining or Finishing—Jointing—Finishing Exterior Work in White Cement—Simla-stone Plastering—Exterior Plastering with Lime Plaster and Keene's Cement—Lime Plaster—Roman Cements—Medina Cement—Mastic Cement—Portland Cement Moulding—Mitres and Returns—Rough-casting with Shingle, Marble Chippings, Brick-hogging, etc.—Throwing on the Rough Shingling, etc.—Joinings in Rough-cast Finish—Angles and Arrises in Reveals, etc.—Coloured Rough-cast—Crushed Gravel—Marble or Granite Chippings—Broken or Crushed Bricks—Derbyshire Spar—Economy in Rough-casting—Damp-proofing—Specialities—Old Work.

EXTERIOR decorative plastering is now rarely executed in so elaborate a style as was commonly practised during the past century. Such plastering is an art, which can only be carried out by the experienced, well-trained craftsman. We have only to examine the old work to appreciate what was involved. There are plain and fluted columns and pilasters, which had to be worked by hand, with the aid of the Entasis rules and wooden floats of different shapes and sizes. Pediment heads, closed or broken, with their raking mouldings and return ends, also had to be worked by hand. The radius rod (gig-stick) had to be used for the segmental and circular or semicircular heads over window and door openings. Rusticated bases and quoin stones also had to be worked by hand. Balustrading calls for turned models, piece moulding and casting and nice handling in the erection ; whilst modelling and carving is needed for the execution *in situ* of swags, drops, scrolled leaves and complicated mouldings on capitals, and for various ornaments, including the pressing in cement of plain capitals, truss blocks and other cement decorative details. Although we rarely see this now, nevertheless there is a return to fine work.

PLASTERING ON BRICKWORK

Plastering on external brickwork is usually carried out by floating with a cement mortar, composed of 1 part of Portland cement to 4 parts of sand. The quantities should be measured out in a box, and carefully gauged in batches just sufficient for immediate use.

Before putting in the dots as a preliminary to commencing the floating, it is necessary to damp the brickwork thoroughly, using clean water from the gauge pot. The amount of moisture required will vary, partly according to the nature of the bricks (some being very absorbent, others hard), and partly according to the atmospheric conditions. Great heat and steady northerly or easterly winds tend to dry up the surface.

The great point is to keep the walls thoroughly damp throughout the operations, otherwise the suction may be great enough to deprive the plaster of its own moisture before it has time to set, and so cause it to disintegrate or crumble.

Exterior Dots

Readers should understand that there is a considerable difference between dots used on interior and exterior work. For interior jobs wood is the basis; for exterior work the dots are formed with nails. There is also a variation as to the screeds. On interior work the screeds are placed horizontally, but on external work they are run vertically. This variation is due to the difference of the materials used. Coarse

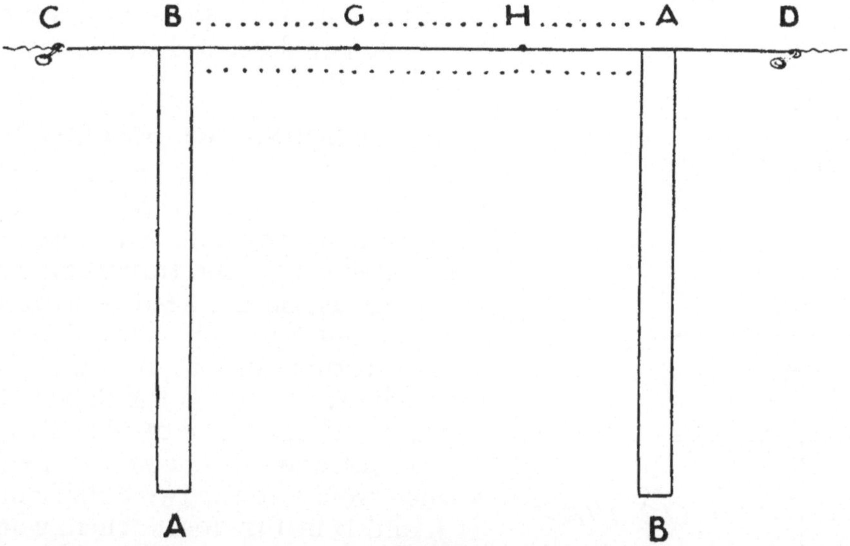

Fig. 102.

stuff used on internal walls being composed of lime and sand is a much fatter material than cement and sand, and so the former is easier to float on across the walls. Apart from this, the construction of the scaffolding would not allow of the cross screeds, though these are the most convenient. Then again, the amount of stuff falling in the ruling off would be considerable, and would be wasted by falling through the scaffolding. By forming upright screeds, when the plasterer is filling in the bays between the screeds in the ruling-off process, the plasterer is able to catch the surplus stuff on the floating rule, and with the use of the laying trowel he can take it off and put it on his board for use again. In this way material is saved which would otherwise be lost by falling through the scaffolding, unless it was possible to get a save-board against the building.

If the building which has to be plastered has a plain façade with no external angles at either end, such as one of a row of buildings, the nail dots are put in as follows : Drive a nail in the joint of brickwork (clout-headed nails in preference), so as to project about $\frac{5}{8}''$, and about 6″ away from the outside limit of brickwork to be plastered. If a cornice is to be run, the nail should be driven in just under the position of the bottom member. Repeat this operation at the opposite end of façade, and then stretch a line from end to end, fastening it at each end with a longer nail driven in the joint. Fig. 102 shows A and B the dot nail, C and D the line fastened to larger nail. Intermediate dot nails can be put in to the line where required, which should be in between windows.

The next process is to plumb from each nail with the plumb bob and two gauges, as described in " Interior Floating." It is not necessary to plumb the whole height of the façade, as two or three scaffolds, or any convenient length, is quite suitable, and any intermediate dot nails can be put in with the line hanging vertically, as previously described in the method of plumbing dots.

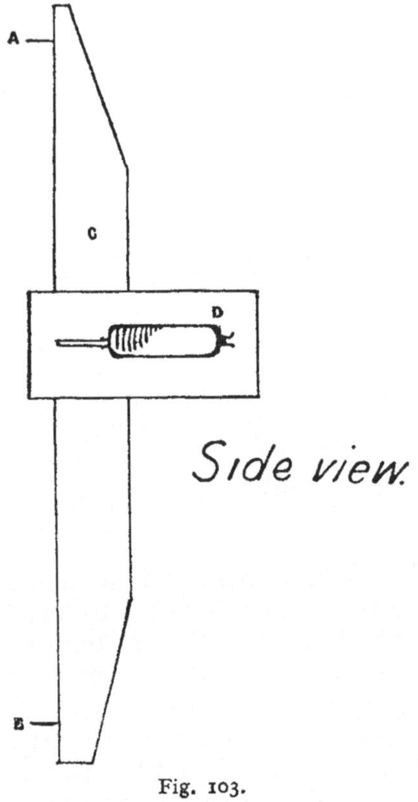

Fig. 103.

Forming the Screeds

When the required number of dots have been put in, the screeds are formed as narrow as possible, not more than 3″ wide, seeing that the traversing rule has to be worked on the head of a nail, and not on a dot lath as in interior screeds.

If a cornice has to be run, a screed would be traversed as the dotted line in Fig. 102. If there is no cornice, this would not be necessary. The first screeds to be formed would be the two outside ones, as at A and B in Fig. 102. Then, when any windows or other openings of whatever height occur, the line can be stretched through from the two outside screeds, and the reveals ranged through. What is meant by ranging is that after the line is stretched through, the distance is measured, from the frame or back of each opening, to ensure the reveals being of the same width. Another reason why these external screeds should be as narrow as possible is, that after being traversed they are cut off and should not be more than 2″ wide.

A very useful way of cutting off the screeds after traversing is to hold the traversing rule central and cut each side of the rule by drawing down the edge of the laying trowel, as in Fig. 103, where A and B are the two dot nails, C the traversing rule, D the laying trowel. Having arranged the position of the intermediate dots so as to come where no openings or

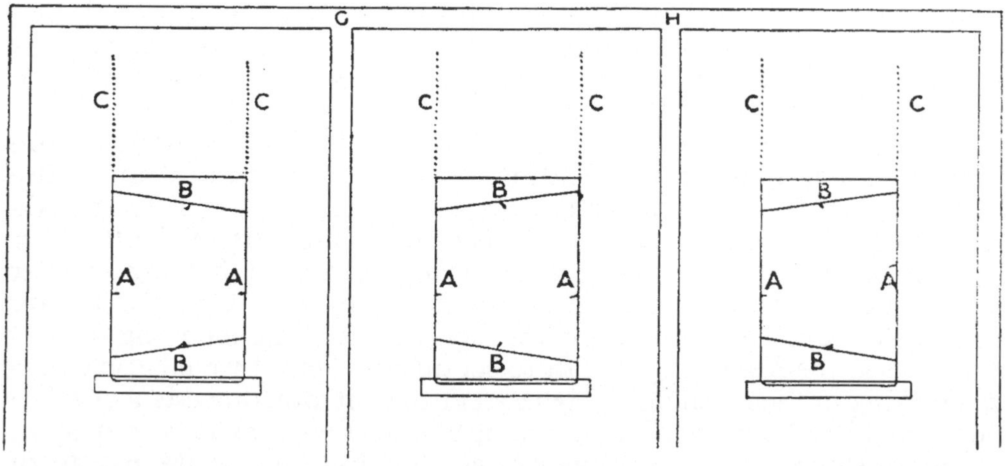

Fig. 104.

windows occur, as G and H in Fig. 102, these screeds are ranged through as previously described.

Where windows or other openings occur, rules are tacked up inside the reveals or openings, or wedged across, and ranged through, as in Fig. 104, where A are the rules fixed, with B wedged across. After the rules have been ranged top and bottom, it is necessary to try them with the floating rule to see that they are not hollow or round in the centre. When straightened, additional screeds are carried over and under if required, as in the dotted line C in Fig. 104. If the windows have a projecting sill, a notched rule, such as a piece of floor board, would be required to obtain the same line of work underneath, and a dot is put in, and another one lower down, if required. Fig. 105 shows the notched floor board against the rule in reveal and against the dot nail A, rule in reveal B, notched board C, sill D, nail under sill in Fig. 104.

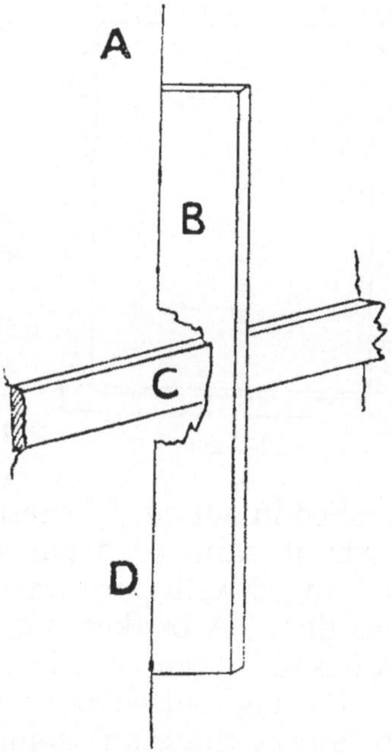

Fig. 105.

Floating

When all the screeds have been formed and allowed to set hard and the rules fixed to opening, etc., the spaces in between are filled in as previously described in floating. It will be understood from what has been written as to the running of the screeds, that it is customary to start floating on the top scaffolding, working downwards. This is partly due to the nature of the material used. Moreover, it permits of the scaffolding being struck as the work progresses. One very important point is that the materials for the screeds are gauged in the same way as the remaining part of the floating.

It is not necessary to tack up rules to the face of the head of the windows or openings for the floating, as after the face is floated the rules are taken down and nailed on the face of the heads, and the reveals are floated as previously described. The remaining part of the floating is continued, and the most important thing in cement floating is that only enough material is gauged as can be used each day; for if any material left over from the previous day was used the following day, it would cause the fresh gauged stuff to set quicker, and it would also cause fire-cracking in the finishing. Apart from that, it would create a suction which would make it very bad for finishing, or fining, as it is termed in cement finishing. This fining, moreover, would dry out white instead of the natural cement colour. In cement work the slower the set the more natural the Portland stone colour obtained and the harder it becomes. So the plasterer must always bear in mind that the most important points in cement work is to well wet the brickwork and the floating, to accurately measure the proportions of cement and sand, and always to use fresh stuff.

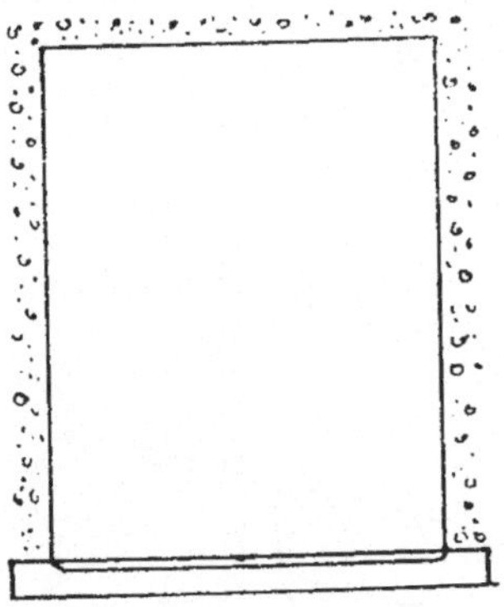

Fig. 106.

Fining or Finishing

When sufficient floating has been done and allowed to stand to set dry and hard, the next step is the fining or finishing, which is done after all the mouldings, if any, are run and completed. When all is ready for finishing, the floating is well soaked with water, using the gauge pot. This preliminary soaking of the whole surface is necessary, for if carried out piecemeal it would disturb, if not destroy, any fining laid on. The finishing material is composed of sand washed in a fine $\frac{1}{16}''$-mesh sieve, in the proportions of not more than 5 parts of sand to 2 parts of cement. These are mixed dry, and should be gauged with pure water in a banker, so as to keep it free from stones and dirt. A banker is a gauging board about 4′ square, with three side ledges to it about 6″ high.

Having wetted the whole surface and reveals, the plasterer proceeds to lay on the stuff, commencing, if there are any openings or reveals, by laying a small quantity about 3″ or 4″ wide all round the openings or reveals, as in Fig. 106. This is traversed off with the feather-edge or short rule. When this has been completed, one man holds the feather-edge against this face about $\frac{3}{16}''$ in front of the face of floating of the reveal, and the plasterer then lays the stuff on the reveal, drawing it with the trowel from the inside up to the feather-edge rule. Whilst the rule is being held, the reveal is wiped out with a feather-edge, any hollow places being filled in with the cross-grain float. After this has been completed, the feather-edge is drawn off and away from the angle as carefully as possible, so as to leave a good arris on the angle. The same process is followed for the other side of reveals, and also the head. When reveals and heads to windows are finished before the front face is done, the joint on the face should be cut back to within a ¼″ of the finished arris of the reveal to disguise the joint.

Having filled out the reveals and brought the arrises as forward to a finish as possible, the remaining part of the floating is laid on with the laying trowel. When laid on, it is gone over with the feather-edge, and any hollow places that the feather-edge does not touch are filled in with more stuff till all the hollows are filled out to a fair surface. When this has been completed, and the face of the work is fairly steady or firm, it is laid in with the skimming float, which should be made of soft pine wood. This in turn fills in any remaining small hollows or rough places. After being treated with the skimming float and a fairly even surface has been obtained, the face is rubbed up with the cross-grained float, using it in a circular manner, working from left to right. Why it is necessary to finish the work from left to right is, that while the work is being rubbed up to a finish by the cross-grain float, the plasterer must keep on working away from the face he has finished; otherwise, if the float was continuously being rubbed over the surface, it would bring the face out patchy and rough. When the floating begins to suck the water out of the finishing, the rubbing up with the float should be stopped and the surface patted with the float. When the process of patting is being done, the water brush is held in the disengaged hand, and as the float becomes dry water is shaken on it from the brush. On no account must water be thrown on the face when rubbing up with the float. This patting process is intended to bring the grit in the sand to the face, so as to give it the appearance of a stone-like finish.

Jointing

As nearly all external cement work is jointed, the next process would be to set out the joints, usually in accordance with the dimensions specified by the architect. The depth of the stones is marked off (on a rod of convenient length) at each end of the finished work, and a chalked line is snapped through. Then the length is marked out as near the given size as possible, and set out to suit accordingly. When the vertical lines have been set out, the plumb bob should be used to see the marks are upright. When this has been checked, the vertical lines can be snapped with the chalk line, and then the marking out can be proceeded with. For this process a feather-edge rule is very suitable. It is held against the face of the work by two men, one at each end, provided with a jointer.

This is an S-shaped tool, with a shoulder towards the centre and a blade at each end, about 2" long and $\frac{1}{8}$" thick (Fig. 107). This is held against the rule and to the shoulder of the jointer, and drawn along the line previously snapped. When the finishing cannot be done all in the one day, the finishing must be cut off about $\frac{1}{4}$" below the lowest horizontal joint, so that when the next part is done it is not noticeable, for the joint of the stone would disguise the joint of new finish. Here, again, the importance of measuring the cement and sand to exact proportions must be remembered, otherwise the second part of the finished work would be of another colour.

When a façade is being finished down, and the scaffolding struck progressively, it is not advisable to bed in the bricks where the pud-logs are taken out, as it causes a bad place in the finished surface. Apart from that, when the work is dried out there is a possibility of the place where the pud-log hole has been filled showing up, as these places would have a different finishing suction to the other parts of the floating. The usual course adopted, which I consider is the best, is to wedge a brick or part of a brick as required to fill the hole, and keep it a trifle back from the face of the floating. After the remaining part of the floating has been soaked with water the bricks, as before described, are covered to the floated surface with some of the gauged finishing stuff, and then the remaining part of the finishing is carried on as before described.

Fig. 107.

Finishing Exterior Work in White Cement

With white cement the finished appearance is very similar to a lime stucco finish. When the external part of any floating is required to be finished in White Atlas cement, the preliminary process is the floating, which is done in the ordinary manner (as previously described), with cement and sand gauged in the same proportions as if the face was to be finished in the ordinary Portland cement.

Before commencing the floating for a white cement finish have the brickwork all well soaked and rendered. When white cement is sprinkled with water after it has been finished (rain, for instance), map cracking is very often noticeable. I think failing to well soak the brickwork is the chief cause of this. I have tried the soaking of the floating, but that does not rectify the defect, though if Majauca powder, or Colmanite, is added to the finishing coat, it helps to counteract this cracking. The laying on of the white cement finish is done in the same way as with the Portland cement; the only difference being that the white cement is gauged with silver sand and mixed 2 parts of cement to 1 part of sand. As the silver sand is so much sharper than the ordinary sand, it requires gauging fatter, as it works very short (that

TEMPORARY BUILDINGS IN FIBROUS PLASTER.

TEMPORARY BUILDINGS IN FIBROUS PLASTER.

is, not so sticky or fatty as with pit sand). Some experts in white cement prefer to add a small amount of lime putty when running cornices, but I have not adopted that method. If the plasterer uses white cement on the same lines as Portland cement, I do not think he will have any need to add lime.

Whilst I am touching on the process of running in white cement, it should be stated that it is carried out in the same way as I have described for the running of cement cornices.

In finishing in white cement by jointing, it will be found that, owing to the much harder nature of this material, the jointer is not suitable; a file is far better, using the handle end.

Before I conclude with the finishing of Portland or white cement, I must point out the usefulness of the cork and felt float faces, and also variously shaped small floats, in the running of mouldings and cornices. All these greatly assist towards the forming up of the members and save much time.

Simla-stone Plastering

Simla-stone finish gives a stone-like surface closely resembling the best stucco. It can be used for interior and external plaster decoration. The treatment of this material is the same as for white cement, with the exception that sometimes interior work is required to have a smoother surface. If such is the case, instead of rubbing up with the float after being laid on with the feather edge, it should be laid on with the float, and in addition a short piece of rule should be dragged over the face. A piece of three-ply wood about 15″ long is a very useful thing for this work. After this treatment, the laying trowel passed over on its edge will pay the face down to a very good finish, a smooth-faced stone.

This Simla-stone mixture sets very hard, and the jointing must be carried out as soon as possible after the face is finished, otherwise great difficulty is experienced in jointing the stone.

This material is composed of white cement, silver sand and a small amount of lime. When required for interior plastering, it can be toned down with umber and yellow ochre. The method of running of mouldings with this material is the same as for white cement. The use of the small floats is a big factor in the successful handling of this material.

External Plastering with Lime Plaster and Keene's Cement

Keene's cement is not used a great deal for external plastering, though I have done a fair amount of external work with this cement, especially in finishing gables and other parts not directly exposed to the weather. I have also used it in running mouldings and cornices. In such cases the background must be floated with cement and sand, and the mouldings cored out with the same material, though a little fatter with cement, and then run off and finished in Keene's. But in these cases the weathering of the cornices was covered and slightly turned down over the top member with sheet copper, and the whole of the work when finished was painted. I may add that this work is as good now as when it was done nearly thirty years ago. Indeed, I see no reason, if Keene's is painted, why it should not be used a good deal more for external work than it is, especially in large towns.

Lime Plastering

Lime plastering for external work has been used from very remote times. In England it was very general for lath and plaster work in half-timbered houses and for covering brick nogging and rubble filling in frame buildings, examples of which are still to be seen plentifully enough in the country, and even in a few instances in London. It was in general use before Portland cement finish was introduced for this purpose.

For lime plaster the lime is run or slaked in sand as already described. It is then left for some considerable time before being used. If slaked lime, after being mixed with sand and gauged, is allowed to stand, the hardening properties are increased and it also gets fatter. If a lime plaster finish is required for covering lath work, it will be necessary to prick up the first coat with 3 parts of sand to 1 of lime, and to every 3 cubic feet of the mixture 1 lb. of well-beaten hair is added. After the first coat has been allowed to stand until nearly dry, the surface is scored over with the lath scratcher. It is then floated with a mortar of the same proportion as before. After the floating has been completed, the face is finished in the same way as described under skimming, with this difference, that after the skimming has been laid down with the float, it is scoured up with the skimming float, instead of being laid down with the trowel. This kind of work is left to stand as the float leaves it. The finishing coat is gauged up 1 part of lime to 2 parts of sand.

In some districts of the South of England, in the early part of the last century, it was a common practice for external plastering to be floated in lime and sand, finished with lime and washed sand, and jointed to imitate stone. The floating stuff was made in the following manner: the lime was slaked by sprinkling water over it, and the sand thrown over to cover it. Having stood for a day or two, it was turned over, well mixed and passed through a screen. This had a frame about 6′ × 2′ 6″, having two sides and a head about 6″ in depth. On this frame a wire mesh was nailed. It was propped up at one angle of 60°, the material being thrown up against the iron mesh. In Fig. 108 A is the iron mesh, B the sides and head, and C the prop. After the stuff had been well screened it was wet up and used in the ordinary manner of floating.

The finishing stuff was made up in the same manner, but, instead of being screened, it was sifted through a fine sieve. In some cases, when Portland cement began to get known, the floating was done with lime and sand, and finished with lime and sand with a small quantity of cement.

This kind of work stood the weather for a good many years, but the face of the finishing became scaly, entailing a lot of repairing. I have hacked off this kind of lime-plaster finish, and though the face was slightly weather-worn, the floating was very tough.

Roman Cements

Before the introduction of Portland cement external moulding and cornices were run and finished off in Roman or Medina cement. Roman, or black cement, as it was sometimes called, dates back to the seventeenth century. It is not of the same strength as Portland cement, but if mixed with Portland it sets very quickly, and is quite useful for casting purposes.

When running mouldings, it should not be gauged less than half and half. When running off cornices or mouldings, they are cored out in the same way as in Portland cement mouldings, though in the running off for the finishing no driers are required. Roman cement, being of a fatty nature, runs itself out to a smooth surface, so instead of having to use driers the plasterer keeps on filling out the members and passing the mould over till they are formed. The finish can be assisted by the use of the gauging trowel and smaller tools.

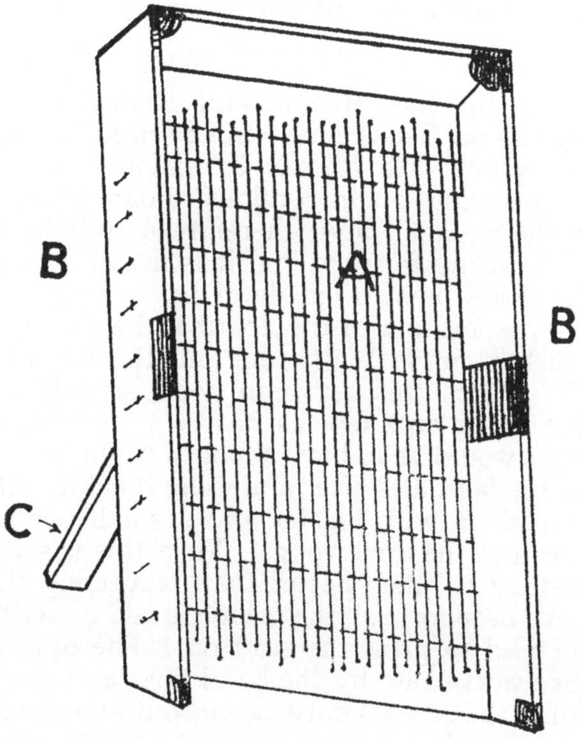

Fig. 108.

Roman cement is now very rarely used, except for repair work. However, it will be found very useful for this purpose, as it can be followed on by painting. With Roman cement it is much easier to run mouldings than with Portland cement, as in the forming of Portland mouldings, the running off with the driers is a very tedious operation. Nevertheless, though I have done a good deal of running in Roman, I prefer Portland for working.

Medina Cement

This cement is of a buff colour, very much like a dark shade of yellow ochre. In working, it is very similar to Roman cement, though it has twice the strength and sets very rapidly; therefore it is useful for casting purposes. Here again we have a cement with which it is quite easy to form mouldings, as it is of a fatty nature. Like Roman, it is run off to a smooth face, and it does not require much labour in running off. This cement, though stronger than Roman, has not the strength of Portland, and when gauging twice the amount is required as compared with what one would be taking if using Portland cement with sand.

Medina cement appears to have gone right out of fashion. I have not used it myself, but have seen buildings that have been plastered with it and have stood a good many years, presenting quite a good appearance at the present time.

Mastic Cement

Mastic, like Medina and Roman cement, has nearly gone out of existence. This cement can be used for running mouldings, but requires coring out with Roman to within a $\frac{1}{4}''$ of the profile of the moulding. Mastic cement is impervious to damp and is not affected by the changes in the weather, and, therefore, is worth attention for special work.

In using this material it is necessary to have very clean trowels, and the floats must be made of boxwood (which is a very hard wood). In the preparation of the floating, the joints of the brickwork must be well brushed out with a broom and the brickwork brushed over with linseed oil. When this has been done, screeds are formed as before described in floating and kept as narrow as possible. A good method of forming these would be to have a long straight-edge about 1″ wide and to grease or oil the face, and after plumbing the dot nails, fix the rule against the nails and fill in with plaster and a small portion of lime putty between the rule and the brickwork. When this is set, the rule is taken away. This method is much better than traversing the screeds. When the screeds have been formed, the mastic is laid on with the trowel, and the hands can be used to press the stuff on. The bays having been filled in, and the face worked up by the hand float and the floating rule passed over carefully, the screeds must be cut out and the space filled in with stiff mastic, and the whole surface finished off with the hand float to a smooth face.

Mastic is composed of ground or powdered stone and silver sand, twice the amount of stone dust being required to the amount of silver sand, and about the fourth part of vitrified lead or a mixture of copper to the amount of silver sand, and the same amount of red lead can be added. This when mixed should be passed through a fine sieve and put in a cask till required for use. When it is needed, it is knocked up (moistened) with equal proportions of raw linseed and boiled oil, and the more it is knocked about the better, as it requires to be well mixed to get it to hold together.

Portland Cement Moulding

Running a cement moulding or cornice with a tooled stone-like finish (to imitate the face of stone which has been tooled up with chisels after being sawn in a block) is sometimes required. The process of fixing the rules and coring out the moulding is carried out exactly in the same way as in the ordinary manner when running mouldings, but a different method is adopted for the running off for the finish. After the moulding has been cored out and the muffle taken off, the running off is proceeded with. Washed sand and cement is gauged up to the proportion of 2 parts of sand to 1 part of cement; sufficient being gauged as estimated to fill out all the members of the moulding, with a little over. When the mould has been passed over several times, additional stuff is gauged half of sand and half of cement, which is the extra amount worked up allowed in the previous gauging. This causes the stuff to work up fatter, but not too much should be gauged at a time, otherwise the mixing of the previously gauged stuff with the new will cause it to set quicker, though it also causes it to work fatter. Cement and sand, when once set and used again, acts very differently than killed plaster, as cement and sand mixed up with fresh gauged stuff sets rapidly and gets hard; though if it is used in too big a quantity it will crack, showing fire or map cracking described previously. In cement work of this kind it is very important that only a small proportion of a previous gauging should be used in a new gauging for filling out the members. If there should be many large places in the drip, etc., the method of using the dry material as described under the running cement cornices can be adopted, but no other driers are used. When the whole of the members have been formed, the stuff is thrown on as liquid as possible and as it is nearing completion less sand is gauged with the cement, for in the final finishing of this kind of cement running the members have to be faced up to a smooth surface with the trowel and other steel tools.

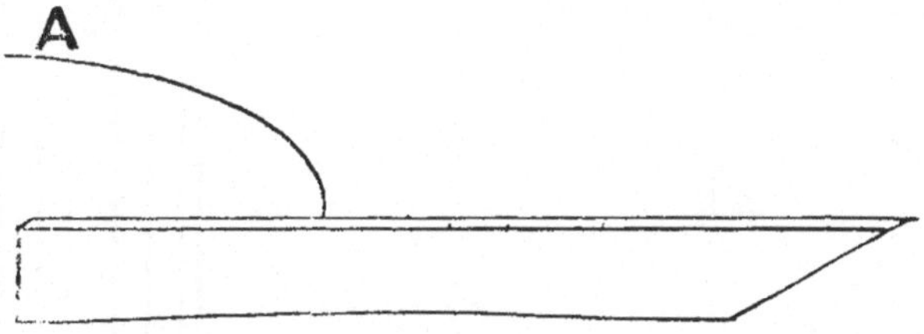

Fig. 109.

Treatment of Mitres and Returns

In running cement architraves, panels and other mouldings, it is necessary when filling out the moulding to fill in the mitres at the same time, working these out with a wood-joint rule. This is made of a thin piece of planed wood, about 12″ long, 2″ wide and about $\frac{1}{4}$″ or $\frac{3}{16}$″ thick, chamfered off both edges of one side (Fig. 109), and used in the same way as a steel-joint rule; working the heel of it (which is the back part, at A in Fig. 109), on the members of the moulding. This is very important, and it is necessary when filling out the moulding that every time the mould is passed over to take off any surplus stuff the mitres should be filled out with the same stuff and wiped out with the wood-joint rule.

A mitre being the joining of an angle (here the intersection of the member of a moulding), it will be seen that there are both internal and external mitres, and that these may be at different angles. With a panel moulding it is merely described as a mitre. In Fig. 110, A is the profile of the panel moulding, B the position of the moulding forming the panel, and C the intersection of the members. When the mitres have been well brought forward with the wood-joint rule, the steel-joint rule is used, as it requires the sharp edge for the finishing parts. It matters not whether the mouldings are finished off as a sawn-stone finish or a tooled face, the mitres must be dealt with at the same time as the running of the moulding, even to throwing the driers on the mitres in addition to the moulding when the mould is passed over. The joint rule would have to be used and passed over the members of the mitres, and in the finishing to a smooth face finish the mitres would have to be faced up in the same manner as the running of the moulding. This procedure has to be adopted with mitres because if the mitres were formed after the moulding is run, they would show up a different colour, however careful the plasterer was in mixing and gauging the materials. Here again we see a different procedure is adopted for the treatment of mitres in external work to that for interior work. In the latter, putty and plaster being used, the moulding and cornices are run off first and the mitres finished afterwards.

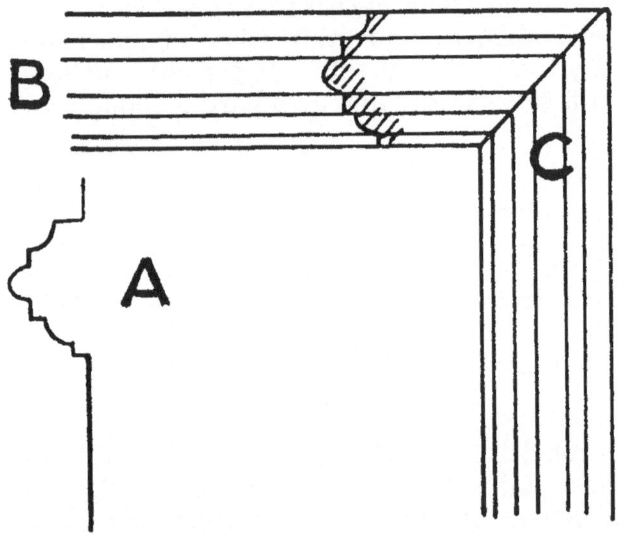

Fig. 110.

When using the joint rule for forming the mitres, the heel must be worked very lightly and carefully on the newly formed members, otherwise it would damage the finished mouldings. When running window heads, sills or cornices, etc., with returns, one very important matter is to mark off the finish of the wall line above the moulding and from this mark off the projection of the various members, so that when the running is proceeding the projections of the return is being filled out at the same time as the running. If each projection is marked on the wall with a pencil or piece of chalk, it will greatly assist the plasterers, acting as a guide as to how far the return needs to be filled out with stuff.

When the moulding is finished and the returns have not been completed, the members are worked out by the various length joint rules and the drag (a steel joint-rule with a saw edge), as very often the returns take some time to form and when left from the day previous become very hard, so that it requires the drag to comb it out. Another important point when working in returns is that they return in square from the face, either by holding a square on the face and trying it with a short piece of rule or an L-square (such as used by the masons), on the face. Fig. 111 shows the square and short rule in position for use (as was described for squaring in piers), also the mason square. In addition to the squaring,

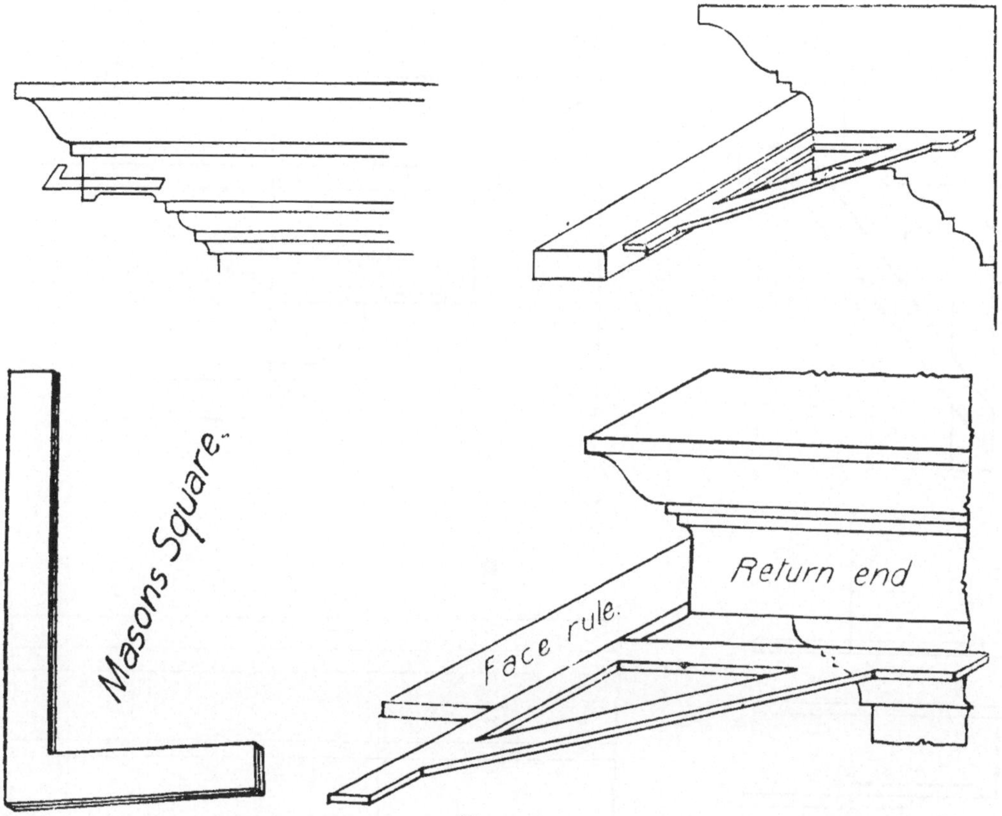

Fig. 111.

the members must be kept level, which can be done by placing the square against the upright face. In Fig. 112, A is the upright face, B the square, C the profile of the moulding at the external mitre for the return. In addition to being square and level, it must also be looked to that the mitre cuts correctly, as unless the projection and the members are worked out exactly the mitre would not cut (meaning the members would not show in the same line from the top members to the bottom at wall face). So it is necessary to hold up the plumb bob and line and sight the mitre cut. Fig. 113 shows section of cornice; the dotted line A is the upright wall. The finish of bottom member and the line with bob is held up to top member at B, and sighted down (that is, by closing one eye and boning the plumb line to the cut of the mitre). When all members of the return have been formed, soft stuff is covered all over, either by throwing it on with the gauging trowel, or if the members are hard the stuff can be laid on. Then with the joint rule the whole of the stuff is taken

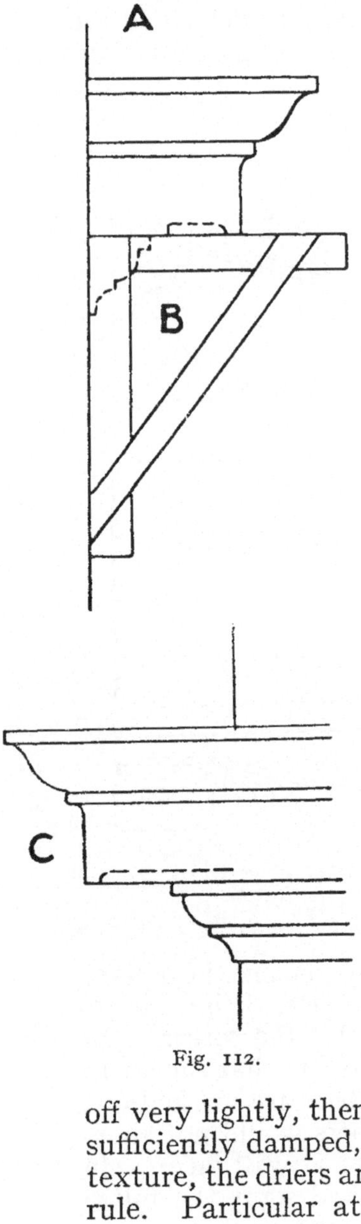

Fig. 112.

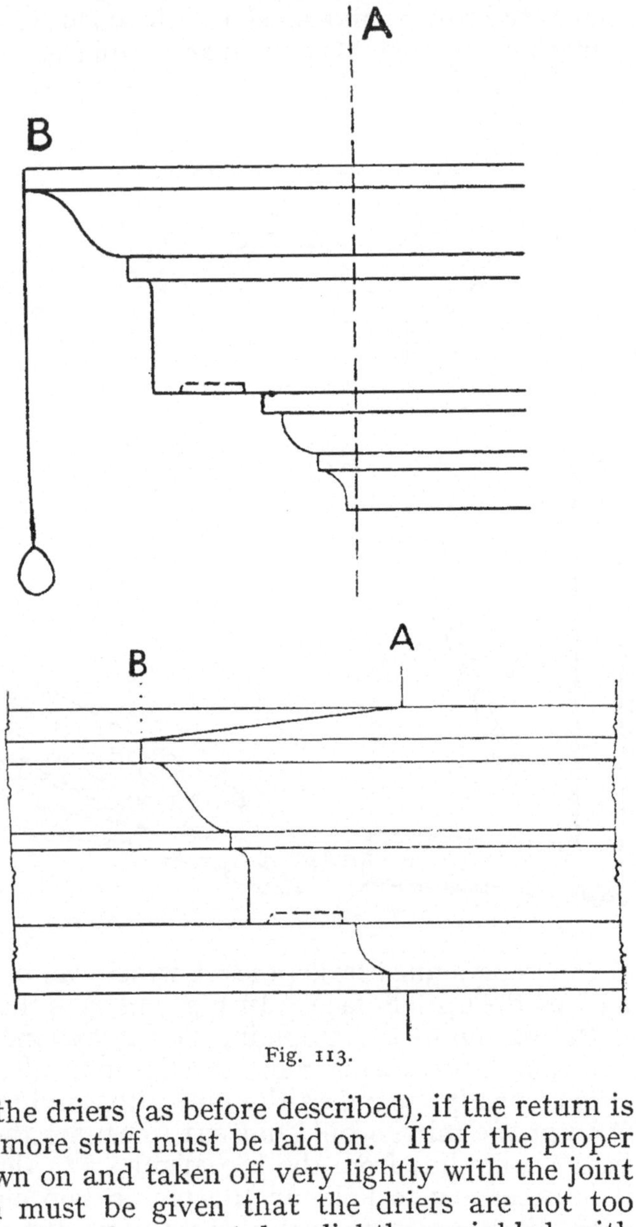

Fig. 113.

off very lightly, then with the driers (as before described), if the return is sufficiently damped, if not more stuff must be laid on. If of the proper texture, the driers are thrown on and taken off very lightly with the joint rule. Particular attention must be given that the driers are not too damp (though the sand is dry, they must be slightly sprinkled with water). This also applies to gauging of any materials for any returns which have to be put in after the front part of the moulding has been run, otherwise the return would show of a different colour.

Rough-casting Clean and Coloured Shingle, Marble Chippings and Brick-hogging, etc.

Rough-casting is a method of external finishing which is much in use. It is capable of considerable variation, chiefly due to the materials used, such as natural coloured or tinted sand, shingle, stone or marble chippings, brick-hogging, etc. The background preparation in all cases is, or should be, the same, though in speculative and cheap contract work not so much care is given to the preliminary stages.

As a rule, the brickwork or concrete having been well soaked, the walls are plumbed, screeded and floated in the ordinary way and allowed to get dry. If it is desired to rough-cast with clean shingle or pebble-dash (as it is sometimes called), gravel of anything up to ½″ is washed through a coarse sieve to remove earth and dust and to throw out larger stones not required. It is then laid out on clean sacks or boards to drain.

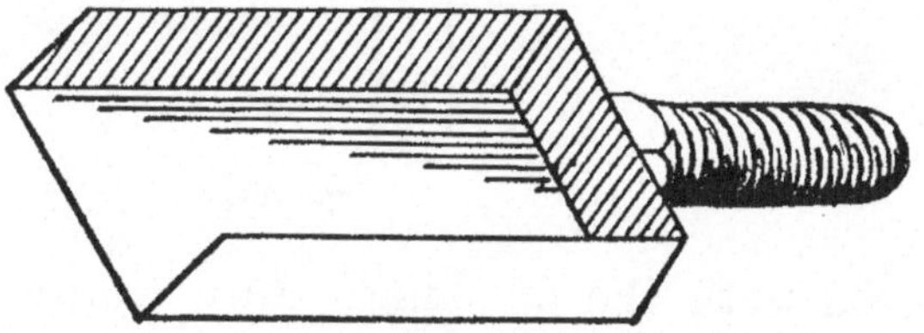

Fig. 114.

Throwing on of the Rough Shingling, etc.

When the floating is ready for the finishing with the clean shingle, it is necessary to gauge the laying or bedding coat the previous afternoon, because it is found that it clings so much better when knocked up again next morning than if fresh gauged stuff is used, and as the floating has to be well saturated with water before laying on, if fresh cement and sand were used the plasterer would find the laying on stuff would start running down, then great difficulty would be experienced in throwing on the shingle to make it hang together. Enough material should be gauged to complete the job. It should be gauged 4 parts of sand to 1 part of cement and 1 part of lime coarse stuff. This should be well mixed and gauged very soft, seeing it will be lying by till the next morning, when it is softened up and an additional ½ part of cement is added.

Everything being ready, and the floating sufficiently wetted so that the face appears to have a damp surface, a fair thickness is laid on and a thin rule is drawn over the surface to get a regular thickness. This should not be less than ½″ thick. When a sufficient amount has been laid on and squared up, the shingle can be thrown on in several ways. The plasterer can hold it in his apron, or a sieve, or a small, shallow box, and use a small scoop for picking up and throwing. Such a scoop is easily made with a piece of zinc, about 6″ × 5″, with the sides turned up about 1½″. This can be nailed on to a piece of batten 2″ × ½″ and shaped off at the end for a handle as in Fig. 114. Many plasterers use their laying trowel, as in Fig. 115, which shows the thrusting in the trowel and the use of the hawk. This method is adopted when using the wet coloured shingle or crushed or broken brick. As there is no jointing in rough-casting, it is necessary to complete the whole part required to be dashed in the one day. In throwing on the shingle, assuming it is done by a right-handed plasterer, it is thrown on from the left to right.

There is a certain knack in throwing on, for it is desirable that each scoopful or trowel load should be widely but evenly spread. This is attained by a twist of the wrist. If this is not carefully done, the dash will be unevenly distributed, which is disastrous to appearance and protective properties. Before beginning, as much scaffolding as possible must be dispensed with, so that no time should be lost in dropping scaffolds to proceed with the finishing. Any pud-log holes should be filled in, as previously described in the finishing of cement facades, before dashing. If all the foregoing instructions are carried out there should be no trouble in having a good finish to the rough-cast.

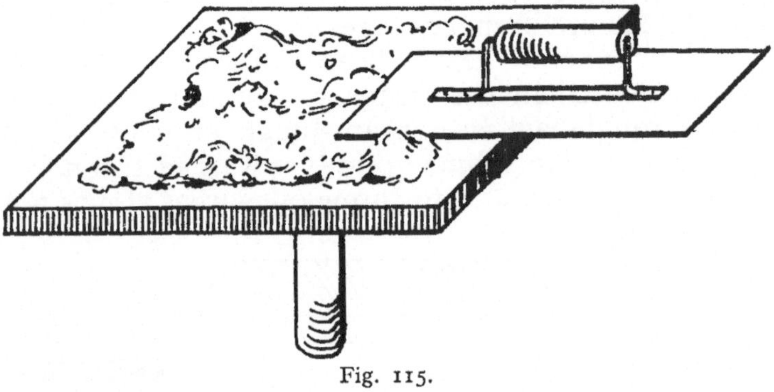

Fig. 115.

Joinings in Rough-cast Finish

When rough-casting with shingle or any other material, one very important point is the joining up (or the shet, as it is sometimes called in cement work), needed after lowering the scaffolding, or the break for meal-time, or any other cause. Unless it was absolutely impossible to finish, it is always best to complete rough-casting without a joint in the finish. However, in the ordinary joint which occurs after short delays, the proper thing is repeat the soaking of the floating as before and to carefully regulate the throwing on of the first scoopfuls. If necessary, a brush of water is thrown on at the joint of the previous work. Another tip in joining is to cut the bedding off with an irregular edge, as then the joint is not so noticeable. Fig. 116 shows the irregular jointing cut off.

Fig. 116.

Angles or Arrises to Reveals, etc.

When reveals to windows or external angles require to be done, the reveals are treated in the same manner as in the finishing with the ordinary cement work. A small band of stuff should be laid round the outside of the reveal and a feather-edge held up on this face. The laying on stuff should be laid on the reveals, and before the rule is taken down the rough-cast material should be thrown on. Then the rule should be drawn away very carefully. After the rule has been taken off, if necessary a little water can be thrown on with the brush to soften the band of stuff put on. When this has been done, the rough-cast is thrown on to complete the angle on the face and the edge would be cut off to form the irregular joint for the remaining part to finish up against. For external angles other than to reveals, etc., this will be found a very good method, as though material used for rough-casting is rough, it is necessary for the arris of the angle to have a regular line.

Coloured Rough-cast and Other Materials

It is sometimes desired that the pebble or shingle dashing should have a grey or yellow or a white colour. Then the gravel or shingle should be washed to clean it, and when well drained it is measured. A mixture of 4 parts of shingle to ½ part of Portland cement will give a greyish tint, and more cement can be added to darken it. It is then thrown on in the usual way. If a white tint is required, instead of cement being added, ground or air-slaked lime is added ; 1 part of lime to ¼ part of Portland to the 4 parts of shingle ; the whole being well mixed and well damped, but not to excess, otherwise the lime and cement would drain away from the gravel.

Crushed Gravel

When crushed gravel or Thames or other river ballast is used, it is necessary to have two different-sized mesh sieves. A coarse sieve to the required dimensions, and a fine sieve to take out the flour (as called), the dust caused by crushing it, and what remains in the sieve would be the required size for the rough-cast.

Marble or Granite Chippings

Crushed marble makes a very nice rough-cast finish for the outside of buildings. Only this material is of a larger size than gravel. I have used this material in black and white about ¾″ rough, and mixing 2 parts of white to 1 part of black, as I found it gave a much better appearance when finished, having more white than black. When using this material it is well to wash it and have it in a sieve for throwing on, so that it can be continually wetted before using. The cleaner it is kept when being used the better the appearance it has when dry. The same applies to any clean dashing. Apart from appearance, the cleaner it is the better it adheres to the bedding coat. With coarser material, such as this kind, the laying on or bedding coat should not be less than ½″ thick.

Broken or Crushed Bricks

Broken or crushed bricks (sometimes called hoggin) is used for external finish to buildings of the bungalow type, but when this is done it is usually mixed with lime or putty. This is a larger material than gravel or most other material, as bricks (generally old ones) are broken or crushed. If crushed brick is used, it is sifted with the two sieves, as in the case of crushed gravel, to take out the dust, etc. It is well soaked with water and measured out 4 parts of crushed brick to 1 part of air-slaked lime or lime putty. This is well mixed and is thrown on in a semi-fluid state. The hawk can be used for this operation as in Fig. 115, or a small light box of a convenient size could be substituted.

As this dash is a kind of a slurry form, it is a cleaner method to use the hawk or a box than a sieve. This material being of a coarse nature, the laying on or bedding coat to receive it should be $\frac{1}{2}''$ to $\frac{5}{8}''$ thick, to provide a good bed for it to adhere to.

Derbyshire Spar

This mineral takes its name from the county of Derbyshire. In appearance it is much like quartz (which is a rock crystal), and has a fine glitter. It is used a good deal in the Midlands, intermixed with other materials. Usually it is applied as crushed bricks or marble chippings ; though if it is added to shingle, etc., it is treated as described for Shingle Dashing.

Economy in Rough-casting

When using marble chippings, broken bricks or any materials of a coarse nature, it is not necessary to float the walls. All that is required is to render a thin coat of cement and sand, gauged 4 parts of sand to 1 part of cement, all over the brickwork and leave it rough with the trowel, or brushing with an old birch broom and then apply the laying-on coat as previously described.

Damp-proofing

It is often necessary for plaster on walls, ceilings and floors to be damp-resisting, if not waterproof ; in some cases as a preventive against damp to wood floors, etc. For this purpose concrete floors are screeded with cement and sand with various kinds of anti-damp or damp-proof patent materials, some of which are in a powder and others in a liquid form.

Colmanite

In the liquid form we have Colmanite, which is easily handled ; a tenth part of which is added to a three-gallon pail of water for ordinary work, though more can be added in bad cases. When running cornices with white or other cements, Colmanite, apart from acting as a damp-proof, greatly assists in the hardening of cements.

In powders we have Majauca, Pudlo, Prove-it and Toxcement. Powders require mixing with cement and sand before they are gauged up, as they are of a very light nature. A small box or a quart beer-can will serve for measuring out the powder, as a quarter of a pint of powder is sufficient for two bushels of gauged materials.

Majauca

Majauca is a white powder. When mixed with coarse stuff and cement it makes a good damp-proof plaster for internal walls which are below the ground level, and greatly assists in keeping back any dampness that might percolate through.

Toxcement

This is another powder. When being used it should be sprinkled over the cement and sand by means of a small and very fine hair sieve, such as is used by a painter for straining paint. This distributes the powder much better than dumping it on, as it is very fine, much like snuff and of the same colour.

Pudlo

Pudlo is similar to Majauca in colour and texture, and of the same strength. It is of a light nature, and should be well sprinkled over the material for mixing purposes, otherwise amalgamation will be difficult.

Prove-it

This is another material having damp-proofing properties. It is not unlike Pudlo. In some tests it has been proved highly damp resisting.

Old Work

In the majority of cases *damp-proof* plastering is called for in the attempt to remedy defects—that is to say, in order to arrest the infiltration of moisture through soft bricks, or the rise of ground water in walls that have not been provided with *damp-proof* courses. As a rule, such walls need the attention of the architect and builder, so that measures may be adopted to stop the evil at the source of origin. As far as plastering is concerned, if the walls are already covered with plaster, which will naturally be defective, the first thing to be done is to hack off the old plaster, and well clean out the joints, especially at the lower and top parts. The walls should be well brushed with a hard broom. Then take 2 parts of cement, 1 of sand and 1 of coarse stuff, adding the necessary amount of damp-resisting powders. This amount will vary according to the dampness of the wall. When well mixed, the materials should be pressed into the joints with the laying trowel, and then the whole surface rendered over and allowed to remain till it is hard, but it should not be left too long to lose the power of suction. When it has been floated out and allowed to get hard, the surface not being left too rough (as that would incline to make it porous), it can be finished off with Keene's cement, adding a small amount of damp-proofing powder. As soon as the Keene's cement is sufficiently set to take the brush, it should be painted with good white lead and turps. This is for interior work. If damp-proofing is required externally, cement and sand should be used with the powder or liquid, and the plastering finished, if possible, in the one coat, after the rendering has been allowed to get hard.

For interior damp-proofing cement and sand are not at all suitable materials, as such a mortar is so soon affected by atmospheric changes. Being a hard face plaster, the least condensation causes the surface to appear as if bathed in moisture.

Ironite Cement

Ironite cement is similar in colour and weight to iron moulders' sand. After having been used it is of a pale black colour and is heavy. It can be mixed dry with Portland cement, in the proportion of 25 lb. of Ironite to 75 lb. of Portland cement, ready for gauging in small quantities when required. To achieve success, about 1 lb. of Ironite to 3 lb. of cement should be used for each square yard covered. On old walls the joints should be well raked out, the brickwork well brushed and the cement applied as a thick grout, making sure that all surfaces to be covered are well brushed and damped before the material is applied.

CHAPTER VIII

FIBROUS PLASTER

Origins—Widely Applicable—Methods of Manufacture—Enrichments—Application of Enrichments—Fixing—Combining Modelling and Moulding.

ORIGINS

IT is always difficult to state with certainty who was the originator of any particular process of construction which is universally adopted in building. An idea germinates, and is tried out by its author who, as a rule, will not be long in discovering that a similar idea was already in being many generations, or perhaps even many centuries, before it occurred to him. And the history of the original invention is probably also the history of his own—no sooner has the crude idea taken practical form than some other eager inventor sees its imperfections, improves upon it, and forthwith claims the whole credit for the invention for himself.

This was so in the case of the famous stucco composition used with such delightful effect by the brothers Adam. The medium, which formed the basis of their wall and ceiling decoration, is reputed to have been invented by a Swiss clergyman named Liardet; and yet it is claimed that when Robert Adam brought Pergolesi's designs from Italy he brought his " pastiglio " or paste composition with him. So it is with fibrous plaster. It was a French modeller, Leonard Alexander Desachy, who first patented the process of fibrous plaster work, in the year 1855. But fibrous plaster was known to the ancient Egyptians, who wrapped their mummies in canvas steeped in burnt gypsum; and it was doubtless through reading the French writers on Egyptian arts and architecture that Desachy took up the idea of reviving the process and of introducing it into England.

This enterprising French modeller established a business in London for the manufacture of fibrous plaster decorations, and enlisted the patronage and financial support of Owen Jones, the author of *The Grammar of Ornament*. Desachy, who employed a large number of hands, male and female, carried on this London business for some time, but at length was compelled to relinquish it and return to Paris. The business was then carried on for a while by two of his former assistants, J. M'Donald and R. Hanwell; but it was eventually taken over by Messrs. G. Jackson & Sons, who acquired the then existing patents, and have since introduced many improvements and brought fibrous plaster work to a high degree of perfection.

About sixty years ago Messrs. Jackson & Sons successfully defended their rights, which they considered had been infringed during the building of the old Oxford Music-hall. Fibrous plaster, however, is now open to all, and many firms who specialise in this class of work have their own methods of construction.

Widely Applicable

Fibrous plaster received its first great impetus at the Paris Exhibition of 1878, when the ceiling panels of the principal edifice were constructed in this material. In this country it was brought into prominence by, among other well-known decorators, Mr. William Millar, who constructed the ornamental parts of various architectural details in Shakespearean scenes for the productions of leading managers in this medium, carried out much plastering work in public buildings and private building, and notably executed fine repair work on ancient plaster ceilings in Scotland. It was also introduced into the Exhibition buildings at Earl's Court and elsewhere; while its latest adaptation to this class of building on a large scale was at Wembley in 1924 and 1925, in the Indian, Australian, Canadian and several other important buildings.

It is, in fact, a branch of the architect's, builder's and decorator's art that grows in demand every day. Theatres, music-halls, cinemas, and large hotels and restaurants throughout the country afford very fine examples of this branch of craftsmanship. Fibrous plaster lends itself to such uses because it can be cast and finished in workshops, but specially owing to its lightness, which makes it easy of transport, and the rapidity with which the job can be carried out. The popularity of period rooms on steamships and in large public and commercial buildings is also tending every day to extend the use of this material; and the reproductions that one so frequently encounters of fine examples of the craftsmanship of the past, while being beautiful in themselves, serve to keep alive an interest in that fine solid work in plastering which is now but too rarely seen.

By the use of fibrous plaster, the ordinary lath and plaster can be dispensed with, as ceilings with elaborate mouldings and enrichments are cast in large sections and screwed on to the joists. Old ceilings may be covered or enriched in this way, any joints being screened by fibrous mouldings or ribs, which will form a plain panelled ceiling at small cost.

No matter what form or design may be required, fibrous plaster will meet the need both readily and economically. Its manufacture, as we shall see presently, is to a large extent mechanical, and the production of good effects is therefore rather a matter for the designer than for the actual craftsman—although, of course, the personal element enters into this branch of the plasterer's art, as in all forms of handicraft. The architect who designs a fine interior, whether on classical or original lines, is sure that his ideas will be carried out without his having to search for a craftsman of special artistic ability who is capable of turning out work equal to that of the old plasterers of the sixteenth and seventeenth centuries.

Fibrous plaster is a matter of pre-casting which is carried out in the workshop, and is therefore economical in production. In many cases models of classic design are kept in stock, so that it is often possible, in the case of a rush job, to order columns for the clothing of steel stanchions, and these fibrous plaster columns, which are supplied in sections, can be fixed with the greatest rapidity. Ceiling and wall panels, ceiling roses, enrichments in the forms of swags, wreaths, etc., can also be supplied in this matter, and it is therefore possible to finish the interior of a building in fibrous plaster in a mere fraction of the time that would be occupied by craftsmen working *in situ.*

Such headway has the use of fibrous plaster made in the past generation that it may not be too much to say that the interiors of the great majority of large buildings erected nowadays are finished in this form of plaster work. For temporary work, such as exhibition buildings and crush rooms which are sometimes built as an annexe to some large hall or other public building, fibrous plaster is invaluable. In many cases buildings, whose framework is constructed of timber or steel, are finished both as regards the interior and exterior in this manner.

The influence of the weather is, of course, more noticeable on the exterior of such buildings than on the interior, thus necessitating the keeping of a watchful eye on the structure—but the cost of maintenance is not excessive. For interior use, fibrous plaster may be reckoned upon to last as long as plaster work carried out in the ordinary way.

The fine buildings at Wembley that were finished in this manner, notably the Indian Pavilion, testify to the beautiful effects that may be obtained by the external use of fibrous plaster ; and it has this additional advantage, that when the first purpose of the buildings so treated has been served, they may be taken down, removed, and re-erected on some other site.

Buildings are still standing intact and as new after many years, having received only the ordinary care and repair necessary for buildings styled permanent. The main point to ensure the life of fibrous plaster is to protect cornices and similar projections by suitable weatherings—such as zinc, asbestos sheets, felt, etc. Providing fibrous plaster is painted periodically, it is proof against decay.

Methods of Manufacture

In early books on the subject, fibrous plaster has been placed in the same category with carton pierre and papier mâché. Papier mâché as a medium for the internal decoration of buildings has to a very large extent been superseded, but carton pierre is still used, although not to such a great extent as in the past.

During the last generation there has been a simplification in the variety of materials used in the construction of fibrous plaster. Each workshop, of course, has its own formulæ and its own methods, but whereas two or three decades ago the workers in fibrous plaster were using a dozen or more different kinds of material in the construction of their work, the present-day practice, as a rule, is to use little more than plaster of Paris, canvas, size, and wood laths or reeds, with tow for small and fine work.

As we have said, every studio or workshop in which the manufacture of fibrous plaster is carried on, will have its own methods of workshop practice, but most of these will follow very much the same procedure. Working drawings of the mouldings, or sections of mouldings, to actual scale having been received from the architect, tracings of these are made at the works. In the case of what may be described as straightforward work, such as panels, columns, etc., the procedure is fairly simple.

For panels, whether for wall or ceiling, a zinc moulding is made from the tracing, and this is strengthened by a wooden template, the laths being placed at intervals according to the depth of the moulding. The panels are usually run out flat on the bench.

In the case of columns, the work of moulding is more complicated The tracings having been prepared, the first step in the process of manufacture is to make a wood template in accordance with the architect's design for the purpose of forming a mould. This mould, which in shape is the half section of a column, is then made in plaster from the wood template. Rods, also cast in plaster, to form the flutings are then laid in the bed of the mould, the flutings being formed by this means in the column sections by reverse moulding.

When the mould, with its fluting rods in position, is dried off sufficiently for use, the scrim, or canvas, generally of $\frac{1}{4}''$ mesh, is laid in, and the plaster is then poured. The plaster is evenly and lightly distributed by the spinning action of the mould which revolves by means of a shafting attachment in a similar manner to the rotatory motion of the ordinary lathe. The spinning action is continued until the plaster is evenly spread, the mould then being brought to rest and the plaster left to set, when the finished section of a column is removed.

Enrichments

In the making of large enrichments, such for instance as an Adam, medallion for wall or ceiling decoration, the process is more intricate and calls for greater artistic skill. In such a case the design is first modelled from the drawing in clay by a modeller. It is interesting to watch the development of a model. The modeller first draws an outline of the design on a slab of clay, and then proceeds to build it up by hand. Taking small lumps of clay as required, a touch here, a pinch there, and the design gradually evolves under the skilful fingers of the modeller from a shapeless mass to a finished work of art.

From the modeller the design goes to the moulder. A gelatine mould is taken if the design is for high relief work, or a plaster mould if the design is low relief. The model is put into a bed of clay, and liquid gelatine or plaster is carefully poured over it. From this mould fibrous plaster castings are taken.

A coating of plaster is first applied to the mould, then a layer of canvas and laths, and finally another coating of plaster. The casting is allowed about half an hour to set, and is then removed from the mould, and any little blemishes caused by air bubbles are filled in.

Application of Enrichments

Enrichments made in this manner are applied to fibrous plaster panels or columns, as the case may be, by roughing the surface at the back of the enrichment, and making fast by screws to joists or studs or plugs previously inserted in the wall, as may be necessary.

Fixing

Fibrous plaster slabs, cornices, etc., are fixed by means of French nails or screws, driven through the material into wood, soft brick or porous concrete at the back. Nailing is permissible for temporary work, although even then it has the disadvantage of possible damage to background or ornament as the result of concussion from the hammering. Screws are much better. They should be countersunk, the heads varnished with shellac, to prevent rusting, and the holes stopped with fine plaster.

When two sheets have to be joined they should be joined at an angle in opposite directions on each edge, so that they fit together closely, the edges being glued and the joint stopped with a thin wash of plaster.

Nails and screws must, of course, be driven into woodwork (joists, studding, etc.) or soft bricks. When the ordinary wood framing is not sufficient to provide adequate fixing points, fillets must be fixed where required, and for heavy cornices, covings and similar sections appropriate wood brackets or cradles must be provided. Hard brick, stone and concrete walls will require suitable plugging to receive the screws, or to which fillets may be attached.

Combining Modelling and Moulding

Of course the highest art in decorative plaster is hand modelling. This, however, is often too expensive, as it demands much experienced labour, particularly when heavy and intricate designs are involved. Such work can be carried out far more expeditiously and cheaply by means of carefully prepared fibrous plaster, similar to that which we have been considering, and this will satisfy most customers. In some cases a superior style is required, although the full cost of hand work is not possible, and this can be secured by combining hand modelling with moulded work. We give in Plate V examples of hand-modelled plaster panels and ornament, which were inset, or planted, in elaborately moulded plaster ceilings. In this way individuality is introduced and the unavoidable hardness of cast repeat ornament avoided. The panels, flowers, leaves, fruit, festoons or other ornaments are modelled separately on the bench as distinct items. When heavy, copper wire, bent double to form a loop, is embedded in the plaster from the back, with the ends left protruding. The plaster of the cast fibrous slabs is cut back where the ornament is to be placed to the canvas and the ornament fixed in the depression, either by gluing alone or by gluing plus the fastening by means of the copper wires, which are driven through the canvas, bent over and sealed with plaster, thus protecting the fastenings. On the face of the slab the jointing of the planted ornament and the background is smoothed over with fine stucco of the same quality as used in the main work. This admixture of workmanship can be carried out both with specially designed and with stock sheets of fibrous plaster. Thus one good modeller can give the finishing touches which lend originality to the decoration in a short time and at a comparatively low cost.

In the examples given, the Sutherland coat of arms was modelled in stucco applied to a plain fibrous plaster panel. Obviously this modelling could be in high, medium or low relief as desired. The companion panel shows combined hand and cast ornament. The rowan leaves and berries were cast in plaster as separate items and planted in position on the plain fibrous plaster panel. Then stucco was applied and the flowers, leaves, fruit, birds, etc., modelled direct *in situ*. These panels which are in Dunrobin Castle, measure 12″ by 8″.

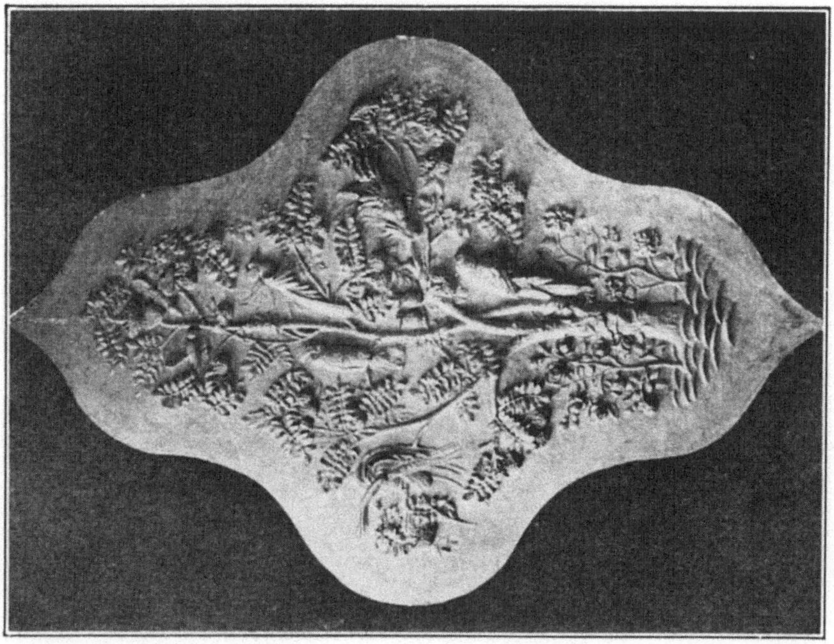

MODELLED AND CAST PLAN ENRICHMENTS COMBINED WITH FIBROUS PLASTER SHEETS.

(By courtesy of *Building*.)

PLASTER CEILING BY SIR CHRISTOPHER WREN, ST. CLEMENT'S CHURCH, EASTCHEAP, LONDON, BEFORE RESTORATION.

CHAPTER IX

REPAIR WORK

Interior Work: Repairing Ceilings or Partitions on Lath-work—Filling in the Patch—Material for Lath-work Patching and Floating Out—Finishing Patches—Finishing Patches in Putty and Plaster—Expanded Metal or Lath-work—Materials for Patching on other than Lath or Wire—Floating Patches with Keene's and Sand—Repairing Walls with Cement and Sand—Cutting Out Cracks on Lath-work, Brickwork, Concrete and other Solid Ceilings—Cutting out and Filling in Cracks in Walls or Concrete Ceilings—Filling in Cracks with Wads—Repairing Mouldings or Setting Coat only to Walls and Ceilings—Repairs to Mouldings—Taking the Squeeze—Fixing the Running Rule—Taking a Squeeze off a Cornice having a Cove—Taking a Squeeze off a Moulding having an Enrichment—Casting Enrichment for Repair Work with Wax Moulds—Bed Mould Enrichment—Heating and Pouring the Wax Mould—Taking Off the Mould—Repairing Wax Mould for Casting—Taking Out Casts from Wax Moulds—Repairs to External Plastering—Repairing Cement Work with a Jointed Surface or Plain—Cracks in Mouldings—Repairing Cement Floors.

THE repairing of plaster work requires more attention and time than the average person imagines. In repair work the plasterer has many minor details to attend to preparatory to the actual filling of the plaster work. It is attention to these small details that tends to make the repairs a success, whether it be on interior work on wall, on lath-work to ceilings, stud partitions, concrete ceilings or floors. The same thing applies to exterior work, though with this there is not the variety of jobs that the plasterer meets with as in interior work.

INTERIOR WORK:

REPAIRING CEILINGS OR PARTITIONS ON LATH-WORK

In repairing patches in ceilings or partitions on lath-work of any size, the plasterer must use a certain amount of discretion as to how much round the patch he should take down. He must not think because it appears weak it should be taken down, as very often work that has been plastered on thin laths naturally has a certain amount of give or spring in it when pressed with the hand. Sometimes, too, the joints or studs are wide apart, which causes the laths to be weak, owing to the weight of plaster and the long bearing. It is not advisable to have the joint to the old work on the joist or stud, because the material would not get such a good key through the laths. In patching or repairing, an irregular joint is not so noticeable as a round or square patch, which some plasterers insist on adopting.

FILLING IN THE PATCH

All loose plaster round the patches should be taken down, and the spaces between the laths which are to form the key should be well cleaned out with the gauging trowel, at the same time knocking the lath-work with the flat of the trowel to shake off any dust. The edges of the old plaster should be well trimmed back from the front edge to form a dovetail for the material in the new patch. Fig. 117 shows the outline of a patch, also the section, A being the face of old work, B the lath-work, C the dovetail from the front edge. Any loose laths should be renailed and all broken ones replaced. The whole of the patch should be well brushed with a dry duster, also well dusted out round the edge. In first-class repair work—that is to say, other than repairs in the ordinary way—the edges should be painted with any ordinary paint. This helps to prevent the dampness from the new patch spreading out on the existing old work. Apart from this, it stops the absorption of moisture from the new material by the old plaster at edges, and helps to make a better joint to the old. New patches will be apt to fall out, or crumble at the edges, if the moisture in them is absorbed by the old plaster.

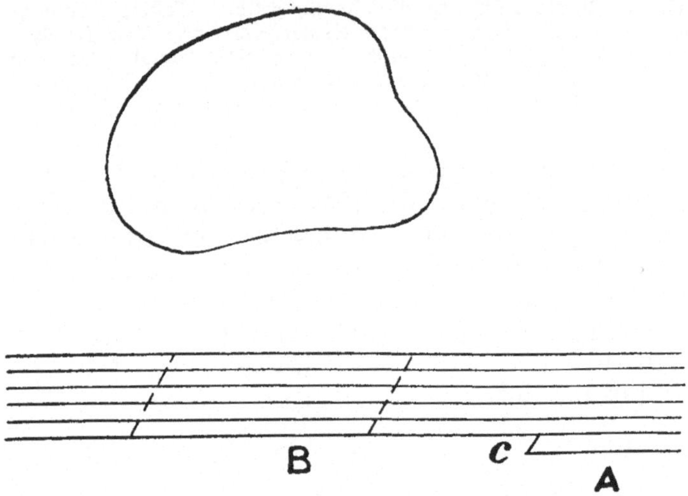

Fig. 117.

MATERIAL FOR LATH-WORK PATCHING AND FLOATING OUT

Lime and hair coarse stuff is gauged strong with plaster, as small patches are filled out with the one coat. The coarse stuff is opened out into a ring, and water is poured to the amount estimated to be absorbed by half the quantity of plaster. When gauged and not too soft, it is spread over and across the laths. A notched rule is placed about $\frac{1}{8}''$ above the old surface for the ruling off of the stuff. After laying on the amount gauged, it should be scratched before it sets, under-cutting the scratching. This gives a good key for any further material required to fill out the patch. The patch being filled out, it is combed or devilled.

Finishing Patches

In the finishing of patches, Keene's cement and putty and plaster are mostly used. For a quick job, where decoration is to follow, it is necessary to use Keene's, and this must be followed on immediately the plasterer has finished by the painter, as it requires a good sharp coat of colour. When finishing patches in Keene's, if a small place the plasterer gauges the material on the spot board, the Keene's being opened out in a ring, water poured in, and it is then mixed up with the trowel. If a large quantity is required, the labourer gauges it in a pail. About half a pail of water is needed and the Keene's is sprinkled in by hand, and when soaked it is stirred up with a piece of batten 2″ × ¾″ by about 2′ long, and shaped to a handle at one end. The Keene's is laid on with the laying trowel, keeping the edge of the joint well down to the old work. A rule is passed over the stuff, working the two ends over the old work. Several patches should be laid on with the Keene's, and when well filled out with the rule and allowed to get firm the long joint rule should be passed over. This is to make sure the finishing coat is flush with the old work. The patch is then laid down with the float and rubbed up with the cross-grained float, assisted by the use of the stock water brush. The patch having been rubbed up, it is laid in tight with the laying trowel, and trowelled in, the water brush being passed all round the edges. The joint rule should again be passed over and round the joint to ensure the joint being flush.

Finishing Patches in Putty and Plaster

In ordinary repair work lime-putty and plaster is generally used. This applies to any work which is likely to remain for some length of time before decorating. The putty being a fine material, the plasterer is able to make a good job of the joint, which is one of the chief matters in finishing patches. The putty should be gauged strong with plaster, and used before it sets. No more than can be used before it sets should be gauged. When using putty and plaster for finishing, the floating should not be much over $\frac{1}{16}$″ less than the old work, as with putty and plaster it is laid on tight with the trowel and trowelled off. No rule is necessary if the patches are properly floated, as putty and plaster when gauged of the consistency of cream can easily be manipulated with the skimming float.

Expanded Metal or Lath-work

In all classes of work, whether it is on lath-work to ceilings or partitions, or breeze or other kinds of partitions, or concrete ceilings, the method of finishing the patches, whether in Keene's, putty and plaster, or any other material, is the same as previously described. A little variation can be adopted on filling out of the patches in the floating, also in the material.

The following method is usually adopted. The patches are prepared (as previously described), and coarse stuff (gauged 3 parts to 1 part of cement) is rendered over the laths and scratched with the lath scratcher. The rendering is allowed to stand to become dry, and when dry it is floated out with 3 parts of coarse stuff, 1 part of sand and 1 part of cement. This material is applied to lath-work, though the method is suitable for any patching.

Materials for Patching on other than Lath or Wire

When repairing old walls the best job is made by rendering the brickwork and letting it stand before floating out for finishing. For this the following materials should be used. To 3 parts of sand add 1 part of coarse stuff and 1 part of cement. When rendering the walls, a lath scratcher should be used. The same materials used for floating Keene's can be applied in the finishing, or if it is ordinary repair work, a small amount of setting stuff, such as 1 part to 2 of Keene's, can be used (providing the work has not to be painted). In repairing old walls, all loose plaster should be hacked off and the joint trimmed round (an old wood chisel is very useful for this operation). The patch should be well brushed with a stiff brush, and afterwards well wetted. When patches are required to be finished immediately, the best method to adopt is to gauge coarse stuff and plaster in the same manner as described for lath-work and finish in Keene's. The gauged coarse stuff answers all purposes for the repair of inside work, whether it be on lath-work or brick or expanded metal, and it can be relied upon to make a good job, but it must be gauged strong with plaster.

Floating Patches with Keene's and Sand

Repairs to walls or ceiling are sometimes specified to be floated in Keene's and sand. If such is the case, lath-work should be rendered with not more than 2 parts of sand to 1 part of Keene's and a small quantity of well-beaten hair. On walls not more than 2½ of sand to 1 part of cement. In finishing these patches neat Keene's only would be used.

Repairing Walls with Cement and Sand

In repairing patches with cement and sand, the mortar should be gauged 4 parts of sand to 1 part of cement. When floating with cement and sand, it should always be finished in Keene's, though a great deal of Sirapite plaster is used as a finish on a cement and sand background; but Sirapite is not such a good material for repair work as Keene's. In repairing walls in sculleries, larders, vaults, etc., with cement and sand, the best finishing material is cement and setting stuff, gauged 2 parts of cement to 1 part of setting stuff.

Cutting Out Cracks on Lath-work, Brickwork, Concrete and other Solid Ceilings

When repairing cracks on lath ceilings or partitions they should be carefully cut out, but not more than 1½″. The repairing of cracks is very often unsatisfactory, because the cracks are caused through vibration from the floors above, owing to wide joists or long spans between walls. Another cause is a certain amount of settlement taking place in the building. Unless the cracks are carefully filled in, the various causes just mentioned would, in all probability, cause the new work to crack again each side at the joint.

The best method of cutting out cracks is to use a short, stumpy knife and hammer, using them as a glazier cuts out old putty when putting in new glass. The cracks should be cut well back each side from the edge. This gives a better key, and is called birdsmouth. Joints must be dusted out and well wetted and filled in with strong gauged coarse stuff, the stuff pressed to each side and through any part of the laths visible. In cutting and repairing cracks, the same sort of material is used for filling out and finishing as is used in repairing patches. In the filling in of cracks less than 1½″, the best way is to use putty and plaster with well-beaten hair, cleaning out the cracks with the trimming knife. Very important points when filling in cracks are to allow the roughing-in coat to stand as long as possible before finishing, and to paint the edges if decoration is to follow immediately after.

Cutting Out and Filling in Cracks in Walls or Concrete Ceiling

Cracks on plaster walls, or concrete or other materials, should be treated in the same way as for lath-work, and the same materials used as described in patching, other than lath or wire, in the previous chapter. In cracks of this description clout-headed, galvanised nails can be driven in the joints of the brickwork, and where possible in the concrete, and then filled in. This forms a good key and, if the material is well pressed in the sides, this method makes a very good job.

Filling in Cracks with Wads

Another method for filling in small cracks on solid ceilings or walls is, after treating them as before with the trimming knife, brushing and well wetting, to cut some narrow strips of canvas and soak them in a bowl of plaster (gauged round), and tuck this well into the crack, and then drive in clout-headed nails. This is called wadding, and the strips of canvas when soaked in plaster are called wads. These cracks should be finished off with neat plaster gauged in a bowl.

Repairing Mouldings, or Setting Coat only, to Walls or Ceilings

Sometimes in repair work the plasterer finds the setting coat is loose, or scaly, and the more he tries to finish the patch, the more it comes away. I, myself, have had to take the whole of the setting coat off large ceilings, scaling it off with a stripper like stripping paper. When a job of this description is met with, the whole surface must be well-brushed and then the water brush passed all over. A wash of Portland cement is made just thick enough to be washed with the brush over the surface of the old floating. Cow-dung is also very useful. But the most convenient material, one which I recommend, is cement. Size water can be used, but it leaves too much of a soft surface, and when the wet material is used it softens the face of the size. The finishing coat can be done in putty and plaster, and kept as thin as possible; or in Keene's gauged 2 parts of Keene's to 1 part of putty and a small quantity of sand. It should be laid on as tight as possible with the laying trowel and rubbed over with the float, and then laid down with half and half of Keene's and putty. In repairing work of this kind two plasterers are needed, as the dry old work has a great deal of suction and a small portion only should be done at a time, as it is much the better and easier way to finish this kind of work as it is laid on.

Scaly walls should be treated in the same manner as ceilings.

There are several causes for the scaling of the setting coat, leaving the floating exposed. This is most noticeable in old work, though I have known cases of setting coming off before the job has been completed. In this case the ceilings, being concrete, were floated in Portland and coarse stuff, and whilst green (still soft and not set) were finished in putty and plaster. After a month or so the trouble began. I had all the ceilings stripped and hacked, then finished with Keene's and setting stuff. They are now quite sound. The same thing has occurred in other cases, both in new and old work, when floated with raw coarse stuff and while green (to make it easier and a quicker job) set with setting stuff and plaster, which is putting a strong material on a soft background.

Other cases, again, I have noticed that the frost had got hold of the floating which had been set before the frost entirely thawed; but in all cases in re-finishing old floating, the plasterer cannot do better than use the cement wash.

Repairs to Mouldings

Cracks in plaster mouldings should be cut out with a dovetailed line from face line and filled in with putty and plaster with a small quantity of hair, and kept back to allow of a thickness of putty and plaster for finishing. In repairing mouldings, all loose parts should be taken down and treated as in dealing with cracks. Gauged coarse stuff and plaster is used for the roughing out, and the finishing off is done by putty and plaster, using the joint rule for the working off of the members. If it is long, it would be necessary to take a squeeze off the old moulding and cut a mould so as to run the portion to be made good.

Taking the Squeeze

The taking of the squeeze is not a very difficult job, but there are several things which must not be overlooked. Sometimes the section of the moulding may have a small cove in it; if such is the case, the squeeze must be made in two or more pieces. Then, again, part of the moulding or cornice may have an enrichment. But whatever the section to be taken, it can be dealt with in the following way. Fig. 118 shows a cornice which has to be repaired, and a squeeze is required. Clean off a part of the old moulding, about 2″ or 3″ wide, and brush a little linseed oil over the part cleaned, or a little tallow. If these are not available, a little lime-putty can be used. Gauge a small quantity of neat plaster (in a bowl or gauge pot, as the plaster soaks better when gauged this way), and lay on the part prepared from ceiling to wall line, pressing the plaster in the various members with a small tool or gauging trowel, and keeping the width about 1½″ to 2″ wide, according to the depth of the moulding. It must be borne in mind that the squeeze should be of sufficient thickness to take down without breaking. The sides are cut down straight, as shown in Fig. 118. This is to allow the squeeze being laid flat when taken down, as it is laid on a piece of zinc and the profile marked round with a pointed knife or pencil, and the mould cut out and horsed up in the ordinary manner.

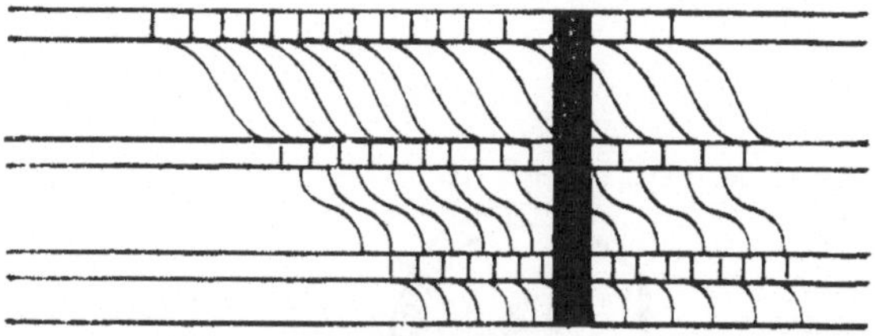

Fig. 118.

Fixing the Running Rule

When fixing the rule for running the cornice, the position for the rule is taken from the under-side of the mould to the bottom of the member of the old cornice. This distance is taken from the dotted line at A, Fig. 119, and marked off on the wall and the rule fixed. This should give the proper intersection of the new to old.

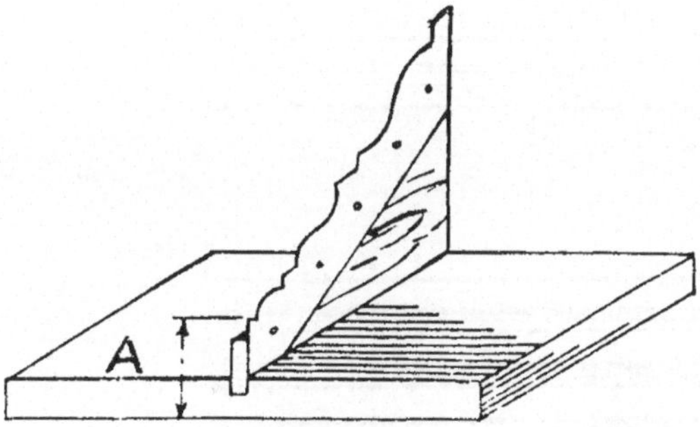

Fig. 119.

Taking a Squeeze off a Cornice having a Cove

With a coved cornice, as Fig. 120, it would be necessary to have three pieces to enable the squeeze to free itself, unless the cornice was to be extended or made up. In this case the plaster would be laid on and dealt with in the same manner as in Fig. 118, and when set the squeeze would slide off the broken end of the cornice as in Fig. 120, where the broken end is shown at A. If there is no broken end and a section is required, as in Fig. 120, the first piece of plaster would be put on to form the piece at A. The joints would be varnished with shellac, and the next piece formed as at B, and the third piece at C. The three pieces are shown as when taken away.

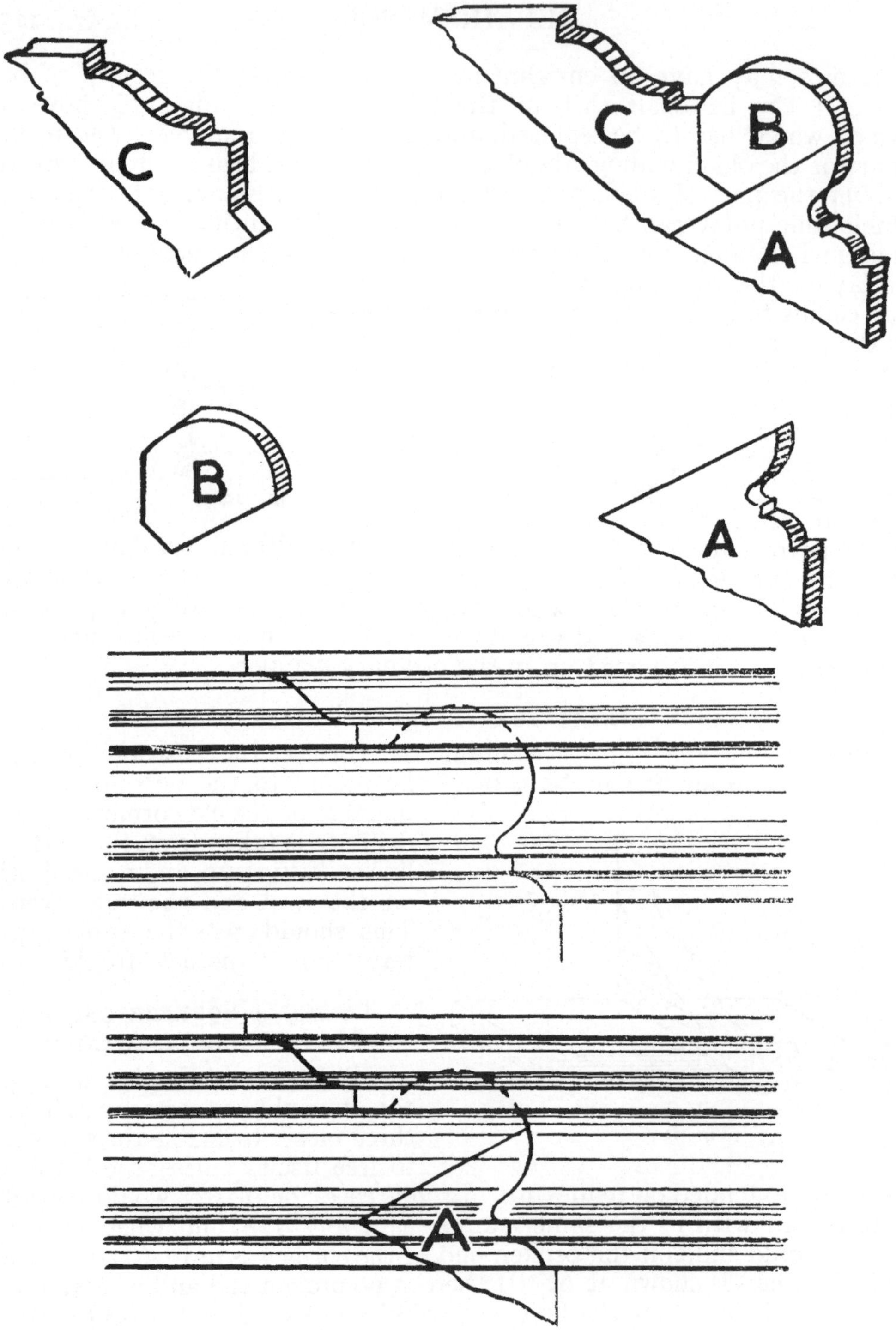

Fig. 120.

Taking a Squeeze off a Moulding having an Enrichment

When repairing mouldings with bed moulds or other enrichments, modelling clay or " plasticine " is necessary for the taking of the squeeze. The remaining parts of the mouldings are dealt with as before described, but the ornament required has to be treated in a different manner. The part required is dusted out with french chalk, and the clay or " plasticine,"

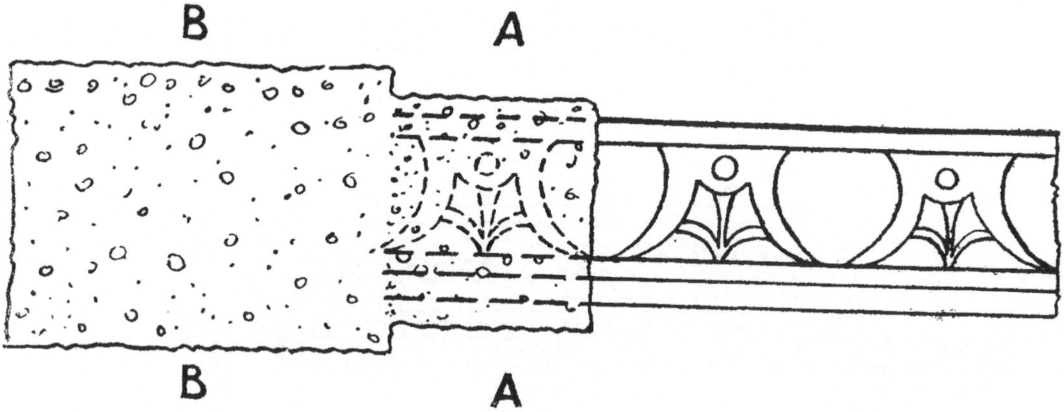

Fig. 121.

whichever is used, is pressed into the ornament a little over ½" thick, leaving the back part rough and undercut, so as to form a key for the plaster backing. Fig. 121 shows an egg and tongue enrichment, the section A being the back view of the squeezing, B the plaster backing. Sufficient plaster is laid on to hold the squeezing material together and to allow of its being drawn away from the original enrichment. Fig. 122 shows the squeeze taken away; A being the reverse egg and tongue ornament, B

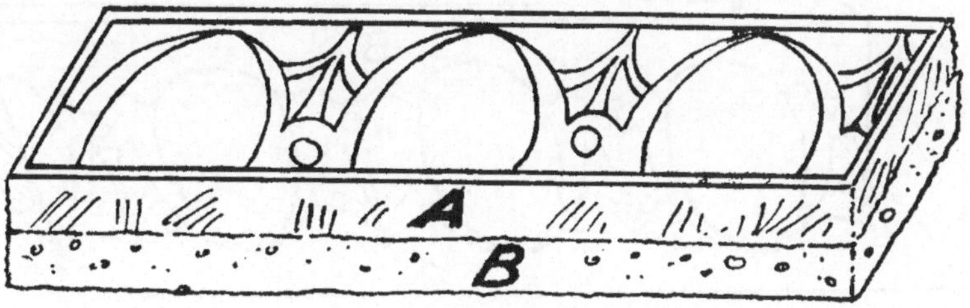

Fig. 122.

squeezing, C plaster backing. This forms the mould, and it is then filled in with neat plaster, shaking the mould so as to ensure every part being filled. When set, the backing is taken away as well as the squeezing material, the cast cleaned up, remoulded and the required number of casts taken.

Casting Enrichments for Repairs with Wax Moulds

When repairing enrichments sometimes the plasterer is able to get a short piece down, of sufficient length for moulding; in this case it would not be necessary to take a squeeze of the enrichment. The piece for moulding is put in a bowl or pail of water to soak, to clean off any dirt or old whitewash. When clean, it is bedded on a piece of 1″ plaster plate, a little wider and longer than the piece to be moulded. If it is a flat enrichment the one plate will do, as all that is necessary is for a clay

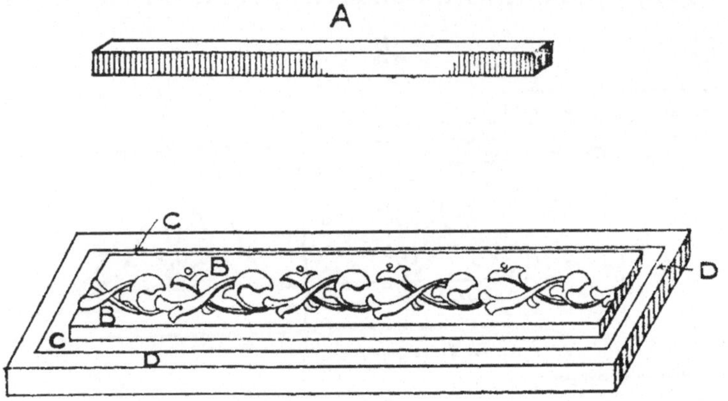

Fig. 123.

fence about ⅜″ higher than the top part of the enrichment to be placed round the plate. The clay fence is a strip of clay about 1″ thick the length of the plate, the width being governed by the need of its being ¾″ higher than the enrichment to be moulded. Fig. 123 shows a flat enrichment bedded on a plate or ground; A is the clay fence, which is kept

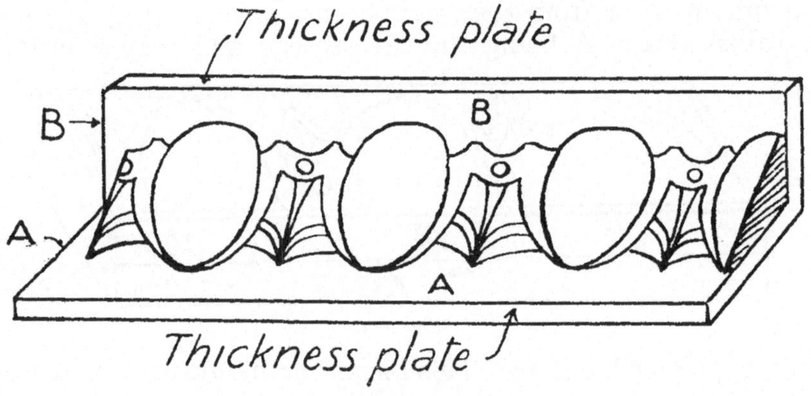

Fig. 124.

about ⅝″ clear all round the cast on the plate, B the edge of cast, C ⅝″ spaced clay fence. This space forms the sides to the mould after it is poured. The sides of the cast should be slightly out of square upwards, to make it easier to remove from the mould. Before the clay fence is put around, the model should be well soaked in clean water till it cannot absorb any more. Take off any extra wet places with a clean brush and then put round the clay fence, holding it in position and putting a small portion of plaster round it.

If a bed mould enrichment, such as an egg and tongue moulding, has to be cast, the model or original is bedded on a plaster ground flush at the back edge and about 1¾″ inside the two ends and front. Fig. 124 shows the

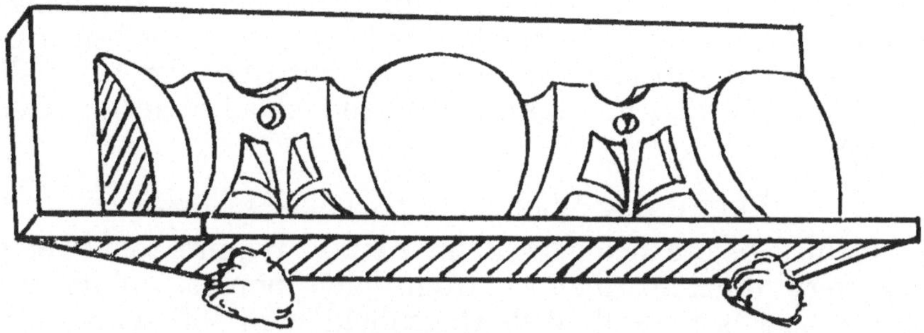

Fig. 125.

egg and tongue ornaments on the plate; A is the plate, B the back plate, which is bedded against the back of the egg and tongue and against the plate A. The plate B should come above the enrichments of this kind. The model is turned up to the angle of 45 degrees, as Fig. 125. A piece of clay is put up at each end about ¾″ away from end of cast, Fig. 126, the

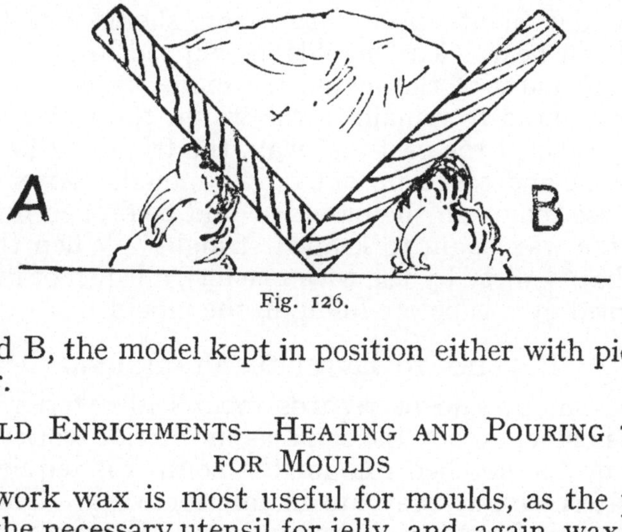

Fig. 126.

clay end A and B, the model kept in position either with pieces of clay or a little plaster.

Bed Mould Enrichments—Heating and Pouring the Wax for Moulds

In repair work wax is most useful for moulds, as the plasterer very often has not the necessary utensil for jelly, and, again, wax is less trouble and can be used immediately it is cold. The wax is heated in a pail or any iron vessel which will stand the fire. The utensils must be clean; an oily brush or a little grease passed over it helps to prevent the wax adhering. A slow fire or gas can be used to melt it, and it must not be allowed to get too hot. As soon as it is dissolved it should be taken off the fire and allowed to stand a short period. When pouring the wax, it should be poured from one end of the framing, covering the model when pouring to the other end. Sufficient wax should be poured on at the one operation to fill up to the top of the clay fencing.

Taking off the Mould

When the wax has been allowed to get cool (but not cold), the clay fence is taken away and the wax lifted up by placing the trimming knife under and round the edge. The mould should be laid on the bench, or on a flat piece of wood, and straightened with the joint rule; for, being slightly warm, there is a possibility in taking it off the model of getting the back out of shape.

Preparing Wax Moulds for Casting

Wax moulds before being used for casting require to be prepared, but in a very different manner from jelly moulds. With wax all that is necessary is to well wash the mould with soft soap, rinsing off with clean water. The mould is filled with a small filling-in brush and slightly shaking the mould. Care must be used, as wax is very brittle.

Taking Out Casts from Wax Moulds

After filling in the wax moulds, they should be immersed in warm water for a few seconds, just long enough to soften the wax and permit the cast to slip out easily. A pail is most suitable, and being deep several moulds can be put in at once. The water should not be any hotter than the hand can bear. The moulds are dipped in and turned end for end, so that both parts of the mould are made warm by the water. If the mould was allowed to remain in the water, the wax naturally would melt. Having warmed the cast and wax, the trimming knife can be put in the end of cast and eased up out of the mould. After this has been done and the case removed, the mould should be dipped into clean water (cold) to cool the wax again to keep it straight. When the wax begins to get firred (that is, dirty by use with plaster), a brush of linseed or sweet oil can be brushed over it before filling in the mould.

Repairs to Exterior Plastering

Little more can be said as regards repairs to exterior plastering, as the treatment is much about the same as in interior work.

For patching, the notched rule for the floating is required and cracks are repaired externally, as they are in interior work. This also applies to the making good of cornices and mouldings. However, some variation as to materials and the shape of patches is usually desirable. If repairs are being done to a cement façade which is jointed, the patch to be repaired would, if possible, be cut to fit in with the jointing. Then, again, if the repair work is to be painted, Roman cement would have to be used instead of Portland; for if Portland is used and painted soon after, the cement destroys the character of the paint, and the first shower of rain washes it off, thus disfiguring the remaining parts.

Repairing Cement Work with a Jointed Surface or Plain

In repairing exterior work, cement and sand, gauged 3 parts of sand to 1 part of cement, is used for the floating, and for finishing, 2 parts of washed sand to 1 part of cement. The patch is cut out as neatly as possible and, unlike interior patches, it should be square or oblong if plain, or as near joints as possible if jointed. Fig. 127 shows a jointed façade; A is the patch to be repaired, which is cut neatly to the joints, and when finished the joint carried across at the dotted, line and the remaining parts jointed to match.

Cracks in Mouldings

Cracks in mouldings or cornices are cut out, not more than 1″ wide, and well dusted out and wetted. Gauged cement and washed sand. half and half, is well tucked in and kept back to allow for finishing, Cracks when filled in should be allowed to stand several days before being finished off, to allow for any slight shrinkage that may be caused by the suction of the old work. When finishing, the same sort of material is used and the joints well rubbed up with the float and then trowelled in, cleaning off each side of the joint with the water brush. Any other repairs, or re-forming mouldings, are treated in the same manner as for interior repairs. In cement repairs, however, large places in mouldings can easily be done with a thin feather-edge, using it as a joint rule, working on the members of old moulding and facing up the new part with the small narrow floats and steel joint. If the work is to be painted, Roman cement should be used, but if colour washed, Portland cement is suitable.

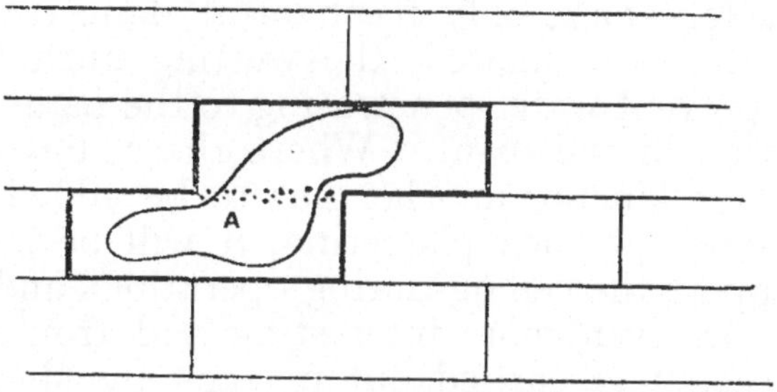

Fig. 127.

Repairing Cement Floors

In repairing bad places in cement floors, the places should be broken up or cut out with the hammer and chisel, though in breaking up, if the underneath part is not very solid, this method tends to drive the existing old parts lower. In such cases, too much force must not be used. But by breaking up, a key is formed for the new mortar. After breaking up or cutting out the patch, the place should be cleaned up and consolidated. Then 3 parts of broken bricks or fine ballast and 1 part of cement are mixed, wetted slightly and levelled in the patch, and flattened down with a wooden rammer, leaving about 1″ for a cement and sand finish, gauged 2 parts of sand to 1 part of cement. It is ruled off with a straight-edge, bearing on the old floor level. When the initial set has begun, a hand float is pressed over it, and the surface trowelled off, but not too smoothly.

CHAPTER X

THE RESTORATION OF PLASTER CEILINGS

Repairing in Situ—Relathing—Back-plastering—Removal and Re-erection.

IN certain cases where fine and heavy old plaster ceilings are concerned, restoration may mean something far more extensive and difficult than making good surface defects, such as filling cracks or filling in with new plaster and modelling missing sections.

At Greenwich Hospital, Hampton Court, Monmouth Castle and elsewhere, it has been found that valuable modelled plaster ceilings have been imperilled by the giving way of the mechanical parts, viz. either the rotting of laths or the crumbling of the plaster squeezed up between and over the laths. In some cases the trouble has been overcome by relathing and adopting back-plastering methods. The danger here is that the old plaster requires damping and the new plaster laid on fairly wet in order to secure adhesion of the old and the new. If it is plain plaster work, only mechanical dangers have to be faced, but if the ceilings are painted, the wetting must be carried out very cautiously to prevent water penetrating to the paint and causing flaking or chemical action in the paint. Where the ceiling is extensive and metal lathing of one kind or another has to be added, which will need a substantial amount of back plastering, it will be necessary to provide centring to support the ceiling during operations and setting.

An even more interesting and troublesome case of restoration was successfully carried out in 1925 by Mr. G. E. Nield, F.R.I.B.A. The work can be seen in St. Clement's Church, Eastcheap, London. The old church was destroyed in the Fire of London, 1666, and was rebuilt by Sir Christopher Wren and completed about the year 1684. The church underwent restoration at various times, but in a more or less perfunctory manner, and about 1924, after a thorough survey, it was found that drastic measures were necessary to save the building. For one thing, the great solid oak beams, supporting the heavy roof above and the no less heavy plaster ceiling below, were discovered to be eaten away by the "death-watch" beetle. These had to be removed and replaced with a modern form of construction, as the timbers were too much destroyed to be retained. At first it was thought that Wren's plaster ceiling would have to be sacrificed, but a way to save it was found.

As the drawings show, the ceiling consisted of a large cove springing from the architrave at the sill level of the clerestory windows, the cove being groined for the window openings. This cove supported a

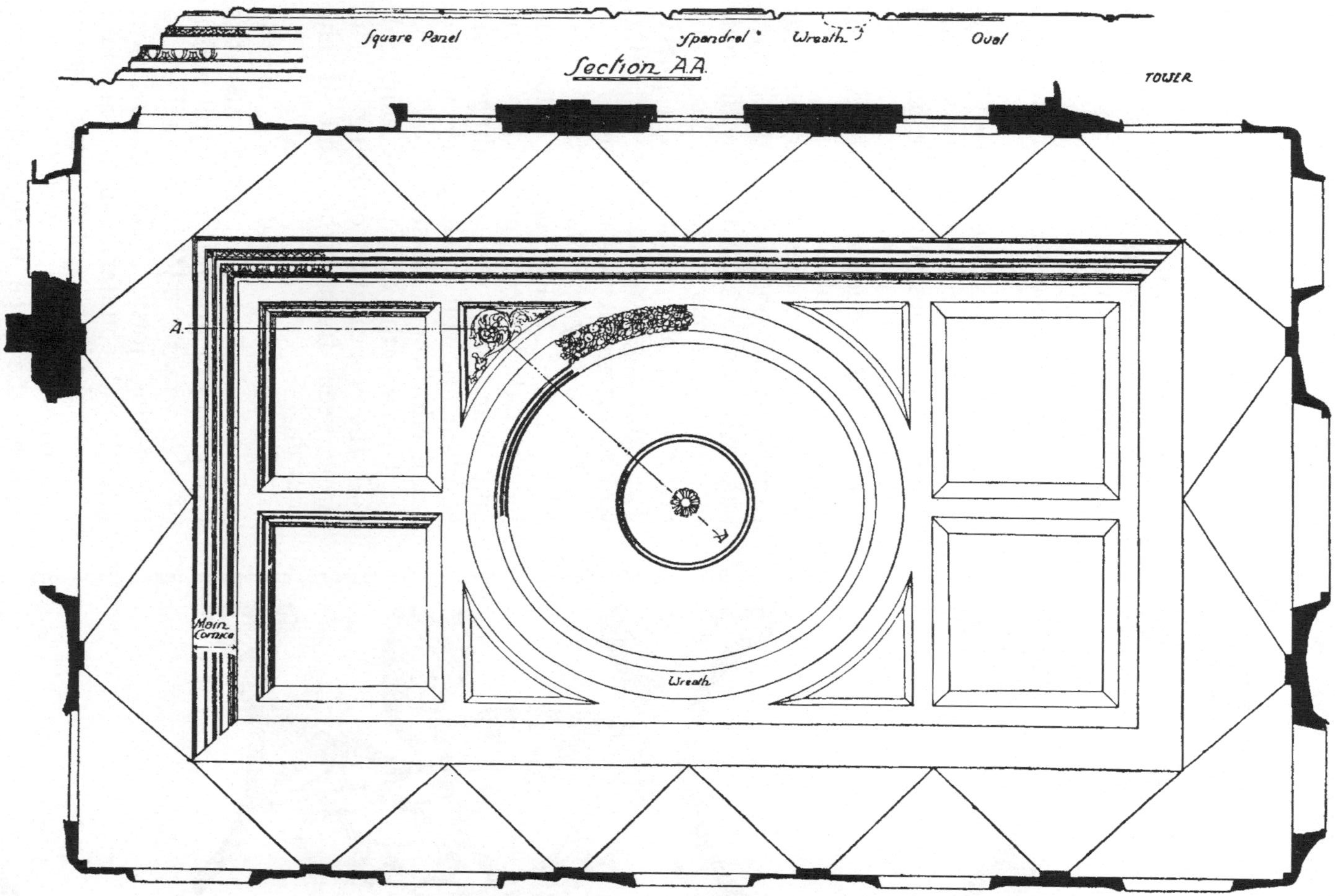

Fig. 128.—Plan of Ceiling, St. Clement's, Eastcheap.

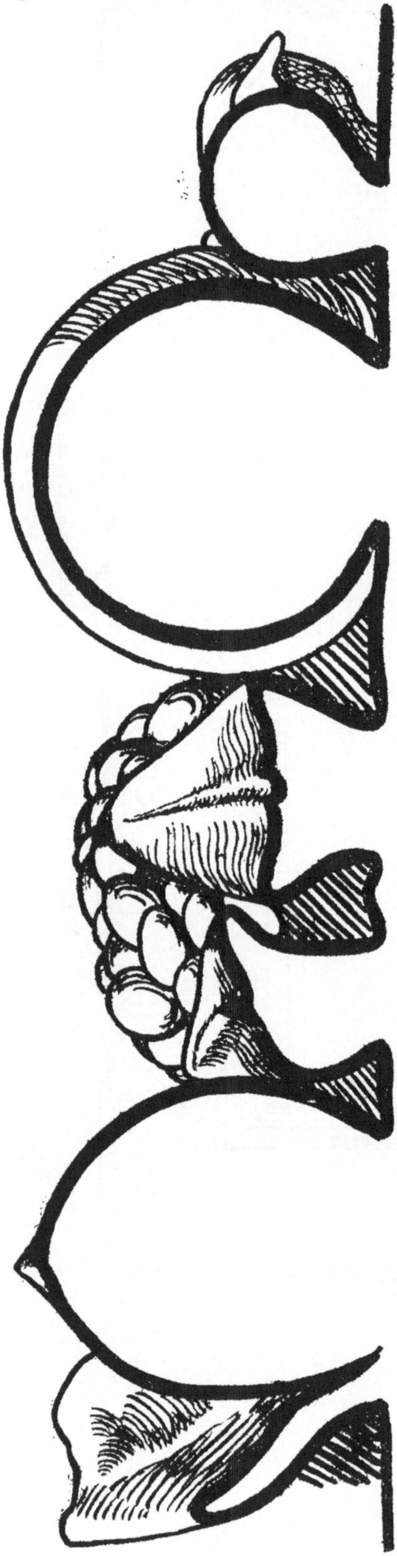

Fig. 129.—A Section of Plaster Wreath, St. Clement's, Eastcheap, showing method of construction. (*See plan.*)

deep and heavy cornice which was enriched by a bold egg and dart moulding.

The soffit enclosed by the cornice was divided into four panels with a large wreath in the centre, the space between the wreath and the four panels being filled with four enriched spandrils.

In the centre space enclosed by the wreath was a large centre flower, but from its construction and the material from which it was formed it was evidently of a much later date and not the work of Wren.

The cove above mentioned was formed with wooden ribs, lathed with oak laths, rendered with lime and hair (a very generous supply of long hair being used) and finished with putty. The groining was a little irregular, widths often varying one inch.

The heavy cornice was formed with timber bracketing or blocking pieces, oak lathed, rendered in lime and hair, and the mouldings and enrichments run with putty with a small addition of plaster.

The wreath, which was elliptical, had the appearance of being a true circle seen from either end of the church.

The form of construction of the wreath adopted by Wren was the fixing of a fir ground of the full width of the wreath, 1½″ thick, spiked to the ceiling joists; this ground was lathed and counter-lathed. Six and 7″ wrought-iron spikes were driven into the fir ground, which formed the core for the large fruits and flowers. The plaster backing for the wreath varied from 2″ to 2½″ in thickness.

The wreath consisted of laurel leaves through which projected the fruit and flowers, and it is to be par-

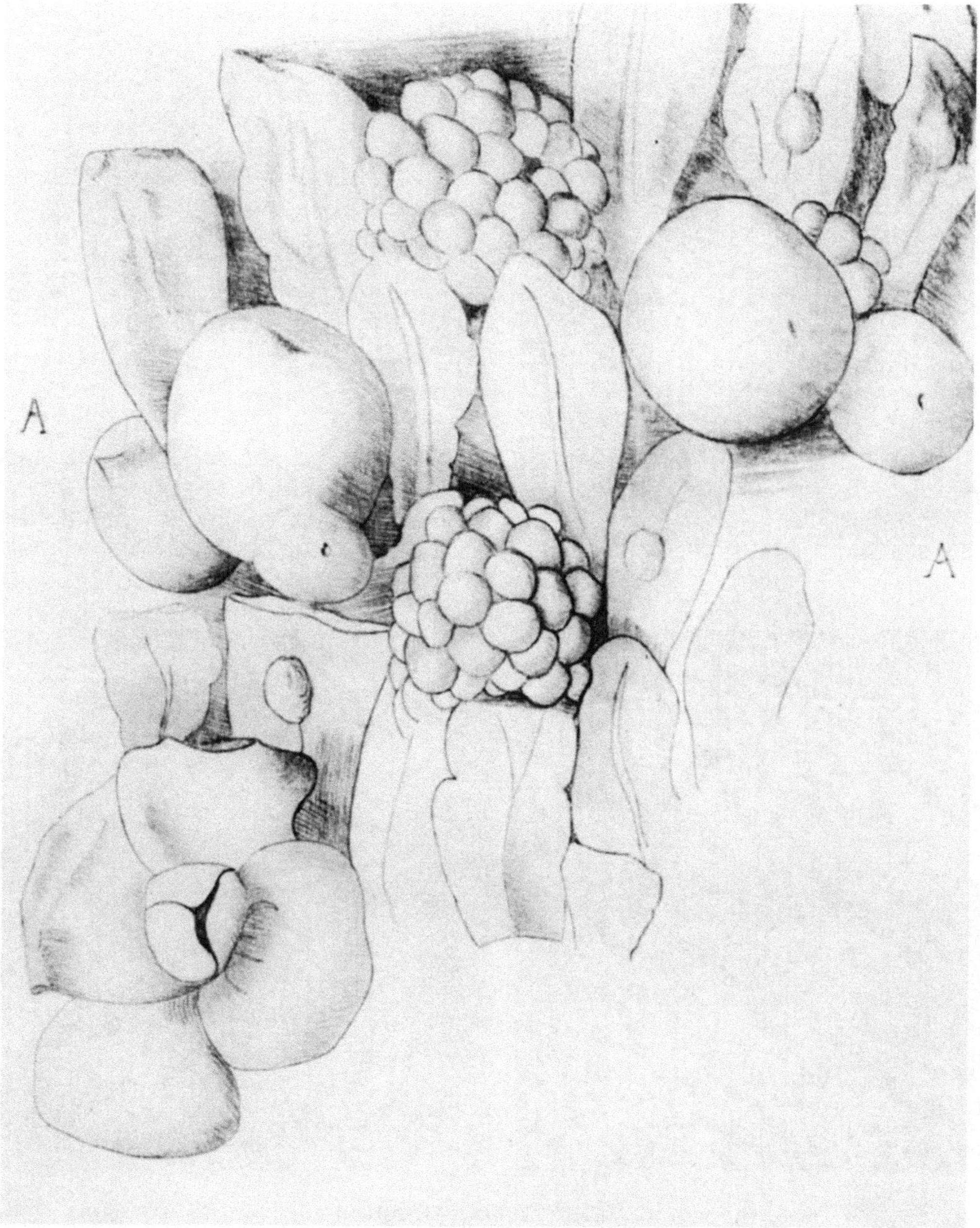

ENRICHMENT OF PLASTER CEILING BY SIR CHRISTOPHER WREN, ST. CLEMENT'S CHURCH, EASTCHEAP, LONDON.

PLATE VIII

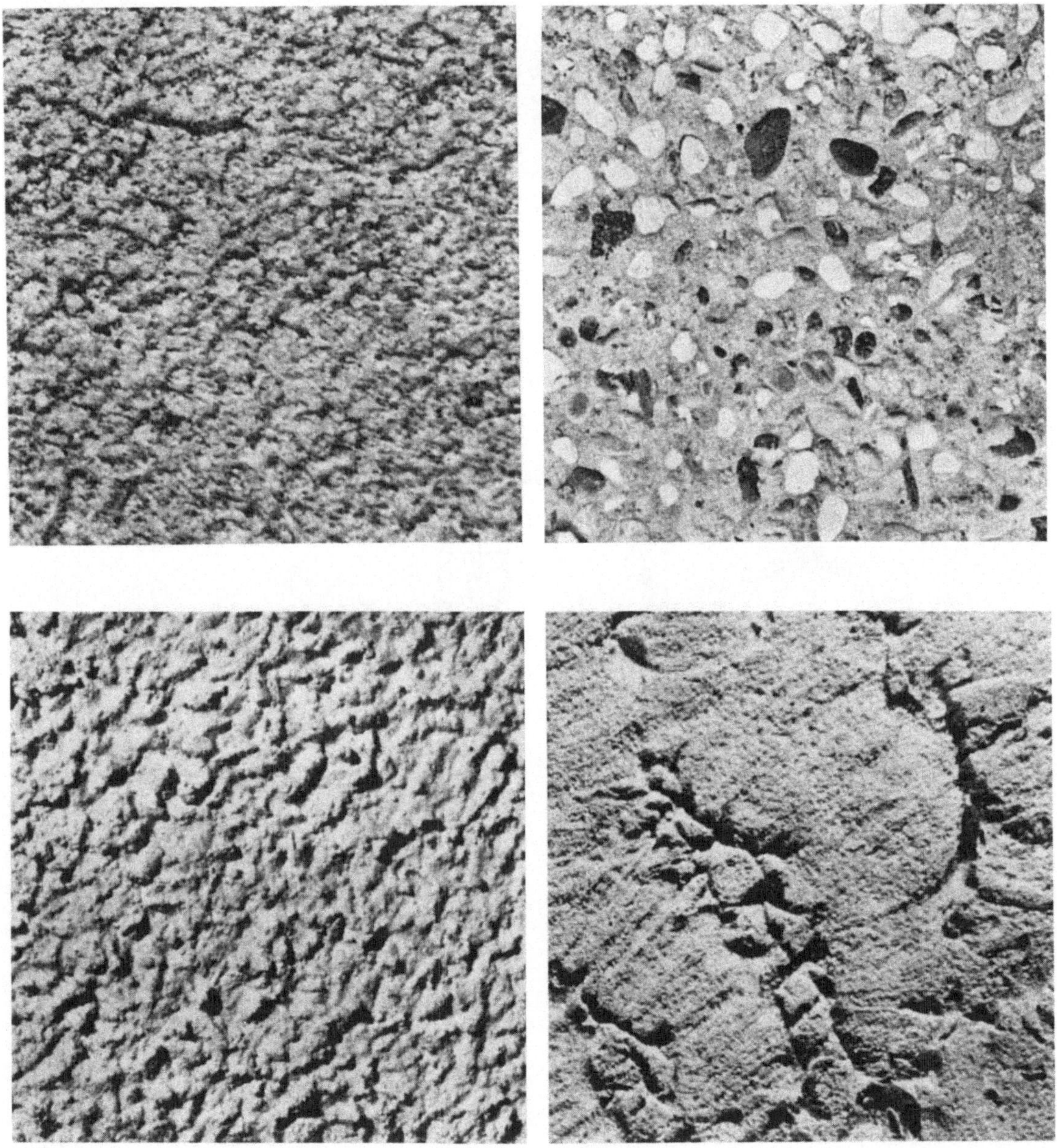

CEMENT STUCCO FINISH.

Sponge Finish. Pebble Dash.
Stippled Finish. Spatter Dash.

ticularly noted that there was no repeat of design, the general opinion being that it was modelled upon the scaffold. The fruit consisted of apples, pears, pomegranates, grapes, what appears like the end of a vegetable marrow, and pea-pods. The flowers are lilies and very flat-modelled, open roses.

Consideration appears to have been given to scale, for the lemons, which are 7½″ long, viewed from the floor of the church, appear no larger than the usual size.

The modelled work was formed with ordinary sand, lime and long ox-hair, finished with putty and a very little plaster.

Preparatory to retimbering, the wreath was removed in sixteen sections. The spandrils were each lifted out entire. Each section was carefully cleaned, the plaster background removed, the whole reinforced and wadded to new plaster ground, and refixed to the fir ground in the original position. All the mouldings and enrichments of the cornice were carefully reproduced, and the whole work as it now exists is practically Wren's ceiling, the wreaths and spandrils being his original work. The only work taken down and not replaced was the centre flower, which, as mentioned above, did not belong to the original plaster work.

The plaster work was in very good condition, and would probably have endured for a further hundred years, had it not been for the destruction of the supporting timbers and also the oak laths by the death-watch beetle.

The removal and replacing of the plaster was undertaken by G. & A. Brown, Ltd., of Hammersmith Road, London, under the direct supervision of Mr. Nield, the general contractors for the restoration being John Greenwood, Ltd., of London.

CHAPTER XI

PLASTERING BY SPRAYING

Claims as to Compressed-air Spraying—The Machine—The Nozzle—The Air Compressor—The Material—Operating—Care of the Machine—Applications—The Eight Points.

THE use of machinery is one of the outstanding characteristics of the age. It has invaded most handicrafts, and plastering has not escaped the innovation, though so far its influence has not been greatly or widely felt over here. Mechanical plastering is not altogether new, having exercised the thoughts of many practical men.

The idea of applying cement for rendering or plastering by means of sprinkling by compressed air originated in Europe, but has been perfected in America, where the "cement-gun," brought forward in 1914, is now extensively used, both in engineering work and in house construction. This machine, known as the "cement-gun," has been introduced into England, and the system of applying cement mortars with its aid appears to be gaining ground, so far mainly for external work on housing schemes and in big factory construction. It is a development which requires attention, and is, therefore, briefly described in this place.

CLAIMS AS TO COMPRESSED-AIR SPRAYING

The claims made as to the advantage of applying cement mortar by means of propulsion by compressed air through a distributing nozzle are the saving of (1) time, (2) labour and (3) material, also (4) exceptionally strong adhesion, increased (5) waterproofing and (6) fireproofing properties.

Apart from compressed air supply, a crew of five men will be required for a gun, and they can cover 140′ super 1″ thick in eight hours. There will, of course, be variations according to the mixture used, the surface to be covered, the air pressure and the experience of the men. It is possible to minimise and simplify scaffolding, as a good height can be reached with a properly handled nozzle and suitable air pressure. As the mix is applied evenly and there is little loss by rebound, the waste is very small, a good covering average being obtained. Adhesion depends partly on the nature of the mix, the character of the surface covered and the pressure used. Data as regards waterproof qualities will also give some idea as to adhesion. In some independent laboratory tests it has been proved that cement-mortar shot slabs, $\frac{5}{8}''$ to $1\frac{1}{2}''$ thick, when "tested with water pressures ranging from 700′ to 1,600′, for several hours, showed no moisture coming through the slab. One 1″ slab held a load of 1,610′ for 2½ hours without showing moisture; then the water pressure was raised gradually to 3,400′ and the specimen broke in bending, having leaked just a little just before breaking."

Further evidence as to adhesion is afforded by experiments carried on at the University of Toronto in coating metal with cement. The report states:

"Neat Portland cement is known to be an effective preventative of rusting when applied to steel surfaces, and on occasions has been employed as a paint for this purpose. To afford a satisfactory protection for steel reinforcement, it is advisable that the metal be surrounded by a mortar rich in cement applied preferably wet, as this ensures that all parts be coated. 'Gunite' is a concrete essentially of this character, but, since it is applied under pressure, it possesses greater density than ordinary concrete. When the raw materials are forced under pressure against a hard metal surface there is considerable rebound—mostly of the sand. This means that the cement adheres to the surface as a matrix into which the particles of sand are subsequently driven. This matrix is a preventative of corrosion."

As regards fireproofing, a slab 1½″ thick, 6′ 3″ span, loaded with 40 lb. per sq. ft., was subjected to a temperature of 1,700° F. for three hours. Subsequently this slab when cooled was loaded with 200 lb. per square foot without failure. Further tests were carried out on two 2″ slabs with 8″ air space, and concrete studs spaced approximately 7′ apart. One sample was subjected to a temperature ranging from 900° F. at five minutes to about 2,100° F. at 257 minutes. A second test was made on a sample subjected to 1,700° F. at one hour, after which water at 50 lb. pressure was shot against the heated surface for five minutes. Subsequent tests were made to show the stability of this type of wall against breaking down from the impact of falling steel.

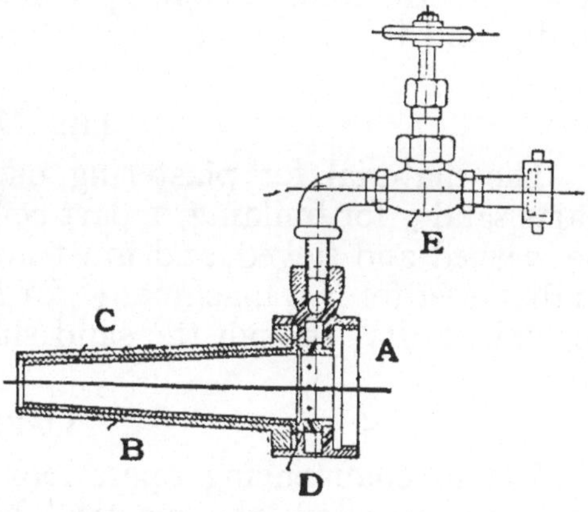

Fig. 130.

The Machine

The mechanical contrivances consist of three parts—the machine proper and, attached to it by means of a rubber hose, a nozzle with valves, and the air compressor. The "cement gun" itself is composed (Fig. 131) of a hopper (*a*), upper chamber (*c*) and lower chamber (*d*), closed by two bell valves (*e*) and their corresponding hand levers. By means of a system of pipes and valves, the two chambers may be brought under equal pressure. At the bottom of the lower chamber is placed a feed wheel (*g*), connected to a vertical shaft which is driven by an air motor (*i*) through a worm gear (*h*). The compressed air enters through the so-called goose neck I, and empties the pockets of the feed wheel as these rotate under it, blowing their contents through the outlet valve I into the material hose. See pages 140, 141.

There are three standard sizes, whose respective capacities, air and power requirements are given in the table below:

—	Air pressure in lb.	Free air per minute of operation.	Power for compressor.		Capacity per day in terms of sq. ft. 1" thick.
			kw.	h.p.	
N-0	35-50	105-125	18	25	950
N-1	35-50	150-165	25	35	1,500
N-2	35-50	225	35	45	2,000

Although the machines are tested for a safety limit of 75 lb., 50 lb. should never be exceeded.

The Nozzle.—Two types of nozzles are used; but the main features are the body (Fig. 130) A, the tapered tip B, with rubber lining C, water ring with its internal notes D, and a water-regulating valve E. The nozzle is connected to the machine by a large diameter rubber hose, and the water valve has a smaller hose connected with water main or other source of supply under pressure, which should be rather higher than the pressure from the machine, in order to procure a good spread.

The Air Compressor.—The question of compressed air will be solved according to circumstances, but whether supplied from a central source or a separate compression, special care is necessary that the air should be dry and free from oil.

The Material

The material for plastering usually consists of 1 part cement to 3 parts sand; for building, 1 part cement to 4 parts sand. The sand must be washed and sieved, and must not contain any particles larger than $\frac{1}{8}''$ in diameter for N-0 machine, $\frac{1}{4}''$ for N-1, and $\frac{3}{8}''$ for N-2. Cement and sand are mixed dry, though the sand should be slightly damp.

Operating

Before commencing operations all cocks must be closed, the compressor connected, the material hose coupled to the outlet valve and the water hose to the main, and both cone valves opened. The mixed material is fed into the machine through the hopper, the upper cone valve closed, and the compressed-air inlet cock opened until a pressure of between 10 to 15 lb. is reached. Then the cock *m*, to the outlet of material hose, is opened. The cock to the air motor is next opened, and the feed wheel begins to revolve, pushing the material towards the material outlet, where it is taken up and propelled forward by the rush of the compressed air. By increasing the speed of the air motor, the volume of material ejected is increased. It is also possible to regulate the supply of air to the nozzle. When the material supply cock is opened, the water valve is also turned on, and as the material comes forward in a finely divided powder mixed with air, it meets the water stream percolating through the holes in the water ring and hydration takes place. The amount of water supplied is under control.

The spread depends partly upon the rate at which the air motor is run and partly on the water supply, but is also influenced by the distance at which the nozzle is held from the surface to be operated upon. The nozzle must not be held too near, or the spread will be reduced, and there will be loss, owing to excessive rebound. On the other hand, if too far away, the spread may be reduced in effectiveness on its outer ring, there will be unevenness and possible interference from wind.

The surface to be treated should be carefully cleaned and in repair, all loose and damaged parts cut away. If the surface is absorbent, it should be thoroughly wetted. In practice, the nozzle is found to give the best results if held from 30″ to 35″ from the surface under treatment. It should not be held straight, but slanting, so that the material hits the surface at an angle of 90°. By approaching the nozzle nearer to the surface the rapidity and strength are increased; by withdrawing it farther they are diminished. Corners should be treated by shooting at the deepest point. No deposit should be less than ½″ or more than 2″ thick; if more is required, the floating should be in two layers, but no great time need elapse between the two operations. If too much water is supplied, adhesion will be bad, and there will be loss of time and material through running. Running, of course, leads to unevenness in the thickness of the layer, a defect which it is difficult and comparatively costly to remedy. The essence of mechanical plastering is to secure rapidly a coat of uniform thickness and strong adhesion. It should be remembered that the action of the compressed air on the cement and sand is to produce intimate mixture prior to introducing moisture, and so while the cement is still inert. The method adopted for hydration ensures the mortar being quite "live," but the hydration is thorough, and, provided that hydration is not overdone, the setting is scarcely retarded. Indeed, it is found that a second coat can follow on a first application fairly quickly, as the whole mass is thoroughly aerated.

At frequent intervals, two to three minutes according to rate of operating, the lower chamber will require refilling with material. This is done without stopping the machine. It is first necessary to ascertain that the upper chamber is quite empty. Then the lower cone valve is closed and the cock *g* (Fig. 131) opened to allow the air to escape; this causes the upper cone valve to open, when the material is fed in, and, this operation accomplished, the upper cone valve is pressed down into its seat and compressed air admitted to the upper chamber. When pressure in this upper chamber reaches an equality with that in the lower chamber, the lower cone valve opens automatically and the material falls down and comes under the influence of its feed wheel. The aim is to refill as soon as the upper chamber is empty, so that there shall be no interruption.

In case of need, the machine can be quickly stopped by shutting down the air motor and then the compressed air supply. Hoses can then be disconnected for inspection. Should, through any neglect, material of too large a diameter be fed in, it will be caught between the feed wheel and outlet. In such a case the air motor should be shut down and the compressed air supply cut off, the material hose disconnected and the obstruction removed.

As a rule, a crew of five men are needed ; the nozzleman or plasterer, the gunman or machine controller, the man running the air compressor, and two labourers for filling in.

Care of the Machine

Before starting operations, the water valve and the water ring in the nozzle should be inspected to see that they are clear. Accumulations may occur here and interfere with free working. At the same time, all lubricators and grease cups should be examined, to see that they work well and if they need refilling. The drop oiler should feed at the rate of five drops per minute. The air pressure should never be allowed to exceed the 50-lb. limit. Care should be taken that the sand used is not too dry or too large. If stoppage takes place through an obstruction in the material outlet, the air motor and inlet of compressed air should be stopped promptly. On closing down for the day, the machine must be completely emptied, the hose disconnected and cleaned, and the feed wheel slowly revolved so that each pocket should come under inspection and be cleaned out. A special air hose is attached to enable the machine to be cleaned inside and out. On stopping the machine, the drop oiler is shut off. A thorough overhaul once a week is advisable.

Both material and water hose must be kept free of kinks, and protected from rough wear as much as possible.

Applications

What more particularly interests us here is the use of the " cement-gun " for plastering work, and this under special conditions it appears to do well and economically. It will be clear that all surfaces to be treated must be clean; if absorbent well wetted; that the mortar should be applied neither too dry nor too wet, and in coats of not less than ½" or more than 2". Where applying for the purposes of protecting wood or iron, less than ½" coating is often permissible. In repair work, the surface must be clean, all " perished " material cut away and the old work wetted before being operated upon.

It should be borne in mind that the "cement-gun" can be used as a sand-blasting machine.

The "cement-gun" has also been successfully used in constructing walls of framed buildings, or partitions, on expanded metal or similar types of reinforcement. The standard specification lay down that:

"'Gunite' for walls shall be mixed in the proportions of 1 part cement to 4 parts sand, and shall be applied to such thickness as shown on the plans.

"The mesh shall consist of expanded metal of no less opening than 1½", with a sectional area of not less than ·07 square inch per foot of width, or of square welded mesh of 2" × 2", or larger openings with similar area of steel, or of such dimensions and weights as shown on the plans. No mesh shall have a larger opening than 4" × 4".

"The mesh shall be fastened firmly to the supporting members at intervals of about 4', and shall be lightly attached to the intervening supporting members. It shall be furred out by rods, chairs or other satisfactory support, so that the mesh shall maintain a plane about in the centre of the slab."

Floor and roof slabs are also made by means of the "cement-gun." The specifications demand that:

"'Gunite' for floor and roof slabs shall consist of 1 part cement to 3 parts sand (by volume).

"The reinforcing mesh shall be laid in sheets of sufficient size that they may, wherever possible, form a continuous reinforcement past at least one supporting beam. Adjacent sheets shall have a side lap of at least 4" and the end laps over beams shall be not less than 6". Sheets shall be tied together at frequent intervals at the points of the lap, and shall in no case approach nearer than ¾" to any exposed surface."

In very heavy repair work to external walls, etc., it is recommended that after cleaning down the surface, wire mesh should be fixed to the wall and embedded in the mortar, a reinforced casing thus being provided. The procedure suggested by an official architect being as follows:

1. All loose concrete should be removed from the face by hand or pneumatic picks.

2. Holes should be drilled at about 30" (75 cm.) centres, in which short expansion bolts should be set for the purpose of attaching reinforcing mesh.

3. The layers of "Gunite" should not be more than 3" thick, and a sheet of mesh, consisting of galvanised wire fabric with openings about 4" square, should be placed in each layer.

4. The concrete should be thoroughly cleaned by sand-blasting and washing before the application of a layer of "Gunite," and each layer should be washed down with fresh water sprayed on under pressure prior to applying a succeeding layer, in order to remove all traces of salt water.

5. After all disintegrated portions have been brought to proper lines, a layer 1½" thick, reinforced with a 2" × 2" mesh, should be placed over the entire face and top of the wall.

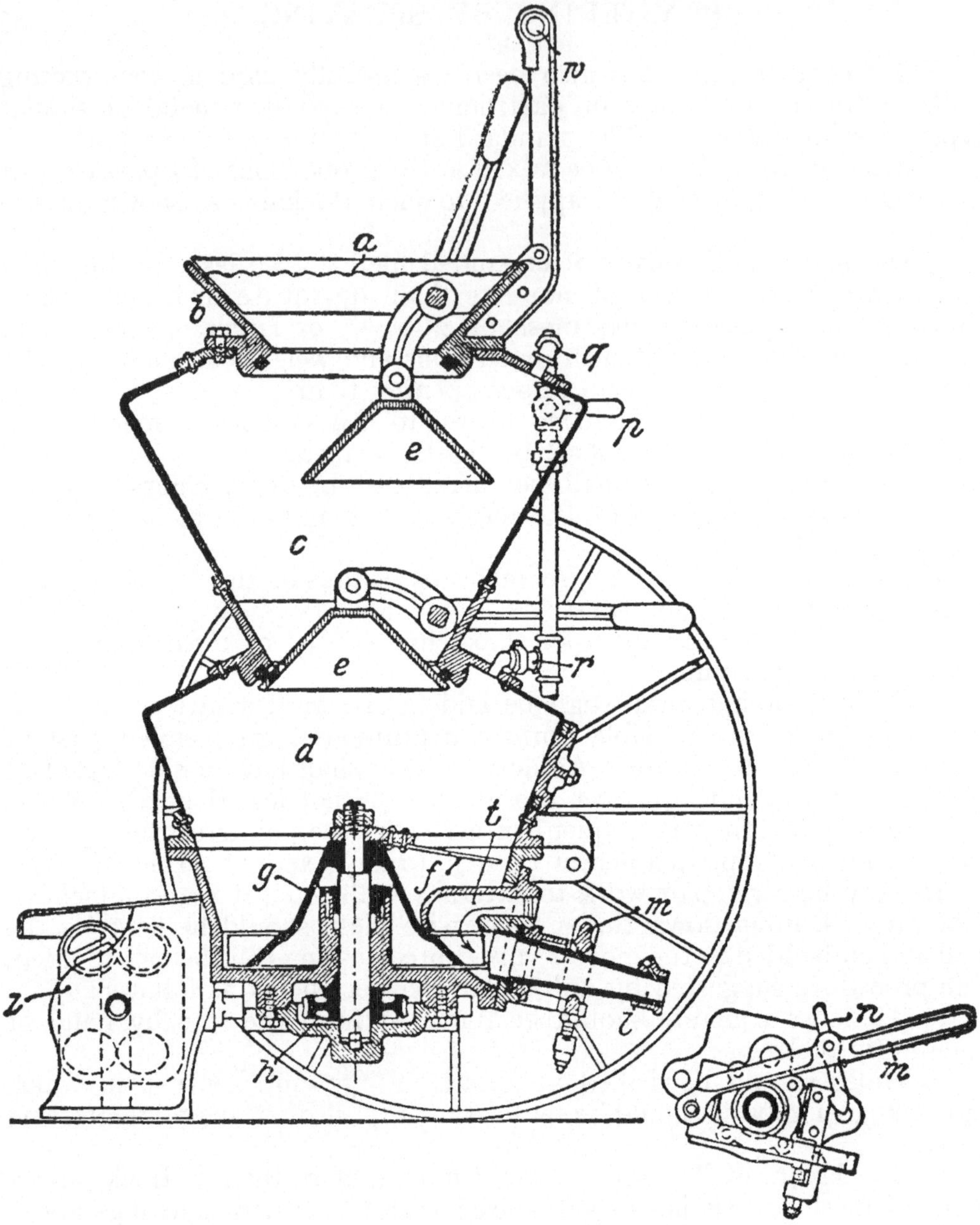

Fig. 131.—Section of Cement Gun with detail.

The Eight Points

The manufacturers of the machine make the following eight points :

1. Sufficient air must be provided. The N-00 machine requires a minimum of 60 cubic feet of actual free air per minute at 35 to 50 lb. pressure. The N-0 machine requires a minimum of 100 cubic feet of actual free air per minute at from 35 to 50 lb. pressure. The N-1 machine requires a minimum of 175 cubic feet actual free air per minute at from 35 to 50 lb. pressure. The N-2 machine requires about 225 cubic feet actual free air per minute at from 40 to 60 lb. pressure.

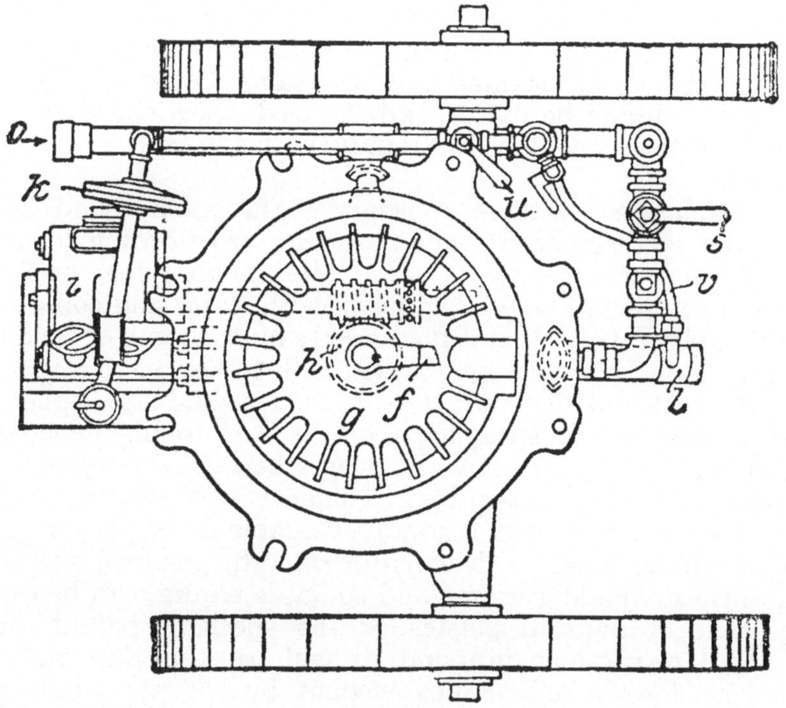

Fig. 132.—Plan of Cement Gun.

2. For successful operation with least loss of time the air should be thoroughly dried ; an air dryer is required for this purpose. With wet air there is always a tendency of the hose to become clogged, due to the cement adhering to the walls of the hose. This adhesion causes stoppages, which must be located and removed. As a prevention of such stoppages, it is always wise to use the by-pass blow-off to clear the hose during any short periods when machine is not operating.

3. Successful operation of a " cement-gun " demands that the machine be cleaned out thoroughly each night when work stops. Otherwise the cement left in the machine will set up.

4. In order to produce true surfaces, the nozzleman must keep the

stream of mortar continuously moving, so that the "Gunite" is placed on the wall in thin layers, thereby avoiding lumpiness.

5. The nozzleman should hold the nozzle at as near right angles to the surface being covered as possible, and the nozzle should be held at about 3′ from the surface.

6. The best results are obtained with from 50′ to 150′ of hose from the machine to the nozzle. Excellent results, both in character of "Gunite" and progress made, have been obtained with as much as 450′ of hose, raising the material 80′. In this case air pressure of about 75 lb. at the machine was needed. "Gunite" has been elevated to as great a height as 250′, but in such cases it is advisable to use a "booster jet" of air.

7. The "cement-gun" is frequently employed as a sand-blast machine. In case of such use, in cleaning steel the sand should be "bone dry"; a special nozzle should be used; and the feed wheel should be operated very slowly. If the machine is run rapidly with dry sand, static electricity is developed.

8. In cleaning concrete or masonry surfaces, the sand blasting is done with the smallest-sized regular nozzle; with ordinary sand; and with the addition of water to the sand. Before applying the "Gunite," the surface must always be washed down with water and compressed air.

The principles involved, and the underlying economy of such a system and machine, will be easily understood. But they have a special field. Owing to the cost and the character of the material used, plastering by means of compressed-air spraying is only suited to big jobs, or a close multiple of small jobs, as in town housing schemes, where great expanses of plain surface dressing are required. Under these conditions the "cement gun" has its uses, and it is certainly valuable in engineering. But it does not supersede hand work where a smooth, polished surface, fancy textured surface, mould running and so on, is required to be carried out with fine lime or gypsum plaster or the special "patent" cements. Bearing these restrictions in mind, it will be seen that the gun and "Gunite" need to be taken into account by the up-to-date plasterer and contractor.

CHAPTER XII

PORTLAND CEMENT STUCCO

Composition of the Stucco—Application: First, Second and Finishing Coats—Finishing: Smooth Finish; Sand Floated; Stippled Finish; Sand Sprayed; Exposed Aggregate; Dashing; Sponge Finish; Modelling; Sgraffito; Other Variations—Colouring—Constructional Value.

IN the Introduction of this book much has been said on the use of stucco, using the term in its modern acceptance—that is to say, plain plastering, both for interior and exterior work, and the practical side of the matter has been dealt with at length in the body of the book. In this chapter we shall confine the subject to Portland cement stucco work, but at the same time give attention to the finishing of surfaces. Methods will be detailed which, although evolved in connection with cement mortar rendering, are also applicable to ordinary plaster work. It will be seen that they offer many charming contrasts and openings for the display of individual skill, so call for careful study.

PORTLAND CEMENT STUCCO

By Portland cement stucco is meant something different from, and superior to, an ordinary mixture of cement and sand. The aim is to produce a material which will set with moderate rapidity, dry hard and give good protection, but which will be more plastic than ordinary cement mortar, and possessing reasonable retardation properties of setting, can be manipulated over a sufficiently long period to give freedom in obtaining decorative effects and without interfering with ultimate or definite setting.

MATERIALS AND MIXING

For this purpose a fine-grained, slow-setting Portland cement is chosen. It is mixed dry, with not more than one-fifth of its volume of hydrated lime, the desired amount of washed sand, together with any coarser aggregate, such as screenings of crushed stones or crushed ballast. The usual proportions are 1 part Portland cement to 3 parts aggregate. Any colouring matter, which should be of a mineral nature, such as those mentioned in the table printed on another page, is next added, but never in a greater quantity than 10 per cent. of the volume of the cement. The whole should be intimately mixed. This is best attained by shovelling the materials from one heap to another through a sieve of appropriate mesh, until the mass is of a perfectly uniform colour. The mixture is then ready for gauging, working up only just enough for immediate requirements. The water should be clean and only enough used to produce a workable mix.

A mortar-mixing machine is a good mixer. It should be worked for about five minutes after adding the water. If to be hand-mixed, use a water-tight box, sprinkle in the water, and turn over with a shovel until thoroughly hydrated. It should feel like well-made dough, not wet or crumbly.

Application

The method of application is much like that adopted in ordinary plastering, only it is found that the best results are obtained if the foundation is wire-netting or metal lathing. If applied direct to brickwork or masonry, the jointings should be raked out and the surface scratched to provide a key. The surface must be brushed down and well wetted.

First Coat.—The first coat, which need not be coloured or contain aggregate of a decorative value, is applied about ¼″ thick. It should be well trowelled into lathing, and if it is to be back-plastered should contain hair in about the same proportion as for plasterers' rough-stuff. It must be made quite level, and then deeply cross-scratched, using the pointed lath or other implement at an angle, so as to improve the key.

Second Coat.—If it is to be two-coat work, the second coat should be applied as soon as the first coat is completed—that is, levelled and scratched. It is laid on ⅜″ to ½″ thick, and levelled with a straight-edge and smoothed with wooden float kept wet.

On the other hand, for three-coat work the first coat should be allowed to stand for 12 to 14 hours. It is then wetted and the second coat applied, about ⅛″ thick, levelled with the straight-edge, and when sufficiently set, smoothed with a wooden float kept wet.

The Finishing Coat.—The finishing or setting coat is not to be applied until a lapse of at least eight days after the second coat. This coat varies in thickness according to the finish adopted. Apart from rough castings, the finishing coat consists of cement, hydrated lime and fine sand. This is the coat which requires tinting, though for some finishes it is also essential that the second coat should be coloured.

In all cases it is well when possible to finish a wall in one job. If this cannot be done, the day's job should be carried to a convenient point where joining up can be easily concealed. The edges should be left rough and well wetted, and on resuming work the surface of the edges should be brushed off and again wetted, to secure a good key and adhesion.

The different coats should be encouraged to set and dry slowly. This can be done, keeping the surface sprinkled with water and, if necessary, covered with wet sacking.

Finishing

So far the method presents no great difficulties. The art resides in the finishing processes. These are capable of very wide variations, and here we shall touch upon only the leading types, which contain the underlying principles that will suggest other developments.

Smooth Finish.—When a smooth finish is desired, the sand should be sharp, but as fine as possible, and the full percentage of hydrated lime added. It is trowelled and finished with the wet float, but the less it is worked the better.

Sand Floated.—A sand-floated finish is obtained by rubbing the surface in a circular manner with a wetted wooden float, sprinkling on a little sand to secure whatever degree of roughness is aimed at.

Stippled Finish.—This finish is obtained by going over the surface with a suitable brush of bristle, wire, etc. The closeness and depth of the stippling can be controlled, but it is advisable to secure an even effect.

Sand Sprayed.—This is a mild form of dash applied to a smooth or sand-floated surface. A grout of equal parts of cement and sand (the coarseness varying according to taste) is prepared and applied to the surface by dashing on with a whisk broom. The dash can be light and close, or heavy and open. Excellent effects can be obtained by adopting different colours or shades for the finishing coat and the grout.

Exposed Aggregate.—Another way of securing broken colour is to use strongly coloured aggregate in the finishing coat. Twenty-four hours after trowelling smooth, the surface is washed with plenty of water and a stiff bristle or wire brush. In dry weather it may be necessary to add 1 part of dissolved hydrochloric acid in 6 parts of water. After exposing the aggregate, the acid solution must be promptly and thoroughly washed off.

Dashing.—Any kind of dash can be applied before the finishing coat is quite dry. The simplest method is to throw against the damp cement-mortar sand stone screenings or fine gravel. If a strong dash is desired, the aggregate can be mixed in a thin grout.

Sponge Finish.—The sponge finish, perhaps, gives the best opportunity for introducing novelty and expressing individuality as to touch. The finishing coat should contain the full percentage of hydrated lime and gauged rather soft. It is laid on $\frac{1}{2}''$ to $\frac{3}{4}''$ thick, lightly straight-edged to secure a true level, and while still wet pressed into and slightly wiped up or down in graceful sweeps with a moist sponge. The sponge will have to be washed out from time to time and squeezed dry. It will be readily understood that by selecting the texture of the sponge, and adopting the degree of pressure and boldness of sweep, suitable and beautiful surfaces can be obtained.

The finish may be given an old ivory tint, and when dry and hard given a wax polish, or the setting may be left white and when dry given a coat of scumbling colour, wiped off with the sponge and then wax polished, the colouring being deepest in the depressions, and the high polish in the elevated portions.

Modelling.—If the finishing coat is fine and about $\frac{3}{4}''$ thick, it is possible to carry out modelling in low relief. An admirable instance of this consisted of the lower part of the wall being sponge finished in a bluey-green tint of uneven surface sprayed to represent waves ; on these was lightly modelled a galleon under full sail ; while the upper part of the wall was sponge finished to give cloud effects in a pale blue tint, with fleecy white clouds and an albatross or two. Another example gave vague suggestions of human and animal figures, partially concealed by the sponging.

A type of modelled ornament which has great charm is the making of maps, on which symbols can be introduced. This is quite suitable for a hall in a country house, a panel over a chimney-piece, or walls in schools and some other descriptions of public buildings.

Sgraffito.—Where the upper or finishing coat differs in colour from the undercoat, it is easy to introduce incised decoration, emphasis being given by deepening and widening the cuts as desired. The two colours or shades are shown in the designs.

Other Variations.—It is also possible to mingle two or more of the finishes described, so that you can have sponged panels within smooth or sand-floated borders. Other changes will suggest themselves.

Colouring

The colouring agents should be of a mineral nature and mixed dry in the proportions given in the table below :

TABLE OF COLOURS FOR PORTLAND CEMENT STUCCO

Colour.	Pigments.	Pounds of colour required for each bag of cement.	
		Light shade.	Medium shade.
Greys, blue-black and black	Lampblack	½	1
	Carbon black	½	1
	Black oxide of manganese	1	2
Blue shade	Ultramarine blue	5	9
Brownish-red to dull brick-red	Red oxide of iron	5	9
Bright red to vermilion	Mineral Turkey red	5	9
Red sandstone to purplish red	Indian red	5	9
Brown to reddish-brown	Metallic brown (oxide)	5	9
Buff, colonial tint and yellow	Yellow ochre	5	9
Green shade	Chromium oxide	5	9
	Greenish-blue ultramarine	6	—

Constructional Value

In America Portland cement stucco is largely used, not merely as a rendering, but as a structural element in frame building, the stucco being applied to metal lathing. A notable instance of this is a large three-storied building in the " Colonial " classic style just built by Mr. Clarence E. Shepard, Architect, in Kansas City. It is a frame construction with Portland cement stucco on metal lathing. Interior portions are wood studding, with wood laths plastered and given a hard finish.

Here are a few constructional details :

Stucco on sheathed frame construction is not intended to add greatly to the rigidity of the building. Diagonal bracing should be used at corners of walls, and the details of the framework given attention to reduce the vibration in the structure to a minimum. Bridging should always be used for the back-plastered type for which sheathing is omitted.

Sheathing should be horizontal and not diagonal. It should be firmly nailed to the studs and covered with a good quality tar or asphalt paper, with plenty of lap between sheets and at baseboard and flashing. Furring should be $\frac{3}{8}$" crimped, and should be fastened over the sheathing paper directly along the lines of studs with 1$\frac{1}{4}$" staples 12" apart. Furring should be placed entirely around all openings and around curved surfaces.

Where metal furring for back-plastered construction is used, it should be stapled to studs at intervals not greater than 12" with 1$\frac{1}{4}$" by 14-gauge staples. Self-furring lath should be rigidly fastened to the studs with staples.

Metal lath should be galvanised or painted; expanded metal should weigh not less than 3·4 lb. per square yard, or galvanised or painted woven wire be not lighter than 19-gauge 2$\frac{1}{2}$ meshes to the inch, and stiffeners placed at 8" intervals.

Vertical laps of metal lath should be at supports and fastened with staples 4" apart. Horizontal joints should be locked or butted, or tightly laced with 18-gauge galvanised wire. Sheets of metal lath should be folded around the corners and firmly stapled down. Corner beads should not be used.

Additional insulation may be given to back-plastered walls by fastening felt fibre or cork insulating sheets to the studs between the back-plaster coat and the interior finish. A heavy grade of good building paper may be used for this purpose, but should be so placed as to leave an air space between it and the stucco.

Old frame structures to be stuccoed should be made strictly sound and rigid before applying the stucco.

CHAPTER XIII

MODELLING IN PLASTER AND PLASTIC CONCRETE

Types of Modelling—Coring or Backing Out for Modelling in Clay—Modelling in Plaster—Method of Gauging—Modelling in Cement Mortar—Method of Proceeding—Construction of the Core—Gauging the Material—Plastic Concrete—Gauging and Using the Material required for Plastic Concrete—Forming Caryatides, or Work of Similar Description, in Plastic Concrete—Cavo-relievo.

MODELLING is an art which one can safely say dates back to an unknown period. Whether executed in clay, plaster, cement or any other material, the art of modelling is one of importance in plastic decoration.

The young plasterer will require a great amount of study to make himself proficient in this particular kind of craftsmanship. It is essential for him to have a good knowledge of drawing. The distribution of light and shade has to be studied as it affects the low or high relief of the work. It is necessary to cultivate freedom in interpreting a design to be modelled, so as to obtain the following lines when dealing with foliage.

Proportion has also to be taken into consideration, as the various parts of the modelling must have the same relationships, for unless the design is evenly balanced, the finished model will have none of that beauty which a well-proportioned design shows.

TYPES OF MODELLING

There are several kinds of modelling. Modelling in the round, such as figures, which are usually taken from life, or from scale models and also from photographs, and high and low relief modelling—such as we see in plaques, panels, bridges, etc. Ornament is usually modelled from a detail drawn on paper or cardboard in pencil and shaded to show any particular relief required. Now, to give the needed roundness or projection to these in the plastic copy, requires not only skill, but a thorough knowledge of the objects represented. Thus in figure modelling and moulding it is most necessary to acquire a thorough familiarity with the figure form, anatomy and the proportions of the various parts of the human or animal body. In modelling ornaments of all kinds, the study of light and shade is of very great importance, for this must be taken into consideration and varied in accordance with the position, especially the height, the finished work is to occupy. Thus, if modelled work is to be placed high above the line of vision, emphasis must be given to the upper part, while if it is to be below the line of vision, the emphasis must be laid on the lower part, the upper portion then being in lower relief.

Coring or Backing Out for Modelling in Clay

When preparing the background for modelling in clay, a full-size detail drawing, or half section full size, is required to enable the plasterer to get the proper profile and depth of the various parts of the design.

The measurements are taken from a vertical line drawn in the centre of the position of the design, then setting off a horizontal line at right angles to it through the other centre or some principal part of it. If the design has any large prominent places, wood cradling (a framework of wood) must be fixed at these parts. Such cradling can be covered with expanded metal or laths, and should be kept back at least 1″ from the fullest place. This allows sufficient room for a thickness of clay. In the parts where, owing to the unevenness of the design, it cannot be framed out, the following method is adopted. Drive in on the skew, or drive in and bend over, nails of various lengths to suit thickness, leaving sufficient length protruding to form a key, but not more than 1″ to the finished face of design. When this has been completed, plaster is required for the next process; also tow, hemp, straw or wood shavings. The plaster is gauged up in a bowl or pail, according to amount required. When gauged the tow, or the material being used, is dipped in the plaster (which should be fairly round), and when well soaked is pressed round the nails, filling out the part required. If a large amount of plaster is required, size water is added to the gauging water for the plaster, to retard the setting (size such as is used by the painters). This gives the plasterers more time to form up and pack out places where required.

This method of coring out saves a great deal of weight and clay, as the modelling is modelled upright and then lowered down on the bench for moulding.

Modelling in Plaster

When modelling in plaster, the same procedure is carried out as described in the previous paragraph in coring out for modelling in clay, with the difference that the one in clay is done on a large board in the shop or studio and the other *in situ* on the job. The nails for the job *in situ* must be galvanised to prevent rust. Having carried out all the foregoing instructions as to the setting out and the cradling and the method of gauging and coring out, the whole of the surface of the design is covered with a fair coat of plaster, mixed with a small quantity of well-beaten hair (the hair being soaked in water before use). The horizontal and vertical lines are re-marked and the design sketched in with a piece of coloured crayon or charcoal, deepening the colour for the prominent parts. Any small prominent places too small for coring out with shavings, etc., should have a few nails driven in for a key. The design having been sketched out, everything is ready for the gauging of the plaster. In most cases the first part of the design requires a big gauging, so as to form the most prominent parts. If such is the case and the design is large, a pail of plaster can be gauged and used quite easily before it sets.

Method of Gauging

In half a pail of water carefully sprinkle in plaster with the hand, allowing it to well soak, as this is the chief thing in gauging plaster. The plaster being thoroughly soaked, it should be well stirred up with the hands, and should then be like thin cream. The plasterer should grease his hands with tallow before mixing. By this method time is gained by using a large quantity of plaster before it sets (without the use of size); the plaster is much better to use and put in position where required with the hands and small tool. If plaster is gauged in this way, when it is stiffening up, by adding more water and keeping on manipulating with the hands and continually adding water, the plaster can be kept soft and workable for some considerable time. But this method of manipulating with the hands should not be put in practice till the design has been well filled out and only the finishing touches are required. Wire tools are necessary to drag off any plaster that sets before being modelled in shape. However, it is much the better way to gauge in small quantities and model in shape before the plaster sets.

If the weather is warm, which affects the setting of plaster, size water can be added to retard the setting; but if this is the case the plaster can be opened out in a large bin several hours before using.

By being exposed to the air, setting is retarded to a great extent. Scratch and wire tools are very useful in plaster modelling, as many small parts of the design can be finished off much quicker by adding soft creamy plaster where required and dragging off with the wire tools. It is very important when gauging plaster for the finishing touches that it is gauged very milky and allowed to stand before using, as it soons stiffens up and reaches a creamy consistency. This allows plenty of time for use.

Modelling in Cement Mortar

Modelling in cement is usually carried out on external work. The material generally consists of cement and mortars. The mortar consists of lime and sand, and is made up in the manner described for external plastering, or it can be made up as coarse stuff without the hair. The amount of cement added to the mortar depends on the size of the design. If the design has large projecting parts, it requires to be gauged half and half of cement and mortar. Cement mortar requires time to set, therefore it must be gauged strong, otherwise it would be continually falling off; apart from that, it would cause much time to be wasted waiting for the material to set.

Method of Proceeding

The method adopted for modelling in cement is similar to that used in plaster. The design is sketched out from the vertical and horizontal lines and any prominent parts noted. But the next proceeding is quite different, as regards the coring out. The cement being a much heavier material, and being used for external work exposed to all weathers, the construction must be made strong, otherwise when filling out the design there would be a possibility of the whole design falling off with its own weight.

Construction of the Core

The prominent parts of the design being marked, holes are cut in the brickwork or concrete (that is, assuming the background is brick or ferro concrete), and short lengths of iron bars, two or three together if necessary, are cemented in the background, protruding sufficiently to provide a support, but so as to keep clear of the face of design. Roof tiles are split in half and are used in place of the iron bars. The tiles, being split the long side, are put in on their edge. They are better to use than iron bars, as the iron has a tendency to rust in time.

Gauging the Material

Having fixed the irons or tiles, cement and coarse stuff, with a small quantity of hair added, gauged 2 parts of cement to 1 part of coarse stuff well mixed, is covered over the irons or tiles, and, after well soaking the brickwork or concrete with water, the whole of the part to be modelled can be covered with the same sort of material. This should be well scratched after being rendered with the front corner part of the trowel. The design should be sketched out roughly again and the modelling proceeded with, scratching each portion of the design after each coat. After the first coat cement and mortar, or sand only, is used, and as the modelling is proceeding less mortar is used and more sand, which should be well washed through a fine sieve. The material should be freshly gauged for each coat, and should be shaped out as much as possible to the design when being put on with the trowel or small tool. Here again the use of the wire or scratch tools will be found very useful. The design having been well formed and modelled into shape, cement and washed sand (half and half of each) is gauged in whatever amount required, and the work is covered, starting from the top and working downwards. It can be laid on with the gauging trowel or small tool, according to the size of the ornament. This process is greatly assisted by the wire tools, which are made in various shapes and sizes. When the whole of the work has been modelled and allowed to get hard, it greatly improves the look if a little cement wash, not too thick, is brushed over it.

Plastic Concrete

Plastic concrete can be used to great advantage when forming concrete mouldings, such as described in mouldings in coloured cements. With reverse moulds for concrete enriched mouldings required to be done *in situ*, plastic concrete can be adopted when filling in the mould. Work of this description requires to be executed as soon as the moulds are taken down, so that the ornament can be carved out while the material is soft or workable. The face of concrete usually remains green on the surface till exposed to the atmosphere; therefore it is necessary to follow this kind of work very speedily.

Modelling plastic concrete is carried out on similar lines to those used in cement, though in carving in relief the cement is laid on the surface to be covered in a thickness or body, and then the design carved out. In concrete, moulds are necessary for the filling in of the material generally in a large body.

When artificial stone is being made with a cement face and concrete backing, and is made and transported to the various jobs, the use of reverse moulds, either in wood or plaster, for enrichment, or dentil or other blocks, are used to a great advantage, as imitation stone not made *in situ* is made in sections; therefore, the reverse mould for enrichments or blocks, etc., would only be of short lengths. If enrichments are required to be done *in situ*, as described in coloured cement, with a reverse mould, plastic concrete is the best material. Referring to Fig. 61 in that chapter, the dotted section at E is required to be enriched. When filling in this section, extra thickness of material of a moist substance must be laid to allow for the carving out and the depth of ornament.

Gauging and Using, and the Material Required for, Plastic Concrete

Sections of enriched moulding that are required to be executed in plastic concrete must be of a very fine substance, otherwise when forming the ornament required great difficulty would be experienced in manipulating, if the material used were coarse. Washed sand is necessary and should not be used too wet. When gauging for this work, 3 parts of sand to 1 part of cement should be well mixed, dry, as it is not mixed up with water as in gauging the ordinary sand and cement. Having well mixed the material, it is tested by the plasterer taking some up in his hand and squeezing it together as though making a snowball. If it is too dry, it must be sprinkled with water; if too wet, dry sand must be added with the proportionate part of cement. The material being of the right texture, it is pressed in the section of the mould as described in Chapter III. It should be well pressed till the required thickness has been put in, then the remaining part of the filling in is proceeded with as described. When the moulds are taken down, each length of ornament should be completed before another length of mould is removed. The filling in of boxes for concrete trusses or cantilevers which have to be carved is carried out in the manner described. A plain box or shuttering is made up, filled in and taken down as the work proceeds, the design being carved out in the plastic concrete just as though it was a block of stone, though the plastic concrete, if properly made, is much easier to carve.

To give the finished concrete an aluminate appearance, colours can be added as described, mixing the tints required with the cement and sand before wetting up. The desired plasticity and retardation in setting is attained by the addition of hydrated lime, which produces an almost unctuous mass, making moulding, modelling and carving all the easier.

Forming Caryatides, or Work of Similar Description, in Plastic Concrete

If several caryatides have to be formed, it is much quicker and better to rough out the figure in clay and prepare the model as for moulding (leaving it free from undercut), then casting several moulds from it. These can be fixed up in the position required and filled in as described, not forgetting that the prominent parts require iron supports let in the wall for reinforcing. When sufficient time has elapsed, the plaster moulds are taken down and the carving out proceeded with. If only one subject is required, whatever it may be, the box or shuttering shaped near to the design is the cheapest method. Any similar work that requires to be executed in plastic concrete can always be carried out with very little expense as regards shuttering by profiling the framing to as near the shape of the subject as possible.

Cavo-relievo

Where elaborate modelling is contemplated, the cavo-relievo method, a cameo-like treatment, is excellent. In this, though the modelling may be in high relief, with rounding almost amounting to under-cutting, yet the highest point always remains on a level with the surrounding wall. That is to say, panels are traced out on which the design is drawn, and then the background of the panel is cut away, leaving the design standing out, which then can be finished with chisel and hammer to any degree of fineness desired. It is very suitable for figure and emblematic designs and also for lettering.

This treatment is in happy relation to the material itself, and so is æsthetically true. Though this is not to say that plain relief modelling should be debarred. An example of work which may be mentioned is the fine modelling in plaster of heroic-sized nude figures, by Acke, seen on the embrasures of a long series of windows in the Prince's Gallery of the new Stadhus, Stockholm.

CHAPTER XIV

MISCELLANEOUS DECORATIVE PROCESSES

Sgraffito—Preparing the Walls for the Designs—Depeter—Pebble Pictures—Pargeting —Coloured Cement Plaster and Imitation Stone in Ferro-Concrete—Coloured Cement Plain Face Decoration—Coloured Inlays in Cement Slabs and Tiles.

Sgraffito

Sgraffito, or graffito, plaster work is a very ancient method of surface decoration. Although having some affinity to the old English pargetting, this type is more popular on the Continent, but it has certain merits which entitles it to attention.

Strictly speaking, sgraffito is an incised form of ornamentation, a scratching through the finishing coat in thin lines. Occasionally, however, it is carried out in deeper cutting, the incisions having sloping sides, forming V-shaped grooves. It is also sometimes associated with raised ornaments in low relief.

It will thus be seen that sgraffito work calls for some artistic taste and manipulative ability.

Sgraffito in interior work is usually carried out on a fine stucco surface, or on a finish of Keene's or Parian. The cement should be exposed to the air for twenty-four hours before using, in order to delay the setting. This allows more time to execute the design. If for any reason the Keene's or Parian has not been exposed to the air, a small quantity of old lime or putty can be added to it when gauged.

Preparing the Walls for the Design

The walls are rendered with the ordinary coarse stuff, with a fifth part of Portland cement added, and allowed to stand to get dry. Before rendering the walls they should be well soaked with water, as then the rendering adheres so much better. If this were not done the suction of the bricks would drain a great amount of the moisture out of the rendering material, and so, instead of it setting, it would merely dry—becoming weak and powdery. The finish need not be in Keene's or Parian cement, it can be lime with a very small quantity of washed sand ; this last assists in the hardening and prevents a certain amount of fire-cracking. But in the final coat pure lime putty should be used, so as to get a sharper edge for the finished design. When the rendering has been allowed to become dry, the next process is to prepare for the finishing of the design. The rendering should be well soaked with as much water as it will absorb.

Then the design is spaced or set out to the required dimensions. Sufficient materials must be mixed to complete the job.

During the mixing, colour is added: if red, Venetian red; if black, lamp black; if white, chalk lime putty. But whatever material is required, sufficient coloured stuff must be gauged to complete the job, otherwise there is sure to be a difference in tint. Assuming a frieze design is being executed with a white background, a sufficient quantity of lime, with a small amount of sand added, is gauged with 3 parts of white Keene's or Parian cement. This is laid on the whole length and width of the frieze, in sufficient thickness to cover the rendering coat. This, when stiffened up, is stippled over with a stiff brush to give a key for the next coat. This next coat is floated on immediately, so that it adheres to the background. This coat should not be more than $\frac{3}{16}''$ thick (unless it is a design in low relief), and when the design is being cut out it should be cut on the splay, which gives it the appearance of being thicker than it really is. Neat Keene's or Parian (which has been exposed to the air), with the desired colours added, should be used. This is traversed, stuff filled in where required, and then rubbed up lightly with the float.

After this has been completed, the design, which of course should be of a full-size sketch, is fastened over the face of the work. It is then either pierced through with a fine bradawl or the design traced lightly with a soft lead pencil. If it is desired to show another colour in the depth of the cut lines, it simply means that another coat, not more than $\frac{1}{8}''$ thick, is added. But in all cases the thickness of each coat depends entirely on the design; the thickness of the coats being varied to suit the colours and the design. The beauty of this kind of work greatly depends on the treatment of design, the clearness of the lines and the blending of the colours. The character of the line is also important. If a smooth surface finish is required (after the float has been passed over, as previously described), the final coat is trowelled in, and then the design traced or pierced as before mentioned, showing up the colours to better effect and strengthening the light and shade of the finished design.

With external sgraffito decoration the method is much the same as with interior work. The brickwork, or whatever may be the backing, is rendered in cement and sand, measured out 3 parts of sand to 1 part of cement.

The rendering coat should be well roughened to form a key, and should be allowed to become thoroughly dry before laying on the finishing coats. When the rendering is in fit condition to proceed, the design is nailed up temporarily in the correct position, and if the detail is in low relief and in different colours, those parts are pounced or pricked through sufficiently to be seen on the rendering coat. The sketch is then removed and the spaces for the various colours or thickness marked in with chalk or charcoal pencil. Care must be taken that the position of the design is noted when fixing it up temporarily, so that it can be put back again when desired. This can be assured by fixing up the sketch with nails and replacing them when taking the sketch down. Having pierced the design and sketched in the detail or outline, the colour coats are laid on according to the pencilled outline. This saves a lot of unnecessary material being used and wasted (which otherwise would have to be cut away). If the coloured background coat is put on after the rendering, the next coats are only required for the extra thickness in the detail. Having laid on the background and the colour coats for detail (which should be left rough), the final surface coat is laid on, and as much as can be cut and cleaned up in the one day is put on. This coat can be left with the float or trowel, but much the better finish is obtained if left with the float. When the final coat is sufficiently firm, the design is again fixed up in position, guided by the nails previously mentioned. The detail is then pounced or pencilled through, the drawing removed and the outline cut out in the surface coat with a short stumped and pointed knife. It is next cut through to the various colours according to the design. With Portland cement for the colour coats, it should be laid on a dry floor and exposed to the air for at least twenty-four hours before using, and should be gauged 3 parts of cement to 2 parts of colour, according to tints required, a small quantity of washed sand can be added (either pit or silver sand). If a black background is required, the rendering coat can be gauged with cement and ashes instead of sand ; if red, red oxide can be added up to 10 per cent. Portland cement gives grey colours, and whiting or lime, white. Other colours, such as ochres, lime blues or brunswick green, and all such found in distempers, can be used.

Depeter

Depeter is a similar plaster decoration to rough-casting, though rough-casting generally denotes shingle or other materials being thrown on to make a surface decoration of a uniform size. Depeter work is more decorative. The brickwork or background (which may be lathwork) is rendered over with a fair thickness of gauged cement, coarse stuff or Portland cement and sand. As soon as it is laid it is straightened up to a uniform thickness with the skimming float, and whilst it is still soft various kinds of materials are pressed in the face, such as broken flints, crockery, spar glass, seashells, marble or broken pottery. This kind of plaster decoration would be very effective in gables to high-pitch roofs, and, if used in conjunction with rough-cast and coloured cement bands, makes a good finish. Much of the effectiveness of this finish depends on the colour and treatment of the bands or margins.

Pebble Pictures

A form of art allied to some of the above processes that was practised during the eighteenth century in England might well be revived. I refer to pebble pictures, which would be suitable for both interior and exterior decoration, and might well be used in the designs of memorials, as well as forming panels on walls of inns, shops and so on.

Pebble pictures consist of large or small slabs of stone or slate which are partly covered with a thin layer of cement, in which pebbles are embedded so as to form pictures, portraits or decorative designs. The procedure is simple enough. Having made a drawing of the size desired, it is traced on the slab; or the drawing may be done in freehand direct on the slab, using black or white paint or chalk, as may be more convenient. The part to be covered with cement and pebbles should then be roughened, to give a good key. The border may be left rough, dressed smooth or polished, as desired; or there may be bands of matt and polished surfaces.

Pebbles must be carefully selected, assorted according to colour and then as to size. In some cases one can go down to coarse sand. Split flints and stone chips are also admissible. If monochrome effects are desired, it will only be necessary to collect different shades of grey, brown, yellow or red. But polychrome pictures are quite possible, for to the above colours can be added white, black, green and certain shades of blue. The more elaborate the design and the colour scheme, the greater precautions must be taken to have various sizes, shades and colours arranged in conveniently accessible heaps or trays. The slab is placed flat on the ground or on a table, and a moderately slow drying cement laid on with a trowel and smoothed to gauge. Only a small amount should be applied at a time. Then the pebbles and sand are laid on the cement in accordance with the drawing, using small or large, dark or light, so as to secure the necessary effect of roundness, of light and shade. It is preferable to keep the picture flat, but bossing or relief effects are quite possible. But where relief is attempted, the work will be less durable, being more exposed, not only to rubbing, but to weathering. The pebbles should be pressed in gently, so as to cover the whole surface. With the point of the trowel, or a flat, flexible-bladed knife, any corrections can be made. Any bad mistake will make it necessary to remove the pebbles and cement, clean up and lay down a fresh coating of cement. So the work demands both dexterity and celerity. The work is carried out bit by bit.

The main difficulty is with the cement, as only a small portion of the design can be proceeded with at a time; for when an extensive surface is in question, this requires considerable management.

Pebble picture slabs can be treated as panels, rectangular, circular or oval, or as friezes. In the first case these can be mounted in frames. But the effect is better if they are treated as accessories for decoration and built into stone or brick walls, or set into concrete structures. In the last case the pebbled slabs would be of great use in softening the monotonous hardness of lines and dullness of tone so characteristic of concrete.

Pebble pictures, like rough-casting, afford an outlet for utilising stone-yard waste and local pebbles or sand deposits.

Pargeting

This is a very old system of floating, and formerly was generally applied to exterior plaster decoration in England, especially in the eastern counties and in the south-east, on the plastered spaces of half-timbered houses. The ornamentation is either stamped with an engraved wood or metal stamp, having a repeat design to form panels, or with a running pattern engraved on a roller. This was often associated with hand-worked foliage of low relief. Many patterns, however, are quite simple. For instance, a very old pattern, Fig. 133, can be imitated by making a comb with umbrella ribs, these, having a V-section, provide us with what wood carvers call a parting tool. If the back parts are used, a small V-shaped line is obtained, and by commencing from one angle, and following on in a half circle to whatever space is required to be covered, the design is soon completed. Another design can be made, Fig. 134, by getting a very stiff twig brush, like a gardener's worn birch broom, and when the surface has been covered, stabbing the brush evenly over the surface, starting from one angle and carrying on. Yet another old pattern, Fig. 136, is partly stamped and partlyfinished by hand. The panel is filled out and lightly finished, then the diagonal bands are lightly marked on with chalk or crayon, and the material (which could be of very fat lime and washed sand or Portland and washed sand) laid on where required. A piece of wood is cut to the shape with handle attached, Fig. 135, and a piece of stout zinc or sheet iron is

Fig. 133.

Fig. 134.

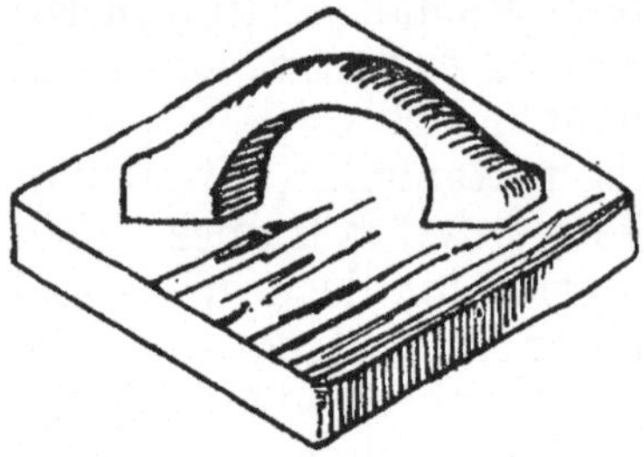

Fig. 135.

tacked round the edge, this being beyond the wood face and the depth that the cut line is desired to be. Before cutting the line, the surface is stippled with the twig brush. The cutting in need not be regular, so long as it is uniform. The reversed fluting pattern, Fig. 137, can either be executed in cement and sand of medium grain or in lime and sand. This design can be roughly drawn on the background with chalk and the material laid up or crossways, somewhat near the design chalked out. This gives a guidance for starting the fluting, and only requires a little practice.

Fig. 136.

A popular form of parget work is to render the brick-work over with a thin coat of cement and sand, and brush it over with a stiff brush to form a key, then to lay a fair thickness (about $\frac{5}{8}''$) and press or stamp in the design with a zinc stamper (made as previously described), setting out the circular pattern from horizontal lines and the cross pattern from the vertical lines, as in Fig. 138, where A and B are horizontal lines, 1 to 2/3 to 4 vertical lines. A chalk line is snapped from 1 to 2 and 3 to 4 over the space required to cover, and then another across from vertical lines. The whole is stamped out, and the intermediate ovals are fitted in and stamped according-ly. Many of the present-day houses of the bungalow type are being ornamented with a similar pattern, with

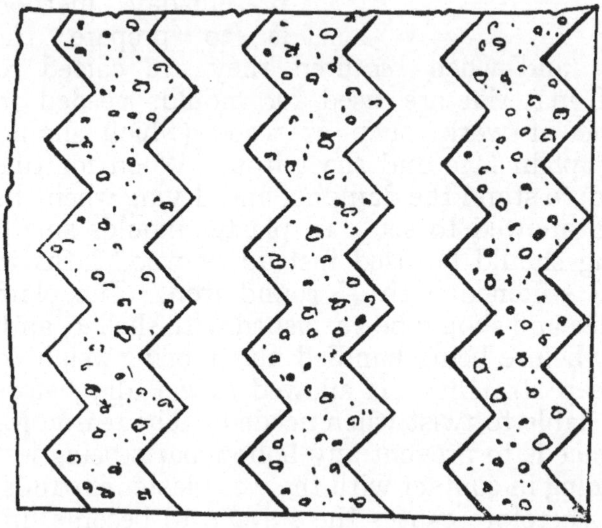

Fig. 137.

a rough-finished surface. The brickwork is rendered over, and when soft a coarse broom or wire comb is brushed indiscriminately over the surface. In a variation, the rendering is daubed when soft with a stiff grass brush. Other walls are rendered, and when set, white cement and sand is gauged up and allowed to get lumpy and then laid over with the float and left. The last few styles are coloured washed over to suit taste.

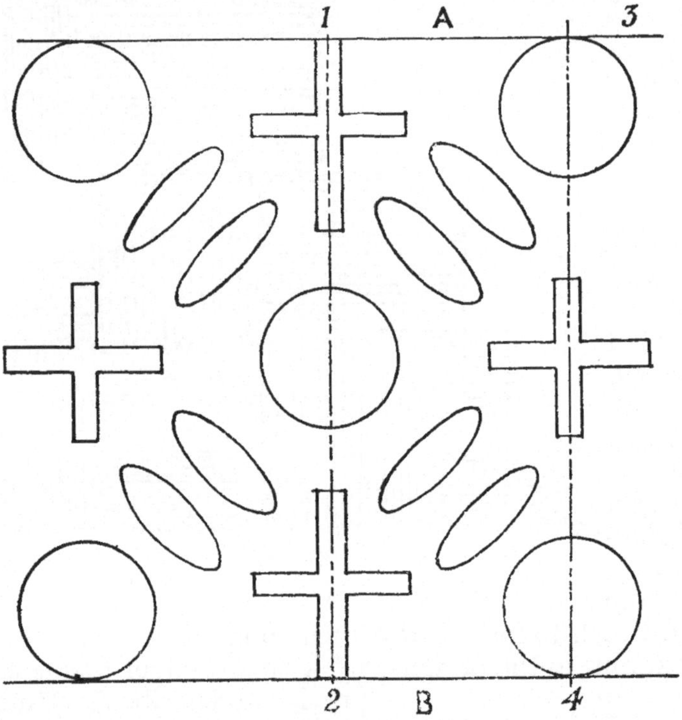

Fig. 138.

Coloured Cement Plaster and Imitation Stone in Ferro-Concrete

Coloured plaster (with cement and sand) are used for the imitation of stone. At one period in London this was largely adopted for great blocks of flats and dwellings, either in plain or ornamental work, and in the majority of cases this work was done *in situ*. The present-day method is to prepare the blocks away from the job, and when required they are carted to the site. Several kinds of material are used for moulds needed in the preparation of this kind of work, such as wood (which should be planed smooth), plaster plain face and sheet iron. Wood for this purpose must be well seasoned, to stand the damping and drying when the wet material is placed in and allowed to set. In plaster moulds small-meshed ordinary wire-netting should be used instead of wood laths in addition to the canvas, and stiffened with $\frac{1}{4}''$ round iron. The plain face is made to the required size and should be varnished with shellac, and, if possible, allowed to get dry before being handled much, being weighted down on a flat surface to keep straight. If allowed to get dry before being used, the slab is not so liable to twist when needed. The reason for not using wood in the plain face is to prevent any hollow parts being left in the face. Wood when coming in contact with the wet plaster expands, and contracts when dry, and therefore causes the surface to become uneven, with the result that the face of the finished work would become

COLOUR INLAID CEMENT SLABS BY MR. THOMAS BEATTIE.

(By courtesy of *Building*.)

very uneven. With iron the foundation is the same as wood, only instead of the wood being planed, it is left rough and the sheet iron tacked on, with sufficient flat-headed nails to keep it in position, the holes for the heads of the nails being counter-sunk, to allow the heads to become flush. A coarse file is passed over to take off any rough edges. Red oxide and the natural grey of Portland cement and sand are the colours most in demand.

Coloured Cement Plain Face Decoration

If only a plain face decoration is required and mouldings run *in situ*, the mouldings are cored out in the usual manner, allowing $\frac{1}{4}''$ thickness muffle for finishing coat. The finishing is carried out in the same manner as in running colours. The mouldings have to be worked up with the floats as much as possible to a finish, to help the running off of the moulding. In plain face work finished with a coloured surface the background is floated out in the ordinary way with cement and sand, and the finishing coat proceeded with as described in finishing cement face. A sufficient quantity should be mixed dry with the Portland cement. The sand should be washed, but it is desirable to let the sand get dry before mixing.

Coloured Inlays in Cement Slabs and Tiles

Everybody admires the old brilliantly coloured inlaid Hispano-Moresque tiles, copied in much more sober tones but with bold and delightful effects by the Norman and Gothic builders. A somewhat analogous system of decoration has been applied to cement tiles and slabs by Mr. Thomas Beattie and colleagues of Edinburgh, some of whose work, reproduced from *Building*, we illustrate on Plate IX. The tiles and slabs are moulded, usually with a bottom plate embossed with the design, which therefore appears sunken on the finished tile. Colours in powder are mixed with dry cement, gauged in small batches and used to fill in the depressions according to the colour scheme. Smooth or rough textures are produced by using pure cement or different qualities and quantities of sand in the mix. The secret of the process resides chiefly in the selection and mixing of the pigments and the subsequent drying. The tiles, slabs or other moulded and decorated architectural parts are set up and fixed as with any other kind, but it is found that they become covered with efflorescence, and so they require to be liberally washed down periodically until thoroughly cured.

CHAPTER XV

GESSO

Preparation—Mixtures—Application—Cast Ornament—Colouring—Gilding—Applicability.

Gesso is a form of decorative work which partakes at once of the nature of painting and modelling. It is a choice type of stucco in which gypsum, marble dust and occasionally other materials are mixed with size water or dilute glue, to form a soft, plastic mass or a stiff, semi-liquid paste, applied either with the palette knife or with brushes.

Preparation

The Greek and Roman mural and ceiling decoration, revived by the Italians in the fifteenth century under the name of stucco, was composed of 3 parts of white marble dust and 1 of air-slaked lime, gauged with water. When gypsum was added in order to increase hardness, the substance was known as *gesso duro* or *gesso grosso*, hard or coarse gesso. Fine gesso, according to early practice, was prepared by slaking white marble dust in a damp atmosphere, repeatedly chopping up the cake thus formed and turning it over. The process lasted some months. It was finally reground and mixed with glue made from hides, diluted as needed for applying with the palette knife or the brush, the latter mixture being rather more fluid, though not enough so to run.

Mixtures

A more speedy way of preparing gesso paste is to slake gypsum or plaster of paris in water, stirring it vigorously for half an hour or more and setting aside for thirty days, but stirring once every day until quite smooth and "fat." It is then shaped in small cakes, placed in muslin and squeezed dry; or the water is drained off and the broken-up mass air-dried. When required, the gypsum is pounded fine and mixed with pure hide or parchment size.

A mixture much used was prepared by soaking fine glue in cold water, diluting the jelly with linseed oil and scraping into this enough gilders' whiting to make a thick cream.

Walter Crane's recipe was to boil 1 part of resin in 4 parts of linseed oil, and subsequently adding 6 parts of dissolved glue. Whiting was soaked in water and part of it mixed in the liquid to produce a cream; only enough being prepared for immediate use. Caution is necessary with the resin, for though it retards setting, if used too liberally it is apt to cause flaking. It is not a suitable mixture to be used in thick layers or for modelling in high relief, and is only advisable when applied with the brush to produce low relief painting.

Mixtures of gypsum with size or glue and a very little linseed oil are the fest for general purposes.

Application

The splendid workers of the Italian Renaissance, and others in their wake, used both fine and coarse gesso in combination, employing the former for delicate brushwork and the latter for modelling. They can be applied direct on wood or plaster, or on skrim dipped in glue thickened with a little gypsum and pressed smooth against wood or plaster. The glued skrim method is commonly adopted for friezes and panels, but is equally fitted for ceiling work, and if properly laid on proves a very valuable precaution against cracking.

When using brushes, the gesso is gradually built up by repeated layers. The grounds may be applied with broad, flexible badger brushes, using finer camel hair and other brushes for the modelling. The design can be helped up by scraping with a sharp knife ; but the aim is to leave a smooth, rounded surface.

When applying thick paste and high relief is desired, it is usual to boss out the prominent parts in a design with tow dipped in glue thickened with whiting. The paste is then laid on either with a small trowel or a broad, flexible-bladed knife. Here, too, the design must be built up gradually, in many layers, the upper one always being applied before the lower one has quite set. A sharp knife and other tools are used for modelling, so that elaborate high relief, with some degree of under-cutting, is attainable.

Fine gesso may be used for surface enrichment on, or round about, the coarse gesso ornament.

Cast Ornament

Another method of introducing strongly projecting gesso ornament, used during the Renaissance, but not much since, was to cast the enrichments and then plant them. This, of course, necessitates the modelling of the ornament, clay or otherwise, which is then moulded. If straightforward work, the mould can be in plaster, in two or more sections, which are lacquered with shellac and greased with tallow and oil before use. If elaborate modelling is in question, a jelly skin mould is necessary. In moulding, a common practice was to pour first a little fine gesso, which was swirled round to form a thin layer inside, and following on with a larger amount of coarse gesso. The thickness of the inner layer would depend upon the size of the enrichment. It is desirable that the casting should be as light as possible ; but a large piece must not be a mere thin shell. In large pieces, the hollow is usually filled with a core of tow steeped in glue thickened with whiting.

The coats must be so contrived that they have a flat, moderately thick rim projecting all round them. If the piece is large this rim should be reinforced with a strip of canvas steeped in the gypsum thickened glue. This rim is for fixing or planting the enrichment in the ornamental work. This is done by cutting out sufficient of the background to receive the enrichment. A thin coating of gesso is applied, the rim of the casting roughened, and the ornament pressed home firmly. Gesso paste is then smoothed over the joint, to produce a kind of sealing effect.

While this method is very suitable for bosses and drops, it is equally applicable for enrichments of lesser projection, such as rosettes, pateræ, shields and the like, whether these are introduced on walls or in ceilings. It, of course, reduces the cost of repetition work, as a number of ornaments can be more quickly cast than modelled, and after allowing for the process of fixing, there is a considerable saving of time, while with proper care there will also be a reduction in weight.

Colouring

Gesso work is usually associated with colour and gilding. Even backgrounds were seldom left white. The Italian and French work was very brilliant, glowing both with rich tints and gold. Sir Edward Burne-Jones and Walter Crane in their gesso work used quite gay colouring, with no large untinted spaces.

The colouring can be effected by mixing pigments in powders with the gesso, and this is the method generally adopted with the creamy mixtures. In other cases, more particularly with the thicker pastes, the modelled designs are painted; in such instances size or glue being used as a medium in which the colours are mixed. If this latter process is adopted, the colours must be laid on in thin washes, as one of the beauties of well-mixed gesso is that it has a certain brilliancy, which would be lost with heavy pigments. The Italian idea of using powdered marble was to enhance this brilliancy, due to the semi-translucency of the crystalline formation. Paints should not only be applied in thin washes, but the tints should be clear, light in tone, at all events avoiding the tertiaries and dark, light-absorbing colours.

Gilding

Gilding is, of course, applied to the finished surface. It requires great care. The surface, when quite dry and firm, is dusted and then given two or three coats of Armenian bole dissolved in yolk of egg. When ready the priming is damped, but not really wetted, and the gold leaf applied in the usual way, left matt or burnished as desired.

Gilded background or bandings can be covered with diaper patterns by impressing them with engraved dies.

Silvered effects can be obtained with aluminium paints. It is preferable not to use oil paints, but the powdered metal ground up in clear nitro-cellulose varnish. The varnish dries rapidly and is not affected by lime. Aluminium paint does not tarnish readily, as true silver paints will.

Applicability

Gesso decoration is necessarily expensive, not so much on account of the materials employed, though these must be of the finest and carefully prepared, but because it is a delicate branch of art craftsmanship. It may be widely applied, but is specially suitable for choice enrichments of ceilings, friezes, panels, etc., where coloured designs in relief are desired, or when more finished relief modelling on a small or medium scale than is readily obtained in ordinary plaster is aimed at. It is only adapted to interior decoration, but while delicate, if well made it sets hard and is more durable than ordinary plaster.

CHAPTER XVI

SURFACE TREATMENT OF PLASTER AND CEMENT

Rough-textured Surfaces—Other Surface Decorations—Cavo-relief Embossing—Painting on Plaster and Cement—Painting on Keene's Cement and Hard Plasters—Paint and Enamel on Plaster : Painting Schedule—Enamelling Schedule—Water Colours on Cement—Painting on Firbo-cement—Water Paints over Distemper on Plaster—Papering and Painting over Repaired Plaster—Waterproof Plaster—Cement Washes and Rough-cast.

THE surface dressing of plaster and cement is often a matter presenting considerable difficulty, while it also affords opportunities for adding beauty and distinction to a " stuccoed " job. Outside walls may be left plain, with trowelled or brushed surfaces. But this is rarely the case with interior work, except where Keene's, Sirapite or similar hard-drying white cement is used, which may be left with a smooth trowelled surface or a surface polished with a small float.

ROUGH-TEXTURED SURFACES

A good deal of ingenuity has lately been shown, especially in the United States, in imparting distinctive character to the texture of finished plaster surfaces. While a perfectly smooth, hard surface is aimed at for hospitals, schools and similar buildings, a finish with a marked texture is largely adopted for both the exteriors and interiors (corridors, halls, bathrooms, etc.) of private houses. This texture is obtained by a dexterous manipulation of the trowel or brush. The semi-roughed surface can be finished by a series of semicircular sweeps of the trowel, held at a slight angle ; or much the same effect can be obtained by using a stiff painter's brush. With the brush a " dash " finish is obtained, the sweep of the brush on the wet plaster being given an upward and diagonal sweep. A pitted finish is secured by means of a wire brush.

Such surfaces can be left plain or can be painted with thin paints—that is, paints with not sufficient body to obliterate (though they may tone down) the texture.

OTHER SURFACE DECORATIONS

The moulding and modelling of both cement and concrete have, of course, been dealt with in detail in other chapters. The running of mouldings by means of good templates is fairly easy and certainly quite effective. If the mouldings are at all elaborate, means should be adopted to retard setting, as it will be necessary to run the templates just about the stage of "initial setting," as the harder the cement is the more likely is there to be dragging, which may disturb a fairly wide area and lead to disintegration later on.

For concrete, the templates should be as simple as possible, though much will depend upon the fineness of the aggregate. Sometimes the parts of concrete to take a moulding receive a sufficiently thick coating of cement mortar or pure cement. But this, besides adding to expense, also detracts from the effect desired from exposing the character of the concrete.

When cement or concrete surfaces are to receive mouldings, the bands are first bossed up and roughly shaped. This should be done so that the bands or panels form an integral part of the body, not added on afterwards.

The same method of bossing up is adopted when surfaces are to be carved or modelled. The work must be carried out rapidly on slow-setting materials.

If parts are dragged away by templates, moulding stamps or by the chisel in carving, all loose parts should be removed, cutting away well on each side and downwards, the surface thoroughly dusted and wetted, and the part made good with freshly gauged cement, mortar or concrete, as the case requires. Promptness and boldness in cutting away are the only means to safety.

Exterior mouldings of considerable projection should have the top surface weathered and the lowest far projecting throated. This by way of precaution against the effects of rain and frost.

Cavo-relief Embossing

Cement and fine concrete surfaces will take embossed decoration quite well. This can be applied by using the chisel or by employing stamps or rollers of wood or metal having raised or sunken designs on, much in the same fashion as in old pargeting. These stamps and rollers must be constantly dipped in water while in service, for their faces and sides must be thoroughly wet whenever they come into contact with the cement.

Sunk patterns have their merits ; not only are they effective artistically, but in many positions have a mechanical advantage over raised ornament. For instance, wherever the ornament is less than 6′ above ground level, whether in or out of doors.

A style worthy of note, because looking particularly well when applied to cement or concrete, is the combination of sinking and embossing which we find in cavo-relief. This is a method much used by the ancient Egyptians, sometimes seen in Celtic work, and associated with gem carving.

The design, whether it be figure or floral work or conventional pattern, is carved so as to give more or less relief and rounding, but no part of the design is on a higher level than the general surface. That is to say, the design is traced on the surface, then panels are traced round the designs, and the material within the panel cut away to the depth desired, taking care not to injure the pattern. When the superfluous material has been removed, it is possible to attack the pattern, carrying out the design by the use of hammer and chisel, or other suitable instruments, and giving it all the prominence of detail and roundness necessary.

It is possible to execute elaborate carving in this way, also very striking lettering, but also quite simple patterns can be supplied with the help of templates and stamps.

Cavo-relief is much used for monuments. It is also well adapted for panels on large or otherwise uninteresting wall surfaces, for small rectangular or fan-shaped panels over door and window openings, and also for banding as a frieze or a string course.

Painting on Plaster Cement

The primary difficulty with regard to painting on plaster or cement is the present-day craving for speed. Now, all fresh plaster and cement contains free lime, and free lime has the effect of saponifying the oil in linseed oil in paint; on the other hand, the free lime absorbing carbon dioxide from the air, while the fresh plaster exudes moisture, results in the disintegration of water paints, the free lime at the same time attacking and modifying many of the pigments. We must make an exception as regards Keene's and the like cements, as we shall see presently.

Mr. Arthur Seymour Jennings, Editor of *The Decorator*, has favoured me with the following abstract of a report by Mr. W. E. Emley, of the National Bureau of Standards of the United States Department of Commerce, which relates to this subject and puts forward many useful hints:

"Several precautions must be observed if an oil paint is to be successfully applied to plaster: Do not apply an oil paint to plaster which is in immediate contact with an exterior masonry wall. There is a strong probability that water will find its way through the wall and into the plaster and make the paint peel off. Be sure that the plaster is air-dry when it is painted. As an extra precaution, it is better to have a circulation of air behind the plaster, in order that any water trapped in it at the time of painting may find a way out. In other words, it is safer to use oil paint when the plaster is on lath rather than on masonry. A sufficient interval of time between the plastering and the painting should be permitted to elapse so that the reactions involved in the hardening of the plaster will not be stopped before they are reasonably complete. This time will vary with the kind of plaster and the weather conditions. It should never be less than three months, and probably need not be more than one year.[1]

"Too much emphasis cannot be laid on the thorough drying of all

[1] *B.S. Circ.* 151, "Wall Plaster, Its Ingredients and Properties."

plaster, concrete, cement, etc., before painting. It must be remembered that such structures, when new, contain much water, and unless sufficient time, under proper conditions of temperature and humidity, is allowed for thorough drying, the water remaining in the wall will eventually seriously injure any paint coating. It should be pointed out that a wall may appear dry on the surface but be still damp in the interior. Even when properly dried, new structures of this kind may contain free lime, hence it is advisable when the structure is less than a year old, even if it is known to be dry, to apply before painting an aqueous solution of zinc sulphate (3 to 4 pounds crystallised zinc sulphate to 1 gallon of water). Allow this to thoroughly dry before applying paint.[1]

" The painting of plaster is important from an artistic as well as from a constructional standpoint.

" In some localities the customary building stone may be of such poor quality that it would not long withstand weathering. It is then protected by a coat of lime plaster, which is in turn protected from the heavy rain by several coats of paint. In this case it is literally true that the paint does hold up the wall.

" For present purposes, it is perhaps well to consider separately the chemical and the physical properties of plaster as they affect paint.

" All fresh plaster contains free lime. In some localities the base coats of plaster are made of lime and sand ; in others, of gypsum and sand. The customary smooth white finish is composed of two parts of lime and one of gypsum. Practically all gypsum or hard-wall plasters have some lime added to them at the mill if not on the job. Even Portland cement, which is less commonly used as a plaster, is alkaline because of the lime which is liberated as the cement sets.

" All fresh plasters contain free lime, and will surely destroy linseed oil paints. But when the plaster is exposed to the air for a while, the lime absorbs carbon dioxide fron the air and is converted into the neutral and harmless calcium carbonate. It is not safe to paint plaster until after the free lime has been destroyed. How long does this take ? It all depends.

" Air contains only a very small amount of carbon dioxide (about four-hundredths of 1 per cent.), so that even if all of the carbon dioxide reaching the plaster were immediately absorbed, it would be necessary for an enormous quantity of air to pass over the surface of the wall before the lime would all be neutralised. For complete carbonation, the the figures are approximately 2,000 cubic feet of air for each square yard of plaster. Ventilation or free circulation of air in a freshly plastered room is, therefore, vitally essential.

" As usual, the theoretical figures can never be attained. The temperature and humidity of the air have a great deal to do with the speed of the reaction. Increasing the carbon dioxide content of the air, by means

[1] *B.S. Tech. Paper* 274, "Use of U.S. Government Specification Paints and Paint Materials."

of salamanders, will speed things up, provided the air is not at the same time rendered too hot or too dry.

"The carbon dioxide must all be taken in through the exposed face of the plaster. Consequently, the absorption grows progressively slower, as the face is carbonated to great and greater depths. Unless the plaster is exposed on the back also, as in lath partitions, it is likely that it will contain some free lime after many years. However, this will not affect the paint, unless there is a chance for water to get in and bring the lime to the painted surface.

"In discussing the physical conditions which affect painted plaster, let us dispose first, very briefly, of the structural question. Of course, the paint will not stay on the wall or ceiling unless the plaster does. It is obviously essential to see that the building is properly constructed, so that structural movements which might crack the plaster are prevented. The plaster must be permanently and rigidly attached to the structure, for the same reason. The flashing is very important, for water getting into or in back of the plaster will be sure to cause trouble.

"Next, there is the much-discussed question of damp walls—referring now to walls which are old enough to have become thoroughly dry. Does the water come through from the outside, or is it deposited on the inside? Probably a little of both. Ordinary building bricks are slightly porous and absorb rain. It must, however, be a very severe and long-continued storm to drive the water clear through. Vitrified bricks are not porous, but the rain gets into the joints at the end of the bricks and is apt to go through the wall, because the brick cannot absorb and hold it.

"There are many other places from which water can get into plastered walls. It may come up from the foundations, it may come from daily mopping the floors, if the base-boards are not water-resistant. All such means of access should be carefully guarded against if the paint is expected to stay on the plaster.

"Free lime can be taken care of simply and easily; extraneous water can be obviated by a little care in construction; it is the gauging water used in placing the plaster which is the most frequent case of trouble with paint.

"We frequently hear it stated that plaster should not be painted until it is dry. When is it dry, and how long does it take to get that way? The gauging water used in placing lime plaster will amount to perhaps 10 lb. per square yard. This quantity, plus a little more liberated by the carbon dioxide, must be carried off by air circulating over the surface. Air at 70° F. can carry about ·001 lb. of water per cubic foot, so that about 10,000 cubic feet of perfectly dry air at 70° F. are required to evaporate the water from each square yard of plaster. If the air already contains some moisture, as it always does, or if it is colder than 70°, more air will be required. Circulation is obviously the most important factor.

"Scientifically 'dry' means 'containing no moisture.' In speaking of plaster, the best we can hope to attain is 'air-dry.' This means that the water in the plaster is in equilibrium with the water in the air, so that there is no tendency for water to evaporate from the plaster or to condense from the air. With the hourly fluctuations of humidity and temperature in which we live, the density and the thermal capacity of the plaster cause the plaster to lag behind the air, so that even air-dry conditions can never be quite attained. A plaster is practically air-dry when a slight increase in the humidity of the air will cause water to go from the air to the plaster.

"Plasters may appear dry on the surface and still contain large quantities of water inside. If a paint coat is applied under these conditions, no harm will be done, provided the water has a chance to get out the back way. That is, if the plaster is on lath, and the space back of the lath is ventilated, the water will go out that way and will not bother the paint. If, however, the plaster is applied directly to masonry, or more particularly over a coat of waterproof paint, the water will be trapped in the plaster and will surely cause trouble. It will gradually find its way to the surface, where the normal changes in temperature will cause great changes in its vapour pressure, pushing the paint off as the temperature goes up, pulling it in as the temperature goes down. While these movements may be small, they are almost certain to destroy the bond between the paint and the plaster if they are permitted to continue over a period of years.

"Most plaster contains some free magnesia—perhaps as much as 5 per cent. This material will not react with water under normal conditions, but if it is permitted to soak in water for five or six years, it probably will combine with water and expand. A few cases of plaster failure have been traced to this cause. Obviously this prolonged soaking can occur only when water is trapped in the plaster, between an impervious wall back of the plaster and a coat of paint on the exposed surface. Such a failure cannot occur if the plaster is dry when painted.

"With a free circulation of artificially heated and dried air, it is possible to dry plaster completely in forty-eight hours. If the air contains 100 per cent. humidity, the plaster will never dry, no matter what the circulation. With our normal conditions, a minimum of three months should be allowed in warm, dry weather, with a proportionately longer time if the weather is damp or cool, care being taken to maintain free circulation of air at all times.

"Just one more precaution: Plaster which is dried too quickly is apt to map-crack. Small cracks, running in all directions over the surface, are practically invisible before the paint is applied. But along each crack it will be found that the plaster has curled away from the wall. This curled plaster reflects the light differently from the flat surface, and when a gloss paint is applied, the cracks stands out in a startling manner.

"Plaster on lath can be painted while still green, provided that there is no free lime on the surface, and that the space behind the lath is open and ventilated. As an additional insurance against the presence of free lime, a priming coat of 3 lb. zinc sulphate per gallon of water can be used, although it is far better to allow sufficient time before painting for the lime to be carbonated in the usual way.

" Plaster on masonry will normally require at least three months to become dry enough to paint. Any attempt to hasten the drying by artificial means will probably result in cracked plaster. Painting the plaster before it is dry will trap the water which it contains, and this trapped water will certainly cause trouble."

Painting Keene's Cement and Hard Plasters

Keene's cement and other "patent" hard plasters, unlike the ordinary run of cements and plasters, must be painted as soon as they leave the trowel. This is because all of them dry rapidly, are very compact and become exceedingly hard, some drying with a glossy surface. This prevents the paint securing a key and so adhering closely. But even when Keene's is painted directly after trowelling the result is not always satisfactory, as it is sometimes patchy, showing water blisters and later scaling off. This appears to be due to the fact that average trowelling leaves the surface more or less uneven. The paint will not easily adhere to the places compacted and left glossy by the pressure of the trowel (which also retards the evaporation of water in the plaster), but does to the parts more lightly touched.

Mr. Jennings informs me that this difficulty has been overcome by a quite simple expedient. In a series of experiments, the first fact to be established was that the harder and smoother the Keene's was trowelled, the more difficult it became to induce the paint to penetrate, and the reason the paint peeled or perished in patches was that the wall was patchy in its degree of hardness according to the manner in which the trowel caught the heights and missed the hollows. The consequent hard spots acted in a doubly adverse way towards the paint. Being much smoother than the rest, they afforded less key, and being more compressed they imprisoned more moisture, or perhaps it would be more correct to say that it had less chance to escape because the surface was sealed. The other fact which threw considerable light upon the subject was the manner in which plasterers in certain parts of the country finished off their Keene's cement and hard plasters. This was done with the water brush instead of the trowel, and it was found that no difficulty at all was experienced in getting paint to key and to stay on such a surface. There was, if anything, rather too much porosity, and the paint had to be used very much on the thin side. There was also a certain roughness of the surface which might be objected to from a hygienic standpoint; or when a glossy finish was desired, as is usually found in hospitals, institutions and public buildings. In other situations such a surface as that left by the water brush would be considered quite desirable on account of its wonderful texture and old-world appearance.

The problem resolved itself into the question of obtaining the open surface of the water brush finish with the smoothness of the trowelled surface, and in the tests just completed this was accomplished in the following manner, and with the results described.

The walls were backed with Portland cement, and floated with Keene's in the usual way. As the Keene's " set " it was worked level with the wooden " float " and then finished off with the steel trowel or float, just as in ordinary practice.

Paint and Enamel on Plaster

Working schedules in detail for finishing new plaster walls in flat, gloss and enamel are given for the following working methods :

Painting Schedule

1. Brush the walls down to remove any loose dirt, plaster or other substance.

2. Apply a coat of first-class paint, coloured to match the finishing coat to be used. This paint may be white lead mixed with about three-fourths boiled linseed oil and one-fourth turpentine. Or any first-class prepared flat wall paint may be used.

3. *Size.*—If glue size is to be used, mix it just strong enough to feel slightly tacky when the fingers are dipped into it and allowed to dry. If you get it too strong it will make the paint scale off. If you use one of the several first-class factory-prepared wall sizes made to stop suction and also to kill any hot lime spots, follow the directions on the package, which often call for putting the size on before the first coat of paint. Glue size should go on between coats to protect it from moisture. It is used because it is cheap, not because it is good. Very little size is needed for a room, and so the best ought to be used. Some prefer varnish size made by thinning a quart of the best floor varnish with a quart of turpentine. Then add a very little of the paint to it, just enough to offset the colour of the varnish and make the size hide the surface to some extent. If the paint is not put in, a small handful of fine pumice-stone should be added to the varnish size to give it a " tooth," so the next coat of paint will take hold and cover well.

4. Second coat of paint. This may be white lead thinned with flatting oil or turpentine only and tinted to suit with colours ground in oil. It is best to stipple this coat with a wall stippling brush. If a semi-gloss is wanted, use one-third boiled linseed oil and two-thirds turpentine with the lead and stipple the coat. If a full gloss is wanted, use two-thirds boiled linseed oil and one-third turpentine. If you use ready-mixed paint for this coat, the regular flat wall paint will give you the flat

finish. For gloss, use an interior gloss paint of this type. Either of these materials may be stippled if desired.

5. As a rule two coats of paint and a coat of size will do a pretty fair job, but there is no doubt that a third coat is well worth the little extra money. When the old surface is very dark in colour or very dirty a third coat must be put on.

Enamelling Schedule

1. Clean off the wall as suggested for painting. If any repairs are to be made, fill the cracks and holes with prepared patching plaster, plaster of paris or other putty. Let the fillings dry and coat over with paint mixed to dry flat after they have been sandpapered lightly to make them smooth.
2. First coat of paint. Same as for any painting job as scheduled.
3. Size coat. Same as scheduled for painting job.
4. Second coat of paint. White lead thinned with flatting oil or turpentine. Or any good interior paint that will dry flat. When bone dry, sandpaper lightly to remove dirt nibs, dust, etc. Clean off with duster brush and damp chamois.
5. Third coat. Use any high-grade enamel under-coater coloured to match the final finish wanted. Or use a mixture of one-half white lead in oil and one-half zinc oxide in oil. Thin with flatting oil or with turpentine. Let dry hard, sandpaper lightly with No. $\frac{1}{2}$ paper and clean up as before.
6. Fourth coat. Same as the third coat.
7. Fifth coat. Any high-grade, factory-made white or tinted enamel. It may be one made to dry with a gloss, with a semi-flat or matt effect. If white enamel is used and you want to tint it, use tinting colours ground in Japan. Thin the colour with turpentine and strain it before adding to the enamel. Sandpaper enough with No. 0 or $\frac{1}{2}$ to remove gloss. Clean up thoroughly.
8. Sixth coat. Same as the fifth, but do not sandpaper.

When a cheaper job is wanted, you may omit the fourth coat above, and for a still cheaper job one of the enamel coats.

Water-colours on Cement

The decoration of cement walls still presents difficulties to many, who dislike the dull grey tones peculiar to the untreated material. However, there is no reason why fresco effects should not be obtained if due precautions are taken. The chief thing is to see that only fine, cool cement or cement mortar is used for the finishing coat, which should be left with a medium rough surface. The colours chosen should be the earth pigments with a few other oxides. These ground fine are mixed with from one-third to one-fifth of the bulk of cement. This cement should be very fine in grain and as light in colour as possible. The powders are mixed dry, and then water added when required for painting. They should be quite fluid and should be applied to the cement while still green. It is necessary to apply the colours swiftly and in thin washes, which will preserve a certain transparency, more or less approaching that of frescoes according to the "whiteness" of the cement. Retouching is not advisable, as that will produce muddy effects. White can be obtained by the use of lime, which must be thoroughly slaked, finely ground and sifted to remove all unslaked particles.

If pictures or other elaborate patterns are attempted, the outlines must be formed on the cement walls, and the colours then applied in thin layers, with a dull, soft brush. Stencilling is also possible.

Treated thus, the colours sink in and the painting will be found fairly permanent, resisting atmospheric influences. If too much cement is mixed with the paint powders, or not enough water added, this sinking in will be very superficial, and then the colour is more likely to fade and to be rubbed off.

While not possessing the much-appreciated luminosity of true fresco painting (earth colours and other oxides mixed with fine, well-slaked lime, applied to fine lime mortar—thoroughly slaked lime 1 part, fine silver sand 2 parts), it will be sufficiently pleasing and permit of a great variety of treatment.

In this paragraph, we are not discussing here the application of oil paints or any of the special colours prepared for work on cement. Our immediate aim here is to give information for fine art painting on cement, which requires something different from the old water-colours or tempera which suited lime-plaster.

Painting on Fibro-cement

Apply to the fibro-cement panels a weak solution of sulphuric acid and allow to stand for an hour or two ; then wash off with plenty of water, repeating the washing at least twice. When quite dry apply an under-coat, which should be glossy if the finishing coat is to be matt, or matt if the surface is to be enamelled.

Sulphate of zinc can be used, both on new cement and on asbestos sheets ; the proportion should be three-quarters of a pound to a gallon of hot water, but the solution should be cold before the application is made. Two coats should be given with an interval of twenty-four hours, and then either distemper or oil paint may be applied with safety. The Indestructible Paint Company make a special paint called "Kewato" for application to cement, asbestos sheets, etc., and it is very successful, but must be applied direct to the sheets or cement without any intervening preparation. "Duresco" is also successfully used for the same purpose, and there are many other primings.

Water Paints over Distemper or Plaster

Water paints cannot be placed directly over old or new distempers or plasters. Mr. Ransome W. Cooper explains that the cause of the trouble is the spongy, hungrily-absorbent, washable distemper surface, which at once absorbs with avidity the binder of the water paint applied over it. This leaves the water paint film with less binder than it needs and impairs its durability. The film then proceeds to harden with age in the normal way, and ultimately becomes too strong for the surface over which it was applied. The latter, having no binder of its own to properly support its own weight, to say nothing of that of a film superimposed upon it, breaks away from the wall surface in large or small flakes and unending trouble begins. It is an absolute impossibility to completely remove all traces of this unsatisfactory material, and it is, therefore, highly desirable to go over the ground with a coat of sharp colour to bind the lurking remnants, capable otherwise of giving rise to flaking.

Papering and Painting over Repaired Plaster

Much trouble is often experienced when it is necessary to paper or paint over repaired plaster surfaces. This is owing to the fact that it is extremely difficult to secure a perfectly even surface by ordinary methods. The cracks or patches filled up are apt to show through the surface dressing and so spoil the job. The best way is to prepare a plastic material by taking equal parts of zinc white paste and fine whiting and making into a thinnish paste by the addition of three parts boiled linseed oil (in which a little glue has been melted), one part turpentine and liquid driers. This is applied to the wall wherever patching has been carried out and a smooth surface obtained. When dry paper or paint can be applied with the certainty that the mended parts will not show through. They can also be prepared by mixing the zinc and whiting in flatting varnish, but this is not quite so easy to manipulate.

Weather-proof Plaster

As a weather-proof plaster for outside walls, the following has been suggested. Mix cement and sand and gauge with rubber solution extended with the necessary quantity of water : ½ to 1 pint of solution to go to every cwt. of cement. The difficulty is that the rubber solution and water would not mix well, so it would be better to adopt rubber latex alone as the moistening agent.

An advantage of such cements is that they would be non-absorbent and non-cracking. It is for these properties that rubber solution or rubber latex is already being used in the manufacture of certain classes of paints and varnishes.

Rubber solution and latex, however, are not very easy to use, though effective if carefully applied; so readers should refer to other methods described in this book.

Cement Washes and Rough-cast

Cement washes and rough-cast are much in favour for the dressing of outside walls, whether for dwelling-houses or other buildings, and rightly so, for they possess decided merits. Both methods of treatment provide a pleasing, easily varied and not too costly a finish, usually blending well with any other architectural feature, such as timber framework, exposed brickwork, stone masonry, tile or slate hanging, etc.

Apart from this, both methods, if a sound cement forms the basis, furnish, when carefully carried out, a large measure of protection against dampness. Not only does the cement in some degree add to the imperviousness of building materials, but it greatly helps water-shedding properties, a condition much enhanced by, and proportionately with, the degree of the roughness of the surface. Actual tests, carried out under strict scientific supervision, have demonstrated that the harder and smoother the surface the greater will be the amount of condensation moisture resulting from a change of temperature in a water-laden atmosphere. A smooth surface seems to encourage and retain a film of moisture, which, therefore, has the greater chance of penetration, and in the case of heated dwelling-houses is assisted by the higher interior temperature. This argument for outer faces of walls applies with special force to houses in damp or rainy situations or exposed to high, moisture-laden winds.

Among the advantages of cement washing and rough-casting are that both colour and texture can be varied almost indefinitely. We can now select colours in practically any tone which will mix with cement and not be bleached. Moreover, fine to coarse ground cement can be used, or a small amount of clean sand added, so the granulations of the surface can be regulated to a nicety.

The same applies to rough-casting. Here the colouring can be obtained by means of the purity of the lime, the addition of ochre or other earth colour, or the special selection of sand, crushed stone or pebbles. Rough-casts may range from a cement-sand wash, giving a slightly granulated surface, through a coarse rendering of cement with stone chips, to pebble dash, the degree of roughness and tint values varying endlessly to suit the character of the building or fit in with the natural surroundings.

Within reason, the damp-protecting and insulation value increases with the coarseness of the rendering.

Success is not always attained with these methods; cement washes especially being apt to peel off in flakes or powder. This is due to imperfect hydration, the water either being absorbed or evaporated too quickly before the cement has had time to set and gain adhesion to the wall surface. A too rapidly setting cement should be avoided, and the surface, if absorbent (brick, limestone, plaster or breeze concrete slabs), should be thoroughly wetted before applying the washes. Hot, bright, sunny days or days when there is a high wind are also bad, because likely to rob the wash of its moisture before setting. If this takes place, the cement will not form a continuous hard sticking skin, but rather a more or less disconnected mass of dry particles, having a feeble adherence to the wall, and so sooner or later falling away in particles or powder.

When hydration is perfect and dehydration slow, the setting of the cement will take its normal course, and so the cement or cement-sand layer dries, not in an uneven manner, leaving a powdery mass, or at best an independent pellicle, but as a closely adhering if not penetrating skin. So we see that it is by means of properly controlled hydration and dehydration that we can secure the desired full measure of protection and durability.

It will thus be clear why we dwell in the first place, and particularly, on atmospheric conditions, and secondly on the physical state of wall surfaces. Both have their influence on results, but the first, being too often overlooked, requires emphasising. It is a pity that an excellent practice should fall into disrepute through neglect of detail.

The principal points leading to success have been touched upon above. A few further details are necessary.

Even when the atmosphere is thoroughly damp and the wall surface well wetted, it is wise to have the washes fairly fluid (stirring the container vigorously to maintain uniformity) and the rough-cast quite moist. In the latter case, if the sand or crushed stone is at all absorbent, it should be first steeped in water before adding to the cement-wash, the whole being kept in the condition of a semi-liquid mortar.

When circumstances compel the work to be carried out in hot summer weather, or during high winds, it is possible to afford protection by hanging wet sacking before the wall. This precaution is also useful when frost is to be feared, which is equally destructive of good work, interfering, as it does, with proper hydration, though in a different way.

CHAPTER XVII

ARTIFICIAL MARBLES AND GRANITE

Uses of Artificial Marbles—Characteristics—Varieties—Scagliola—Materials: Plaster, Glue, Pure Water, Hair, Pigments—Mixing—Applying—Polishing—Marezzo—Other Marbles: Magnesite, Cement, Gypsum, Granite—General Precautions.

ARTIFICIAL marbles were produced in quite early times and have retained esteem ever since, though the modern revival began with the Italian Guido Sassé in the sixteenth century, and in this country by Wyatt about the middle of the eighteenth century. A fresh awakening as to its merits began on the Continent at the opening of this century, and is fast gaining ground in this country.

Its merits are, indeed, conspicuous. When properly made, artificial marbles are very durable, while the material is easily applied *in situ* to walls, pillars and columns, or in the form of slabs prepared in the workshop. It can also readily be moulded and modelled. Its possible uses are, therefore, manifold. Artificial marbles are very suitable alike for exterior and interior work; for plinths, panels, string courses and pilasters on external elevations, and inside for lining walls of halls, passages, staircases, bathrooms and the like, whether in public or private buildings, and are for several reasons far preferable to genuine marble for lining restaurants. These mixtures are admirably fitted for columns, chimney-pieces, balustrades and other ornamental work, which must look handsome and yet be able to resist hard wear. One strong argument in their favour is that they are much more easily and quickly repaired than natural stone.

CHARACTERISTICS

When carefully prepared, artificial marbles are lighter in weight and decidedly less brittle than true marbles, and therefore, whether applied as a plastic material *in situ* or as cast slabs, are more stable. The stuff, too, can be applied in thick or thin veneers, on flat or curved surfaces, over brick, stone, concrete, wood or metal, without waste. It dries hard, takes a very fine polish and, as we have already said, is very durable under all atmospheric conditions. Decoratively it can be varied almost indefinitely, both as to colour and form, being easily tinted in polychrome designs, moulded and modelled. Thus the surface can be plain and smooth, panelled and banded according to any colour scheme, fluted or carved. Moreover, the tinting can be so carried out that it becomes permanent, whether the colour penetrates only just beneath the skin or permeates throughout the mass, the latter method being preferable when the surface is decorated with mouldings or carvings. Thus artificial marbles are of such a nature that they can withstand hard wear, and while their smooth surface diminishes the tendency to soiling, it can be subjected to constant routine cleaning without fear of diminishing its brilliancy, and so calling for periodical redecoration. So, as regards upkeep, this type of finish is much cheaper and satisfactory than painted surfaces. While this finish sets hard and takes on a brilliant polish, it is less cold than natural stone, so is more suitable for interior work in such a climate as ours. Artificial marbles of this kind, moreover, lend themselves readily to the insertion of metal ornament, such as lighting brackets, friezes and so on.

These are the main characteristics which, together with other advantages, will become more apparent as the processes of production are described.

Varieties

There are several types of artificial marble, differing in the materials used, the processes of manufacture and the results obtained. While the earlier forms were compounds of carbonate of lime (slaked limestone mixed with crushed marble or marble dust), such as the famous scagliola, to-day those with a sulphate of lime base (including plaster and "patent cements," of which Keene's is typical) are more common. Others, again, are magnesite or cement compounds. Each has its peculiar qualities; those with a carbonate or sulphate of lime basis being remarkable for their brilliancy, while the magnesite are notable for their hardness. Those with a Portland or kindred cement basis are apt to be coarse and are the most difficult to colour.

Scagliola

When artificial marble produced by plasterers is mentioned, most people mean scagliola. As its name implies, it was originally a conglomerate, a compound of small fragments in a matrix. In practice, crushed marble and refuse from the marble dressing yards was mixed with carefully slaked fat lime, to which sometimes marble dust was added. Marble, either in the form of small fragments or dust, has, however, long ceased to be used in the production of scagliola.

Materials

The materials required by the modern plasterer in the production of scagliola are plaster, glue and pigments.

Plaster.—The plaster selected should be pure, very fine, as nearly white as possible, and one which setting slowly will dry hard. For white marbles, Keene's cement is chosen, though others having similar characteristics are suitable. They should be of good colour. All plasters must be thoroughly dry and powdery before gauging. For high-class work, the plaster should be sifted before use, to get rid of lumps and coarse grains.

Glue.—The plaster is gauged with size. For this purpose the best fish glue should be used. " Strong water " is made by melting 1 lb. of glue in 7 gallons of water. The glue is first placed in cold water to swell, then the rest of the water is added and the pot put over the fire. Only sufficient glue for immediate requirements should be made at a time, as it weakens when kept. Size water is used chiefly, because it has the effect of retarding setting, but adding to the ultimate hardness. It also tends to increase polishing properties. But this is only the case if the glue is pure and strong and the size fresh. If stale, retardation and hardness are lowered, and the surface of the finished product is apt to be dull. If any scum is formed in melting the glue, or a skin forms over the size, it should be removed. The pot or pail should be protected from dust.

Water.—The water should be clean, pure (especially free from iron) and fairly soft.

Hair.—Well-washed hair, such as used in ordinary plastering, is sometimes added in the proportion of not more than 10 per cent. by weight. While this adds to toughness and cohesion, and so may do well for exceptionally heavy work, it is liable to give a certain coarseness to the finished surface and also to militate against the highest polish being obtained. If white or very light marbles are to be imitated, only white hair is permissible, otherwise objectionable dark streaks may appear, which would destroy the pleasing effect.

Pigments.—The general colour of backgrounds is to some extent obtained by the selection of the plaster, but dark and pronounced tints are produced by using special pigments. These are mostly of a mineral character, and in any case must be such as will resist alkaline reaction. These colours include raw and burnt sienna, raw and burnt umber, lamp and vegetable blacks, vermilion, copperas, ultramarine and indigo blues, lemon and other chromes. Various colours and shades are produced by mixing different pigments, while intensity is increased or diminished by adding more or less colour to the gauged plaster. The colours are not used as powders, but in the form of duly prepared water-colours. They are always used for tinting the plaster, not as surface washes, and this intimate mixing of colour and plaster makes the careful selection of the pigments all the more necessary.

In the best class of work the luminosity associated with veining and speckling in natural marbles is obtained by the admixture of fragments of alabaster, coloured marbles or tiny flakes of mica, even threads of silky asbestos, with the gauged plaster.

Mixing

Both the floating coat and the decorative veneer of scagliola (commonly abbreviated to "scag.") are prepared by gauging good plaster with strong size water, so as to obtain a stiff paste. If well made, with good materials, it will take from twelve to eighteen hours or more to set.

The true test of skill comes with the tinting of the stuff, of craftsmanship in the blending of the variously coloured lumps of plaster.

The desired quantity of stiff-gauged plaster is placed on a marble or slate slab, the colour added and thoroughly mixed by kneading with the hands. It is then rolled out, cut into strips about 1″ thick and 1″ wide. These strips are pinched off and formed into balls about the size of large walnuts. These lumps are rolled in "crumbs," made by adding a little dry plaster to a reserved lump of the gauged plaster and breaking it up into crumbs. At this stage any fragments of marble, alabaster or specks of mica are added to the crumbs. Meanwhile a little plaster is gauged in a bowl fairly liquid; a 2″ wide fairly soft brush is dipped in this to smear over the balls. These are knocked up into a large lump, which is flattened, chopped up with a knife, knocked up once, chopped, again knocked up, and then finally sliced ready for applying. Dark-coloured gauged plaster is known as "rich mixed," and light shades as "plain mixed." As a rule, plain mixed stuff is gauged less stiff than rich mixed, so as to obtain the natural blending associated with light veining.

The choice of pigments, the intensity of colouring, the relative proportions of different coloured stuffs, are matters of taste and skill, based on a close study of choice examples of natural stone. Each man will have his own ideas as to these matters and his little "dodges" in mixing and blending. It is here only possible to lay down general principles. As we know, different specimens of the same type of marble vary among themselves, so there is a wide range for the display of taste and skill. It may be laid down as a general rule that it is well to avoid exaggeration both as to colouring and veining. However, when small panels or narrow bandings of a richly coloured marble are used in a lighter tinted mass, emphasis is not so disagreeable, and may be necessary in large public halls, theatres, cinemas, etc. Veining should not be too hard, especially in outline; it will be noticed that in almost all cases, even dark-coloured veins soften towards the edges and blend into the backgrounds. This varies with different kinds of marbles, and a study of these should guide the plasterer in the gauging of each coloured lump. Veining must not be too stiff; straight lines (except in the case of stained cracks) are not natural or graceful. It is by exaggeration and stiffness that the artificial is usually betrayed.

While it would be impossible to give directions for the proper colouring of every kind of marble, or even of those most commonly imitated, the following will serve as examples of how scagliola is built up to represent any particular marble which may be specified, or which the fancy of the craftsman may suggest.

Devonshire Marble.—Plain mixture: one shade of grey (compounded with a little lamp-black, and a tinge of indigo, to be used sparingly); one shade lemon chrome; one shade purple brown (a careful blending of burnt sienna, vermilion and indigo); one shade of flesh colour (a very little vermilion with a touch of chrome yellow). Veining: burnt sienna. Rich mixture: one shade light purple brown (slightly more indigo); one shade of dark purple brown (a touch of lamp-black added); one shade middle purple brown; one shade of grey; one shade of lemon chrome. Veining: burnt sienna with a sprinkling of alabaster splinters.

Rouge Royal.—General body: light purple brown. Veining: sienna, umber, ultramarine, blue-black, verd antique, dark green, dark brown, light brown.

Swiss Red.—Rich mixing: light purple brown, dark purple brown (with indigo), Venetian red. Veining: black for ground, white and green for mixings, with small alabaster spots.

Applying

As stated, scagliola may be applied *in situ*, or used in the shop for casting slabs or ornaments.

If it is to be applied to brick or stone walls, the joints should be raked out and the surface well wetted. Should the walls be new, it is well to repeat this wetting several times, leaving the walls exposed as long as possible. This is for the purpose of removing saline matter, which otherwise might work through the "scag.," producing unsightly efflorescence. The process of curing can be hastened by coating the walls with a wash of freshly slaked hot lime, which when dry is scrubbed off with a hard brush and warm water. Wood and lathing should be well wetted. Iron or steel pillars, etc., should be well brushed to remove rust or scale, and then painted with a suitable close-adhering metallic paint, containing a fair admixture of fine sand or cement to provide a key. Or they may be sheathed with expanded-metal lathing.

The surfaces are then ready to receive the floating coat, which is prepared as before stated. It is laid on with the trowel, trowelled or gone over with a float, and when nearly set is well scratched with a single pointed lath, the cuts being fairly deep, given at a slant and diagonally. This is usually of the same thickness as ordinary first coating.

The finishing or scagliola coat is laid on in slices by hand and pressed down with a square-pointed gauging trowel. Before laying on the strips, the floating coat is brushed over with soft-gauged plaster. When the whole face is laid and compacted, it is gone over with a toothed plane, with irons set at an angle of 70°, such as cabinet makers use for veneering. This produces a smooth surface and rectifies any distortioning caused by pressing with the trowel. The slices are usually 1" thick, reduced by pressing and planing. Additional thickness may be needed if raised or sunk panels are required in parts where mouldings are to be run by means of templates, or where carving and modelling is to be carried out *in situ*.

When columns are to be covered *in situ*, usually two wooden templates are cut corresponding to the upper and lower diameters, and these are fixed at top and bottom to work from. The "scag." is applied about $\frac{1}{8}''$ thicker than the true outline, to allow for the loss of thickness in pressing and planing. Special provision is made for the bases and capitals. When finishing with the plain toothed or the fluting plane, the outline is tested with the rule applied to bottom and top template. This method is also adopted with pilasters. Templates are also used when building up projecting parts, such as the framing of a doorway.

Carving and modelling is carried out on the moist plaster with the usual carving and modelling tools.

Scagliola slabs can also be formed in the workshop (using for this purpose a slate bed), much in the same way as described above. Such slabs, if not too large, are easily transported and handled. They are screwed in position to plugs or battens, the sunk screw holes being stopped with freshly gauged and fittingly coloured stuff, so as to produce an unbroken surface.

Columns, balusters, vases and ornaments generally can be formed of scagliola in the workshop. There are two types of column, the "independent," intended to stand by itself, chiefly as an ornament and without serious structural function, and the "case," designed to form a casing for an iron stanchion or brick pier already in position. For the former a skeleton frame is built, roughly of the shape of the finished column, either of expanded metal (for light work) or of strong timber (when desired to have some supporting value). This is rendered with a floating coat, which is treated in the usual way to receive the scagliola slices. Cases can be prepared in two ways. Either they are made in two longitudinal sections, or a hollow shell is formed which is then cut into two long sections. In the first instance long moulds are used, at least $\frac{1}{8}''$ greater in diameter than that desired for the finished column. The scagliola slices are pressed into the oiled mould, and then the floating coat is applied fairly stiff; or a muffle core (larger in diameter than the permanent core to be cased) may be used and floating coat poured in the usual manner. A second way is to build up a skeleton frame of timber, slightly larger in diameter than the permanent core, this is gauged with plain plaster gauged stiff and scratched to receive the scagliola veneer. Such columns are planed and polished before they are cut in two lengthwise. The cutting is done with a thin, fine-toothed saw, either with a straight or a wavy cut. The latter is preferred, as such a joint is easier to disguise. The joint is concealed by means of suitably coloured stuff. As a rule, the bases and capitals are moulded separately. The fixing is generally done by means of brass screws or clamps.

Vases and other small pieces are moulded much in the usual way, though the outer coating (and the inner face of vases or bowls) have to be lined by hand with the scagliola and then backed with plain plaster.

Small columns, vases and the like in good-class work are often turned on the lathe, which gives the best finish, but naturally is more expensive.

Polishing

Polishing truly puts the finishing touch to scagliola, showing off its beauty to advantage. It is a lengthy, somewhat tedious, work, but well worth while, indeed necessary. The process differs slightly according to whether plain white Carrara or coloured marbles are being imitated.

Imitation white marble is usually carried out with the best quality Keene's or similar white cements. Great care is demanded throughout to prevent the work being soiled. In polishing, perfectly clean materials only must be used, and steps taken to keep them clean, and to keep away dust, and especially smoke, during the process. Once polished, it is safe enough.

White marble must be allowed to thoroughly set. It is then rubbed down with fine pumice-stone, rubbing lightly and evenly, and using a fine, soft sponge dipped in clean water. It is next gone over with snake-stone, again using the sponge and water freely. It is then wiped dry with the sponge, preparatory to stopping. For stopping, Keene's cement (or any other cement used for the body) is gauged in a clean bowl to a thick cream, which is dubbed over the object with a sponge and brushed in. When nearly dry, the superfluous stopping is scraped off with a fine-edged, hard-wood scraper. The brushing on of the stopping and scraping is repeated until a perfectly smooth surface is secured. It is then left for several hours, when it undergoes another rubbing with snake-stone and a wiping down with clean, soft cloths. It is allowed to dry thoroughly, when it is polished with oxide of tin (putty powder) and clean, soft cloths dipped in water and wrung out.

Coloured or ordinary scagliola receives slightly more elaborate treatment. When very nearly set it is rubbed down with marble grit on a soft cloth and a sponge dipped in clean water. Columns and large flat slabs in the workshop are usually planed as before described. The grit and dust are removed with a soft sponge and plenty of water, and a dry sponge passed over to remove all moisture. A soft paste of plaster (gauged with weak size water and tinted to the ground colour) is used for stopping, being dubbed on liberally and then scraped off with a steel scraper. These processes are repeated twice more. The actual polishing is carried out with putty powder and pure, raw linseed oil. It is sometimes found that polished work becomes dull after a short time. This is the result of faulty work, due to too much haste. The work must be allowed to become thoroughly dry after each stopping and stoning, which will take perhaps as much as two or three days. No salamanders or other artificial means of hastening drying must be used, though a hall where work is being carried out, or a workshop, should be kept comfortably warm.

Polishing is important, not only because it enhances the value of colour and figure in the finished work, but also because it adds to its hygienic qualities (by not harbouring dust) and its durability.

Marezzo

Marezzo is a variation on scagliola, and may also be used for the imitation of porphyry and granite.

It can be made with ordinary high-grade plaster, gauged with strong water as for scagliola, or with Keene's cement gauged with plain water.

Slabs are made on polished marble, slate or plate-glass. The last is to be preferred, as it permits of the work being viewed as it proceeds, and also leaves the surface with a good polish, so reducing the time necessary for finishing. Columns, balusters, mouldings, vases, etc., are produced in well-seasoned, well-oiled moulds, such as are used for other plaster casts. It is essential that they should be hard and non-absorbent.

The plaster is gauged to a thick cream and is divided up into a number of pots, one for each colour or shade required. The pigments should be similar to those used for scagliola, the powders being ground in water. They are added to the plaster until the right shade is obtained.

Strands of floss silk (for high-class work), manila grass or flax are needed. They should be from 12″ to 20″ long and of varying breadth, and some should be knotted at one end. These strands are dipped into the coloured slip (creamy plaster), opened out with the fingers if necessary, and flipped firmly against the face of the mould at suitable angles. This process is repeated again and again, constantly dipping the fibres in the slip, and using new sets of fibres for each colour. This is according to the colouring and nature of the veining. With certain marbles, where there are blotches and specks, it will be necessary to dip the fingers in the slip and flick them hard over the mould so as to cut into the veins. The size of the drops can be regulated by taking up more or less slip and using more or less force in flipping. If the marble has a white ground, a thin wash of plain plaster or Keene's is applied ; or the wash may be tinted to suit the case. The back surface is then consolidated with a trowel or spatula, and coarse canvas pressed down on it. Over this coarse dry plaster is spread to absorb moisture. The coarse plaster and canvas are then removed and a coating of stiff gauged plaster run by way of backing. With large pieces, the canvas may be left in to give stiffness and toughness to the cast, which facilitates handling before fixing.

Panels and slabs are produced differently. In this case the fibres dipped in the coloured slips are disposed on the moulding slab in accordance with the figuring on the marble, the fibres being so disposed that the knots hang loose outside. Specks and blocks are put in as with case work. The background is then run, smoothed with the trowel and, when the body is nearly firm, the fibres are carefully withdrawn. Excess moisture is absorbed as described above, and the backing finished with or without the canvas.

Porphyries are imitated by a stiff gauge of plaster or Keene's tinted as desired, and formed into slabs about ⅛″ thick. When quite dry these slabs are broken up into small pieces to represent the stone to be imitated. They are sized by means of a suitable sieve, dust being reserved for gauging for stopping purposes. These pieces are mixed with the tinted plaster, which is then pressed into the moulds or trowelled flat on the moulding slabs. This process will explain how *rouge royal, verd antique,* or granites can be imitated, mica specks or the like being added when needed.

Marezzo is stopped, stoned and polished as with scagliola, but the processes must not be so drastic, owing to the nature of the colouring.

Other Marbles

Magnesite.—Magnesia artificial marbles are made with plaster prepared from magnesite (magnesium carbonate, $MgCO_3$), or from dolomite (calcium magnesium carbonate, $CaMg\ (CO_2)_2$). These minerals are washed, roasted in kilns and reduced to a very fine powder. The cements are gauged with an equal volume of a solution of sulphate of magnesia (density 1·9), which may be coloured with any suitable pigment.

For panels, slabs and the like the gauged plaster is rolled out on marble, slate or glass slabs to the required thickness. Shaped objects are cast in moulds, those of vulcanised rubber giving the best results.

Veining and figuring is imitated with the brush. Yellow is obtained with a 3 to 4 per cent. solution of bichromate of potash in water. When dry it is flooded with, or steeped in, a 5 per cent. solution of acetate of lead in water. The reaction gives a yellow chromate of lead. When dry the excess of the acetate is washed off with water. Red is obtained with an alkaline solution of dragon's blood, followed when dry by a wash of a 1 per cent. water solution of chloride of tin, which when dry is washed off with clear water. For choice work a half per cent. solution of nitrate of silver can be used, the reactive being 1 per cent. solution of bichromate of potash, which gives a red chromate of silver. Blue is obtained with ferro-cyanide of potassium, the reactive being an ammoniacal citrate of iron. Black is obtained with a solution of hematite, followed by a 14° solution of pyrolignite of iron. After drying, a solution of hydro-fluosilicate of aluminium or of zinc is washed over.

All reactives must be washed off after drying.

Polishing is carried out, after lightly stoning with fine pumice-stone or tripoli, with a little wax dissolved in turpentine.

Cement.—Portland or similar cement can also be used for imitating marbles. It should be white and very finely ground. A dry mix is made of 20 parts by volume of cement, 10 parts of fine white sand, 4 parts of chalk, 6 parts of sulphate of zinc and 5 parts of pulverised white or coloured marble, according to the stone to be imitated. A small quantity, not more than 5 per cent., of hydraulic lime may be added to make the mass more plastic. A small part of this is separated from the rest. The batches are gauged medium stiff, colouring matters being added at the same time. For stones of very mixed colouring differently tinted balls are formed, and these are thrown, in the proportion desirable, in a mill, ground up, moistened, and the mix either placed in moulds or rolled flat on tables. When nearly set, the veining batches are thinned with water, and the slip is applied by means of a funnel or other suitable instrument to represent the veins and specks. When thoroughly dry, the surface is stopped with slip, lightly stoned and polished with wax dissolved in turpentine.

Aluminate of Gypsum—On the Continent artificial marble is largely made in the workshops with an aluminate of gypsum. The plaster is prepared thus: The limestone is dried in kilns, pulverised and then treated in water so as to produce a sulpho-aluminate of calcium. This is mixed with quantities of water, and crystallisation obtained by precipitation. The sediment is ground after drying in kilns. The gauging is done with water, when the desired mineral colours are added. The mixture is poured in moulds or rolled out on polished slabs. When quite dry and hard, the finishing touches are put by stopping, stoning and wax polishing.

According to official tests, such artificial marble is nearly as strong as fine Carrara, possesses a high resistance to compression and gains in hardness with age.

Another method is to gauge the plaster with weak size water in which a little resin dissolved in turpentine is added. The plaster can be tinted with mineral colours, and veining is carried out by depositing threads, thin or flattened as may be necessary, in moulds or on moulding slabs, and then covering with the body plaster. When quite dry it can be stopped and stoned to a smooth, even surface, and is then ready for varnishing. For white or very light-coloured marbles, the varnish is prepared by dissolving 2 lb. of white gum lac in 6 quarts of alcohol and adding about a half pound of plaster. For dark-coloured marble, orange-tinted gum lac is used and rather more plaster. Before applying the varnish the marble is rubbed with a rag dipped in alcohol.

Granite.—A granite finish for plaster is sometimes carried out as follows: The walls are floated with hydraulic lime and when nearly dry are rendered with another coat tinted light brown. At once the moist surface is splashed with white slip, then with black slip, and finally stippled with a brush dipped in red slip. Success depends upon the art with which the splashing and speckling is carried out. The colouring may, of course, be varied. The surface may either be left rough, or smoothed and polished with the trowel kept just moist. When quite hard, the walls are set out and jointed to represent blocks.

An improvement on this was introduced by William Millar in the eighties of last century. External walls are floated with Portland cement or hydraulic lime ; interior walls with Keene's, Parian or other white cement and duly scratched for keying. For the surfacing finish, the same materials are gauged rather stiff, one batch being left white, another tinted black and a third red. These are rolled flat, about $\frac{3}{16}''$ thick, and cut into small irregular cubes. The walls are then floated with a thin coat of cement, tinted light brown. On this wet surface the cubes are pressed according to fancy, gauged with the light brown cement and pressed down with a hand float, but not sufficiently hard to break the cubes. When set and dry, the surface is rubbed down and polished. Here, again, mica specks and alabaster fragments may be introduced with advantage. This process lends itself to the introduction of plain mouldings by means of templates, applied while the surface material is still soft. The surface can also be panelled, the colouring of the granite varied and jointings introduced. Polishing adds to the durability of the finishing.

General Precautions

Several other methods of producing artificial marbles and granites are practised, and numerous patents for their manufacture have been taken out from time to time. Some are still in use, both protected and unprotected. It is, however, scarcely necessary to describe these, as they are mostly rather complicated and often expensive, while the processes described are open to any skilled craftsmen. They differ in many respects, so that the needs of any case can be met. Workshop methods have the advantage in that stock can be produced in slack seasons.

It is as well to repeat that success in the art depends upon a close study of nature, the avoidance of exaggeration and the capacity for taking infinite pains. Such work cannot be left to pure fancy or be unduly rushed, for scamping leads to poor effects and to work which soon deteriorates. Necessarily entailing much labour, it is unavoidably expensive work, so only the best of materials should be used. If this is done, the work is really economical, because it is lasting, and it looks well as long as it lasts.

Certain points must be emphasised.

All plaster, lime and cement should be passed through a fine-meshed sieve, not only because fineness of grain is desirable, but because it is essential to remove all lumps. Such lumps often contain free lime and in any case they prevent perfect hydration, with the result that there may follow, after a more or less lengthy lapse of time, unequal expansion, giving rise to cracks or blisters ; or the surface may be blemished by efflorescence, very difficult to remedy.

If canvas is left in as a permanent reinforcement, as is sometimes the case with marezzo marbles, care should be taken that the textile is completely embedded, not showing itself at the edges ; and both surfaces, back and front, should be well compacted and polished, to make them non-absorbent. If this is not done decay may set in and destroy the work. When well covered with hard, glossy plaster the textile will last for ages, as Egyptian examples, 3,000 years old, testify.

When artificial marble panels or slabs are fixed to frames (as with pillar casings) or wooden battens, not much is to be feared. But if scagliola or the like is applied *in situ* to walls, great care must be taken that perfect adhesion is secured. Thus, brick and stone walls must have all mortar in joints or elsewhere removed. All old plaster must be cut away. The bricks or stones must be thoroughly wetted to stop absorption. Wood and iron must be cleaned, dried and then painted with strongly adhering paint, made gritty by the addition of very fine sand or coarse cement.

CHAPTER XVIII

CONCRETE AND PLASTIC FLOORS

Chunam—Cement Floors—Concrete Floors—Joist Concrete Floors—Concrete Floors on Metal Lathing—Tubular Floors—Slab Floors—Wood Flooring—Some Experimental Floors—Solid Concrete Floors—Concrete Floors for Dancing—Fire-resisting Properties—Terrazzo—Specification for Topping Terrazzo Finish—Plastic Floorings—Magnesia Cement-Sawdust Flooring—Magnesium Oxychloride Composition Floors—Use of Magnesium Oxychloride Cements—Constituent Materials: Caustic Magnesia, Magnesium Chloride—Fillers and Pigments—Methods of Mixing and Laying—Advantages and Disadvantages—Summary.

ROUGH plaster, with sand, has been used for floorings, and until the middle of last century was still often seen so used in Lincolnshire and Cambridgeshire. It was used almost entirely for upper floors, and laid on laths, or the older rushes, supported by the floor joists. It does not, however, provide a good, lasting surface, and has been generally superseded by cement mortars. Nevertheless, rough plaster still forms the principal ingredient of at least one patent fireproof flooring.

CHUNAM

In India chunam is still widely used as a flooring material. As mentioned in the Introduction, chunam is a plaster composed of quicklime and fine sand. It is pure white, takes on a high polish, hardens well and is durable. But it must be remembered that Indians in the home are a bare-footed or slippered people, so floors do not sustain rough wear.

CEMENT FLOORING

Cement flooring is composed of a mortar of 1 part Portland cement and 5 parts sand, gauged rather on the dry side, and laid about 3″ thick on laths or expanded metal and the like, supported on flooring joists. Such flooring can be finished with a $1\frac{1}{2}''$ layer of 1 part Portland cement and 3 parts fine sand, with hardening and waterproofing materials added, or can form the base for boarding or parquet. When boarding or parquet is to be laid, as a rule fillets of wood dipped in pitch are embedded in the cement mortar, round the room against the walls, and at suitable intervals, to enable the top flooring to be fixed to them. Wood fillets may be replaced by screeds formed of coke-breeze concrete. If the situation is at all damp, the cement mortar should receive a coat of tar applied hot to the dry mortar. In any case, the mortar should be perfectly dry before the wood flooring is applied. If the mortar is damp, rot may set in, and in the absence of under ventilation there is also danger of dry-rot.

CONCRETE FLOORS

There are two main types of concrete floors. Those laid on the ground, or other solid foundation, and those forming upper floors, whether over basements or higher up.

The latter may be divided into : (1) Joist floors, cast *in situ* ; (2) floors on metal lathing ; (3) tubular floors, in which terra-cotta or pre-cast reinforced concrete tubes of various shapes are embedded in concrete ; and (4) slab floors. The two last comprise most of the patent fireproof floors, with which we do not propose to deal here, as such work is usually carried out by specially trained concretors.

Joist Concrete Floors.—In this type of floor the joists (practically always of steel) are embedded in the concrete, and for these floors centring is necessary. This centring or sheathing should be of well-seasoned planks, closely adjusted and suspended from the joists at the right distance by means of wires. The strength of the centring and suspending links will depend upon the area and thickness of the flooring, as the centring will have to support the weight of the concrete until it has set and hardened. By way of guidance, it should be mentioned here that 1 cubic foot of average 1 : 2 : 4 concrete will weigh, according to the nature of the aggregate : coke breeze, 100 lb. ; clinker, 110 lb. ; brick, broken, 125 lb. ; limestone, 135 lb. ; ballast, 145 lb. ; reinforced 150 lb. or more. In certain cases, therefore, the centring may require supporting from below. This further support is obtained by placing planks under the centring and nailing to this upright timbers resting on the lower floor. These are struck when the concrete is set and the centring removed.

The concrete is usually a mix of 1 part Portland cement, 1 part sand, and 4 parts aggregate, which may be coke breeze, clinker, broken bricks or fine gravel. Before pouring, the centring should be well wetted. The concrete is then carefully laid so that it should be packed close under and round the girders. For ordinary floors, $\frac{1}{2}''$ below the under part of girders, and $\frac{3}{4}''$ over the top of girders will be sufficient. Where heavy loads or traffic have to be provided for, the top layer should be 1″ thick, as the depths of the girders, and consequently the total thickness of the flooring, will be greater. Three to four weeks should be allowed for the concrete to harden. Then the wires can be cut and pulled out as the centring is removed, the holes being filled with cement grout.

If the flooring is to be left with concrete finish, it is usual to leave the concrete with rough surface, and as soon after initial set as possible to lay on a top dressing about $\frac{1}{2}''$ thick, of 1 part Portland cement. 2 parts sand, 2 parts fine aggregate, which is smoothed with the trowel. As ordinary concrete floors are apt to " dust " or powder, the top dressing should contain some hardening agent, or the surface receive special treatment, a point to which we shall return presently.

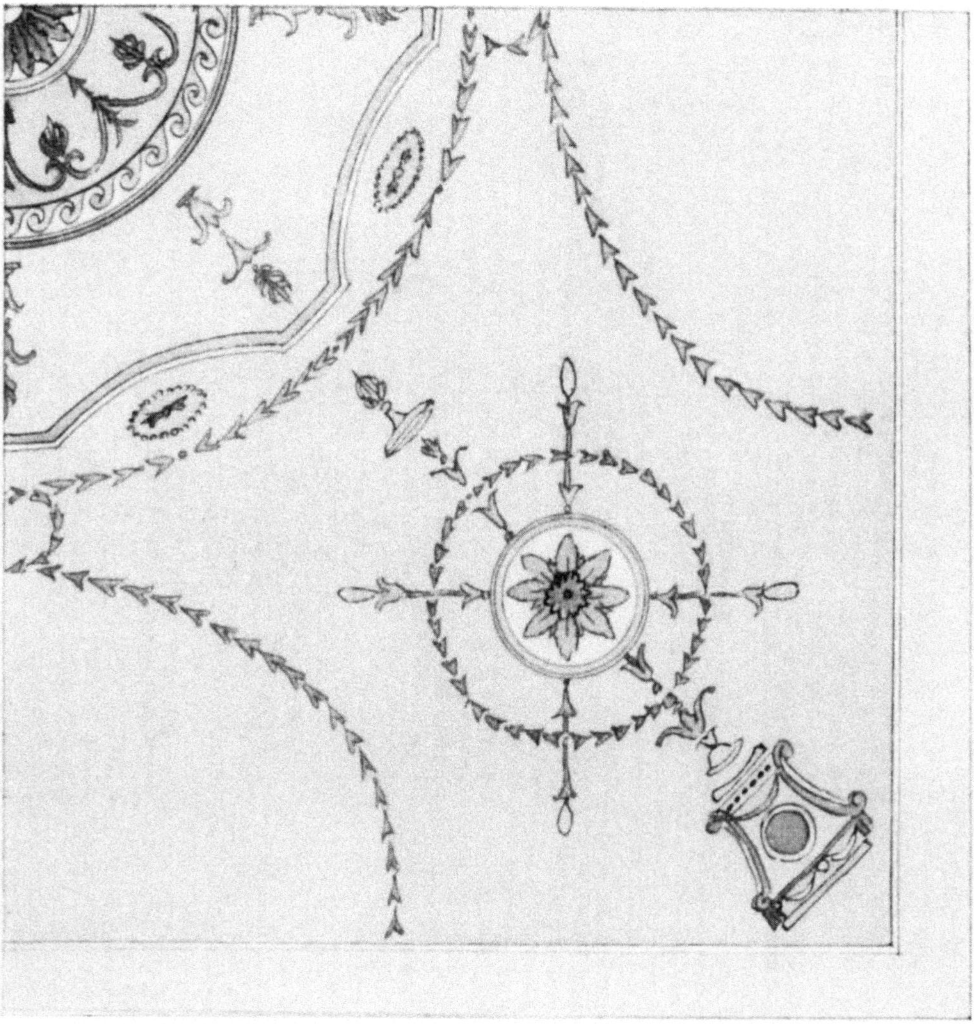

SECTION OF CEILING IN THE LATER STYLE OF ROBERT ADAM
Ornament usually in composition or plaster in low relief.

Concrete Floors on Metal Lathing.—Metal lathing or expanded metal may be wired or bolted to the upper or lower parts of joists. If the floors are of small area and light, the former position is permissible, the actual thickness of the concrete being kept to 1″ or 2″, and the lower parts of the joists lathed for the ceiling. If, however, the span is considerable, heavier floors will be necessary, and then the lathing should be bolted to the lower part of the joists, these and the lathing being completely embedded. The nature of the concrete and the methods of laying are the same as for the "joist floors," though as a rule the centring can be somewhat lighter, and as this deck is reinforced throughout, the depth and width of the joists can also be slightly reduced.

It will be readily understood that the underside of both kinds of floor can be left rough to receive a plaster ceiling, the only precaution necessary being to wet the concrete before applying the plaster. If, owing to the nature of the centring, the underside is too smooth, the surface may have to be hacked before applying the plaster.

It should be remembered that if concrete floors are adopted, it is impossible to have concealed pipes between flooring and ceiling, as the surface cannot be broken up to get at them when necessary, so other methods of piping and wiring must be adopted.

Wood Flooring.—Concrete being hard, noisy and cold, a top finishing of plank flooring or parquet is often adopted. Several methods are in use. Wooden fillets, partly embedded in the concrete or coke breeze screeds, are allowed to project, these being placed at suitable distances so that the boards can be nailed to them. If this method is selected, the space between the concrete and boards should be ventilated, so as to prevent dry-rot. In any case, the concrete should be thoroughly dry before the boarding is commenced.

A flooring of this description is expensive, as there is the concrete deck *plus* the ordinary board flooring and the cost of forming the screeds. With proper precautions it is possible to lay the boards or parquet on the concrete, when there will be no necessity for screeds, and the wood dressing can be quite thin. When this method is chosen, particular care should be taken with the drying of the concrete. It should not be gauged too wet, and soon after the initial set has begun the surface should be covered with dry sawdust, which will absorb the extra moisture. This may have to be renewed more than once. Drying may take anything from six days to three months, according to the thickness of the concrete, the nature of the aggregate and climatic conditions. The underside of the boards or parquet should be coated with a mixture of tar and pitch, applied hot, and if in a damp climate the concrete surface should also be painted. Or a thin layer of slag wool may be laid down, or felt dipped in pitch or creosote. Slag wool is the better material and the easiest to handle.

Whichever type of top flooring is adopted, close tonguing and grooving is essential, as dust and water must not be allowed to percolate to the hard concrete deck.

Many specifications demand that "concrete under boarded floors of basement (or ground) level shall be composed of 1 part cement to 6 parts of aggregate," and that "the concrete shall be 6″ in thickness, lightly rammed with wood beaters until the moisture rises to the surface, floated to a smooth and even face, and finished to proper falls or levels."

Some Experimental Floors

In an experimental cottage built by the Department of Scientific and Industrial Research, Mr. W. R. Jaggard, F.R.S.B.A., Architect to the Department, the whole of the upper floor was of concrete composed of 1 part of Portland cement, 1 part of clean sand and 4 parts of clean granite chippings cast *in situ*, the edge resting upon the inner half-brick walls of the external walls. The floor was calculated to carry a super load of 70 lb. per square foot. The floor consisted of a series of reinforced concrete beams supporting a concrete deck, at bearings of from 5′ to 5′ 3″. The method of construction was:

The centring for the beams was erected with 5″ × 2″ soffit bearers and 7″ side casings. The soffit bearers were carried by scaffold poles about 4′ apart, resting upon 1½″ boards laid upon the ground. Lengths of "Hy-rib" metal reinforcement were then laid upon the edges of the side casings of the beams, and temporarily strutted in the centres of the bays. The reinforcement of the beams was then placed in position, and the floor was ready for concreting. To avoid temperature cracks, ¼″ circular steel rods 4′ long by 9′ apart were placed over each of the beams, the ends resting upon the "Hy-rib" filling, and similar rods were placed in the direction of each part of the slab resting upon the "Hy-rib." The concrete was carefully placed in position, the whole floor being completed in one operation to avoid joints. The side casings of the beams were removed fourteen days after the beams were filled, but the soffit bearers and the slab struts were left some weeks longer. When these latter were removed, the underside of the "Hy-rib" was covered with a ½″ thickness of 3 to 1 cement rendering, making a total thickness of 3″ for the floor slab and providing a ceiling. The upper surface of the floor was finished with a ½″ floating of 3 to 1 cement rendering. This flooring was 35′ 10″ × 25′ 10″.

Some tests of concrete beam and slab floors by the same Department are worth recording here.

Slab D (reinforced concrete beam and slab) was made up of units between two concrete moulded beams, wedge-shaped in section with blunt point, reinforced with ⅝″ diameter steel rods near the bottom and two ¼″ diameter steel rods near the top. Ten-gauge steel wire connected the bottom and top rods, being wound round them, thus keeping them in position. The wires were continued upwards, sticking out, to be turned down into the joints between the concrete slabs before the fine concrete rendering, thus forming ties. Between the beams, and between beams and walls, the space was bridged by 1½″ concrete slabs, 2′ 8″ long by 1′ wide, made on the ground in order to obtain a rough face. Each slab was reinforced with three $\frac{3}{10}$″ diameter steel rods. The top edges of the slabs were chamfered all round. Each slab was bedded on both the beam and the wall. When finished, the floor received a layer ¾″ thick of fine concrete (¼″ shingle, sand and cement, 4 : 2 : 1). Beams and slabs were 159 days old, and the complete floor 79 days old, when tested. There was no deflection under a load of 71·18 lb. per foot super, but under a load of 287·93 lb. per foot super there was gradual deflection to 1·60″ after fifteen minutes, when there was complete collapse.

Slab F (reinforced concrete beam and slab). This floor, measuring 13 ′3″ × 12′, was built of six concrete beams (of blunted T-shape), reinforced with a ¾″ diameter rod at bottom and a ¾″ diameter rod at top, connected together by steel wire stirrups. These beams, 5½″ at bottom, 6″ deep and 4″ wide at top, with shoulders on each side, were made in moulds. About 1′ above the shoulders holes were made to receive a ¾″ diameter steel rod. These beams were spaced at 2′ centres and filled in between with 2¼″ coke breeze blocks, chamfered on the top edge to admit a ⅝″ diameter rod passing through the beam and running transversely therein for the whole length of the slab. The breeze blocks were filled over with fine concrete, brought to a level with the top of the beams and finished smoothly. This floor was thus reinforced in both directions. Its weight was 50 lb. per square foot. Tested when the units were 87 days old and the floor 54 days old, there was a deflection of ·08″ under a load of 69·78 lb. per foot square, which reached ·10″ after two hours, with a permanent set of ·05″, though with no apparent ill-effect. Loaded to 139·57 lb. per foot super, after two hours there was a net deflection of ·26″, and the permanent set reached ·10″ under a load of 279·13 lb. per foot super, the deflection reached ·75″, little increased the next day, though under continuous load ; when the load was raised to 367·21 lb. per foot super, the deflection was 1·20″ and the floor settled slowly and broke. This floor behaved as a monolithic structure.

As a rule, plasterers would be concerned only with the top finishing of such floors and the floating of the ceilings beneath, the latter being carried out in the usual way.

Solid Concrete Floors

When concrete floors are laid on the ground or other solid foundation, the mix is generally one of 1 part Portland cement, 1 part sand, and 4 parts aggregate, the latter usually being broken brick, broken stone, or ballast. The ingredients should be thoroughly mixed while dry, and just before being required sprinkled with water, quickly mixed with the spade and shovelled into position.

If the paving or flooring is to be laid on the ground, the position should be excavated to a depth of about 12″ below floor level, the bottom consolidated and drained, and then filled in to a depth of 6″ to 8″ with hard core or other suitable material. On this the concrete is shovelled and beaten lightly with the back of the spade to a depth of about 3″. Over this is floated a mixture of 1 part Portland cement to 1 part of sand gauged stiff and smoothed with a float.

Concrete floors should not be poured in one mass, but laid in squares, not more than 6′ to the side; kind of coffers being formed by means of battens which serve as guides for the straight-edge when levelling the surface, and also as expansion joints. The battens are removed when the concrete has set, and the space filled in with a similar mixture. Slow drying is advisable, and it is well to cover the surface with wet sand, which can be damped as required and removed when the concrete has hardened.

As already mentioned, concrete floors are liable to "dust" or crumble under hard wear. To overcome this, some hardening ingredient, which is usually also waterproofing, is added, either to the whole body of concrete, the top dressing, or both. There are many of such substances which have proved efficacious in practice. Some are in the form of powders, generally mixed dry with the cement; others are pastes or liquids to be dissolved and diluted in the gauging water. These agents act as hardeners and binders. Their use varies considerably, and it is essential that the directions given by the manufacturers should be followed scrupulously.

If 5 per cent. of hydrated lime is added to the cement, a more plastic, slow, but hard-drying and wearing material is obtained.

It is found that if Portland cement floors, particularly those in passages, halls, engine-houses and other places subject to hard wear, are painted with a special cement paint, they will last much longer and give no trouble from dust arising from the surface. There are many of such paints, but with some it is important that they be applied direct to the cement without any intervening primer.

Concrete Floors for Dancing

A concrete floor can be made quite suitable for dancing purposes. It must, of course, be perfectly level, and, if necessary, this must be done by repairing broken surfaces, cutting out loose material, dusting, wetting thoroughly, and filling in with fairly moist concrete to make good the deficiency. If the concrete is of a very coarse description, it may require a thin dressing of cement mortar, say, 1 part Portland cement to 2 parts sand. The floor must first be hard brushed, and then thoroughly wetted before floating on the mortar and smoothing with the float. A ½″ layer should be sufficient.

To dress this surface when dry (or any other clean, sufficiently smooth surface) for dancing, proceed as follows: Prepare a thick lather of common household soap, and scrub with a brush. When dry, sprinkle with powdered soap and then polish with a pad or mop. A dressing of this kind lasts long, only needing occasional polishing to keep it in fine condition.

Another way is to dissolve paraffin wax in turpentine. The wax should be powdered or finely shaved, added gradually and stirred; not more than the turpentine will take up to form a viscid fluid should be used. This is brushed into the cement and then polished.

In both cases, while a sufficient amount of dressing material should be employed, this must not be overdone. The quantity required will depend upon the nature of the concrete and the quality of the work. The object is to fill the pores of the floor and leave a mere film over it, which will give the desired slidable surface. If overdone, there will be a thick film, apt to become sticky and soon damaged by the dancers.

If the dancing floor is in the open air, the paraffin wax and turpentine mixture is to be preferred.

Fire-resisting Properties

The temptation to use coke breeze as an aggregate in floor making is very great, for it is light in weight and provides the decided advantage that nails or screws can be driven into it, thus facilitating the laying down of wood floorings, boards or parquet. But it must not be overlooked that coke breeze has its disadvantages, for it is absorbent and produces a porous concrete, apt to become wet in damp situations. Moreover, if not exactly inflammable, it is greatly influenced by heat, and in case of a serious fire offers practically no resistance, and by crumbling away quickly disintegrates the whole mass of concrete. Thus in many cases coke breeze is not a suitable aggregate, especially for suspended or upper floors.

Where fireproof and waterproof qualities in concrete are demanded, special care is needed in the selection of materials, with a view to produce a compact, close-textured mass. For this purpose the aggregate must be hard, not too large or irregular in shape, yet not too smooth and of varying sizes within a narrow limit. Thus for average purposes the aggregates should be in about equal proportions of $\frac{3}{4}''$, $\frac{5}{8}''$, $\frac{1}{2}''$, $\frac{3}{8}''$, $\frac{1}{4}''$. The sand should be washed, sharp, not too fine, and free from dust. It should be well mixed in dry state, gauged rather wet, but not too much, and firmly but gently consolidated by tamping.

Ballast of the gravel type does not make good aggregate, for under long and intense heat it disrupts with an explosive force, which disintegrates the matrix or cement-sand mortar.

Well-made lime or gypsum plaster, if laid in two coats and compacted by trowelling, while not fireproof, is a good non-conductor and strongly fire-resistant. It is even an effectual protection for wood. But this protection is lost to a large extent if the plaster contains a high percentage of coarse sand.

A suspended floor with a top coating of asbestic plastic material and a good plaster ceiling below will prove an effective means of resisting the vertical spread of fire. This form of construction, therefore, makes the use of coke breeze in a concrete case permissible, at all events in small- or medium-sized dwelling-houses or shop property.

Terrazzo

Venetian mosaic, the so-called Terrazzo mosaic, may be included in the class of semi-plastic floorings. It can be successfully tackled by anyone who is willing to take pains. It is made up of fragments of marble (which may be all of one colour, black and white, or varicoloured) or of granite, having a rough polished face, rough base, and are of irregular shape, but should be small, not more than $\frac{3}{4}''$ in any direction, and not more than $\frac{5}{8}''$ deep.

The foundation for flooring, terrace or steps should consist of a rich, slow-setting concrete. On this is floated a $1\frac{1}{2}''$ layer of liquid mortar, composed of 1 part of Portland cement and 2 parts of marble dust in the case of marble work, or silver sand for granite. The paving cubes are bedded in this at random, but so as to pack tightly together, the whole then being floated with grouting and well rubbed down with a float.

A cheaper grade of this type of paving is carried out with irregular chips of marble, or granite, thrown down broadcast on a $\frac{1}{2}''$ layer of mortar, pressed down by means of a small float, then covered with liquid mortar and finished with a large smooth-faced float. As the mixture is fairly plastic, it can be easily shaped to corners, etc.

In both cases the matrix of mortar will be fairly conspicuous, so colour is of some importance. This will partly depend upon the quality of the cement and the nature of the sand or marble dust. A splendid, sparklingly white, hard-setting mortar can be made with well-slaked fat lime, fine ground white marble lime, and a little medium ground white marble. Ground slag and brick dust give greys, browns and reds, while adding to hardness and making the top dressing heat-proof. Forge ashes will give black mortars ; ashes and marble dust, pearl greys ; litharge or red oxide of lead, reds ; 2 parts marble dust, 1 part lime and $\frac{1}{2}$ part oxide or carbonate of copper, blues ; green enamel and marble dust, greens.

It is possible, by using skeleton templates embedded in the cement, to form patterns, the templates afterwards being lifted out ; but this is going away from the nature of the work.

Specification for Floor Topping, Terrazzo Finish

1. The concrete contractor shall pour the flour slab and screed the concrete with a straight-edge, bringing the surface of the concrete $1''$ ($\frac{3}{4}''$) below the finished floor grade.

2. As soon as possible after the concrete slab is poured, the surface shall be thoroughly broomed to expose the coarse aggregate and remove the loose scaling or other foreign material, and allowed to cure.

3. Before applying the wearing surface mixture, the slab shall be washed and sprinkled with neat cement, the cement grout to be brushed over the surface with a broom.

4. A traprock wearing surface mixture, in the proportion of 1 part cement to 1 part and sevenths (1·7) parts of stone (mix 3 sacks of cement to 5 cubic feet of coarse and fine traprock), ranging in size from ½″ down to dust, the mixture to be put down reasonably dry.

(*Note.*—¾″ by ¾″ angle irons, with the heel of the angle turned up, are recommended for screed guides.)

5. The wearing-surface mixture, after being screeded, shall be compacted with water-weighted rollers. (Two rollers are required, one light weight and the other heavy weight.) Rolling to be done in both directions and continued until water is brought to the surface (the surface before rolling may be sprinkled with a dry mixture 1 part cement to 1½ parts traprock, in order to absorb surplus water so as to permit rolling earlier than would otherwise be possible. Atmospheric conditions should be taken into consideration.

6. After rolling, the surface to be swept with soft brooms to remove the water and superfluous cement which have been brought to the surface by the rollers. The surface shall then be lightly steel trowelled, but no attempt made to remove all trowel marks, as is usual (one trowelling sufficient).

7. About a week or more after the surface is trowelled, it should be ground down with an approved type of grinding machine, shod with No. 20 carborundum brick, to expose the coarse aggregate. The material removed in grinding to be washed off the surface.

8. All holes, pits, etc., developed in the grinding process shall be filled with a thin grout composed of 1 part Portland cement to 1 part of No. 80 grain-powdered carborundum grit spread over the surface and worked into pits with a steel straight-edge, to be followed with rubbing process with the grinding machine. Excess material shall be promptly removed after the grinding-machine operation. Filler shall be allowed to set for seven days.

9. The floor surface shall receive a final grinding to remove the film of grout and to give it a polish. It shall then be thoroughly washed and all surplus material removed.

Remarks.—In this case the finishing was laid 1″ thick on the ground floor of a factory and ¾″ thick on the upper floors. The weight of the finish was 10 lb. per square foot.

Plastic Floorings

Although concrete floors can be cast *in situ*, poured and moulded to a considerable extent, vet even those containing a percentage of hydrated lime, and so slightly more elastic, are hard, somewhat intractable and rather heavy. Moreover, they are apt to expand and liable to crack, which peculiarities impose certain precautions and limitations. It was, therefore, early felt that a more plastic material, which could be spread evenly, but thinly, over any expanse and without joint, was needed. To meet this demand a number of materials, or rather compounds, have been produced. Most of them are characterised by being fairly light in weight and possessing sufficient plasticity to make them easy to manipulate, so that they can be spread over large or small squares, plain or intricate positions in thinnish layers and without joints. Such floors have many advantages. They occupy little depth, are usually impervious to moisture and insects, present a pleasant walking surface and are easily cleaned. The more plastic they are, the better they are, for then they can fill up every crack and crevice, and, without joint, can be carried up the wall, to form skirtings or even deep dadoes, which goes a long way to ensure sanitation.

While these floors differ among themselves, the general principles are much the same, with sufficient distinction to make one or other best adapted for particular purposes or situations.

For domestic purposes and places of public resort, a flooring which is not too highly polished, which is soft to the tread (therefore, comfortable and more or less noiseless) and a non-conductor of heat, so warm in winter and cool in summer, is to be preferred. But it should have a good wearing surface, prove impervious and possess strongly adhesive properties.

For schools, public institutions, hospital corridors and certain classes of shops (fishmongers, butchers, greengrocers, dairies), a harder material with close-grained, polished surface is to be preferred. Such floors are better wearing, less apt to get dirty, but lose a little of their plasticity.

The former usually have a fibrous component or a magnesium chloride basis; the latter generally contain marble or granite, and often may be classed as a modification of terrazzo.

Among the hard-wearing, special resisting floorings are "Granolithic" *in situ* paving. A ½″ topping of this on sound concrete foundation makes a perfect floor for dairies, not being affected by lactic acid. "Marbolith" in one 1″ coat on concrete, or ¾″ in two coats on boards, gives an impervious, resilient but durable surface. "Plastoment" is a plastic fireproof flooring, containing asbestos and cement, but applied in two coats of ⅜″ each. It can be applied on concrete, stone, tiles or boards. "Terrazzolith," an American preparation, is a chemically purified magnesia containing less than ½ per cent. of lime, mixed with long-fibred asbestos and wood pulp, coloured by means of metallic oxides. It is flexible and does not crack.

Most of these floorings are shallow dressings applied over well-laid concrete foundations. The foundations must be quite level, but not smooth; they must be left to afford a key to the top layer. This is the method adopted with a form of "granolithic" first evolved in America, but now fairly common over here. This top dressing is composed (by measure) of 1 part of cement, ½ part of fine aggregate (coarse sand or chippings) and 2 parts of coarse aggregate (preferably hard granite). It is gauged rather stiff, spread evenly to a depth of slightly over 1″ and straight-edged. Any depressions left must be promptly filled up and straight-edged. The whole surface is then worked over and compacted with wooden floats. This must be done before the initial set begins, so slow-setting cement is advisable. After this, the surface is trowelled smooth, but trowelling must cease before there is any risk of interfering with ultimate set. The trowel may be dipped in water, though it is well to avoid a wet surface. This is the way of proceeding if the "grano" dressing is laid on the green foundation soon after initial set. If for any reason the application of the top dressing is postponed, then the surface should be vigorously scratched with a wire brush, brushed, washed and sponged. Before applying the "grano," the foundation must be wetted, but all superfluous water removed. The surface is then floated with a cement and fine sand grout (of the thickness of cream) on which the "grano" is floated and treated as above described. To save the surface, it should be covered with wet sand or sawdust, kept damp for three to four weeks, according to the state of the weather. Fast setting and hardening are not desirable.

Most plastic floors of the first type have a basis of magnesium oxide or magnesium chloride with fillings which range from ordinary sawdust, fine wood flour, cork parings or cork dust, talcum, asbestos fibre or other material.

When well compounded and properly laid, such floors will possess a tensile strength of from 572 lb. to 786 lb. per square inch, and a pressure resistance of from 3,942 lb. to 4,300 lb. per square inch.

Such compounds are commonly laid in one coat ½″ thick, or in two coats ⅜″ thick, according to the quality of the finished work. They can be laid practically on any foundation, provided it is sound, clean and free from dust. If a much-damaged concrete or brick surface has to be covered, it is economical to first level up with mastic. If laid over boards, it should be seen that these are clean and quite free from rot (dry or wet). Mere wearing into unevenness and splinters will not matter, but all rotten wood must be cut out and knot holes stopped.

Magnesia Cement-Sawdust Flooring

The following is a typical magnesia cement-sawdust flooring compound.

Add 2¼ to 2¾ parts (by measure) of chloride of magnesia to 1 part of sawdust (first removing all knots, chips, shavings, etc.). Mix into stiff paste and lay with trowel, smoothing with floats. This mixture should be laid down 2″ to 4″ thick, according to situation and likely wear. It can be carried out to the wall to form skirting, and at the angles of rooms it can be finished with a concave curve. It will dry in from eighteen to forty-eight hours, but should not be walked upon for at least a week. It will then be quite hard and washable.

Magnesium Oxychloride Composition Floors

Mr. J. Turner, member of the National Association of Plasterers, Granolithic and Cement Workers, and former instructor in plastering, kindly sends us the following interesting paper read before his Society :

In 1868 the French chemist, Sorel, discovered that when calcined magnesia was made into a thick paste with magnesium chloride, the mass quickly hardened and set. The possibilities of this property became obvious, although some considerable time elapsed before its use was established, owing to the high cost of the constituents. Research work, however, proved that an economic commercial problem could be solved by the inclusion of aggregates, such as wood flour (a specially prepared fine sawdust), pumice, slate dust, fine granulated cork, asbestos, various fibres and pulps. Their employment soon became universal for floor and wall coverings, the latter more in the sphere of moulded work, such as dado panelling in the form of imitation marble panelling. Their setting powers being thus unimpaired by the additional ingredients, the commercial outlook became much clearer, and to-day their employment is general, inasmuch as officers' quarters, saloons, alley-ways, etc., on shipboard, alike with floors in buildings, are laid under various trade names. Magbestic, Wirolithic, Athena, Linolite, and many other firms advertise themselves as jointless floor layers. The strongest cement available—which is cheap enough to be used on a large scale—is resilient ; a good qualification when used to replace lino, and may be laid with successful results upon wood flooring, and in addition thereto upon prepared concrete floors. Two coats are usually specified of a total thickness of $\frac{3}{4}''$, the top coat alone of which is coloured. The colouring matter for flooring mixtures is iron oxide, yellow ochre and other umbers.

When magnesia or magnesite in a proper physical condition is mixed with a sufficient quantity of a dilute solution of magnesium chloride to form a plastic mixture, the whole sets in a few hours to a hard, stone-like solid. The formula expressed for the composition of the compound is $3MgO.MgCl_2.10H_2O$.

The setting takes place through the combination of the two chief constituent materials—magnesium oxide, magnesium chloride—with water. It will be seen, then, that this cement differs from the commonly used Portland cement and plaster of Paris, in that a chloride compound is required to complete the setting in addition to the usual water. The neat cement, which resembles a hard white stone similar in aspect to meerschaum, may be made nearly twice as strong as Portland cement, while its binding powers are so great that it can be made to hold together very large volumes of filling materials, such as those aggregates aforementioned. The resultant composition is a solid and strong mass, and its important use is to produce a hard floor covering which has some of the properties of wood, in that it is not too cold to the feet, while nails, etc., may be driven into it.

The magnesite, as it is better known, is first mixed with the fine sawdust in the proportion of about 3½ to 1, and when required, the colouring matter in the proportion of 10 lb. of, say, red oxide is incorporated with 1 bushel of the magnesite to make a rich red. These ingredients are now thoroughly mixed while in the dry state by means of sieve and rake. The magnesium chloride as obtained is a solid and necessitates being broken up when removed from its tub or other container and dissolved in water. It is very soluble and easily deliquesces in the air. It is of the utmost importance that the specific gravity of this liquid chloride should be correct, because, if too strong, expansion which will result in large mounds will take place during the setting process at least, while if it is too weak the composition will not harden sufficiently and cracks will occur. There are two systems by which the solution may be tested—the Twaddell and the Beaume. With a No. 3 Twaddell the specific gravity should be 1·25, which is 50° Twaddell, while that of the Beaume should be of a specific gravity of 1·188 and 24°.

The specific gravity is controlled by means of a hydrometer. The solid chloride is placed in the tub, when water in the proportion of 5 lb. of the magnesium chloride to 1 gallon of water is added until thoroughly dissolved. The test is made by placing the hydrometer in the liquid, adding more or less water as may be necessary until the hydrometer floats to the right degree. The solution being found correct, sufficient is intermixed with the magnesite, gradually, until a portion squeezed in the hand adheres well together but does not exude any moisture. The mixing should be performed upon a board with raised sides, as the floating sawdust sometimes causes great inconvenience, also by means of the rake the process may be accelerated.

Should the foundation be of concrete, it must be dry and clean, as any dampness or lime mortar adhesions will result in failure. Wood floors require that either flat-headed nails should be driven at close intervals in order to secure the composition, or that the boards should be taken up but relaid with open joints. To further ensure a perfect bond, the surface, whether of wood or concrete, should be slightly damped with some of the chloride solution. The process of laying the floor is performed by first forming screeds, then flanking the whole of the interior space, beating the sawdust mass with a handfloat to level and consolidate it. The topping coat, which will probably be coloured, is applied in a similar way and allowed to harden for about two days, according to the humidity of the atmosphere, before taking part in the final process of scraping and polishing. This process is carried out by means of a steel scraper, when any trowel marks or blemishes may be erased, and meanwhile any cracks or defects may be made good by filling in with some of the scrapings which have been slightly moistened with some of the chloride. These floors may, if required, be further improved by a process of polishing with any good make of polishing wax.

Damp Composition Flooring.—As this flooring is non-absorbent, the least sign of condensation—caused probably by a warm air striking the cold surface after a period of cold, wet weather—may be mistaken as dampness. When this is observed it may be put down to the nature of the materials, one of which is decidedly hygroscopic (magnesium chloride), but this really must be set down to the non-absorbent nature of the flooring. This trouble is not likely to occur in rooms where fires are burning, either in or in close proximity with the new work.

The Process of Setting.—Users generally safeguard their position by seeking to maintain a system of ventilation immediately after laying, because a moisture-laden atmosphere retards the drying, which, with the inevitable expansion that takes place during the setting, is frequently the cause of cracks, the cement not being hard enough to resist them.

Another cause of failure or weakening of the cement is undoubtedly due to the fact that the calcined magnesia may contain as an impurity a proportion of lime which, as it is not in itself a setting material, may, while reacting with the chloride, cause a general weakness. Another annoying factor to floor layers is the knowledge that expansion is always taking place during the setting more or less, and it is proved that, apart from other causes referred to already, a greater expansion may take place from a coarsely ground magnesite than from one finely ground.

The chief source of supply of magnesite to this country is India, where it is calcined in gas-fired shaft kilns, after which it is largely ground under mill-stones.

Analysis of a broken piece of imitation marble removed from a dado panel:

Calcium carbonate	51·45	per cent.
Magnesium chloride	36·10	,,
Silica	1·75	,,
Iron oxide	4·00	,,
Lamp-black	4·10	,,
Moisture	2·60	,,
	100·00	,,

The Use of Magnesium Oxychloride Cements

The subject is of such importance that the Building Research Board of the Department of Scientific Industrial Research issued a report on it, of which the following is an abstract:

About the middle of the nineteenth century, Sorel, a French engineer-chemist, discovered that a hard, dense cement could be obtained by mixing caustic magnesia with an aqueous solution of magnesium chloride. This cement, now known as Sorel's magnesium oxychloride, but more generally and somewhat erroneously as magnesia or magnesite cement, is the strongest cement used in any quantity for building purposes at the present time.

The chemical union of caustic magnesia with magnesium chloride results in the formation of magnesium oxychloride. As is the case with many other cements, the mechanism of setting is as yet imperfectly understood, and there is room for a great deal of necessary research.

The binding qualities of oxychloride cement permit of its use with very considerable additions of filler, and this fact has brought it to the fore as a flooring cement. According to the nature of the fillers used, and by adjustment of the quantity of the filler, a variety of products can be obtained, from a marble-like material to one possessing some of the characteristics of wood.

On the Continent and in America the cement has, since its invention, formed the basis for stucco, artificial stone and bricks, but it has not found much favour in this country in these capacities in the past, nor does it seem likely to in the future.

In view, however, of its very extensive use in jointless floors, and the number of these floors at present on the market as proprietary compositions under trade names, it is important that the properties of the material and the precautions that should be taken to avoid failures, should be fully understood. A fair amount of literature has been published on the subject in recent years, and the object of this bulletin is to present in convenient form such of this information, together with that collected during investigations at the Building Research Station, as is likely to be of assistance to those who have occasion to specify, or use themselves, oxychloride cements for flooring cements.

Constituent Materials

Caustic Magnesia.—Practically all the caustic magnesia used for cement is obtained by calcining magnesite.

The principal deposits of the mineral magnesite, which is a natural magnesium carbonate, occur in Greece, India, North America, Austria and Czechoslovakia. The best of the magnesites, containing at least 95 per cent. of magnesium carbonate and under 1 per cent. of iron, and giving a pure white magnesia, comes from Greece and India. As an example, the following analysis of a good Indian magnesite is given:

Silica	2·20
Alumina, Iron Oxide	0·30
Lime	0·59
Magnesium carbonate	96·08
Water	0·83

The effect of heating magnesite is to drive off the carbon dioxide, leaving magnesium oxide. When broken magnesite is burnt or calcined at very low temperatures, and only a part of the carbon dioxide is driven off, the product when treated with water sets rapidly—much too rapidly for it to be of any use as a structural cement. When calcining is continued, as it is commercially, up to temperatures ranging from 700° to 1,200° C., the resulting amorphous magnesium oxide, known as lightly calcined, caustic or plastic magnesia, is for all practical purposes free of carbon dioxide, and sets when mixed with water alone. The exact temperature of calcination to produce the best results is influenced to a large degree by the nature of the magnesite and the conditions of burning, so the specification of a magnesia burnt at a specific temperature will be of little avail. The presence of under-burnt or over-burnt magnesia will mean erratic behaviour of the cement as regards setting, and considerable volume changes during and after setting.

Calcining at higher temperatures, of the order of 1,700° C., produces an oxide in a different physical state, known commercially as dead burnt, sintered or shrunk magnesia. This magnesia is much denser than the lightly calcined, and not sufficiently active for use as a cement, but it possesses valuable refractory qualities, and is marketed for this purpose. The temperature necessary for obtaining the dead-burnt state is lowered by employing magnesite with a comparatively high iron content, or by adding iron oxide where the natural iron content is particularly low.

After leaving the kiln, caustic magnesia is sorted and finely ground between millstones or in tube and ball mills. The fineness of grinding is of vital importance. Not only does fine grinding assist the chemical combination with the chloride, but it is one of the most important factors governing the physical phenomena of setting and hardening. It has been suggested in an American publication that a residue of 3 per cent. on a No. 100 and 25 per cent. on a No. 200 sieve is a reasonable degree to demand to give satisfactory results.

The behaviour of the cement will be affected if the caustic magnesia has been exposed to the atmosphere, since it readily absorbs moisture and carbon dioxide. In some cases where freshly calcined magnesia has proved unsound, a reasonable amount of aeration has been known to be of advantage, but only where the magnesia was originally defective, possibly containing under-burnt particles or excessive free lime. The apparent corrective effect may have been due to the slowing down of the too rapid setting of under-burnt particles, or, perhaps, to the carbonating of excessive free lime.

A free lime content of over 2½ per cent. in the magnesia is liable to aggravate volume changes in the cement and cause a loss of strength. With magnesia obtained from good Greek or Indian magnesites there is not much likelihood of this figure being exceeded, and something like the following analysis is to be expected :

INDIAN CAUSTIC MAGNESIA

Loss on ignition	1·81
Insoluble residue	1·12
Oxide of iron and alumina	0·63
Lime	1·05
Magnesia	95·39

Magnesium Chloride.—Magnesium chloride is usually imported solid in drums, and an aqueous solution of specific gravity from 1·138 (28° by Twaddell hydrometer[1]) to 1·176 (36°) is ordinarily used. The chloride may contain sulphuric acid, but this can easily be removed by treatment with barium hydrate or barium carbonate. A good method of carrying out this purification is to put the chloride solution into a wooden tub or barrel tapped about one-quarter of the way up the side. The barium hydrate or carbonate should be carefully added, a little at a time, and the whole stirred. When the precipitation of barium sulphate ceases, time should be allowed for settling, and the clear magnesium chloride solution, free from precipitate and from the scum which usually collects on the surface, can be run off. Generally speaking, the amount of barium salt required will be between 6 and 10 per cent. of the weight of magnesium chloride.

The expansion and, within limits, the strength and hardness of the cement increase with the concentration of the solution, but at the same time the set is accelerated ; thus the choice is limited to comparatively dilute solutions in order to allow of time for working the cement to a good surface.

Fillers and Pigments.—It has been mentioned that oxychloride cement can stand the admixture of a high proportion of filler, provided, of course, that this is inert. Where it is not so, the cement would be useless, even dangerous, for building purposes, on account of the great volume changes to which it is subject, without some diluent.

Fillers in common use are sawdust and wood flour, broken and powdered cork, asbestos fibre and powder, quartz sand, leather chippings, kieselguhr, talc, clay, serpentine and peat. Talc, clay and asbestos fibre facilitate trowelling. The selection of fillers depends on whether a warm, resilient and noiseless floor is required, or one more dense and stone-like, with a harder and smoother surface. Such fillers as sawdust, cork and kieselguhr would be employed where the former is the desideratum. There are instances, too, where lightness is essential, as when the composition is laid on ship's decking. Here it is laid over the irregularity of steel plates, and may require to be as much as 3″ thick at plate junctions. An absorbent filler may be saturated before incorporation in the mix, and, if desired, can be coloured at the same time. In practice, saturation with chloride solution appears fairly general.

A large addition of colouring matter is not required. The following earth pigments and metallic oxides are in common use :

Red	Red iron oxide.
Yellow	Yellow ochre.
Browns	Burnt umber and burnt sienna.
Blue	Ultramarine.
Green	Chromium oxide.
Blacks	Lamp-black and manganese dioxide.

[1] A No. 2 Twaddell hydrometer graduated from 24° to 28° and costing from 2*s.* 6*d.* to 5*s.* is suitable.

METHOD OF MIXING AND LAYING

The usual practice is first to mix the magnesia, fillers and pigments together, and then to add the magnesium chloride solution gradually till the right consistency is obtained; that is to say, till the material binds in the hand but does not exude moisture when squeezed.

The following mix, exclusive of the magnesium chloride, for a floor of the resilient type is given in U.S.A. Bureau of Standards Circular No. 135:

	Bottom coat. Per cent. by weight.	Top coat. Per cent. by weight.
Caustic magnesia	35·0	45·0
Sand	7·5	–
Ground silica	25·0	20·0
Sawdust	25·0	–
Wood flour	–	10·0
Asbestos fibre	–	5·0
Clay, talc or sil-o-cel (diatomaceous earth)	7·5	10·0
Pigment	–	10·0

This mix is evidently designed to take full advantage of the properties of various available fillers, but simpler mixes can give perfectly satisfactory results. (See Sherer, *Der Magnesite.*)

It will be noticed that in these specifications two coats are provided for. Where the surface on which the floor is to be laid is uneven, the economical advantages of two coats are obvious, the lower one as a rule being leaner and containing coarser fillers. Further, by using elastic fillers only in the lower coat, and restricting finely ground fillers to the top coat, it is possible to produce a hard-surfaced floor with some measure of resiliency.

The best possible mix cannot be standardised, and experience is necessary to obtain the best results with each batch of materials and for the special object in view. It may even be found that a magnesia which gives unsatisfactory results with some fillers will give good results with others.

Proportioning should be done by weight, on account of the great variation in volume of magnesia handled in different ways.

It is essential that the surface over which the floor is to be laid is clean, dry and fairly impervious. The surface should be washed over with clean water or chloride solution immediately before covering.

The wet flooring mixture, after sieving to break up lumps, is evenly spread and lightly tamped, and then brought to a good surface with a steel trowel or float. Where laid in two coats, the second is applied when the first is stiff enough to work upon, but before it has become non-plastic. A thickness of $\frac{1}{4}''$ to $\frac{1}{2}''$ is ordinarily sufficient for single-coat floors, and, when applied in two layers, $\frac{1}{4}''$ to $\frac{3}{8}''$ for each coat. It is necessary when covering areas of any magnitude to divide them up by screeding into sections. Before the composition in the sections has finally set, the screeds can be removed and the spaces filled.

Patterns are arranged for by temporarily blocking out and afterwards filling with mixtures of the desired colours.

Reinforcement in the form of wire netting is often incorporated when laying over wooden flooring, and nails are sometimes driven into the wood to form a key. These precautions are generally unnecessary and, as explained later, inadvisable when using an oxychloride cement gauged with magnesium chloride solution.

A well-ventilated floor, other conditions being favourable, will harden in two or three days. Floors should not be laid in cold or damp weather, since setting is delayed, and a weak floor may result, though it is not known exactly in what way temperature and humidity affect the subsequent properties of the cement. Before the floor has properly hardened it may be scraped to remove irregularities. If it contains a high proportion of absorbent filler, it should be given one or two coats of boiled linseed oil thinned with kerosene, and, when these are dry, polished with wax in turpentine; the oil treatment can be dispensed with on the harder and denser types of floor.

Advantages and Disadvantages

The advantages of well-made jointless floors may be summarised as under:

(1) They are sanitary and non-inflammable.

(2) They can be laid and are ready for use in a comparatively short time.

(3) They can be used over old floors; their high tensile strength and elastic properties allow them to carry safely over the springiness of a wood under-floor.

(4) Their good appearance gives them a wider scope than equally durable floors that do not possess this asset.

(5) The type of floor best suited for its position can be obtained by the judicious use of fillers.

In comparing a jointless floor with other types of flooring, its possibly shorter life would be a disadvantage to be balanced against corresponding advantages. There is ample evidence, however, that jointless floors can be extremely durable, even when subjected to exceptionally hard wear. On the other hand, there have probably been more failures with this type of flooring than any other. The majority may safely be said to have been due to avoidable causes.

Magnesium oxychloride cement is in no sense a water-resisting material. If a floor containing it is soaked with water, its strength will be materially reduced, and if immersed for any considerable period, it will break up altogether. It has been shown by experiment that the mix is, if anything, more important than the quality of the magnesia where water-resisting properties are concerned, and in this connection a fine aggregate is an advantage. A good floor should, nevertheless, stand periodic washing, and a little care exercised in keeping the surface in good condition by polishing or oiling will minimise disintegration from this cause, and at the same time give added resistance to abrasion.

In an effort to cheapen the floors, there has been a tendency to use unsuitable fillers and in excessive quantities. This is bound to result in poor resistance to both abrasion and water action, cracking and, finally, in complete disintegration.

The cause of cracking, blistering and buckling is sometimes obscure. As already suggested, it may be due to the use of under-burnt or coarsely ground magnesia or too strong a chloride solution. There have been cases where floors have cracked or buckled after some years of use, and a possible explanation is that damp or vibration has gradually lowered their resistance to stresses produced in the hygroscopic material by repeated volume changes. One failure by buckling over ship's plating brought to the notice of the Department appeared to be chiefly due to the use of large quantities of a calcium sulphate plaster as a filler, which naturally added enormously to the volume changes of the composition. Needless to say, the buckling had occurred where the floor was thickest.

A very serious disadvantage of ordinary oxychloride flooring is its tendency to corrode contiguous iron and steel; the other metals ordinarily used in building construction appear to be practically immune from attack. The danger is generally realised in connection with ship's decking, and the metal is first covered with an insulating membrane, e.g. bitumen paint. In buildings the danger is, as often as not, entirely ignored, and the floors are laid in contact with structural steelwork and steel conduits for electric wiring. The elimination of this corrosive tendency has been the aim of an investigation at the Building Research Station and a report published. It seems that a solution of the problem is the substitution of ferrous chloride for magnesium chloride in the mix. Not only is the resulting oxychloride cement non-corrosive, but it has a greater resistance to water action. There are one or two minor limitations as regards colour and filler.

There seems to be a popular belief, unconfirmed so far as is known by scientific data, that oxychloride flooring has an exceptionally poor electrical resistance. A few experiments of a very preliminary nature, carried out at the Building Research Station, have indicated that, generally speaking, its electrical resistance is greater than that of a 1–3 Portland cement-sand mixture. At the same time, the preliminary tests have clearly shown that a reliable estimate of the electrical conductivity of the material, taking into account the numerous influential factors, cannot be made without a prolonged investigation. This has now been commenced.

While it will readily be appreciated that success with oxychloride floors depends very largely on the judgment and skill of the layer of the floor, it is considered that the necessary experience could soon be acquired by anyone capable of applying a Portland cement rendering.

It will be gathered from the preceding notes that in any attempt to guard against failure by specification of the constituent materials, the magnesia presents the only real difficulty. Also that most important factors governing its physical properties cannot be readily ascertained. Chemical analysis is merely helpful in detecting harmful impurities. A certain measure of protection is obtained by purchasing only from reputable suppliers dealing in magnesia calcined from Greek or India ores, and by carrying out, on trial mixtures, a few tests for tensile strength, setting time and volume changes during and after setting. The *Bulletin* proceeds to give notes on the specification of caustic magnesia, and in Part II describes the preparation of non-corrosive oxychloride flooring with corrosion tests and cost.

Summary

The object of this work, to obtain an oxychloride cement which, while retaining all the usual properties of such cements, will not corrode metal work, has, it is claimed, been attained by the substitution of ferrous chloride for magnesium chloride in the mixing. By this means a non-corrosive cement is obtained that sets rapidly, and which, with admixtures, forms a floor that can safely be used three days after laying. A floor of this cement was in constant use for over eighteen months, and showed no signs of wear or cracking, though it was washed regularly and did not have its surface treated in any way. Iron nails and particles kept in this cement showed no signs of corrosion.

The magnesia, filler and water should be well mixed first, preferably in a mechanical mixer, the ferrous chloride being added immediately before using, and the whole well stirred. If too much water be added, the non-corrosive properties of the cement are lost, together with much of its strength. The maximum quantity of water added should be 70 parts to 100 of the magnesia by weight.

Because of this limitation in the amount of water, the nature of the filler must be such that it does not absorb a great deal of water. Ordinary under-waterproofed sawdust cannot be used on this account. Granulated cork forms a very useful filler, and sand, granulated slag, asbestos fibre, etc., could also be used.

The amount of filler that can be added is limited by the water factor, but by suitably grading and mixing different fillers a large amount of the latter can be used.

The non-corrosive cement appears to withstand water to a greater extent than ordinary magnesium oxychloride cement. A disadvantage is that it is coloured so deeply that the addition of colouring pigments has but little effect. On mixing, the whole turns blue, but after exposure to the air the surface becomes a stable brown. With slight limitations, therefore, the non-corrosive cement could replace the ordinary Sorel cement in making floors, and it is considered that the cost of a non-corrosive floor should not compare unfavourably with that of the ordinary type.

CHAPTER XIX

SPECIFICATIONS

Plastering—Materials—Workmanship—Flooring—Concrete Foundations—Removal of Defective Wood.

In all important plastering and flooring work a specification is necessary.

When well drawn up, a specification is a guide and a safeguard to the builder, or plastering contractor, as well as to the architect and the building owner. It should state clearly the quality and description of materials to be used, and, as far as possible, how they should be used. To have a definite understanding on these points is a matter of great benefit to all concerned. At the same time, the contractor should scrutinise the clauses of a specification closely, for once accepted they become binding on him, and it may well be that by adopting stereotyped forms of words, or by setting too high a level than is justified by the kind of work and the contract price, hardship would be inflicted on the contractor. Consequently, it is far better to criticise a clause before accepting it, than to assume that it may be modified later. An architect is within his rights in demanding strict observance of a specification; on the other hand, he should be prepared to consider criticism before signing, because it is his duty to ensure to the best of his ability fairness to both parties—his client and the contractor.

After all, a specification is merely the setting forth in detail of the conditions agreed upon in a contract, and should be so framed as to demand neither more nor less than those conditions.

The following notes are based on a standard form of clauses, with suggested alternatives to meet special cases. But, bearing in mind what has been written above, they should not be taken as definite models, but where necessary should be adapted to each particular case.

Plastering Specification

Materials.—1. All sand shall be fresh water, river or pit sand, free from vegetable matter, earth or saline material; must be well screened and, on demand, be washed.

> *Note.*—Sand should be sharp, free from dust and irregular in size. Hydraulic limes will require fine sand; for fat lime and Portland cement floating coarse is best. But fine decorative plastering needs fine sand. So in high-class plastering it is not unusual to require working to approved samples.

2. Lime for coarse stuff shall be of approved well-burnt grey-stone lime, and be run four weeks [six weeks] before being used, and shall be kept clean.

3. Coarse stuff shall be well mixed, and in such a manner as to keep it free from earth or dust, in the proportions of one part of lime to two of sand [one part lime to two and a half, or three parts, sand]. No more stuff shall be mixed than is needed for immediate use.

4. Coarse stuff for ceilings, lath partitions and for [here specify special work] shall contain 1 lb. of sound, long, curled cowhair, free from grease, loading or other impurities, well beaten in and incorporated with every 3 cubic feet of stuff.

5. Approved chalk lime, free from lumps, flares or core, shall be used for setting putty and other fine stuff; and shall be run four weeks [six weeks] before use.

6. Portland cement shall be of the best quality for plastering work [of brand]; it shall not be used fresh, but properly aerated.

7. In adding waterproofing materials, the maker's directions shall be strictly followed.

8. Plaster of paris, in the proportions of 1 to 4 [1 to 6 or 1 to 20] of ordinary stuff may be used for

Note.—Plaster of paris is used when quick setting is desired, or where expansion (as in stopping holes) is useful.

9. Parian [Keene's, Martin's] cement shall be used for and the maker's directions strictly followed.

10. Laths shall be of fir, straight riven [machine sawn], free from sap, knots or unsound parts [of the strength of single, lath and a half or double]; shall be well nailed with 1" galvanised lath nails, properly spaced for keying, with butt-headed joints; [shall be double-nailed with breaking joints in 3' widths].

Note.—Proprietary wired-wood lathing, "Expanded Metal" or other sheet metal may be specified in place of the above.

WORKMANSHIP.—11. Provide scaffolding, tools, moulds and all other appliances necessary for the proper execution of the work.

12. Lath, plaster, float and set all wood joists, ceilings, soffits and stud [metal lathed] partitions and finish partitions to " in trowelled stucco.

Note.—For one-coat work on laths the clause runs "lath and lay" or "lath and plaster," and on walls "render." For two-coat work the terms commonly employed are "lath, lay and set" or "lath, plaster and set," and on walls "render and set."

13. Concrete ceilings and soffits shall be well hacked for key and set in gauged stuff; concrete partitions shall be floated and set. Or, dub-out concrete ceilings, soffits and partitions in gauged stuff.

14. Cover all chases containing pipes, wires, etc., with wire lathing

[gauge usually specified] suitable for plastering on, and secure in thorough manner. Wet wire lathing in lime water before plastering.

15. Render, float and set all walls not otherwise specified.

16. Walls at to be finished in trowelled stucco.

17. Render and float walls and lath and plaster and float ceilings and finish with a grit texture by freely scouring with float [or describe other finishes required].

18. Cornices and moulded work throughout shall be run clean and accurately to sections shown.

Or, Mitres and returns shall be truly worked and all enrichments and modelling to be carried out to architect's designs and approval.

Or, Cornices to shall be cast in fibrous plaster, fitted and fixed with proper galvanised nails.

Or, Run moulded plaster cornices at girt to rooms, with all mitres, returned, stopped and mitred ends, as required.

19. Narrow reveals, splays and returns shall be finished in Parian [Keene's] cement on a Portland cement backing.

Or, Run Keene's [or other] cement angles and arrises on Portland cement backing to all projecting angles, except the following, which are to be moulded, as here stated

Or, Run ovolo-moulded angles, 3″ girt with 2″ wings to opening, finished with moulded stops and short lengths of angle and arris to detail, all in Parian [Martin's] cement.

Or, Exposed surfaces of concrete lintels and girder casings shall be finished in Keene's [or other] cement internally and Portland cement externally, kept flush with brickwork ; all with arrises and angles, except those otherwise specified.

20. Run Portland cement flush skirtings 9″ high to basement, where plastered, with flush bead to top and trowelled face.

Or, Skirtings to shall be 12″ high and 1″ projection, sunk and twice moulded in Keene's [or other] cement, on Portland cement backing.

Or, Run skirting in cement and to scotia section [or other concave section].

Or, Float off the concrete floors of in Portland cement to the required level to receive mosaic [tile or other] paving.

21. Run necessary quirks, splays, arrises, etc., and make good to all mantelpieces ; cut away for and make good after all other trades, and cut out and make good all cracks, blisters and other defects, and leave plaster work perfect.

22. Ding walls where shown on plans with a coat of Portland cement 1 part, sand 2 parts, pea grit [or chippings] 1 part and ground chalk 1 part. Finish walls where shown with a rough coat of Portland cement 1 part and sand 3 parts, and rough-cast with fine pea grit [sand, chippings etc.].

23. Stop and twice lime white ceilings, soffits and walls of

24. Twice distemper white all ceilings, soffits and cornices, and twice distemper to approved tints walls of

Flooring

Clauses in specifications relating to floorings vary a great deal, according to the nature of the material used. In most cases full directions are laid down as to the concrete foundations, which may then be followed by some such general clause as under :

Provide the *per centum* sum of for super yard for patent composition flooring to be laid, as marked on plan, by the makers upon concrete bed as specified ; add attendance and give all necessary assistance and facilities.

Or, we may have :

Lay flooring, where marked on plan, with special composition in one layer [two layers] ″ thick.

To this may be added :

Run flooring up walls to form skirting 4″ [6″] high, finishing with concave section.

Usually provision is made for the description of finish, polished or matt, and due protection of the surface for a determined period after laying. [See also clauses 20 and 21 above.]

Concrete Foundations

Prepare a mixture of 1 part Portland cement and 5 of aggregate.

> *Note.*—The above proportions are common for concrete floors and foundations required to withstand heavy wear. But there is considerable variation, and usually the quality of sand, nature and size of aggregate are defined.

2. The concrete bed to be 4″ [6 or 8″] thick [finished with rough surface ; or, with granite surface and smoothly floated].

3. Expansion boards shall be provided at intervals of 12′ ; being removed at completion and cavities filled up.

4. Finished surface of concrete bed shall be kept damp for seven days [fourteen days], by covering with wet sand [sawdust], before laying the composition flooring.

Removal of Defective Wood

If the flooring is to be laid on old boards, provision is made for careful inspection, the removal of all parts infected with dry- or wet-rot (and also the removal of joists so attacked), the filling up of knot holes, crevices, etc., and the fastening down of all springy boards. Frequent and reasonable demands are that the woodwork should be treated with preservative composition and the under parts suitably ventilated.

GLOSSARY OF TERMS FOR PLASTERERS

Abacus. The top member of a capital.

Acanthus. The leaf as seen on the Corinthian capital.

Alcove. A recess in a room, usually vaulted or coved.

Annulated Column. Consists of a series of drums united by renewed bands.

Annulet. A ring; the term is applied to the fillets under the Doric capital, and to the bands connecting small detached shafts with the central column or with the wall in Gothic.

Applique. Moulded or carved ornament fastened to a subject as distinguished from planted, which is embedded in its surface.

Arabesque. A fanciful scroll ornament of leaves and animals and human beings. (Renaissance.)

Arcade. A series of arches.

Arch. A structure bridging over an opening with abutmenting imposts resting on piers and having a curved soffit, intrados or archivolt. They are usually semicircular, segmental (one-third of a circle), stilted (with long sides), pointed or triangular (Norman), curved (three parts of a circle, Moorish), or horseshoe form, etc. Gothic arches have trefoil and fretted heads, while the Tudor is a four-centred or flat arch.

Architrave. The lowest and weight-carrying division of the entablature; the moulding round a door or window. (Classic or Renaissance.)

Arris. The sharp edge produced by the intersection of two surfaces.

Ashlar. Squared and regular masonry.

Astragal. Small round moulding surrounding top or bottom of a column.

Atlantes. Male figures used instead of columns. See Caryatides.

Axis. Straight line passing the centre of building, design, etc.

Ball-flower. An ornament-formed Gothic moulding representing four curved and rounded petals enclosing a ball.

Band. A flat moulding.

Bandeler. A plain moulding.

Baque. An annular moulding.

Baquette. A diminutive of the above.

Barrel Ceiling. One that is semicircular in sections.

Base Moulding. One at the bottom of a column or wall.

Batter. When a wall is intentionally built with a sloping face it is said to " batter," and the slope is called " the batter."

Bay. A compartment of a building; the space between two pairs of columns or two roof principals.

Bay-window. A bow-window.

Bead. A small round moulding.

Beaded. A string of balls or beads; often used as a moulding.

Bed-mould. The moulding under the bold projection of a cornice.

Bevel. A slope made by cutting of an angle.

Billet. A Norman enrichment consisting of a succession of short cylinders lying in a shallow or hollow or on a chamfer.

Blocking Course. A plain course of stone over a cornice; a plain string course. (Classic and Renaissance.)

Bolection Moulding. A moulding used for panelling, projecting beyond the face of the framing. (Renaissance.)

Bond. The overlapping of stones or bricks in a wall. In brickwork from the middle of the sixteenth century to the end of the seventeenth century English bond was used. In this system, which is the strongest, one course consists entirely of headers, i.e. bricks showing their ends, and the next course of "stretchers," or bricks showing their sides, and so on alternately. In Flemish bond all courses are alike, and show alternately "headers" and "stretchers." Frequently imitated in plaster finish.

Bowtel. A round moulding. (Gothic.)

Brandering. Fixing small fillets to girders to take plastering laths.

Brattishing. A richly carved cresting on open parapet.

Cable Moulding. In the form of a strand of rope.

Caisson. A sunk panel.

Camber. An upward curve or slope in a beam, such as the tie-beam of a roof. A camber arch is one which tends to flat, as in middle Tudor architecture.

Campanulated. A capital in the form of an inverted bell.

Canopy. Ornamental projection over doorways, windows, niches, etc.

Cant. An obtuse angle.

Canticol. The stalks or anthers seen on a Corinthian capital.

Cantilever. A bracket.

Caps. Abbreviation for the word capital, ornamental mouldings, etc., at the top of columns and pilasters.

Caryatides. Columns in the form of human or grotesque figures. If ending in scrolls or merging into pilasters they are called terminal figures.

Casement. (1) A wide, shallow moulding; (2) a window hinged at the side to open like a door.

Cavetto. A round concave moulding containing the quadrant of a circle. Often found in a cornice.

Cavo-relievo. Sculptured or modelled work in hollowed cavities.

Chamfer. A bevel.

Chapter. The upper part of a column.

Chaplet. A moulding containing pearls or rosettes.

Checquer. A pattern formed of alternate squares.

Cinque-foil. A Gothic ornament representing a trio-leaved flower.

Coffer. A sunk panel in a ceiling, dome or vault in Classic or Renaissance buildings.

Colonnade. A row of columns.

Column. A pillar or shaft placed vertically, usually on a base, and crowned by a spreading mass called the capital. A column may be circular, oval or square in section ; smooth or fluted ; or even dimensions throughout the length of the shaft, diminishing towards the top or provided with an entasis (swelling towards its middle). It may be single or clustered (when two or more are grouped and attached to one another), or engaged, when its column is against, and more or less encased in, a wall. The dimensions and ornamentations of columns vary according to the classic orders and the Gothic period styles.

Concave. A circular hollow, or a rounded hollow moulding.

Console. A bracket supporting the cornice over a doorway or window. (Classic.)

Coved. An arched recess ; a concave moulding.

Convex. A circular rounded boss ; a ball-section moulding.

Cornice. (1) Highest part of an entablature ; (2) any series of mouldings crowning a wall, etc.

Corona. (1) Top edge of cornice, projecting over the bed-moulding to throw off water ; known as drip or weathering ; (2) in ornament, classical coronets, such as the mural, palisade, etc.

Crenelle. A parapet or cornice with battlements or loopholes.

Crocket. A crook-like leaf or bunch of leaves projecting from the slope of a gable.

Cusp. One of a series of points projecting from the soffit or mouldings of an arch, giving a trefoil or multifoil form in the arch.

Cyma. A *cyma recta* moulding is one that is concave at the top and convex at the bottom. A *cyma reversa* moulding is convex at the top and concave at the bottom ; also known as an ogee moulding.

Dado. The lower part of a wall when differently finished to the upper, usually marked off by a moulding or cornice. If only a few inches above the flooring it is called the skirting.

Datum. A fixed point from which levels are taken.

Dentil. An enrichment in a classical cornice consisting of a series of small square projections.

Diaper. A geometrical pattern carved on a wall in a frieze or panel. (Gothic.)

Egg and Anchor. Ornament consisting of an ovular figure and an anchor.

Egg and Dart. An enrichment in a classical cornice consisting alternately of eggs and darts.
Egg and Tongue. Oviformed figures alternately with a tongue.
Enrichment. Added ornament.
Entablature. The horizontal superstructure on the columns in classical architecture. It is divided into three parts: the architrave, or lintel, the frieze, and the cornice or projecting member.
Entasis. The slight swelling in a column. (Classic.)
Epistyle. Architrave on the capitals of columns which unites them.
Extrados. The outer convex side of an arch.

Façade. Exterior surface of a building.
Fan Tracery. Voluted ornaments and form of open fan. (Gothic.)
Festoon. Garlands of flowers, fruit and leaves, arranged in loops, often used on friezes and panels.
Fillet. A band applied principally to mouldings.
Fixing Fillet. A band of wood embedded in a brick, stone or concrete wall to which woodwork is fixed. Sometimes required for attaching the cradle work for plaster coved friezes and ceilings.
Flute. One of a series of channels running up the face of a column or frieze. Sometimes it is filled by a staff which runs up to one-third of the height of the column, and it is then said to be cabled or reeded. (Classic or Renaissance.)
Frieze. (1) That part of an entablature between the architrave and cornice; (2) a band at the top of a wall just below the cornice, or the ceiling in a plain wall.

Glyphs. Perpendicular flutings formed on the Doric frieze.
Gorge. (1) A hollow moulding; (2) narrowest part of a Tuscan or Doric column.
Groin. The edge formed by the intersection of surfaces in vaulting.
Grotesque. The Gothic form of arabesques, consisting of plaited scrolls budding into figures, masks and animals of fantastic character.
Guilloche. A form of plait, consisting of bands of fillets undulating and crossing each other.
Guttæ. Small drops, usually placed under cornices, etc.

Hack. To cut back and roughen a plastered or other surface.
Hood Mould. The horizontal moulding over an arch.

Imbrication. Fish-scale ornament.
Impost. The horizontal projecting member at the springing of an arch.

Jamb. The side of a window or doorway.

Knocked Up. Material after it has been gauged ready for use; and if after it has set it is mixed up again, it is then said to have been knocked up again.

Label. A hood mould.

Linen Fold. A pattern resembling the folds of a napkin, common in wood carving of the mid-Gothic period, sometimes imitated in plaster.

Machicolations (pron. Makikolations). Small arches carried on corbel to support an overhanging parapet of a castle. There were openings in the gutter so that missiles could be thrown upon assailants.

Metope. (1) The space between two triglyphs in the frieze of the Doric Order; (2) the space between two dentils in the Ionic frieze.

Mitre. The line formed by the intersection of mouldings or other surfaces, as at the angles of a picture frame. The internal mitre is that on the receding angle and the external mitre that on the projecting angle.

Modillion. A bracket under a classical cornice, associated with the Corinthian Order.

Mouldings. Projecting or sunken bands on the face of a wall or column, round panels, etc., capable of infinite variations. See under different names.

Mutule. The corresponding name in the Doric Order to the modillion in the Corinthian.

Nail-head Moulding. Moulding composed of a series of pyramidal projections.

Napkin Pattern. See Linen Fold.

Neck Mouldings. Those round the lower part of a capital.

Niche. A recess in a wall, pillar or column. The Norman are shallow and small. Early English Gothic usually of trefoil-headed type and commonly seen in series. Decorated and Perpendicular Gothic very diverse and ornamented. Classic usually deep, semicircular. with arched and coved top.

Nosing. The projecting part of a moulding.

Obtuse. The opposite to an acute or sharp angle.

Ogee. A curve composed partly of a convex and partly of a concave line; applied to the sections of moulding and the outlines of arches.

Orders. The five orders of classical architecture: the Tuscan, the Doric, the Ionic, the Corinthian, the Composite. Their chief parts are the Column, with its base, shaft and capital, and the Entablature, the treatment of which constitutes the particular style.

Oriel. A bow-window, either corbelled out from the wall or rising from the ground.

Oversail. That part of an entablature, cornices, etc., overhanging the face of a wall shaft of a column, etc.

Ovolo. A round convex moulding.

Pamure. A vine ornament.

Pargeting. Plaster work ; the term is generally applied to ornamental outside work ; to parge.

Patera. A small circular ornament in classical architecture. The term is now often applied to any small carved ornament forming one of a series.

Pediment. A gable in classical architecture often placed over doors or other openings. The apex may be open, when it is said to be broken.

Pendant. Bases on other ornaments hanging from ceilings arches. (Gothic).

Pendentive. Triangular portion of a spherical vault. (Byzantine and Gothic).

Peristyle. A colonnade round a courtyard. (Classic.)

Pilaster. A flat, rectangular column against a wall.

Planted. An ornament moulded separate and then fixed in the plastered wall or ceiling is said to be planted.

Plinth. (1) The base of a wall above the ground ; (2) the lower part of the base of a column.

Poppy-head. A carved finial on the top of a bench end.

Profile. Outline of a moulding or other ornament.

Profiling. Forming the outline.

Quarter Round. Moulding a quarter of a circle in section.

Quatrefoil. A four-leaved flower ornament. (Gothic.)

Quirk. Groove separating a convex moulding from the fillet above.

Quoin. An angle-stone.

Radius Rod. Used in the striking out of arches, and also gives the length for the gig-stick in running circular mouldings.

Rebate. A rectangular sinking along the edge of a piece of wood or stone. Sometimes spelt and pronounced " rabbet."

Reeding. A number of parallel beads.

Relief. Sculptured or modelled ornament partly rounded and projecting above the base on which it stands.

Relieving Arch. An arch over a lintel.

Return. An angle, generally applied to mouldings ; thus the hood mould is said to be returned horizontally at the springing of the arch, and to be returned against the wall.

Reveals. The side of an opening such as a window or doorway.

Round Moulding. A double round moulding, the upper part projecting above the lower.

Rustic-work, or Rustication. Ashlar masonry with the surface treated in a particular way. There are several varieties; the face of the stone is left rough or is artificially roughened; or it is smooth, but projects and has chamfered or rebated edges.

Scotia. Hollow moulding at the base of a column or pedestal.
Scroll Moulding. See Roll Moulding.
Sgraffito. Ornamenting plaster walls with scratched or incised lines.
Skirting. Narrow projecting bunch at the base of a wall.
Soffit. The under side of a cornice or arch, etc.
Spandrel. Applied to almost any surface of irregular form; such as the spaces above an arch; between an arch and a cusp; between the ribs of a vault.
Splay. A large chamfer, as to the jamb of a window, etc.
Squinch. An arch built across each angle of a tower to form an octagon to carry a spire.
Staff Bead. A bead at an angle, with a quirk at each side.
Stilted Arch. An arch of which the springing is above the capital.
Strap Work. Ornament formed of intertwined flat mouldings.
String. A horizontal projecting moulding. (Gothic.)
Swags. Festoons of leaves and flowers with drops at each end.

Table. A band or group of horizontal mouldings. (Gothic.) The fillet between the architrave and the frieze of a Gothic entablature.
Talin. A moulding formed of two arcs of a circle, respectively convex and concave.
Tooth Ornament. Serrated decoration found in hollow mouldings of the Norman period.
Torching. Plastering on under-side of roofing slates or tiles.
Torus. A large round moulding.
Tracery. Gothic ornament in panels, etc., consisting of curved flowing lines or straight mouldings intersecting each other.
Trefoil. Ornament based on the clover leaf. (Gothic.)
Trellis Moulding. Intertwined zigzag lines forming diamonds. (Norman.)
Triforium. An upper story over the aisle of a church.
Triglyph. An ornament in a Doric frieze, consisting of a projection with the two vertical edges chamfered and with two vertical grooves.
Trilobe. Ornament having three lobes. (Gothic.)
Tudor Arch. A flat arch.
Tudor Flower. An upright leaf used in cresting on the tops of cornices, etc.
Tympanum. The space enclosed (1) between the lintel of a doorway and the relieving anchor; (2) between the horizontal and sloping cornices of a classical pediment.

Vermiculated. A plastered surface ornamented with irregular channels.

Vesica Piscis. The pointed oval forming the auriole or glory round representations of the Deity and the Virgin.
Vitruvian Scroll. A long series of scrolls joined together.
Volute. A spiral; the characteristic ornament of the Ionic capital.
Vrils. Thin, elongated volutes, as in the tendrils of vines.

Wave Moulding. A succession of wavy lines; a section used in the fourteenth century.
Weathering. A sloping surface of a panel or other projection.
Weather Moulding. A drop, stone or outwardly sloping moulding.
Wreathed. A moulding formed of two twisted members as entwined ribbons. A wreathed column twisted or contorted.

Zigzag. A Norman enrichment. A succession of diagonal lines placed alternately.

APPENDIX

I. Measuring and Pricing—II. Labour—III. Quantities for 100 Square Yards—IV. Covering Capacity per Cubic Yard—V. Equivalents—VI. Mixtures: Coarse Stuff, Fine Stuff, Plasterer's Putty, Gauged Stuff, Common Stucco, Trowelled Stucco, Bastard Stucco, Compositions, Gesso, Papier Mâché, Carton Pierre, Pâte Coulante, Concrete, Fireproof Mortar, Plastic Flooring—VII. Hardening Agents—VIII. Hastening Agents—IX. Retarding Agents.

I. Measuring and Pricing

"The Standard Method of Measurement of Building Works" lays down that:

"The thickness of two-coat work, exclusive of keys (i.e. grooves or open joints in brickwork or spaces between laths) shall be regarded as not exceeding $\frac{5}{8}''$, that of three-coat work as not exceeding $\frac{7}{8}''$, and that of two-coat work in patent plasters as not exceeding $\frac{1}{2}''$. Any requisite dubbing and any additional thickness shall be specially stated in the item in question."

In practice, however, three-coat work is taken as $\frac{3}{4}''$ thick and any additional thickness is charged as dubbing.

Prices are arrived at after taking into consideration labour (and labour insurance), cost of materials, each of which should bear its quota of profit, including allowance for overhead charges.

The greater part of the plasterer's work is measured and priced per square yard or foot run. Repair work is an exception.

Keep interior and exterior work separate.

Measure and price limewashing, colouring, washing, stopping and distempering: rendering and setting; trowelled stucco on brick or lath; lath plaster, render in cement to inside of parapets and party walls, etc., by the square yard.

Cornices in plaster exceeding 12" in girth (to include Scotch bracketing, or similar bracketing, and dubbing-out when required) by the foot run. Soffits bands, margins, etc., under 12" in girth, per foot super.

Cornices 12" in girth and under and skirtings according to schedule; quirks, beads, arises throatings, per foot run.

Cornices 12" in girth or over, take out one of the projections of the cornice and multiply, per foot run.

Girth cornices, including space occupied by enrichments, mouldings, from the nose of the enrichment or moulding to the wall, allowing 1" more for the screed.

Ceilings to be measured from wall to wall; lath and plaster to ceilings and soffits per square yard.

Enrichments, mitres and stopped ends to be counted. Pateras covering mitres not to be counted, unless within frames or coffers.

Measure plastering in brick or stone walls from top of ground to foot of cornice, per square yard. On lath, add to the height of wall one-third of the height of cornice above the moulding for stucco, two-thirds for laid and floated work.

Skirtings per foot run. Reveals and margins, if so specified, per foot run, otherwise per foot super.

Where no wood ground is fixed and the skirting is nailed to plugs, the measurement of plastering is usually half the height of skirting.

Patches and repair work generally to be charged time and materials. Otherwise per foot run of joint, with 1′ over.

The architect, or his clerk of the works, will pay particular attention to see that:

All soffits and walls of cupboards are plastered where specified.

All staff beads are properly fixed to show the same projection on each wall.

All cornices and enrichments are properly finished.

All plastered surfaces are true and even, or given the fancy finish specified.

Plastering reaches below the skirting.

All rough-casting and the like is uniform.

Artificial stone is measured and charged by the cube.

II. Labour

It is reckoned that on ceiling work one plasterer with his labourer will lath, float and set 1½ square yards per hour. On brickwork they will render, float and set 3 square yards per hour.

For detailed calculations as to time of labour required for different items of work, consult *The Building Estimator*, by B. Price Davies.

III. Quantities for 100 Square Yards

The approximate quantities required for 100 square yards are:

One-coat work, or rendering: Lime, 1 cubic yard; sand, 2 cubic yards; hair, 15 to 16 in.

Two-coat work, rough-cast finish: 20 bushels of lime, 2 cubic yards of sand, 16 lb. hair, ¾ cubic yard of gravel.

Three-coat work on brick or stone: Lime, 2⅛ cubic yards; sand, 2½ cubic yards; hair, 28 to 30 lb.

On laths: Laths, 22 bundles; nails, 14 lb.; lime, 2½ cubic yards; sand, 2½ cubic yards; hair, 32 lb.

To render 100 square yards ¾″ thick with neat Portland cement will require approximately 25 bushels. Gauged half cement and half sand will require 13 bushels of each; a 1 : 2 mixture, 9 bushels of cement and 18 of sand; a 1 : 3 mixture, a little over 6½ bushels of cement and 19½ bushels of sand.

For concrete quantities, see Vol. I, p. 88.

IV. Covering Capacity per Cubic Yard

1 cubic yard of chalk lime, 2 cubic yards of sand and 3 bushels of hair will cover 75 yards of render and set on brick, or 70 yards on lath; or render two coats and set on brick 65 yards; on lath, 60 yards. Floated work requires about the same amount of materials as two coats and set.

1 bushel of cement, or cement and sand, will cover 1 yard super 1⅛″ thick; or 1⅛ yards 1″ thick.

1 cwt. of mastic and 1 gallon of oil will cover 2½ square yards ½″ thick.

1 cubic foot of slaked lime will cover 100 square yards; for second coat ½ cubic foot.

12 lb. of whiting, ½ lb. of blue black and 1¾ gallons of size will whitewash

100 square yards. For two-coat work: whiting, 21 lb.; blue black, ¾ lb.; size 2¾ gallons. If the surface is absorbent or very rough, more material will be required.

V. Equivalents

1 bag of plaster (London measure) contains 14 lb. and 7 bags equal 1 bushel.

1 sack of Portland cement equals 2 centals (200 lb.) or 2 bushels.

1 sack of lime equals about 3 bushels or 186 lb.

14 to 15 lb. of hair go to the bushel.

1 bundle of wood laths contains about 125 laths 3′ long to 125 laths 4′ long (making from 360′ to 500′ run).

1 ton of river sand equals 21 cubic feet.

1 ton of pit sand equals 22 cubic feet.

1 ton of Thames ballast equals 22 cubic feet.

1 ton of coarse gravel equals 23 cubic feet.

1 ton of clean shingle equals 24 cubic feet.

VI. Mixtures

Coarse Stuff.—Well-slaked lime, 1 part; sand, 2 to 3 parts; hair, ½ lb. to 1 lb. per cubic foot of stuff.

Fine Stuff.—For first-class work: slaked lime reduced to a cream. Second class: lime 3 parts, fine sand 1 part (a little hair sometimes added).

Plasterer's Putty.—Chalk or stone lime run through a fine sieve.

Setting Stuff.—Made from putty or fine stuff 5 parts to 2 parts fine washed sand.

Gauged Stuff (Hard Finish).—First-class work: fine stuff, 3 parts; plaster of Paris, 2 parts. Second-class: coarse stuff, 3 parts; plaster of Paris, 1 to 3 parts.

Common Stucco.—Hydraulic lime, 1 part; sand, 3 parts.

Trowelled Stucco.—Fine stuff (without hair), 2 parts; sand, 1 part.

Bastard Stucco.—The same as trowelled stucco but with a little hair.

Compositions (Compo).—Mix 2½ lb. of refined glue, 2 lb. of resin, 1 pint of linseed oil and ¼ lb. of Venice turpentine in a hot-water bath. Roll out the paste into slabs.

Gesso.—(1) Fine run slaked lime or plaster of Paris gauged with good size.

(2) Boil 1 part of resin in 4 parts of linseed oil, then add 6 parts of dissolved glue. Use this to gauge whiting into a cream.

Papier Mâché.—Steep torn up tissue paper in glue water. When quite soft, squeeze out the water and mix with strong liquid glue.

Carton Pierre.—Make paper pulp as in previous recipe. Dissolve 2 lb. of fish glue in 2 quarts of water, then add slowly, stirring well, ½ lb. of sifted flour. Add paper pulp to make a thin paste; boil for 1 hour, then simmer for 2 hours until thick. Spread out on a marble slab, dust liberally with whiting and knead into a workable paste.

Pâte Coulante.—Mix equal parts of plaster of Paris and whiting, gauge with weak glue water, adding a little alum.

Concrete.—(1) For foundations of floors, steps, etc.: 1 part of cement, 2½ parts of sand, 5 parts of coarse aggregate.

(2) For rough moulded work: 1 part cement, 1½ parts sand, 3 parts coarse aggregate.

(3) For fine moulding and modelling or carving: 1 part cement, 1 part fine sand, 2 parts aggregate, with hydrated lime ⅛ part of cement (mixed dry with the cement).

Fireproof Mortar.—(1) Slaked lime, 2 parts; blacksmith's ashes, 1 part; gauge stiff.

(2) Mix good stiff clay with cow dung and a little chimney soot. To 4 parts of this add 1 of air-slaked lime. Pass through a sieve.

Quick-setting Mortar.—1 part of air-slaked lime, 1 part of bone ash, 7 parts of fine sand.

Plastic Flooring.—Chloride of magnesia, 2½ to 2¾ parts; sawdust, 1 part.

Screeding to floors for wood blocks, or terrazzo or other materials, ½" to ¾" thick.

VII. Hardening Agents

Plaster gauged with lime water sets harder than if plain water is used.

A little alum or ferric-chloride added when gauging adds greatly to hardness.

Dextrine, obtainable from any oil and colour merchant, is greatly used for hardening purposes when casting pilaster or bases or other exposed parts of fibrous plaster likely to be knocked about.

VIII. Hastening Agents

Sodium chloride and sodium sulphate (Glauber Salts) quicken the setting of gypsum cements and plasters, but cause efflorescence.

Bone ash in lime mortars.

Alum, when added in the gauging, hastens as well as hardens.

IX. Retarding Agents

The addition of 8 per cent. of ground marshmallow root (althæa root) will retard the setting of plaster of Paris (gypsum) by 1 hour.

Gauging with weak size water also retards setting.

Hydrated lime retards the setting of concrete and renders it more plastic.

Sugar dissolved in beer or vinegar.

INDEX